D0086989

$f(t)$	$\mathcal{L}\{f(t)\} = F(s)$
35. $\sin kt \cosh kt$	$\dfrac{k(s^2 + 2k^2)}{s^4 + 4k^4}$
36. $\cos kt \sinh kt$	$\dfrac{k(s^2 - 2k^2)}{s^4 + 4k^4}$
37. $\cos kt \cosh kt$	$\dfrac{s^3}{s^4 + 4k^4}$
38. $J_0(kt)$	$\dfrac{1}{\sqrt{s^2 + k^2}}$
39. $\dfrac{e^{bt} - e^{at}}{t}$	$\ln\dfrac{s - a}{s - b}$
40. $\dfrac{2(1 - \cos kt)}{t}$	$\ln\dfrac{s^2 + k^2}{s^2}$
41. $\dfrac{2(1 - \cosh kt)}{t}$	$\ln\dfrac{s^2 - k^2}{s^2}$
42. $\dfrac{\sin at}{t}$	$\arctan\left(\dfrac{a}{s}\right)$
43. $\dfrac{\sin at \cos bt}{t}$	$\dfrac{1}{2}\arctan\dfrac{a + b}{s} + \dfrac{1}{2}\arctan\dfrac{a - b}{s}$
44. $\delta(t)$	1
45. $\delta(t - t_0)$	e^{-st_0}
46. $e^{at}f(t)$	$F(s - a)$
47. $f(t - a)\mathcal{U}(t - a)$	$e^{-as}F(s)$
48. $\mathcal{U}(t - a)$	$\dfrac{e^{-as}}{s}$
49. $f^{(n)}(t)$	$s^n F(s) - s^{(n-1)}f(0) - \cdots - f^{(n-1)}(0)$
50. $t^n f(t)$	$(-1)^n \dfrac{d^n}{ds^n} F(s)$
51. $\displaystyle\int_0^t f(\tau)g(t - \tau)\, d\tau$	$F(s)G(s)$

A FIRST COURSE IN
DIFFERENTIAL EQUATIONS
WITH MODELING APPLICATIONS

A FIRST COURSE IN DIFFERENTIAL EQUATIONS

WITH MODELING APPLICATIONS

Sixth Edition

Dennis G. Zill

Loyola Marymount University

Brooks/Cole Publishing Company

I(T)P™   An International Thomson Publishing Company

Pacific Grove ■ *Albany* ■ *Bonn* ■ *Boston* ■ *Cincinnati* ■ *Detroit* ■ *London* ■ *Madrid* ■ *Melbourne*
Mexico City ■ *New York* ■ *Paris* ■ *San Francisco* ■ *Singapore* ■ *Tokyo* ■ *Toronto* ■ *Washington*

COPYRIGHT© 1997, 1993, 1989, 1986 by Brooks/Cole Publishing Company, a division of International Thomson Publishing Inc.

I(T)P The ITP logo is a trademark under license.

All rights reserved. No part of this work may be reproduced, stored in a retrieval system, or transmitted, in any form or by any means—electronic, mechanical, photocopying, recording, or otherwise—without the prior written permission of the publisher, Brooks/Cole Publishing Company, Pacific Grove, California 93950.

 This book is printed on recycled, acid-free paper.

For more information, contact:
Brooks/Cole Publishing Company
511 Forest Lodge Road
Pacific Grove, CA 93950
USA

International Thomson Publishing Europe
Berkshire House I68-I73
High Holborn
London WC1V 7AA
England

Thomas Nelson Australia
102 Dodds Street
South Melbourne, 3205
Victoria, Australia

Nelson Canada
1120 Birchmont Road
Scarborough, Ontario
Canada M1K 5G4

International Thomson Editores
Campos Eliseos 385, Piso 7
Col. Polanco
11560 Mexico D.F., Mexico

International Thomson Publishing GmbH
Konigswinterer Strasse 418
53227 Bonn, Germany

International Thomson Publishing Asia
221 Henderson Road
#05-10 Henderson Building
Singapore 0315

International Thomson Publishing Japan
Hirakawacho Kyowa Building, 31
2-2-1 Hirakawacho
Chiyoda-ku, Tokyo 102
Japan

Library of Congress Cataloging-in-Publication Data

Zill, Dennis G.
   A first course in differential equations with modeling applications / Dennis G. Zill. -- 6th ed.
       p.   cm.
     Rev. ed. of: A first course in differential equations. 5th ed. c 1993.
     Includes index.
     ISBN 0-534-95574-6 (hardcover)
     1. Differential equations.   I. Zill, Dennis G.   First course in differential equations.   II. Title.
   QA372. Z54   1997   96-2424
   515′.35--dc20   CIP

Portions of this book also appear in *Differential Equations with Boundary-Value Problems, Fourth Edition,* by Dennis G. Zill and Michael R. Cullen, copyright © 1997 by Brooks/Cole Publishing Company.

Editor: Gary Ostedt
Developmental Editor: *Barbara Lovenvirth*
Production Coordinator: *Monique A. Calello*
Editorial Assistant: *Anna Aleksandrowicz*
Manufacturing Coordinator: *Wendy Kilborn*
Marketing Manager: *Marianne Rutter*
Cover Designer: *Diane Levy, DFL Publications*

Production Services: *Lifland et al.,Bookmakers*
Compositor: *Bi-Comp Incorporated*
Cover Image: Copyright © 1995 "Michael Schimpf"/
    Panoramic Images, Chicago, All Rights Reserved
Cover Printer: *Phoenix Color Corp*
Text Printer: *Quebecor/Hawkins*
Insert Printer: *Coral Graphic Services*

Printed and bound in the United States of America
96 97 98 99 — 10 9 8 7 6 5 4 3 2 1

CONTENTS

A Modeling Application

AZT and Survivability with AIDS   **After page 64**

by Ivan Kramer

A Modeling Application

Wolf Population Dynamics

by C. J. Knickerbocker

A Modeling Application

Decay of Satellite Orbits   **After page 192**

• •

by John Ellison

A Modeling Application

The Collapse of the Tacoma Narrows Suspension Bridge

• •

by Gilbert N. Lewis

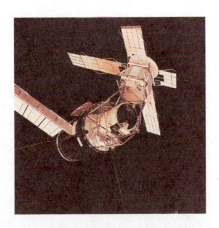

Modeling an Arms Race **After page 352**

A Modeling Application •

by Michael Olinick

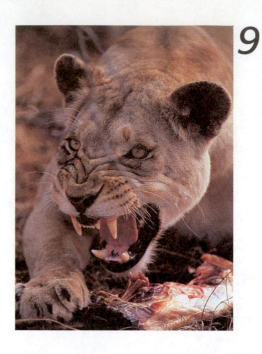

9 Numerical Methods for Ordinary Differential Equations   353

PREFACE

For the Sixth Edition of *A First Course in Differential Equations with Modeling Applications,* the revisions reflect a dual purpose: to ensure that the information is current and relevant to students and yet maintain the rudimental foundations on which the previous editions were constructed. Written with the student in mind, this new book retains the basic level, pedagogical aids, and straight-forward style of presentation of previous editions.

In differential equations, as in many other mathematics courses, instructors are beginning to question certain aspects of traditional teaching methods. This healthy introspection is important in making the subject matter not only more interesting for the students but also more relevant to the world in which they live. Changes in both the content and the style of *A First Course in Differential Equations with Modeling Applications, Sixth Edition*, including the new subtitle, reflect changes the author has observed in the overall approach to teaching differential equations.

Summary of Major Changes

■ *Greater emphasis on differential equations as mathematical models.* The notion of a mathematical model is now interwoven throughout the text, and the construction and pitfalls of such models are discussed.

■ *Five new Modeling Applications.* These applications, illustrated in full color, have been contributed by selected experts in each field. Presented at appropriate points throughout the text, the applications cover timely areas of study, ranging from AZT and survivability with AIDS to the effects on the ecological balance of the reintroduction of the gray wolf to Yellowstone National Park.

■ *Greater emphasis on nonlinear differential equations and on linear and nonlinear systems of differential equations.* Three chapters contain new sections— Sections 3.3, 4.9, 5.2, and 5.3.

■ *Greater emphasis on boundary-value problems for ordinary differential equations.* New to this edition, eigenvalues and eigenfunctions are introduced in Chapter 5.

■ *Greater utilization of technology throughout the text.* Graphics calculators, graphing software, computer algebra systems, and ODE solvers are utilized wherever appropriate in applications and examples, as well as in the exercise sets.

■ *Increase in the number of conceptual problems in exercises.* In many sections, new Discussion Problems have been added. Instead of being asked to solve a differential equation, the student is asked to think about what a differential equation says. In order to encourage students to think, to draw conclusions, and to explore possibilities, the answers to these questions have been intentionally omitted from the end of the book. Some of these problems can serve as individual or group assignments, at the option of the instructor.

Changes in the Sixth Edition by Chapter

Chapter 1 has been expanded to include an introduction to the notions of an initial-value problem and ODE solvers in Section 1.2. The discussion of differential equations as mathematical models, in Section 1.3, has been rewritten for ease of student comprehension.

Chapter 2 now combines discussion of homogeneous first-order equations with discussion of the Bernoulli equation in Section 2.4, *Solutions by Substitution*. Material on the Ricatti and Clairaut equations is now in exercise sets.

Chapter 3 includes a new Section 3.3, *Systems of Linear and Nonlinear Equations*, which shows systems of first-order differential equations as mathematical models. Orthogonal trajectories are now covered in exercises.

Chapter 4 introduces the concept of a linear differential operator in Section 4.1 for the purpose of expediting the proofs of some important theorems. The slightly different manner in which the two equations defining "variable parameters" are presented in Section 4.6 is due to a student, J. Thoo.* The Cauchy-Euler equation is now covered in Section 4.7. Solving systems of linear differential equations with constant coefficients has been moved forward to Section 4.8. A new Section 4.9, *Nonlinear Equations*, begins with a qualitative discussion of the differences between linear and nonlinear equations.

Chapter 5 includes two sections new to this edition. Section 5.2, *Linear Equations: Boundary-Value Problems*, introduces the concepts of eigenvalues and eigenfunctions. Section 5.3, *Nonlinear Equations*, describes modeling with nonlinear higher-order differential equations.

Chapter 6 is now devoted solely to series solutions of linear differential equations.

Chapter 7 now contains, in Section 7.7, the application of the Laplace transform to systems of linear differential equations with constant coefficients. An alternative form of the second translation theorem has been added to Section 7.3.

Chapter 8 covers only the theory and solution of systems of linear first-order differential equations, as material on matrices has been moved to Appendix II. This placement allows the instructor to decide whether to designate the material for a reading assignment or insert it into classroom discussion.

---

* J. Thoo, "Timing Is Everything," *The College Mathematical Journal*, Vol. 23, No. 4, September 1992.

Chapter 9 has been completely rewritten. Error analysis for the various numerical techniques is covered in the appropriate section devoted to each method.

Supplements Available

For Students

Student Solutions Manual (Warren S. Wright) provides a solution to every third problem in each exercise set, with the exception of the Discussion Problems.

For Instructors

Complete Solutions Manual (Warren S. Wright) provides worked-out solutions to all the problems in the text.

Lab Experiments for Differential Equations (Dennis G. Zill/Warren S. Wright) contains an assortment of computer lab experiments with differential equations.

Software

ODE Solver: Numerical Procedures for Ordinary Differential Equations (Thomas Kiffe/William Rundel), for IBM and compatibles and for Macintosh, is a software package that presents tabular and graphical representations of the output for the various numerical methods. No programming is required.

Grapher (Steve Scarborough), for the Macintosh, is a versatile collection of graphing utilities which can plot graphs of rectangular equations, parametric equations, polar equations, interpolating polynomials, series, and direction fields. The program provides an ODE solver for first-order differential equations and systems of two first-order differential equations.

Computer Programs for BASIC, FORTRAN, and Pascal (C. J. Knickerbocker), for IBM and compatibles and for Macintosh, contains listings of computer programs for many of the numerical methods in the text.

ACKNOWLEDGMENTS

A large measure of gratitude is owed to the following persons, who contributed to this revision through their help, suggestions, and criticisms:

Scott Wright, Loyola Marymount University
Bruce Bayly, University of Arizona
Dean R. Brown, Youngstown State University
Nguyen P. Cac, University of Iowa
Philip Crooke, Vanderbilt University
Bruce E. Davis, St. Louis Community College at Florissant Valley
Donna Farrior, University of Tulsa
Terry Herdman, Virginia Polytechnic Institute and State University
S. K. Jain, Ohio University
Cecelia Laurie, University of Alabama
James R. McKinney, California Polytechnic State University
James L. Meek, University of Arkansas
Brian M. O'Connor, Tennessee Technological University

A special thank-you goes to

John Ellison, Grove City College
C. J. Knickerbocker, St. Lawrence University
Ivan Kramer, University of Maryland, Baltimore County
Gilbert Lewis, Michigan Technological University
Michael Olinick, Middlebury College

who were generous enough to take time out of their busy schedules to supply the interesting new essays on modeling applications.

Finally, a personal note. Those with sharp eyes may have noticed that the familiar PWS lion logo is absent from the spine of the book. While I look forward to working with the personnel at Brooks/Cole, a different branch of the parent company ITP, I can't help thinking about the many good people I have been fortunate to meet, work with, and, yes, fight with over the last twenty years at PWS. So to all in production, marketing, and editorial—especially to Barbara Lovenvirth, my *de facto* editor—I wish you godspeed and good luck. Thank you for another—the final—job well done.

Dennis G. Zill
Los Angeles

A FIRST COURSE IN
DIFFERENTIAL EQUATIONS
WITH MODELING APPLICATIONS

INTRODUCTION TO DIFFERENTIAL EQUATIONS

INTRODUCTION

The words *differential* and *equations* certainly suggest solving some kind of equation that contains derivatives. Just as students in a course in algebra and trigonometry spend a good amount of time solving equations such as $x^2 + 5x + 4 = 0$ for the *variable x*, in this course we wish to solve differential equations such as $y'' + 2y' + y = 0$ for the unknown *function y*. But before you start solving anything, you must learn some of the basic definitions and terminology of the subject.

1.1 DEFINITIONS AND TERMINOLOGY

■ *Ordinary and partial differential equations* ■ *Order of an equation*
■ *Linear and nonlinear equations* ■ *Solution of a differential equation*
■ *Explicit and implicit solutions* ■ *Trivial solution* ■ *Family of solutions*
■ *Particular solution* ■ *General solution* ■ *Systems of differential equations*

A Differential Equation   In calculus you learned that the derivative dy/dx of a function $y = \phi(x)$ is itself another function of x found by some appropriate rule. For example, if $y = e^{x^2}$ then $dy/dx = 2xe^{x^2}$. Replacing e^{x^2} by the symbol y then gives

$$\frac{dy}{dx} = 2xy. \tag{1}$$

The problem we face in this course is not "Given a function $y = \phi(x)$, find its derivative." Rather, our problem is "If we are given a differential equation such as (1), is there some way or method by which we can find the unknown function $y = \phi(x)$?"

DEFINITION 1.1   **Differential Equation**

An equation containing the derivatives of one or more dependent variables, with respect to one or more independent variables, is said to be a **differential equation (DE).**

Differential equations are classified by **type, order,** and **linearity.**

Classification by Type   If an equation contains only ordinary derivatives of one or more dependent variables with respect to a single independent variable, it is said to be an **ordinary differential equation (ODE).** For example,

$$\frac{dy}{dx} + 10y = e^x \qquad \text{and} \qquad \frac{d^2y}{dx^2} - \frac{dy}{dx} + 6y = 0$$

are ordinary differential equations. An equation involving the partial derivatives of one or more dependent variables of two or more independent variables is called a **partial differential equation (PDE).** For example,

$$\frac{\partial u}{\partial y} = -\frac{\partial v}{\partial x} \qquad \text{and} \qquad \frac{\partial^2 u}{\partial x^2} = \frac{\partial^2 u}{\partial t^2} - 2\frac{\partial u}{\partial t}$$

are partial differential equations.

Classification by Order   The **order of a differential equation** (ODE or PDE) is the order of the highest derivative in the equation. For example,

second-order $\downarrow$ $\qquad$ $\downarrow$ first-order

$$\frac{d^2y}{dx^2} + 5\left(\frac{dy}{dx}\right)^3 - 4y = e^x$$

is a second-order ordinary differential equation. Since the differential equation $(y - x)\,dx + 4x\,dy = 0$ can be put into the form

$$4x\frac{dy}{dx} + y = x$$

by dividing by the differential dx, it is an example of a first-order ordinary differential equation.

A general nth-order ordinary differential equation is often represented by the symbolism

$$F(x, y, y', \ldots, y^{(n)}) = 0. \tag{2}$$

In general discussions in this text we shall assume that an nth-order differential equation (2) can be solved for the highest derivative $y^{(n)}$, that is,

$$y^{(n)} = f(x, y, y', \ldots, y^{(n-1)}).$$

Classification as Linear or Nonlinear A differential equation $y^{(n)} = f(x, y, y', \ldots, y^{(n-1)})$ is said to be **linear** when f is a linear function of $y, y', \ldots, y^{(n-1)}$. This means that an equation is linear if it can be written in the form

$$a_n(x)\frac{d^n y}{dx^n} + a_{n-1}(x)\frac{d^{n-1} y}{dx^{n-1}} + \cdots + a_1(x)\frac{dy}{dx} + a_0(x)y = g(x).$$

From this last equation we see the two characteristic properties of linear differential equations:

(*i*) The dependent variable y and all its derivatives are of the first degree; that is, the power of each term involving y is 1.
(*ii*) Each coefficient depends only on the independent variable x.

Functions of y such as $\sin y$ or functions of the derivatives of y such as $e^{y'}$ cannot appear in a linear equation. A differential equation that is not linear is said to be **nonlinear.** The equations

$$(y - x)\,dx + 4x\,dy = 0, \qquad y'' - 2y' + y = 0, \qquad x^3\frac{d^3 y}{dx^3} - 4x\frac{dy}{dx} + 6y = e^x$$

are linear first-, second-, and third-order ordinary differential equations, respectively. On the other hand,

$$\underset{\underset{\downarrow}{\substack{\text{coefficient}\\ \text{depends on } y}}}{(1 + y)y'} + 2y = e^x, \qquad \frac{d^2 y}{dx^2} + \underset{\underset{\downarrow}{\substack{\text{nonlinear}\\ \text{function of } y}}}{\sin y} = 0, \qquad \frac{d^4 y}{dx^4} + \underset{\underset{\downarrow}{\substack{\text{power not 1}}}}{y^2} = 0$$

are nonlinear first-, second-, and fourth-order ordinary differential equations, respectively.

Solutions As mentioned before, one of our goals in this course is to solve, or find **solutions** of, differential equations.

> **DEFINITION 1.2**   **Solution of a Differential Equation**
>
> Any function ϕ defined on some interval I, which when substituted into a differential equation reduces the equation to an identity, is said to be a **solution** of the equation on the interval.

In other words, a solution of an ordinary differential equation (2) is a function ϕ that possesses at least n derivatives and

$$F(x, \phi(x), \phi'(x), \ldots, \phi^{(n)}(x)) = 0 \quad \text{for all } x \text{ in } I.$$

We say that $y = \phi(x)$ *satisfies* the differential equation. The interval I could be an open interval (a, b), a closed interval $[a, b]$, an infinite interval (a, ∞), and so on. For our purposes, we shall also assume that a solution ϕ is a real-valued function.

EXAMPLE 1   **Verification of a Solution**

Verify that $y = x^4/16$ is a solution of the nonlinear equation

$$\frac{dy}{dx} = xy^{1/2}$$

on the interval $(-\infty, \infty)$.

SOLUTION   One way of verifying that the given function is a solution is to write the differential equation as $dy/dx - xy^{1/2} = 0$ and then see, after substituting, whether the sum $dy/dx - xy^{1/2}$ is zero for every x in the interval. Using

$$\frac{dy}{dx} = 4\frac{x^3}{16} = \frac{x^3}{4} \quad \text{and} \quad y^{1/2} = \left(\frac{x^4}{16}\right)^{1/2} = \frac{x^2}{4},$$

we see that
$$\frac{dy}{dx} - xy^{1/2} = \frac{x^3}{4} - x\left(\frac{x^4}{16}\right)^{1/2} = \frac{x^3}{4} - \frac{x^3}{4} = 0$$

for every real number. Note that $y^{1/2} = x^2/4$ is, by definition, the nonnegative square root of $x^4/16$.   ■

EXAMPLE 2   **Verification of a Solution**

The function $y = xe^x$ is a solution of the linear equation

$$y'' - 2y' + y = 0$$

on the interval $(-\infty, \infty)$. To show this, we compute

$$y' = xe^x + e^x \quad \text{and} \quad y'' = xe^x + 2e^x.$$

Observe that

$$y'' - 2y' + y = (xe^x + 2e^x) - 2(xe^x + e^x) + xe^x = 0$$

for every real number.   ■

Not every differential equation that we write necessarily has a solution. You are encouraged to think about this and then solve Problem 51 in Exercises 1.1.

Explicit and Implicit Solutions   You should be familiar with the terms *explicit* and *implicit functions* from your study of calculus. Because some methods for solving differential equations lead directly to these two forms, solutions of differential equations can be further distinguished as either explicit solutions or implicit solutions. A solution in which the dependent variable is expressed solely in terms of the independent variable and constants is said to be an **explicit solution.** For our purposes let us think of an explicit solution as an explicit formula $y = \phi(x)$ that we can manipulate, evaluate, and differentiate. We have already seen in our initial discussion that $y = e^{x^2}$ is an explicit solution of $dy/dx = 2xy$. In Examples 1 and 2, $y = x^4/16$ and $y = xe^x$ are explicit solutions of $dy/dx = xy^{1/2}$ and $y'' - 2y' + y = 0$, respectively. Note that in Examples 1 and 2 each differential equation possesses the constant solution $y = 0, -\infty < x < \infty$. An explicit solution of a differential equation that is identically zero on an interval I is said to be a **trivial solution.** A relation $G(x, y) = 0$ is said to be an **implicit solution** of an ordinary differential equation (2) on an interval I provided there exists at least one function ϕ that satisfies the relation as well as the differential equation on I. In other words, $G(x, y) = 0$ defines the function ϕ implicitly.

EXAMPLE 3   **Verification of an Implicit Solution**

The relation $x^2 + y^2 - 4 = 0$ is an implicit solution of the differential equation

$$\frac{dy}{dx} = -\frac{x}{y} \tag{3}$$

on the interval $-2 < x < 2$. By implicit differentiation we obtain

$$\frac{d}{dx}x^2 + \frac{d}{dx}y^2 - \frac{d}{dx}4 = \frac{d}{dx}0 \qquad \text{or} \qquad 2x + 2y\frac{dy}{dx} = 0.$$

Solving the last equation for the symbol dy/dx gives (3). In addition, you should verify that the functions $y_1 = \sqrt{4 - x^2}$ and $y_2 = -\sqrt{4 - x^2}$ satisfy the relation (in other words, $x^2 + y_1^2 - 4 = 0$ and $x^2 + y_2^2 - 4 = 0$) and are solutions of the differential equation on $-2 < x < 2$. ∎

Any relation of the form $x^2 + y^2 - c = 0$ *formally* satisfies (3) for any constant c. However, it is understood that the relation should always make sense in the real number system; thus, for example, we cannot say that $x^2 + y^2 + 4 = 0$ is an implicit solution of the equation. (Why not?)

Because the distinction between an explicit solution and an implicit solution should be intuitively clear, we will not belabor the issue by always saying, "Here is an explicit (implicit) solution."

More Terminology   The study of differential equations is similar to that of integral calculus. A solution is sometimes referred to as an **integral** of the equation, and its graph is called an **integral curve** or a **solution curve.** When evaluating an antiderivative or indefinite integral in calculus, we use a single constant c of integration. Analogously, when solving a first-order differential equation $F(x, y, y') = 0$, we usually obtain a solution containing a single arbitrary constant or parameter c. A solution containing an arbitrary constant represents a set $G(x, y, c) = 0$ of solutions, called a **one-parameter family of solutions.** When solving an nth-order differential equation $F(x, y, y', \ldots, y^{(n)}) = 0$, we seek an **$n$-parameter family of solutions** $G(x, y, c_1, c_2, \ldots, c_n) = 0$. This simply means that a single differential equation can possess an infinite number of solutions corresponding to the unlimited choices for the parameter(s). A solution of a differential equation that is free of arbitrary parameters is called a **particular solution.** For example, by direct substitution we can show that any function in the one-parameter family $y = ce^{x^2}$, where c is an arbitrary constant, also satisfies equation (1). The original solution $y = e^{x^2}$ corresponds to $c = 1$ and thus is a particular solution of the equation. Figure 1.1 shows some of the integral curves in this family. The trivial solution $y = 0$, corresponding to $c = 0$, is also a particular solution of (1).

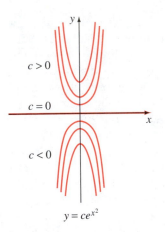

$c > 0$

$c = 0$

$c < 0$

$y = ce^{x^2}$

FIGURE 1.1

EXAMPLE 4   **Particular Solutions**

The function $y = c_1 e^x + c_2 e^{-x}$ is a two-parameter family of solutions of the linear second-order equation $y'' - y = 0$. Some particular solutions are $y = 0$ ($c_1 = c_2 = 0$), $y = e^x$ ($c_1 = 1$, $c_2 = 0$), and $y = 5e^x - 2e^{-x}$ ($c_1 = 5$, $c_2 = -2$).   ∎

In all the preceding examples we have used x and y to denote the independent and dependent variables, respectively. But in practice these two variables are represented by many different symbols. For example, we could denote the independent variable by t and the dependent variable by x.

EXAMPLE 5   **Using Different Symbols**

The functions $x = c_1 \cos 4t$ and $x = c_2 \sin 4t$, where c_1 and c_2 are arbitrary constants, are solutions of the differential equation

$$x'' + 16x = 0.$$

For $x = c_1 \cos 4t$ the first two derivatives with respect to t are $x' = -4c_1 \sin 4t$ and $x'' = -16c_1 \cos 4t$. Substituting x'' and x then gives

$$x'' + 16x = -16c_1 \cos 4t + 16(c_1 \cos 4t) = 0.$$

In like manner, for $x = c_2 \sin 4t$ we have $x'' = -16c_2 \sin 4t$, and so

$$x'' + 16x = -16c_2 \sin 4t + 16(c_2 \sin 4t) = 0.$$

Finally, it is easy to verify that the linear combination of solutions, or two-parameter family, $x = c_1 \cos 4t + c_2 \sin 4t$ is a solution of the given equation.   ∎

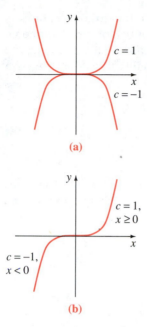

(a)

(b)

FIGURE 1.2

The next example shows that a solution of a differential equation can be a piecewise-defined function.

EXAMPLE 6   **A Piecewise-Defined Solution**

You should verify that any function in the one-parameter family $y = cx^4$ is a solution of the differential equation $xy' - 4y = 0$ on the inverval $(-\infty, \infty)$. See Figure 1.2(a). The piecewise-defined differentiable function

$$y = \begin{cases} -x^4, & x < 0 \\ x^4, & x \geq 0 \end{cases}$$

is a particular solution of the equation but cannot be obtained from the family $y = cx^4$ by a single choice of c. See Figure 1.2(b). ∎

Sometimes a differential equation possesses a solution that cannot be obtained by specializing *any* of the parameters in a family of solutions. Such a solution is called a **singular solution.**

EXAMPLE 7   **A Singular Solution**

In Section 2.1 we will prove that a one-parameter family of solutions of $y' = xy^{1/2}$ is given by $y = (x^2/4 + c)^2$. When $c = 0$, the resulting particular solution is $y = x^4/16$. In this case the trivial solution $y = 0$ is a singular solution of the equation, because it cannot be obtained from the family for any choice of the parameter c. ∎

General Solution   If *every* solution of an nth-order equation $F(x, y, y', \ldots, y^{(n)}) = 0$ on an interval I can be obtained from an n-parameter family $G(x, y, c_1, c_2, \ldots, c_n) = 0$ by appropriate choices of the parameters c_i, $i = 1, 2, \ldots, n$, we then say that the family is the **general solution** of the differential equation. In solving linear differential equations we shall impose relatively simple restrictions on the coefficients of the equation; with these restrictions we can always be assured not only that a solution does exist on an interval but also that a family of solutions does indeed yield all possible solutions. Nonlinear equations, with the exception of some first-order equations, are usually difficult or impossible to solve in terms of familiar elementary functions (finite combinations of integer powers of x, roots, exponential and logarithmic functions, trigonometric and inverse trigonometric functions). Furthermore, if we happen to have a family of solutions for a nonlinear equation, it is not obvious when this family constitutes a general solution. On a practical level then, the designation "general solution" is applied only to linear differential equations.

Systems of Differential Equations   Up to this point we have discussed single differential equations containing one unknown function. But often, in theory as well as in many applications, we must deal with *systems* of differential equations. A **system of ordinary differential equations** is two or more equations involving the derivatives of two or more unknown functions of a single independent variable. For example, if x and y denote dependent variables and t the independent variable, the following is a system of two first-order differential equations:

$$\frac{dx}{dt} = 3x - 4y$$

$$\frac{dy}{dt} = x + y. \tag{4}$$

A solution of a system such as (4) is a pair of differentiable functions $x = \phi_1(t)$ and $y = \phi_2(t)$ that satisfies each equation in the system on some common interval I.

Remark

A few last words about implicit solutions of differential equations are in order. Unless it is important or convenient, there is usually no need to try to solve an implicit solution $G(x, y) = 0$ for y explicitly in terms of x. In Example 3 we can easily solve the relation $x^2 + y^2 - 4 = 0$ for y in terms of x to get two solutions, $y_1 = \sqrt{4 - x^2}$ and $y_2 = -\sqrt{4 - x^2}$, of the differential equation $dy/dx = -x/y$. But don't be misled by this one example. An implicit solution $G(x, y) = 0$ can define a perfectly good differentiable function ϕ that is a solution of a differential equation, but yet we may not be able to solve $G(x, y) = 0$ using analytical methods such as algebra. In Section 2.2 we shall see that $xe^{2y} - \sin xy + y^2 + c = 0$ is an implicit solution of a first-order differential equation. The task of solving this equation for y in terms of x presents more problems than just the drudgery of symbol pushing; *it can't be done.*

SECTION 1.1 EXERCISES

Answers to odd-numbered problems begin on page A-1.

In Problems 1–10 state whether the given differential equation is linear or nonlinear. Give the order of each equation.

1. $(1 - x)y'' - 4xy' + 5y = \cos x$ **2.** $x\dfrac{d^3y}{dx^3} - 2\left(\dfrac{dy}{dx}\right)^4 + y = 0$

3. $yy' + 2y = 1 + x^2$

4. $x^2\, dy + (y - xy - xe^x)\, dx = 0$

5. $x^3 y^{(4)} - x^2 y'' + 4xy' - 3y = 0$ **6.** $\dfrac{d^2y}{dx^2} + 9y = \sin y$

7. $\dfrac{dy}{dx} = \sqrt{1 + \left(\dfrac{d^2y}{dx^2}\right)^2}$ **8.** $\dfrac{d^2r}{dt^2} = -\dfrac{k}{r^2}$

9. $(\sin x)y''' - (\cos x)y' = 2$ **10.** $(1 - y^2)\, dx + x\, dy = 0$

In Problems 11–40 verify that the indicated function is a solution of the given differential equation. In some cases assume an appropriate interval of validity for the solution. Where used, the symbols c_1 and c_2 denote constants.

11. $2y' + y = 0;\quad y = e^{-x/2}$

12. $y' + 4y = 32;\quad y = 8$

13. $\dfrac{dy}{dx} - 2y = e^{3x};\quad y = e^{3x} + 10e^{2x}$

14. $\dfrac{dy}{dt} + 20y = 24;\quad y = \frac{6}{5} - \frac{6}{5}e^{-20t}$

15. $y' = 25 + y^2;\quad y = 5\tan 5x$

16. $\dfrac{dy}{dx} = \sqrt{\dfrac{y}{x}};\quad y = (\sqrt{x} + c_1)^2,\ x > 0,\ c_1 > 0$

17. $y' + y = \sin x;\quad y = \frac{1}{2}\sin x - \frac{1}{2}\cos x + 10e^{-x}$

18. $2xy\,dx + (x^2 + 2y)\,dy = 0;\quad x^2 y + y^2 = c_1$

19. $x^2\,dy + 2xy\,dx = 0;\quad y = -\dfrac{1}{x^2}$

20. $(y')^3 + xy' = y;\quad y = x + 1$

21. $y = 2xy' + y(y')^2;\quad y^2 = c_1(x + \frac{1}{4}c_1)$

22. $y' = 2\sqrt{|y|};\quad y = x|x|$

23. $y' - \dfrac{1}{x}y = 1;\quad y = x\ln x,\, x > 0$

24. $\dfrac{dP}{dt} = P(a - bP);\quad P = \dfrac{ac_1 e^{at}}{1 + bc_1 e^{at}}$

25. $\dfrac{dX}{dt} = (2 - X)(1 - X);\quad \ln\dfrac{2 - X}{1 - X} = t$

26. $y' + 2xy = 1;\quad y = e^{-x^2}\displaystyle\int_0^x e^{t^2}\,dt + c_1 e^{-x^2}$

27. $(x^2 + y^2)\,dx + (x^2 - xy)\,dy = 0;\quad c_1(x + y)^2 = xe^{y/x}$

28. $y'' + y' - 12y = 0;\quad y = c_1 e^{3x} + c_2 e^{-4x}$

29. $y'' - 6y' + 13y = 0;\quad y = e^{3x}\cos 2x$

30. $\dfrac{d^2 y}{dx^2} - 4\dfrac{dy}{dx} + 4y = 0;\quad y = e^{2x} + xe^{2x}$

31. $y'' = y;\quad y = \cosh x + \sinh x$

32. $y'' + 25y = 0;\quad y = c_1\cos 5x$

33. $y'' + (y')^2 = 0;\quad y = \ln|x + c_1| + c_2$

34. $y'' + y = \tan x;\quad y = -\cos x\ln(\sec x + \tan x)$

35. $x\dfrac{d^2 y}{dx^2} + 2\dfrac{dy}{dx} = 0;\quad y = c_1 + c_2 x^{-1},\, x > 0$

36. $x^2 y'' - xy' + 2y = 0;\quad y = x\cos(\ln x),\, x > 0$

37. $x^2 y'' - 3xy' + 4y = 0;\quad y = x^2 + x^2\ln x,\, x > 0$

38. $y''' - y'' + 9y' - 9y = 0;\quad y = c_1\sin 3x + c_2\cos 3x + 4e^x$

39. $y''' - 3y'' + 3y' - y = 0;\quad y = x^2 e^x$

40. $x^3\dfrac{d^3 y}{dx^3} + 2x^2\dfrac{d^2 y}{dx^2} - x\dfrac{dy}{dx} + y = 12x^2;\quad y = c_1 x + c_2 x\ln x + 4x^2,\, x > 0$

In Problems 41 and 42 verify that the indicated piecewise-defined function is a solution of the given differential equation.

41. $xy' - 2y = 0;\quad y = \begin{cases} -x^2, & x < 0 \\ x^2, & x \ge 0 \end{cases}$

42. $(y')^2 = 9xy;$ $y = \begin{cases} 0, & x < 0 \\ x^3, & x \geq 0 \end{cases}$

43. A one-parameter family of solutions for $y' = y^2 - 1$ is

$$y = \frac{1 + ce^{2x}}{1 - ce^{2x}}.$$

By inspection, determine a singular solution of the differential equation.

44. On page 5 we saw that $y = \sqrt{4 - x^2}$ and $y = -\sqrt{4 - x^2}$ are solutions of $dy/dx = -x/y$ on the interval $(-2, 2)$. Explain why

$$y = \begin{cases} \sqrt{4 - x^2}, & -2 < x < 0 \\ -\sqrt{4 - x^2}, & 0 \leq x < 2 \end{cases}$$

is not a solution of the differential equation on the interval.

In Problems 45 and 46 find values of m so that $y = e^{mx}$ is a solution of each differential equation.

45. $y'' - 5y' + 6y = 0$ **46.** $y'' + 10y' + 25y = 0$

In Problems 47 and 48 find values of m so that $y = x^m$ is a solution of each differential equation.

47. $x^2y'' - y = 0$ **48.** $x^2y'' + 6xy' + 4y = 0$

In Problems 49 and 50 verify that the given pair of functions is a solution to the given system of differential equations.

49. $\dfrac{dx}{dt} = x + 3y$

$\dfrac{dy}{dt} = 5x + 3y;$

$x = e^{-2t} + 3e^{6t}, \quad y = -e^{-2t} + 5e^{6t}$

50. $\dfrac{d^2x}{dt^2} = 4y + e^t$

$\dfrac{d^2y}{dt^2} = 4x - e^t;$

$x = \cos 2t + \sin 2t + \dfrac{1}{5}e^t, \quad y = -\cos 2t - \sin 2t - \dfrac{1}{5}e^t$

Discussion Problems

51. (a) Make up at least two differential equations that do not possess any real solutions.
(b) Make up a differential equation whose only real solution is $y = 0$.

52. Suppose $y = \phi(x)$ is a solution of an nth-order differential equation $F(x, y, y', \ldots, y^{(n)}) = 0$ on an interval I. Explain why $\phi, \phi', \ldots, \phi^{(n-1)}$ are necessarily continuous on I.

53. Suppose $y = \phi(x)$ is a solution of the differential equation $dy/dx = y(a - by)$, where a and b are positive constants.
(a) By inspection find two constant solutions of the equation.

(b) Using only the differential equation, find intervals on the y-axis on which a nonconstant solution $y = \phi(x)$ is increasing; on which $y = \phi(x)$ is decreasing.

(c) Using only the differential equation, explain why $y = a/2b$ is the y-coordinate of a point of inflection of the graph of a nonconstant solution $y = \phi(x)$.

(d) On the same coordinate axes, sketch the graphs of the two constant solutions found in part (b) and a graph of the nonconstant solution $y = \phi(x)$ whose shape is suggested in parts (b) and (c).

54. The differential equation $y = xy' + f(y')$ is known as **Clairaut's equation.**

(a) By differentiating both sides of the differential equation with respect to x, verify that the family of straight lines $y = cx + f(c)$, where c is an arbitrary constant, is a solution of Clairaut's equation.

(b) Discuss how the procedure in part (a) leads naturally to the discovery of a singular solution of Clairaut's equation.

(c) Find a one-parameter family of solutions, as well as a singular solution, of the differential equation $y = xy' + (y')^2$.

1.2 INITIAL-VALUE PROBLEMS

- *Initial-value problem* ■ *Initial condition* ■ *Existence and uniqueness of a solution*
- *Interval of existence* ■ *ODE solver*

Initial-Value Problem   We are often interested in solving a differential equation subject to prescribed side conditions—conditions that are imposed on $y(x)$ or its derivatives. On some interval I containing x_0, the problem

$$\textit{Solve:} \qquad \frac{d^n y}{dx^n} = f(x, y, y', \ldots, y^{(n-1)})$$

$$\textit{Subject to:} \quad y(x_0) = y_0, \quad y'(x_0) = y_1, \quad \ldots, \quad y^{(n-1)}(x_0) = y_{n-1}, \tag{1}$$

where $y_0, y_1, \ldots, y_{n-1}$ are arbitrarily specified real constants, is called an **initial-value problem (IVP).** The given values of the unknown function $y(x)$ and its first $n - 1$ derivatives at a single point x_0: $y(x_0) = y_0$, $y'(x_0) = y_1, \ldots, y^{(n-1)}(x_0) = y_{n-1}$ are called **initial conditions.**

First- and Second-Order IVPs   The problem given in (1) is also called an ***n*th-order initial-value problem.** For example,

$$\textit{Solve:} \qquad \frac{dy}{dx} = f(x, y)$$

$$\textit{Subject to:} \quad y(x_0) = y_0 \tag{2}$$

and

$$\textit{Solve:} \qquad \frac{d^2 y}{dx^2} = f(x, y, y')$$

$$\textit{Subject to:} \quad y(x_0) = y_0, \quad y'(x_0) = y_1 \tag{3}$$

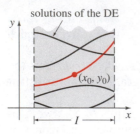

FIGURE 1.3
First-order IVP

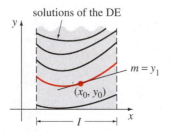

FIGURE 1.4
Second-order IVP

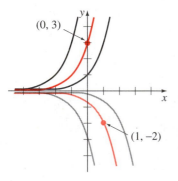

FIGURE 1.5
Solutions of IVPs

are first- and second-order initial-value problems, respectively. These two problems are easy to interpret in geometric terms. For (2) we are seeking a solution of the differential equation on an interval I containing x_0 so that a solution curve passes through the prescribed point (x_0, y_0). See Figure 1.3. For (3) we want to find a solution of the differential equation whose graph not only passes through (x_0, y_0) but passes through so that the slope of the curve at this point is y_1. See Figure 1.4. The term *initial condition* derives from physical systems where the independent variable is time t and where $y(t_0) = y_0$ and $y'(t_0) = y_1$ represent, respectively, the position and velocity of an object at some beginning, or initial, time t_0.

Solving an nth-order initial-value problem frequently entails using an n-parameter family of solutions of the given differential equation to find n specialized constants so that the resulting particular solution of the equation also "fits"—that is, satisfies—the n initial conditions.

EXAMPLE 1   **First-Order IVPs**

It is readily verified that $y = ce^x$ is a one-parameter family of solutions of the simple first-order equation $y' = y$ on the interval $(-\infty, \infty)$. If we specify an initial condition, say, $y(0) = 3$, then substituting $x = 0$, $y = 3$ in the family determines the constant $3 = ce^0 = c$. Thus the function $y = 3e^x$ is a solution of the initial-value problem

$$y' = y, \quad y(0) = 3.$$

Now if we demand that a solution of the differential equation pass through the point $(1, -2)$ rather than $(0, 3)$, then $y(1) = -2$ will yield $-2 = ce$ or $c = -2e^{-1}$. The function $y = -2e^{x-1}$ is a solution of the initial-value problem

$$y' = y, \quad y(1) = -2.$$

The graphs of these two functions are shown in color in Figure 1.5. ∎

EXAMPLE 2   **Second-Order IVP**

In Example 5 of Section 1.1 we saw that $x = c_1 \cos 4t + c_2 \sin 4t$ is a two-parameter family of solutions of $x'' + 16x = 0$. Find a solution of the initial-value problem

$$x'' + 16x = 0, \qquad x\left(\frac{\pi}{2}\right) = -2, \quad x'\left(\frac{\pi}{2}\right) = 1. \qquad \textbf{(4)}$$

SOLUTION   We first apply $x(\pi/2) = -2$ to the given family of solutions: $c_1 \cos 2\pi + c_2 \sin 2\pi = -2$. Since $\cos 2\pi = 1$ and $\sin 2\pi = 0$, we find that $c_1 = -2$. We next apply $x'(\pi/2) = 1$ to the one-parameter family $x(t) = -2 \cos 4t + c_2 \sin 4t$. Differentiating and then setting $t = \pi/2$ and $x' = 1$ gives $8 \sin 2\pi + 4c_2 \cos 2\pi = 1$, from which we see that $c_2 = \frac{1}{4}$. Hence

$$x = -2 \cos 4t + \frac{1}{4} \sin 4t$$

is a solution of (4). ∎

Existence and Uniqueness   Two fundamental questions arise in considering an initial-value problem:

Does a solution of the problem exist? If a solution exists, is it unique?

For an initial-value problem such as (2) we are asking:

Existence	Uniqueness
Does the differential equation dy/dx = f(x, y) possess solutions? *Do any of the solution curves pass through the point (x_0, y_0)?*	*When can we be certain that there is precisely **one** solution curve passing through the point (x_0, y_0)?*

Note that in Examples 1 and 2 the phrase "*a* solution" is used rather than "*the* solution" of the problem. The indefinite article "a" is used deliberately to suggest the possibility that other solutions may exist. At this point it has not been demonstrated that there is a single solution of each problem. The next example illustrates an initial-value problem with two solutions.

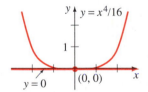

FIGURE 1.6
Two solutions of the same IVP

EXAMPLE 3   **An IVP Can Have Several Solutions**

Each of the functions $y = 0$ and $y = x^4/16$ satisfies the differential equation $dy/dx = xy^{1/2}$ and the initial condition $y(0) = 0$, and so the initial-value problem

$$\frac{dy}{dx} = xy^{1/2}, \qquad y(0) = 0$$

has at least two solutions. As illustrated in Figure 1.6, the graphs of both functions pass through the same point $(0, 0)$. ■

Within the safe confines of a formal course in differential equations one can be fairly confident that *most* differential equations will have solutions and that solutions of initial-value problems will *probably* be unique. Real life, however, is not so idyllic. Thus it is desirable to know in advance of trying to solve an initial-value problem whether a solution exists and, when it does, whether it is the only solution of the problem. Since we are going to consider first-order differential equations in the next two chapters, we state here without proof a straightforward theorem that gives conditions that are sufficient to guarantee the existence and uniqueness of a solution of a first-order initial-value problem of the form given in (2). We shall wait until Chapter 4 to address the question of existence and uniqueness of a second-order initial-value problem.

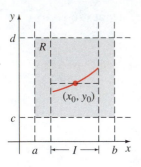

FIGURE 1.7
Rectangular region R

THEOREM 1.1   **Existence of a Unique Solution**

Let R be a rectangular region in the xy-plane defined by $a \leq x \leq b$, $c \leq y \leq d$ that contains the point (x_0, y_0) in its interior. If $f(x, y)$ and $\partial f/\partial y$ are continuous on R, then there exist an interval I centered at x_0 and a unique function $y(x)$ defined on I satisfying the initial-value problem (2).

The foregoing result is one of the most popular existence and uniqueness theorems for first-order differential equations, because the criteria of continuity of $f(x, y)$ and $\partial f/\partial y$ are relatively easy to check. The geometry of Theorem 1.1 is illustrated in Figure 1.7.

EXAMPLE 4   **Example 3 Revisited**

We saw in Example 3 that the differential equation $dy/dx = xy^{1/2}$ possesses at least two solutions whose graphs pass through $(0, 0)$. Inspection of the functions

$$f(x, y) = xy^{1/2} \qquad \text{and} \qquad \frac{\partial f}{\partial y} = \frac{x}{2y^{1/2}}$$

shows that they are continuous in the upper half-plane defined by $y > 0$. Hence Theorem 1.1 enables us to conclude that through any point (x_0, y_0), $y_0 > 0$ in the upper plane there is some interval centered at x_0 on which the given differential equation has a unique solution. Thus, for example, even without solving it we know that there exists some interval centered at 2 on which the initial-value problem $dy/dx = xy^{1/2}$, $y(2) = 1$ has a unique solution. ∎

Theorem 1.1 guarantees that in Example 1 there are no other solutions of the initial-value problems $y' = y$, $y(0) = 3$ and $y' = y$, $y(1) = -2$ other than $y = 3e^x$ and $y = -2e^{x-1}$, respectively. This follows from the fact that $f(x, y) = y$ and $\partial f/\partial y = 1$ are continuous throughout the entire xy-plane. It can be further shown that the interval on which each solution is defined is $(-\infty, \infty)$.

EXAMPLE 5   **Interval of Existence**

For the equation $dy/dx = x^2 + y^2$ we observe that $f(x, y) = x^2 + y^2$ and $\partial f/\partial y = 2y$ are both polynomials in x and y and hence continuous at every point. In other words, the region R in Theorem 1.1 is the entire xy-plane. Therefore through any given point (x_0, y_0) there passes one and only one solution curve. Note, however, that this does *not* mean that the largest interval I of validity for a solution of an initial-value problem is necessarily $(-\infty, \infty)$; the interval I need not be as wide as the region R. In general, it is not possible to find a specific interval I on which a solution is defined without actually solving the differential equation (see Problems 18, 19, and 28 in Exercises 1.2). ∎

ODE Solvers It is possible to obtain an *approximate* graphical representation of a solution of a differential equation or a system of differential equations without actually obtaining an explicit or implicit solution. This graphical representation requires computer software with a utility known generically as an **ODE solver.** In the case of, say, a first-order differential equation $dy/dx = f(x, y)$, we need only supply $f(x, y)$ and specify an initial value $y(x_0) = y_0$. If the problem has a solution, the solver then gives us a solution curve.* For example, in view of Theorem 1.1 we are assured that the differential equation

$$\frac{dy}{dx} = -y + \sin x$$

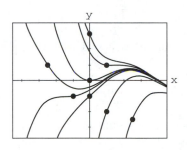

FIGURE 1.8
Some solutions of $y' = -y + \sin x$

possesses only one solution passing through each point (x_0, y_0) in the xy-plane. Figure 1.8 shows the solution curves generated by an ODE solver that pass through $(-2.5, 1)$, $(-1, -1)$, $(0, 0)$, $(0, 3)$, $(0, -1)$, $(1, 1)$, $(1, -2)$, and $(2.5, -2.5)$.

Remarks

(*i*) You should be aware of the distinction between a solution *existing* and *exhibiting* a solution. Clearly, if we find a solution by exhibiting it we can say that it exists, but a solution can exist and we may not be able to display it. In other words, when we say a solution of a differential equation exists, we do not mean that there also exists a method for finding it. A differential equation can possess a solution that satisfies specified initial conditions, but it may be that the best we can do is to approximate it. Approximation methods for differential equations and systems, which constitute the theory behind ODE solvers, will be examined in Chapter 9.

(*ii*) The conditions in Theorem 1.1 are sufficient but not necessary. When $f(x, y)$ and $\partial f/\partial y$ are continuous on a rectangular region R, it must always follow that a solution of (2) exists and is unique whenever (x_0, y_0) is a point interior to R. However, if the conditions stated in the hypothesis of Theorem 1.1 do not hold, then anything could happen: Problem (2) *may* still have a solution and this solution *may* be unique, problem (2) may have several solutions, or it may have no solution at all.

SECTION 1.2 EXERCISES

Answers to odd-numbered problems begin on page A-1.

In Problems 1–10 determine a region of the xy-plane for which the given differential equation would have a unique solution through a point (x_0, y_0) in the region.

1. $\dfrac{dy}{dx} = y^{2/3}$

2. $\dfrac{dy}{dx} = \sqrt{xy}$

---

*From this point on you will be expected to remember that a solution curve generated by an ODE solver is an *approximate* solution curve.

3. $x\dfrac{dy}{dx} = y$

4. $\dfrac{dy}{dx} - y = x$

5. $(4 - y^2)y' = x^2$

6. $(1 + y^3)y' = x^2$

7. $(x^2 + y^2)y' = y^2$

8. $(y - x)y' = y + x$

9. $\dfrac{dy}{dx} = x^3\cos y$

10. $\dfrac{dy}{dx} = (x - 1)e^{y/(x-1)}$

In Problems 11 and 12 determine by inspection at least two solutions of the given initial-value problem.

11. $y' = 3y^{2/3}, \quad y(0) = 0$

12. $x\dfrac{dy}{dx} = 2y, \quad y(0) = 0$

In Problems 13–16 determine whether Theorem 1.1 guarantees that the differential equation $y' = \sqrt{y^2 - 9}$ possesses a unique solution through the given point.

13. $(1, 4)$

14. $(5, 3)$

15. $(2, -3)$

16. $(-1, 1)$

17. (a) By inspection find a one-parameter family of solutions of the differential equation $xy' = y$. Verify that each member of the family is a solution of the initial-value problem $xy' = y, y(0) = 0$.

   (b) Explain part (a) by determining a region R in the xy-plane for which the differential equation $xy' = y$ would have a unique solution through a point (x_0, y_0) in R.

   (c) Verify that the piecewise-defined function

$$y = \begin{cases} 0, & x < 0 \\ x, & x \geq 0 \end{cases}$$

satisfies the condition $y(0) = 0$. Determine whether the function is also a solution of the initial-value problem in part (a).

18. (a) Consider the differential equation $y' = 1 + y^2$. Determine a region R in the xy-plane for which the differential equation would have a unique solution through a point (x_0, y_0) in R.

   (b) Show that $y = \tan x$ satisfies the differential equation and the condition $y(0) = 0$, but explain why $y = \tan x$ is not a solution of the initial-value problem $y' = 1 + y^2, y(0) = 0$ on the interval $(-2, 2)$.

   (c) Determine the largest interval I of validity for which $y = \tan x$ is a solution of the initial-value problem in part (b).

19. (a) Verify that the differential equation $y' = y^2$ possesses a unique solution through any point (x_0, y_0) in the xy-plane.

   (b) Use an ODE solver to obtain the solution curve passing through each of the points: $(0, 0), (0, 2), (1, 3), (-2, 4), (-2, -4), (0, -1.5), (1, -1)$.

   (c) Use the graphs obtained in part (b) to conjecture the largest interval I of validity for the solution of each of the seven initial-value problems.

20. (a) Consider the differential equation $y' = x/y$. Determine a region R in the xy-plane for which the differential equation would have a unique solution through a point (x_0, y_0) in R.

(b) Use an ODE solver to obtain the solution curves for various initial-value problems for (x_0, y_0) in R.

(c) Use the results in part (b) to conjecture a one-parameter family of solutions of the differential equation.

In Problems 21 and 22 use the fact that $y = 1/(1 + c_1e^{-x})$ is a one-parameter family of solutions of $y' = y - y^2$ to find a solution of the initial-value problem consisting of the differential equation and the given initial condition.

21. $y(0) = -\dfrac{1}{3}$

22. $y(-1) = 2$

In Problems 23–26 use the fact that $y = c_1e^x + c_2e^{-x}$ is a two-parameter family of solutions of $y'' - y = 0$ to find a solution of the initial-value problem consisting of the differential equation and the given initial conditions.

23. $y(0) = 1, \quad y'(0) = 2$

24. $y(1) = 0, \quad y'(1) = e$

25. $y(-1) = 5, \quad y'(-1) = -5$

26. $y(0) = 0, \quad y'(0) = 0$

Discussion Problems

27. Suppose $f(x, y)$ satisfies the hypotheses of Theorem 1.1 in some rectangular region R of the xy-plane. Explain why two distinct solutions of the differential equation $y' = f(x, y)$ cannot intersect or be tangent to each other at a point (x_0, y_0) in R.

28. Theorem 1.1 guarantees that there is only one solution of the differential equation $y' = 3y^{4/3} \cos x$ passing through any specified point (x_0, y_0) in the xy-plane. The interval of existence of a solution depends on the initial condition $y(x_0) = y_0$. Use the one-parameter family of solutions $y = 1/(c - \sin x)^3$ to find a solution satisfying $y(\pi) = 1/8$. Find a solution satisfying $y(\pi) = 8$. Use these two solutions as a basis for discussing the following questions: From the given family of solutions, when do you think the interval of existence of the initial-value problem is a finite interval? An infinite interval?

1.3 DIFFERENTIAL EQUATIONS AS MATHEMATICAL MODELS

■ *Mathematical model* ■ *Level of resolution of a model*
■ *Newton's second law of motion* ■ *Kirchhoff's second law* ■ *Dynamical system*
■ *State variables* ■ *State of a system* ■ *Response of a system*

Note to the Instructor   In this section we concentrate only on the construction of differential equations as mathematical models. Once we have examined some methods for solving differential equations in Chapters 2 and 4, we return to, and solve, some of these models in Chapters 3 and 5.

Mathematical Models   It is often desirable to describe in mathematical terms the behavior of some real-life system or phenomenon, whether physical, sociological, or even economic. The mathematical description of a system or a phenomenon is called a **mathematical model** and is constructed with certain goals in mind. For example, we may wish to understand the mechanisms of a certain ecosystem by studying the growth of animal populations in that system, or we may wish to date fossils by analyzing the decay of a radioactive substance either in the fossil or in the stratum in which it was discovered.

Construction of a mathematical model of a system starts with

(*i*) *identification of the variables that are responsible for changing the system. We may choose not to incorporate all these variables into the model at first. In this step we are specifying the **level of resolution** of the model.*

Next is

(*ii*) *making a set of reasonable assumptions, or hypotheses, about the system we are trying to describe. These assumptions will also include any empirical laws that may be applicable to the system.*

For some purposes it may be perfectly within reason to be content with low-resolution models. For example, you may already be aware that in modeling the motion of a body falling near the surface of the earth, the retarding force of air friction is ignored in beginning physics courses. But if you are a scientist whose job is to predict accurately the flight path of a long-range projectile, you must take air resistance and other factors such as the curvature of the earth into account.

Since the assumptions made about a system frequently involve *a rate of change* of one or more of the variables, the mathematical construct of all these assumptions is one or more equations involving *derivatives*. In other words, the mathematical model is a differential equation or a system of differential equations.

Once we have formulated a mathematical model that is either a differential equation or a system of a differential equations, we are faced with the not insignificant problem of solving it. Having solved it, we deem the model to be reasonable if its solution is consistent with either experimental data or known facts about the behavior of the system. If the predictions produced by the solution are poor, we can either increase the level of resolution of the model or make alternative assumptions about the mechanisms for change in the system. The steps of the modeling process are then repeated, as shown in Figure 1.9. Of course, by increasing

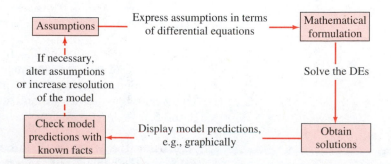

FIGURE 1.9

the resolution we add to the complexity of the mathematical model and increase the likelihood that we must approximate its solution.

A mathematical model of a *physical* system will often involve the variable time *t*. A solution of the model then gives the **state of the system;** in other words, for appropriate values of *t* the values of the dependent variable (or variables) describe the system in the past, present, and future.

Growth and Decay   One of the earliest attempts to model human **population growth** by means of mathematics was by the English economist Thomas Malthus in 1798. Basically, the idea of the Malthusian model is the assumption that the rate at which the population of a country grows is proportional to the total population $P(t)$ of the country at any time *t*. In other words, the more people there are at time *t*, the more there are going to be in the future. In mathematical terms this assumption can be expressed as

$$\frac{dP}{dt} \propto P \qquad \text{or} \qquad \frac{dP}{dt} = kP, \qquad \qquad \textbf{(1)}$$

where *k* is a constant of proportionality. This simple model, which fails to take into account many factors (immigration and emigration, for example) that can influence human populations to either grow or decline, nevertheless turned out to be fairly accurate in predicting the population of the United States during the years 1790–1860. The differential equation given in (1) is still used often to model, over short intervals of time, populations of bacteria and small animals.

The nucleus of an atom consists of combinations of protons and neutrons. Many of these combinations of protons and neutrons are unstable—that is, the atoms decay or transmute into atoms of another substance. Such nuclei are said to be radioactive. For example, over time the highly radioactive element radium, Ra-226, transmutes into the radioactive gas radon, Rn-222. To model the phenomenon of **radioactive decay,** we assume that the rate at which the nuclei of a substance decay is proportional to the amount (more precisely, the number of nuclei) $A(t)$ of the substance remaining at time *t*:

$$\frac{dA}{dt} \propto A \qquad \text{or} \qquad \frac{dA}{dt} = kA. \qquad \qquad \textbf{(2)}$$

Of course, equations (1) and (2) are exactly the same; the difference is only in the interpretation of the symbols and the constants of proportionality. For growth, as we expect in (1), $k > 0$; and for decay, in the case of (2), $k < 0$. The model (2) for decay also applies in a biological setting such as determining the half-life of a drug—namely, the time that it takes for 50% of a drug to be eliminated from a body by excretion or metabolism.

We see the basic model in (1) and (2) next in a business setting.

Continuous Compounding of Interest   Interest earned on a savings account is very often compounded quarterly or even monthly. Of course, there is no reason to stop there; interest could as well be compounded daily, by the hour, by the minute, every second, every half-second, every microsecond, and so on. That is to say, interest could be compounded **continuously.** To model the concept of continuous compounding of interest, let us suppose that $S(t)$ is the amount of money accrued in a savings account after t years and that r is the annual rate of interest compounded continuously. If $h > 0$ denotes an increment in time, then the interest obtained in the time span $(t + h) - t$ is the difference in the amounts accrued:

$$S(t + h) - S(t). \tag{3}$$

Since interest is given by (rate) $\times$ (time) $\times$ (principal), we can approximate the interest earned in this same time period by either

$$rhS(t) \qquad \text{or} \qquad rhS(t + h). \tag{4}$$

Intuitively, the quantities in (4) are seen to be the lower and upper bounds, respectively, for the actual interest in (3); that is,

$$rhS(t) \le S(t + h) - S(t) \le rhS(t + h)$$

or
$$rS(t) \le \frac{S(t + h) - S(t)}{h} \le rS(t + h). \tag{5}$$

Since we want h to become smaller and smaller, we then take the limit of (5) as $h \to 0$:

$$rS(t) \le \lim_{h \to 0} \frac{S(t + h) - S(t)}{h} \le rS(t).$$

And so it must follow that

$$\lim_{h \to 0} \frac{S(t + h) - S(t)}{h} = rS(t) \qquad \text{or} \qquad \frac{dS}{dt} = rS. \tag{6}$$

The point of (1) and (2) and (6) in this last example is simply this:

A single differential equation can serve as a mathematical model for many different phenomena.

Mathematical models are often accompanied by certain side conditions. For example, in (1), (2), and (6) we would expect to know an initial population P_0, an initial amount of the substance A_0 on hand, and an initial amount of the deposit S_0, respectively. If the initial time is taken to be, say, $t = 0$, then we know that $P(0) = P_0$, $A(0) = A_0$, and $S(0) = S_0$. In other words, a mathematical model actually consists of either an initial-value problem or, as we shall see later on in Section 5.2, a boundary-value problem.

Chemical Reactions   The disintegration of a radioactive substance, governed by the differential equation (1), is said to be a **first-order reaction.** In chemistry a few reactions follow this same empirical law: If the molecules of substance A decompose into smaller molecules, it is a natural assumption that the rate at which this decomposition takes place is proportional to the amount of substance A that has not undergone conversion; that is, if $X(t)$ is the amount of substance A remaining at any time, then $dX/dt = kX$, where k is negative constant (since X is decreasing). An example of a first-order chemical reaction is the conversion of t-butyl chloride into *t*-butyl alcohol:

$$(CH_3)_3CCl + NaOH \rightarrow (CH_3)_3COH + NaCl.$$

Only the concentration of the *t*-butyl chloride controls the rate of reaction.
  Now in the reaction

$$CH_3Cl + NaOH \rightarrow CH_3OH + NaCl,$$

for every molecule of methyl chloride one molecule of sodium hydroxide
is consumed, thus forming one molecule of methyl alcohol and one molecule of sodium chloride. In this case the rate at which the reaction proceeds
is proportional to the product of the remaining concentrations of CH_3Cl
and NaOH. If X denotes the amount of CH_3OH formed and α and β are the
given amounts of the first two chemicals A and B, then the instantaneous
amounts not converted to chemical C are $\alpha - X$ and $\beta - X$, respectively.
Hence the rate of formation of C is given by

$$\frac{dX}{dt} = k(\alpha - X)(\beta - X), \tag{7}$$

where k is a constant of proportionality. A reaction whose model is
equation (7) is said to be a **second-order reaction.**

Spread of a Disease   In the analysis of the spread of a contagious
disease—for example, a flu virus—it is reasonable to assume that the rate
at which the disease spreads is proportional not only to the number of
people $x(t)$ who have contracted the disease, but also to the number of
people $y(t)$ who have not yet been exposed. If the rate is dx/dt, then

$$\frac{dx}{dt} = kxy, \tag{8}$$

where k is the usual constant of proportionality. If, say, one infected
person is introduced into a fixed population of n people, then x and y are
related by $x + y = n + 1$. Using this last equation to eliminate y in (8)
gives us the model

$$\frac{dx}{dt} = kx(n + 1 - x). \tag{9}$$

An obvious initial condition accompanying equation (9) is $x(0) = 1$.

Newton's Law of Cooling   According to Newton's empirical law
of cooling, the rate at which a body cools is proportional to the difference
between the temperature of the body and the temperature of the surrounding medium, the so-called ambient temperature. If we let $T(t)$ represent the temperature of a body at any time t, T_m represent the constant
temperature of the surrounding medium, and dT/dt represent the rate at
which a body cools, then Newton's law of cooling translates into the
mathematical statement

$$\frac{dT}{dt} \propto T - T_m \quad \text{or} \quad \frac{dT}{dt} = k(T - T_m), \tag{10}$$

where k is a constant of proportionality. Since we have assumed the body
is cooling, we must have $T > T_m$, and so it stands to reason that $k < 0$.

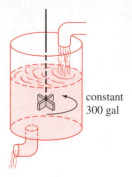

constant
300 gal

FIGURE 1.10

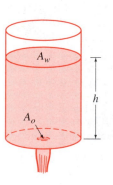

A_w

h

A_o

FIGURE 1.11

Mixtures

The mixing of two salt solutions of differing concentrations gives rise to a first-order differential equation for the amount of salt contained in the mixture. Let us suppose that a large mixing tank holds 300 gallons of water in which salt has been dissolved. Another brine solution is pumped into the large tank at a rate of 3 gal/min, and then when the solution is well stirred it is pumped out at the same rate. See Figure 1.10. If the concentration of the solution entering is 2 lb/gal, determine a model for the amount of salt in the tank at any time.

Let $A(t)$ be the amount of salt (measured in pounds) in the tank at any time. In this case, the rate at which $A(t)$ changes is a net rate:

$$\frac{dA}{dt} = \left(\begin{array}{c}\text{rate of}\\\text{substance entering}\end{array}\right) - \left(\begin{array}{c}\text{rate of}\\\text{substance leaving}\end{array}\right) = R_1 - R_2. \quad \textbf{(11)}$$

Now the rate R_1 at which the salt enters the tank is, in lb/min,

$$R_1 = (3 \text{ gal/min}) \cdot (2 \text{ lb/gal}) = 6 \text{ lb/min},$$

whereas the rate R_2 at which salt is leaving is

$$R_2 = (3 \text{ gal/min}) \cdot \left(\frac{A}{300} \text{ lb/gal}\right) = \frac{A}{100} \text{ lb/min}.$$

Then equation (11) becomes

$$\frac{dA}{dt} = 6 - \frac{A}{100}. \quad \textbf{(12)}$$

Draining a Tank

In hydrodynamics, Torricelli's law states that the speed v of efflux of water through a sharp-edged hole at the bottom of a tank filled to a depth h is the same as the speed that a body (in this case, a drop of water) would acquire in falling freely from a height h; that is, $v = \sqrt{2gh}$, where g is the acceleration due to gravity. This last expression comes from equating the kinetic energy $\frac{1}{2}mv^2$ with the potential energy mgh and solving for v. Suppose a tank filled with water is allowed to drain through a hole under the influence of gravity. We would like to find the depth h of water remaining in the tank at any time t. Consider the tank shown in Figure 1.11. If the area of the hole is A_o (in ft^2) and the speed of the water leaving the tank is $v = \sqrt{2gh}$ (in ft/s), then the volume of water leaving the tank per second is $A_o\sqrt{2gh}$ (in ft^3/s). Thus if $V(t)$ denotes the volume of water in the tank at any time t,

$$\frac{dV}{dt} = -A_o\sqrt{2gh}, \quad \textbf{(13)}$$

where the minus sign indicates that V is decreasing. Note that we are ignoring the possibility of friction at the hole that might cause a reduction of the rate of flow there. Now if the tank is such that the volume of water in it at any time t can be written $V(t) = A_w h$, where A_w (in ft^2) is the *constant* area of the upper surface of the water (see Figure 1.11), then $dV/dt = A_w\, dh/dt$. Substituting this last expression into (13) gives us the desired differential equation for the height of the water at any time:

$$\frac{dh}{dt} = -\frac{A_o}{A_w}\sqrt{2gh}. \quad \textbf{(14)}$$

It is interesting to observe that (14) remains valid even when A_w is not constant. In this case we must express the upper surface area of the water as a function of h: $A_w = A(h)$.

Newton's Second Law of Motion

To construct a mathematical model of the motion of a body moving in a force field one often starts with Newton's second law of motion. Recall from elementary physics that Newton's **first law of motion** states that a body either will remain at rest or will continue to move with a constant velocity unless acted upon by an external force. In each case this is equivalent to saying that when the sum of the forces $\sum F_k$—that is, the *net,* or resultant, force—acting on the body is zero, then the acceleration a of the body is zero. **Newton's second law of motion** indicates that when the net force acting on a body is not zero, then the net force is proportional to its acceleration a; or, more precisely, $\sum F_k = ma$, where m is the mass of the body.

Falling Body

Now suppose a rock is tossed upward from a roof of a building. What is its position at time t? As illustrated in Figure 1.12, let us suppose that its position relative to the ground is $s(t)$. The acceleration of the rock is the second derivative d^2s/dt^2. If we assume that the upward direction is positive, that the mass of the rock is m, and that no force acts on the rock other than the force of gravity, then Newton's second law gives

$$m\frac{d^2s}{dt^2} = -mg \qquad \text{or} \qquad \frac{d^2s}{dt^2} = -g. \qquad (15)$$

Here $W = mg$ is the weight of the rock and g is the acceleration due to gravity. The minus sign is used because the weight of the rock is a force directed downward, which is opposite to the positive direction. If the height of the building is s_0 and the initial velocity of the rock is v_0, then s is determined from the initial-value problem

$$\frac{d^2s}{dt^2} = -g, \qquad s(0) = s_0, \quad s'(0) = v_0. \qquad (16)$$

Although we have not been stressing solutions of the equations we have constructed, we note that (16) can be solved by integrating the constant $-g$ twice with respect to t. The initial conditions determine the two constants of integration.

Falling Bodies and Air Resistance

Under some circumstances a falling body of mass m encounters air resistance proportional to its instantaneous velocity v. If we take, in this circumstance, the positive direction to be oriented downward, then the net force acting on the mass is given by $mg - kv$, where the weight mg of the body is a force acting in the positive direction and air resistance is a force acting in the opposite, or upward, direction. Now since v is related to acceleration a by $a = dv/dt$, Newton's second law becomes $F = ma = m\,dv/dt$. By equating the net force to this form of Newton's second law, we obtain a differential equation for the velocity v of the body at any time:

$$m\frac{dv}{dt} = mg - kv. \qquad (17)$$

Here k is a positive constant of proportionality.

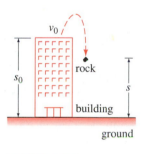

FIGURE 1.12

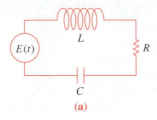

(a)

Inductor
inductance L: henrys (h)
voltage drop across: $L\dfrac{di}{dt}$

$i \rightarrow$ L

Resistor
resistance R: ohms (Ω)
voltage drop across: iR

$i \rightarrow$ R

Capacitor
capacitance C: farads (f)
voltage drop across: $\dfrac{1}{C}q$

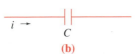

$i \rightarrow$ C

(b)

FIGURE 1.13

Series Circuits   Consider the single-loop series circuit containing an inductor, a resistor, and a capacitor shown in Figure 1.13. The current in a circuit after a switch is closed is denoted by $i(t)$; the charge on a capacitor at time t is denoted by $q(t)$. The letters L, C, and R are constants known as inductance, capacitance, and resistance, respectively. Now according to Kirchhoff's second law, the impressed voltage $E(t)$ on a closed loop must equal the sum of the voltage drops in the loop. Figure 1.13 also shows the symbols and the formulas for the respective voltage drops across an inductor, a capacitor, and a resistor. Since current $i(t)$ is related to charge $q(t)$ on the capacitor by $i = dq/dt$, by adding the voltage drops

$$\text{inductor} = L\frac{di}{dt} = \frac{d^2q}{dt^2}$$

$$\text{resistor} = iR = R\frac{dq}{dt}$$

$$\text{capacitor} = \frac{1}{C}q$$

and equating the sum to the impressed voltage we obtain a second-order differential equation

$$L\frac{d^2q}{dt^2} + R\frac{dq}{dt} + \frac{1}{C}q = E(t). \tag{18}$$

We will examine a differential equation analogous to (18) in great detail in Section 5.1.

Remark

Each example in this section has described a dynamical system: a system that changes or evolves with the flow of time t. Since the study of dynamical systems is a branch of mathematics currently in vogue, we shall occasionally relate the terminology of that field to the discussion at hand.

In more precise terms, a **dynamical system** consists of a set of time-dependent variables, called **state variables,** together with a rule that enables us to determine (without ambiguity) the state of the system (this may be a past, present, or future state) in terms of a state prescribed at some time t_0. Dynamical systems are classified as either discrete-time systems or continuous-time systems. In this course we shall be concerned only with continuous-time dynamical systems, systems in which *all* variables are defined over a continuous range of time. The rule or the mathematical model in a continuous-time dynamical system is a differential equation or a system of differential equations. The **state of the system** at a time t is the value of the state variables at that time; the specified state of the system at a time t_0 is simply the initial conditions that accompany the mathematical model. The solution of an initial-value problem is referred to as the **response of the system.** For example, in the case of radioactive decay the rule is $dA/dt = kA$. Now if the amount of a radioactive substance at some time t_0 is known, say, $A(t_0) = A_0$, then by solving the rule the response of the system for $t \geq t_0$ is found to be $A(t) = A_0 e^{(t-t_0)}$ (see Section 3.1). This solution is unique, and $A(t)$ is the single state variable for this system. In the case of the rock tossed from the building roof, the response of the system, the solution of the differential equation $d^2s/dt^2 = -g$, subject to the initial state $s(0) = s_0$, $s'(0) = v_0$, is the well-known formula from physics $s(t) = -\frac{1}{2} gt^2 + v_0 t + s_0$, $0 \leq t \leq T$, where T represents the time when the rock hits the ground. The state variables are $s(t)$ and $s'(t)$, the vertical position of the rock and its velocity, respectively. Note that acceleration $s''(t)$ is not a state variable since we only have to know any initial position and initial velocity at a time t_0 to uniquely determine the rock's position $s(t)$ and velocity $s'(t) = v(t)$ for any time in the interval $t_0 \leq t \leq T$. The acceleration $s''(t) = a(t)$ at any time is, of course, given by the differential equation $s''(t) = -g$, $0 < t < T$.

One last point: Not every system studied in this text is a dynamical system. We shall also examine some static systems in which the model is a differential equation.

SECTION 1.3 EXERCISES

Answers to odd-numbered problems begin on page A-1.

1. Under the same assumptions underlying the model in (1), determine a differential equation governing the growing population $P(t)$ of a country when individuals are allowed to immigrate into the country at a constant rate r.

2. The model given in (1) fails to take death into consideration; that is, the growth rate equals the birth rate. In another model of a changing population of a community it is assumed that the rate at which the population changes is a net rate—that is, the difference between the rate of births and the rate of deaths in the community. Determine a differential equation governing the population $P(t)$ if the birth rate and the death rate are both proportional to the population present at any time t.

3. A drug is infused into a patient's bloodstream at a constant rate of r g/s. Simultaneously, the drug is removed at a rate proportional to the amount $x(t)$ of the drug present at any time t. Determine a differential equation governing the amount $x(t)$.

4. At a time $t = 0$ a technological innovation is introduced into a community with a fixed population of n people. Determine a differential equation governing the number of people $x(t)$ who have adopted the innovation at any time t.

5. Suppose that a large mixing tank initially holds 300 gallons of water in which 50 pounds of salt has been dissolved. Pure water is pumped into the tank at a rate of 3 gal/min, and then when the solution is well stirred it is pumped out at the same rate. Determine a differential equation for the amount $A(t)$ of salt in the tank at any time t.

6. Suppose that a large mixing tank initially holds 300 gallons of water in which 50 pounds of salt has been dissolved. Another brine solution is pumped into the tank at a rate of 3 gal/min, and then when the solution is well stirred it is pumped out at a *slower* rate of 2 gal/min. If the concentration of the solution entering is 2 lb/gal, determine a differential equation for the amount $A(t)$ of salt in the tank at any time t.

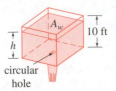

FIGURE 1.14

7. Suppose water is leaking from a tank through a circular hole of area A_o at its bottom. When water leaks through a hole, friction and contraction of the water stream near the hole reduce the volume of the water leaving the tank per second to $cA_o\sqrt{2gh}$, where $0 < c < 1$. Determine a differential equation for the height h of water at any time t for the cubical tank in Figure 1.14. The radius of the hole is 2 inches, and $g = 32$ ft/s^2.

8. A tank in the form of a right circular cylinder of radius 2 feet and height 10 feet is standing on end. The tank is initially full of water, and water leaks from a circular hole of radius $\frac{1}{2}$ inch at its bottom. Use the information in Problem 7 to determine a differential equation for the height h of the water at any time t.

9. A series circuit contains a resistor and an inductor as shown in Figure 1.15. Determine a differential equation for the current $i(t)$ if the resistance is R, the inductance is L, and the impressed voltage is $E(t)$.

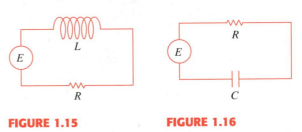

FIGURE 1.15   **FIGURE 1.16**

10. A series circuit contains a resistor and a capacitor as shown in Figure 1.16. Determine a differential equation for the charge $q(t)$ on the capacitor if the resistance is R, the capacitance is C, and the impressed voltage is $E(t)$.

11. In the theory of learning, the rate at which a subject is memorized is assumed to be proportional to the amount that is left to be memorized. Suppose M denotes the total amount of a subject to be memorized and $A(t)$ is the amount memorized in time t. Determine a differential equation for the amount $A(t)$.

12. In Problem 11 assume that the amount of material forgotten is proportional to the amount memorized in time t. Determine a differential equation for $A(t)$ when forgetfulness is taken into account.

13. A person P starts at the origin and moves in the direction of the positive x-axis, pulling a weight along the curve C, called a **tractrix,** as shown in Figure 1.17. The weight, initially located on the y-axis at $(0, s)$, is pulled by a rope of constant length s, which is kept taut throughout the motion. Find the differential equation for the path of motion. Assume that the rope is always tangent to C.

14. For high-speed motion through the air—such as the sky diver shown in Figure 1.18, falling before the parachute is opened—air resistance is closer to a power of the instantaneous velocity. Determine a differential equation for the velocity $v(t)$ of a falling body of mass m if air resistance is proportional to the square of the instantaneous velocity.

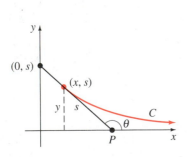

FIGURE 1.17

FIGURE 1.18

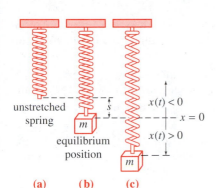

unstretched spring

s

m

equilibrium position

m

$x(t) < 0$

$x = 0$

$x(t) > 0$

m

(a) (b) (c)

FIGURE 1.19

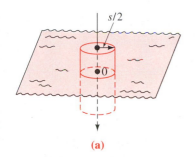

$s/2$

0

(a)

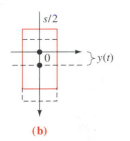

$s/2$

0

$y(t)$

(b)

FIGURE 1.20

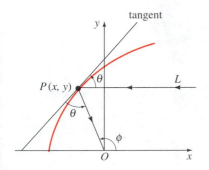

tangent

y

$P(x, y)$

θ

L

θ

ϕ

O

x

FIGURE 1.21

15. After a mass m is attached to a spring, it stretches the spring s units and then hangs at rest in the equilibrium position, as shown in Figure 1.19(b). After the spring/mass system has been set in motion, let $x(t)$ denote the directed distance of the mass beyond the equilibrium position. Assume that the downward direction is positive and that the motion takes place in a vertical straight line through the center of gravity of the mass. Assume, too, that the only forces acting on the system are the weight mg of the mass and the restoring force of the stretched spring, which, by Hooke's law, is proportional to its total elongation. Determine a differential equation for the displacement $x(t)$ for any time t.

16. A cylindrical barrel s feet in diameter of weight w pounds is floating in water. After an initial depression, the barrel exhibits an up-and-down bobbing motion along a vertical line. Using Figure 1.20(b), determine a differential equation for the vertical displacement $y(t)$ if the origin is taken to be on the vertical axis at the surface of the water when the barrel is at rest. Use Archimedes' principle that buoyancy, or upward force of the water on the barrel, is equal to the weight of the water displaced. Assume that the downward direction is positive, that the density of water is 62.4 lb/ft^3, and that there is no resistance between the barrel and the water.

17. As illustrated in Figure 1.21, light rays strike a plane curve C in such a manner that all rays L parallel to the x-axis are reflected to a single point O. Assuming that the angle of incidence is equal to the angle of reflection, determine a differential equation that describes the shape of the curve C. [*Hint:* Inspection of the figure shows that we can write $\phi = 2\theta$. Why? Now use an appropriate trigonometric identity.]

Discussion Problems

18. The differential equation

$$\frac{dP}{dt} = (k \cos t)P,$$

where k is a positive constant, is a model of human population $P(t)$ of a certain community. Discuss an interpretation for the solution of this equation; in other words, what kind of population does the differential equation describe?

19. A large snowball is shaped into the form of a sphere. Starting at some time, which we can designate as $t = 0$, the snowball begins to melt. Assume for the sake of discussion that the snowball melts in such a manner that its shape remains spherical. Discuss the quantities that change with time as the snowball melts. Discuss an interpretation of "melting" as a rate. If possible, construct a mathematical model that describes the state of the snowball at any time $t > 0$.

20. This is a different snow problem—the "snowplow problem." A classic that appears in many differential equations texts, it is due originally to Ralph Palmer Agnew.

One day it started snowing at a heavy and steady rate. A snowplow started out at noon, going 2 miles the first hour and 1 mile the second hour. What time did it start snowing?

Find the text *Differential Equations* by Ralph Palmer Agnew (McGraw-Hill Book Co.), and then discuss the construction and solution of the mathematical model.

21. Suppose a hole is drilled through the center of the earth and a body of mass m is dropped into the hole, as shown in Figure 1.22. Discuss the possible motion of the mass. If possible, construct a mathematical model that describes the motion. Let the distance from the center of the earth to the mass at any time t be denoted by r, let M denote the mass of the earth, let M_r denote the mass of that portion of the earth within a sphere of radius r, and let δ denote the constant density of the earth.

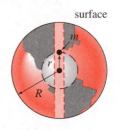

FIGURE 1.22

22. A cup of coffee cools according to Newton's law of cooling (10). Use data from the graph of the temperature $T(t)$ in Figure 1.23 to estimate T_m, T_0, and k in a model of the form

$$\frac{dT}{dt} = k(T - T_m), \qquad T(0) = T_0.$$

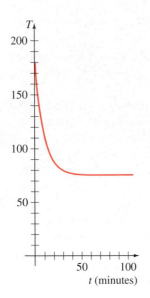

FIGURE 1.23

CHAPTER 1 REVIEW EXERCISES

Answers to odd-numbered problems begin on page A-1.

Answer Problems 1–4 without referring back to the text. Fill in the blank or answer true/false.

1. The differential equation $y' = 1/(25 - x^2 - y^2)$ has a unique solution through any point (x_0, y_0) in the region(s) defined by _____.

2. The initial-value problem $xy' = 3y$, $y(0) = 0$ has the solutions $y = x^3$ and _____.

3. The initial-value problem $y' = y^{1/2}$, $y(0) = 0$ has no solution since $\partial f/\partial y$ is discontinuous on the line $y = 0$. _____

4. There exists an interval centered at 2 on which the unique solution of the initial-value problem $y' = (y - 1)^3$, $y(2) = 1$ is $y = 1$. _____

In Problems 5–8 classify the given differential equation as to type and order. Classify the ordinary differential equations as to linearity.

5. $(2xy - y^2)\, dx + e^x\, dy = 0$

6. $(\sin xy)y''' + 4xy' = 0$

7. $\dfrac{\partial^2 u}{\partial x^2} + \dfrac{\partial^2 u}{\partial y^2} = u$

8. $x^2\dfrac{d^2 y}{dx^2} - 3x\dfrac{dy}{dx} + y = x^2$

In Problems 9–12 verify that the indicated function is a solution of the given differential equation.

9. $y' + 2xy = 2 + x^2 + y^2$; $y = x + \tan x$, $-\frac{\pi}{2} < x < \frac{\pi}{2}$

10. $x^2y'' + xy' + y = 0$; $y = c_1 \cos(\ln x) + c_2 \sin(\ln x)$, $x > 0$

11. $y''' - 2y'' - y' + 2y = 6$; $y = c_1e^x + c_2e^{-x} + c_3e^{2x} + 3$

12. $y^{(4)} - 16y = 0$; $y = \sin 2x + \cosh 2x$

In Problems 13–20 determine by inspection at least one solution for the given differential equation.

13. $y' = 2x$

14. $\dfrac{dy}{dx} = 5y$

15. $y'' = 1$

16. $y' = y^3 - 8$

17. $y'' = y'$

18. $2y\dfrac{dy}{dx} = 1$

19. $y'' = -y$

20. $y'' = y$

21. Determine an interval on which $y^2 - 2y = x^2 - x - 1$ defines a solution of $2(y - 1)\,dy + (1 - 2x)\,dx = 0$.

22. Explain why the differential equation

$$\left(\frac{dy}{dx}\right)^2 = \frac{4 - y^2}{4 - x^2}$$

possesses no real solutions for $|x| < 2, |y| > 2$. Are there other regions in the xy-plane for which the equation has no solutions?

23. The conical tank shown in Figure 1.24 loses water out of a hole at its bottom. If the cross-sectional area of the hole is $\frac{1}{4}$ ft², determine a differential equation representing the height of the water h at any time. Ignore friction and contraction of the water stream at the hole.

24. A weight of 96 pounds slides down an incline making a 30° angle with the horizontal. If the coefficient of sliding friction is μ, determine a differential equation for the velocity $v(t)$ of the weight at any time. Use the fact that the force of friction opposing the motion is μN, where N is the normal component of the weight. See Figure 1.25.

25. By Newton's universal law of gravitation the free-fall acceleration a of a body, such as the satellite shown in Figure 1.26, falling a great distance to the surface is *not* the constant g. Rather, the acceleration a is inversely proportional to the square of the distance r from the center of the earth, $a = k/r^2$, where k is the constant of proportionality.
   (a) Use the fact that at the surface of the earth $r = R$ and $a = g$ to determine the constant of proportionality k.
   (b) Use Newton's second law and part (a) to find a differential equation for the distance r.
   (c) Use the Chain Rule in the form

$$\frac{d^2r}{dt^2} = \frac{dv}{dt} = \frac{dv}{dr}\frac{dr}{dt}$$

to express the differential equation in part (b) as a differential equation involving v and dv/dr.

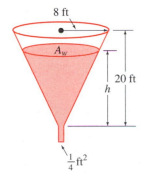

FIGURE 1.24

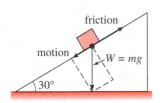

FIGURE 1.25

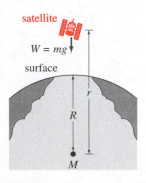

FIGURE 1.26

2

FIRST-ORDER DIFFERENTIAL EQUATIONS

INTRODUCTION

We are now in a position to solve some differential equations. We begin with first-order differential equations. We shall see that if a first-order differential equation can be solved, the method for solving it depends on what kind of equation it is. Over the years mathematicians have struggled to solve many specialized equations. Thus there are many methods of solutions; what works for one kind of first-order equation does not necessarily apply to another kind. In this chapter we focus on three types of equations.

2.1

SEPARABLE VARIABLES

■ *Solution by integration* ■ *Definition of a separable DE* ■ *Method of solution*
■ *Losing a solution* ■ *Alternative forms*

Note

In solving differential equations you will often have to integrate, and this, in turn, may require some special technique of integration. It will be worth a few minutes of your time to review your calculus text or, if a CAS (computer algebra system) is available, review the command syntax for carrying out basic integration, integration by parts, and partial fractions.

Solution by Integration   We begin our study of the methodology of solving first-order equations $dy/dx = f(x, y)$ with the simplest of all differential equations. When f is independent of the variable y, that is, $f(x, y) = g(x)$, the differential equation

$$\frac{dy}{dx} = g(x) \tag{1}$$

can be solved by integration. If $g(x)$ is a continuous function then integrating both sides of (1) gives the solution

$$y = \int g(x)\, dx = G(x) + c,$$

where $G(x)$ is an antiderivative (indefinite integral) of $g(x)$. For example:

If $\dfrac{dy}{dx} = 1 + e^{2x}$ then $y = \displaystyle\int (1 + e^{2x})\, dx = x + \frac{1}{2} e^{2x} + c.$

Equation (1), as well as its method of solution, is just a special case when f in $dy/dx = f(x, y)$ is a product of a function of x and a function of y.

DEFINITION 2.1   **Separable Equation**

A first-order differential equation of the form

$$\frac{dy}{dx} = g(x)\, h(y)$$

is said to be **separable** or to have **separable variables.**

Observe that by dividing by the function $h(y)$ a separable equation can be written as

$$p(y)\frac{dy}{dx} = g(x), \tag{2}$$

where, for convenience, we have denoted $1/h(y)$ by $p(y)$. From this last form we can see immediately that (2) reduces to (1) when $h(y) = 1$.

Now if $y = \phi(x)$ represents a solution of (2), we must have $p(\phi(x))\phi'(x) = g(x)$, and therefore

$$\int p(\phi(x))\phi'(x)\, dx = \int g(x)\, dx. \tag{3}$$

But $dy = \phi'(x)\, dx$, and so (3) is the same as

$$\int p(y)\, dy = \int g(x)\, dx \qquad \text{or} \qquad H(y) = G(x) + c, \tag{4}$$

where $H(y)$ and $G(x)$ are antiderivatives of $p(y) = 1/h(y)$ and $g(x)$, respectively.

Method of Solution

Equation (4) indicates the procedure for solving separable equations. A one-parameter family of solutions, usually given implicitly, is obtained by integrating both sides of $p(y)\, dy = g(x)\, dx$.

Note

There is no need to use two constants in the integration of a separable equation, because if we write $H(y) + c_1 = G(x) + c_2$ then the difference $c_2 - c_1$ can be replaced by a single constant c, as in (4). In many instances throughout the chapters that follow, we will relabel constants in a manner convenient to a given equation. For example, multiples of constants or combinations of constants can sometimes be replaced by a single constant.

EXAMPLE 1   **Solving a Separable DE**

Solve $(1 + x)\, dy - y\, dx = 0$.

SOLUTION   Dividing by $(1 + x)y$, we can write $dy/y = dx/(1 + x)$, from which it follows that

$$\int \frac{dy}{y} = \int \frac{dx}{1 + x}$$

$$\ln|y| = \ln|1 + x| + c_1$$

$$y = e^{\ln|1+x|+c_1}$$

$$= e^{\ln|1+x|} \cdot e^{c_1} \qquad \leftarrow \text{laws of exponents}$$

$$= |1 + x|e^{c_1}$$

$$= \pm\, e^{c_1}(1 + x). \qquad \leftarrow \begin{cases} |1 + x| = 1 + x,\, x \geq -1 \\ |1 + x| = -(1 + x),\, x < -1 \end{cases}$$

Relabeling $\pm e^{c_1}$ as c then gives $y = c(1 + x)$.

ALTERNATIVE SOLUTION   Since each integral results in a logarithm, a judicious choice for the constant of integration is $\ln|c|$ rather than c:

$$\ln|y| = \ln|1 + x| + \ln|c| \qquad \text{or} \qquad \ln|y| = \ln|c(1 + x)|$$

so that

$$y = c(1 + x).$$

Even if the indefinite integrals are not *all* logarithms, it may still be advantageous to use $\ln|c|$. However, no firm rule can be given. ∎

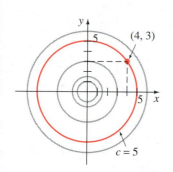

FIGURE 2.1

EXAMPLE 2 An Initial-Value Problem

Solve the initial-value problem $\dfrac{dy}{dx} = -\dfrac{x}{y}$, $y(4) = 3$.

SOLUTION From $y \, dy = -x \, dx$ we get

$$\int y \, dy = -\int x \, dx \quad \text{and} \quad \frac{y^2}{2} = -\frac{x^2}{2} + c_1.$$

We can write this solution as $x^2 + y^2 = c^2$ by replacing the constant $2c_1$ by c^2. The solution represents a family of concentric circles.

Now when $x = 4$, $y = 3$ so that $16 + 9 = 25 = c^2$. Thus the initial-value problem determines $x^2 + y^2 = 25$. In view of Theorem 1.1 we can conclude that it is the only circle of the family passing through the point $(4, 3)$. See Figure 2.1. ∎

Some care should be exercised when separating variables, since the variable divisors could be zero at a point. As the next two examples show, a constant solution may sometimes get lost in the shuffle of solving the problem. Also see Problem 58 in Exercises 2.1.

EXAMPLE 3 Losing a Solution

Solve $xy^4 \, dx + (y^2 + 2)e^{-3x} \, dy = 0.$ **(5)**

SOLUTION By multiplying the equation by e^{3x} and dividing by y^4, we obtain

termwise division → $\quad xe^{3x} \, dx + \dfrac{y^2 + 2}{y^4} \, dy = 0 \quad \text{or} \quad xe^{3x} \, dx + (y^{-2} + 2y^{-4}) \, dy = 0.$ **(6)**

Using integration by parts on the first term yields

$$\frac{1}{3}xe^{3x} - \frac{1}{9}e^{3x} - y^{-1} - \frac{2}{3}y^{-3} = c_1.$$

The one-parameter family of solutions can also be written as

$$e^{3x}(3x - 1) = \frac{9}{y} + \frac{6}{y^3} + c,$$ **(7)**

where the constant $9c_1$ is rewritten as c. Observe that $y = 0$ is a perfectly good solution of (5) but is not a member of the set of solutions defined by (7). ∎

EXAMPLE 4 An Initial-Value Problem

Solve the initial-value problem $\dfrac{dy}{dx} = y^2 - 4$, $y(0) = -2$.

SOLUTION We put the equation into the form

$$\frac{dy}{y^2 - 4} = dx$$ **(8)**

and use partial fractions on the left side. We have

$$\left[\frac{-1/4}{y+2} + \frac{1/4}{y-2}\right] dy = dx \qquad \textbf{(9)}$$

so that
$$-\frac{1}{4}\ln|y+2| + \frac{1}{4}\ln|y-2| = x + c_1. \qquad \textbf{(10)}$$

In this case it is easy to solve the implicit solution (10) for y in terms of x. Multiplying the equation by 4 and combining logarithms gives

$$\ln\left|\frac{y-2}{y+2}\right| = 4x + c_2 \qquad \text{and so} \qquad \frac{y-2}{y+2} = ce^{4x}.$$

Here we have replaced $4c_1$ by c_2 and e^{c_2} by c. Finally, solving the last equation for y, we get

$$y = 2\frac{1 + ce^{4x}}{1 - ce^{4x}}. \qquad \textbf{(11)}$$

Substituting $x = 0$, $y = -2$ leads to the mathematical dilemma

$$-2 = 2\frac{1+c}{1-c} \quad \text{or} \quad -1 + c = 1 + c \quad \text{or} \quad -1 = 1.$$

The last equality prompts us to examine the differential equation a little more carefully. The fact is, the equation

$$\frac{dy}{dx} = (y+2)(y-2)$$

is satisfied by two constant functions, namely $y = -2$ and $y = 2$. Inspection of equations (8), (9), and (10) clearly indicates we must preclude $y = -2$ and $y = 2$ at those steps in our solution. But it is interesting to observe that we can subsequently recover the solution $y = 2$ by setting $c = 0$ in equation (11). However, there is no finite value of c that will ever yield the solution $y = -2$. This latter constant function is the only solution to the original initial-value problem. See Figure 2.2. ∎

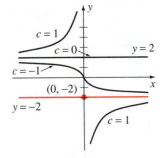

FIGURE 2.2

If, in Example 4, we had used $\ln|c|$ for the constant of integration, then the form of the one-parameter family of solutions would be

$$y = 2\frac{c + e^{4x}}{c - e^{4x}}. \qquad \textbf{(12)}$$

Note that (12) reduces to $y = -2$ when $c = 0$, but now there is no finite value of c that will give the constant solution $y = 2$.

If an initial condition leads to a particular solution by finding a specific value of the parameter c in a family of solutions for a first-order differential equation, it is a natural inclination of most students (and instructors) to relax and be content. In Section 1.2, however, we saw that a solution of an initial-value problem may not be unique. For instance, in Example 3 of that section we saw that the problem

$$\frac{dy}{dx} = xy^{1/2}, \quad y(0) = 0 \qquad \textbf{(13)}$$

has at least two solutions, namely $y = 0$ and $y = x^4/16$. We are now in a position to solve the equation. Separating variables, we have

$$y^{-1/2}\, dy = x\, dx,$$

and integrating gives

$$2y^{1/2} = \frac{x^2}{2} + c_1 \qquad \text{or} \qquad y = \left(\frac{x^2}{4} + c\right)^2.$$

When $x = 0$, $y = 0$, so necessarily $c = 0$. Therefore $y = x^4/16$. The solution $y = 0$ was lost by dividing by $y^{1/2}$. In addition, the initial-value problem (13) possesses infinitely more solutions, since for any choice of the parameter $a \geq 0$ the piecewise-defined function

$$y = \begin{cases} 0, & x < a \\ \dfrac{(x^2 - a^2)^2}{16}, & x \geq a \end{cases}$$

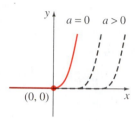

(0, 0)

FIGURE 2.3

satisfies both the differential equation and the initial condition. See Figure 2.3.

Remark

In some of the preceding examples we saw that the constant in the one-parameter family of solutions for a first-order differential equation can be relabeled when convenient. Also, it can easily happen that two individuals solving the same equation correctly arrive at dissimilar expressions for their answers. For example, by separation of variables we can show that one-parameter families of solutions for $(1 + y^2)\, dx + (1 + x^2)\, dy = 0$ are

$$\arctan x + \arctan y = c \qquad \text{or} \qquad \frac{x + y}{1 - xy} = c.$$

As you work your way through the next several sections, keep in mind that families of solutions may be equivalent in the sense that one family may be obtained from another by either relabeling the constant or applying algebra and trigonometry.

SECTION 2.1 EXERCISES

Answers to odd-numbered problems begin on page A-1.

In Problems 1–40 solve the given differential equation by separation of variables.

1. $\dfrac{dy}{dx} = \sin 5x$

2. $\dfrac{dy}{dx} = (x + 1)^2$

3. $dx + e^{3x}\, dy = 0$

4. $dx - x^2\, dy = 0$

5. $(x + 1)\dfrac{dy}{dx} = x + 6$

6. $e^x \dfrac{dy}{dx} = 2x$

7. $xy' = 4y$

8. $\dfrac{dy}{dx} + 2xy = 0$

9. $\dfrac{dy}{dx} = \dfrac{y^3}{x^2}$

10. $\dfrac{dy}{dx} = \dfrac{y+1}{x}$

11. $\dfrac{dx}{dy} = \dfrac{x^2 y^2}{1+x}$

12. $\dfrac{dx}{dy} = \dfrac{1+2y^2}{y \sin x}$

13. $\dfrac{dy}{dx} = e^{3x+2y}$

14. $e^x y \dfrac{dy}{dx} = e^{-y} + e^{-2x-y}$

15. $(4y + yx^2)\, dy - (2x + xy^2)\, dx = 0$

16. $(1 + x^2 + y^2 + x^2 y^2)\, dy = y^2\, dx$

17. $2y(x+1)\, dy = x\, dx$

18. $x^2 y^2\, dy = (y+1)\, dx$

19. $y \ln x \dfrac{dx}{dy} = \left(\dfrac{y+1}{x}\right)^2$

20. $\dfrac{dy}{dx} = \left(\dfrac{2y+3}{4x+5}\right)^2$

21. $\dfrac{dS}{dr} = kS$

22. $\dfrac{dQ}{dt} = k(Q - 70)$

23. $\dfrac{dP}{dt} = P - P^2$

24. $\dfrac{dN}{dt} + N = Nte^{t+2}$

25. $\sec^2 x\, dy + \csc y\, dx = 0$

26. $\sin 3x\, dx + 2y \cos^3 3x\, dy = 0$

27. $e^y \sin 2x\, dx + \cos x(e^{2y} - y)\, dy = 0$

28. $\sec x\, dy = x \cot y\, dx$

29. $(e^y + 1)^2 e^{-y}\, dx + (e^x + 1)^3 e^{-x}\, dy = 0$

30. $\dfrac{y\, dy}{x\, dx} = (1 + x^2)^{-1/2}(1 + y^2)^{1/2}$

31. $(y - yx^2)\dfrac{dy}{dx} = (y+1)^2$

32. $2\dfrac{dy}{dx} - \dfrac{1}{y} = \dfrac{2x}{y}$

33. $\dfrac{dy}{dx} = \dfrac{xy + 3x - y - 3}{xy - 2x + 4y - 8}$

34. $\dfrac{dy}{dx} = \dfrac{xy + 2y - x - 2}{xy - 3y + x - 3}$

35. $\dfrac{dy}{dx} = \sin x(\cos 2y - \cos^2 y)$

36. $\sec y \dfrac{dy}{dx} + \sin(x - y) = \sin(x + y)$

37. $x\sqrt{1 - y^2}\, dx = dy$

38. $y(4 - x^2)^{1/2}\, dy = (4 + y^2)^{1/2}\, dx$

39. $(e^x + e^{-x})\dfrac{dy}{dx} = y^2$

40. $(x + \sqrt{x})\dfrac{dy}{dx} = y + \sqrt{y}$

In Problems 41–48 solve the given differential equation subject to the indicated initial condition.

41. $(e^{-y} + 1) \sin x\, dx = (1 + \cos x)\, dy, \quad y(0) = 0$

42. $(1 + x^4)\, dy + x(1 + 4y^2)\, dx = 0, \quad y(1) = 0$

43. $y\, dy = 4x(y^2 + 1)^{1/2}\, dx, \quad y(0) = 1$

44. $\dfrac{dy}{dt} + ty = y, \quad y(1) = 3$

45. $\dfrac{dx}{dy} = 4(x^2 + 1), \quad x\left(\dfrac{\pi}{4}\right) = 1$

46. $\dfrac{dy}{dx} = \dfrac{y^2 - 1}{x^2 - 1}, \quad y(2) = 2$

47. $x^2 y' = y - xy, \quad y(-1) = -1$

48. $y' + 2y = 1, \quad y(0) = \dfrac{5}{2}$

In Problems 49 and 50 find a solution of the given differential equation that passes through the indicated points.

49. $\dfrac{dy}{dx} - y^2 = -9$

    (a) $(0, 0)$ **(b)** $(0, 3)$ **(c)** $\left(\dfrac{1}{3}, 1\right)$

50. $x\dfrac{dy}{dx} = y^2 - y$

    (a) $(0, 1)$ **(b)** $(0, 0)$ **(c)** $\left(\dfrac{1}{2}, \dfrac{1}{2}\right)$

51. Find a singular solution for the equation in Problem 37.

52. Find a singular solution for the equation in Problem 39.

Often a radical change in the solution of a differential equation corresponds to a very small change in either the initial condition or the equation itself. In Problems 53–56 compare the solutions of the given initial-value problems.

53. $\dfrac{dy}{dx} = (y - 1)^2$, $y(0) = 1$

54. $\dfrac{dy}{dx} = (y - 1)^2$, $y(0) = 1.01$

55. $\dfrac{dy}{dx} = (y - 1)^2 + 0.01$, $y(0) = 1$

56. $\dfrac{dy}{dx} = (y - 1)^2 - 0.01$, $y(0) = 1$

Discussion Problems

57. Before discussing the two parts of this problem, reread Theorem 1.1 and the definition of an implicit solution.

    (a) Find an *explicit* solution of the initial-value problem $dy/dx = -x/y$, $y(4) = 3$. Give the largest interval I of validity for this solution. (See Example 2.)

    (b) Do you think that $x^2 + y^2 = 1$ is an *implicit* solution of the initial-value problem $dy/dx = -x/y$, $y(1) = 0$?

58. Consider the general form of a separable first-order differential equation $dy/dx = g(x)h(y)$. If r is a number such that $h(r) = 0$, explain why $y = r$ must be a constant solution of the equation.

2.2   EXACT EQUATIONS

■ *Exact differential* ■ *Definition of an exact DE* ■ *Method of solution*

Although the simple equation $y\, dx + x\, dy = 0$ is separable, it is also equivalent to the differential of the product of x and y; that is,

$$y\, dx + x\, dy = d(xy) = 0.$$

By integrating we immediately obtain the implicit solution $xy = c$.

You might remember from calculus that if $z = f(x, y)$ is a function having continuous first partial derivatives in a region R of the xy-plane, then its differential (also called the total differential) is

$$dz = \frac{\partial f}{\partial x} dx + \frac{\partial f}{\partial y} dy. \tag{1}$$

Now if $f(x, y) = c$, it follows from (1) that

$$\frac{\partial f}{\partial x} dx + \frac{\partial f}{\partial y} dy = 0. \tag{2}$$

In other words, given a family of curves $f(x, y) = c$, we can generate a first-order differential equation by computing the total differential. For example, if $x^2 - 5xy + y^3 = c$, then (2) gives

$$(2x - 5y)\, dx + (-5x + 3y^2)\, dy = 0 \qquad \text{or} \qquad \frac{dy}{dx} = \frac{5y - 2x}{-5x + 3y^2}.$$

For our purposes it is more important to turn the problem around; namely, given an equation such as

$$\frac{dy}{dx} = \frac{5y - 2x}{-5x + 3y^2}, \tag{3}$$

can we identify the equation as being equivalent to the statement

$$d(x^2 - 5xy + y^3) = 0?$$

DEFINITION 2.2   **Exact Equation**

A differential expression $M(x, y)\, dx + N(x, y)\, dy$ is an **exact differential** in a region R of the xy-plane if it corresponds to the differential of some function $f(x, y)$. A first-order differential equation of the form

$$M(x, y)\, dx + N(x, y)\, dy = 0$$

is said to be an **exact equation** if the expression on the left side is an exact differential.

EXAMPLE 1   **An Exact DE**

The equation $x^2 y^3\, dx + x^3 y^2\, dy = 0$ is exact, because it is recognized that

$$d\left(\frac{1}{3} x^3 y^3\right) = x^2 y^3\, dx + x^3 y^2\, dy.$$

∎

Notice in Example 1 that if $M(x, y) = x^2 y^3$ and $N(x, y) = x^3 y^2$ then $\partial M/\partial y = 3x^2 y^2 = \partial N/\partial x$. Theorem 2.1 shows that this equality of partial derivatives is no coincidence.

THEOREM 2.1   **Criterion for an Exact Differential**

Let $M(x, y)$ and $N(x, y)$ be continuous and have continuous first partial derivatives in a rectangular region R defined by $a < x < b$, $c < y < d$. Then a necessary and sufficient condition that $M(x, y)\, dx + N(x, y)\, dy$ be an exact differential is

$$\frac{\partial M}{\partial y} = \frac{\partial N}{\partial x}. \tag{4}$$

PROOF OF THE NECESSITY   For simplicity let us assume that $M(x, y)$ and $N(x, y)$ have continuous first partial derivatives for all (x, y). Now if the expression $M(x, y)\, dx + N(x, y)\, dy$ is exact, there exists some function f such that for all x in R,

$$M(x, y)\, dx + N(x, y)\, dy = \frac{\partial f}{\partial x}\, dx + \frac{\partial f}{\partial y}\, dy.$$

 Therefore,

$$M(x, y) = \frac{\partial f}{\partial x}, \qquad N(x, y) = \frac{\partial f}{\partial y},$$

and

$$\frac{\partial M}{\partial y} = \frac{\partial}{\partial y}\left(\frac{\partial f}{\partial x}\right) = \frac{\partial^2 f}{\partial y\, \partial x} = \frac{\partial}{\partial x}\left(\frac{\partial f}{\partial y}\right) = \frac{\partial N}{\partial x}.$$

The equality of the mixed partials is a consequence of the continuity of the first partial derivatives of $M(x, y)$ and $N(x, y)$. ∎

The sufficiency part of Theorem 2.1 consists of showing that there exists a function f for which $\partial f/\partial x = M(x, y)$ and $\partial f/\partial y = N(x, y)$ whenever (4) holds. The construction of the function f actually reflects a basic procedure for solving exact equations.

Method of Solution   Given an equation of the form $M(x, y)\, dx + N(x, y)\, dy = 0$, determine whether the equality in (4) holds. If it does, then there exists a function f for which

$$\frac{\partial f}{\partial x} = M(x, y).$$

We can find f by integrating $M(x, y)$ with respect to x, while holding y constant:

$$f(x, y) = \int M(x, y)\, dx + g(y), \tag{5}$$

where the arbitrary function $g(y)$ is the "constant" of integration. Now differentiate (5) with respect to y and assume $\partial f/\partial y = N(x, y)$:

$$\frac{\partial f}{\partial y} = \frac{\partial}{\partial y}\int M(x, y)\, dx + g'(y) = N(x, y).$$

This gives

$$g'(y) = N(x, y) - \frac{\partial}{\partial y}\int M(x, y)\, dx. \tag{6}$$

OUR TIME WILL COME

Finally, integrate (6) with respect to y and substitute the result in (5). The solution of the equation is $f(x, y) = c$.

Note

Some observations are in order. First, it is important to realize that the expression $N(x, y) - (\partial/\partial y) \int M(x, y)\, dx$ in (6) is independent of x, because

$$\frac{\partial}{\partial x}\left[N(x, y) - \frac{\partial}{\partial y} \int M(x, y)\, dx \right] = \frac{\partial N}{\partial x} - \frac{\partial}{\partial y}\left(\frac{\partial}{\partial x} \int M(x, y)\, dx \right) = \frac{\partial N}{\partial x} - \frac{\partial M}{\partial y} = 0.$$

Second, we could just as well start the foregoing procedure with the assumption that $\partial f / \partial y = N(x, y)$. After integrating N with respect to y and then differentiating that result, we would find the analogues of (5) and (6) to be, respectively,

$$f(x, y) = \int N(x, y)\, dy + h(x) \qquad \text{and} \qquad h'(x) = M(x, y) - \frac{\partial}{\partial x} \int N(x, y)\, dy.$$

In either case *none of these formulas should be memorized.*

EXAMPLE 2   Solving an Exact DE

Solve $2xy\, dx + (x^2 - 1)\, dy = 0$.

SOLUTION   With $M(x, y) = 2xy$ and $N(x, y) = x^2 - 1$ we have

$$\frac{\partial M}{\partial y} = 2x = \frac{\partial N}{\partial x}.$$

Thus the equation is exact, and so, by Theorem 2.1, there exists a function $f(x, y)$ such that

$$\frac{\partial f}{\partial x} = 2xy \qquad \text{and} \qquad \frac{\partial f}{\partial y} = x^2 - 1.$$

From the first of these equations we obtain, after integrating,

$$f(x, y) = x^2 y + g(y).$$

Taking the partial derivative of the last expression with respect to y and setting the result equal to $N(x, y)$ gives

$$\frac{\partial f}{\partial y} = x^2 + g'(y) = x^2 - 1. \qquad \leftarrow N(x, y)$$

It follows that $g'(y) = -1$ and $g(y) = -y$.

The constant of integration need not be included in the preceding line because the solution is $f(x, y) = c$. Some of the family of curves $x^2 y - y = c$ are given in Figure 2.4.

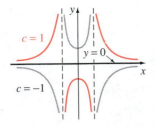

FIGURE 2.4

Note

The solution of the equation is *not* $f(x, y) = x^2 y - y$. Rather it is $f(x, y) = c$ or $f(x, y) = 0$ if a constant is used in the integration of $g'(y)$. Observe that the equation could also be solved by separation of variables.

EXAMPLE 3   **Solving an Exact DE**

Solve $(e^{2y} - y \cos xy) \, dx + (2xe^{2y} - x \cos xy + 2y) \, dy = 0$.

SOLUTION   The equation is exact because

$$\frac{\partial M}{\partial y} = 2e^{2y} + xy \sin xy - \cos xy = \frac{\partial N}{\partial x}.$$

Hence a function $f(x, y)$ exists for which

$$M(x, y) = \frac{\partial f}{\partial x} \quad \text{and} \quad N(x, y) = \frac{\partial f}{\partial y}.$$

Now for variety we shall start with the assumption that $\partial f / \partial y = N(x, y)$;

that is,

$$\frac{\partial f}{\partial y} = 2xe^{2y} - x \cos xy + 2y$$

$$f(x, y) = 2x \int e^{2y} \, dy - x \int \cos xy \, dy + 2 \int y \, dy.$$

Remember, the reason x can come out in front of the symbol $\int$ is that in the integration with respect to y, x is treated as an ordinary constant. It follows that

$$f(x, y) = xe^{2y} - \sin xy + y^2 + h(x)$$

$$\frac{\partial f}{\partial x} = e^{2y} - y \cos xy + h'(x) = e^{2y} - y \cos xy, \qquad \leftarrow M(x, y)$$

and so $h'(x) = 0$ or $h(x) = c$. Hence a family of solutions is

$$xe^{2y} - \sin xy + y^2 + c = 0. \qquad \blacksquare$$

EXAMPLE 4   **An Initial-Value Problem**

Solve the initial-value problem

$$(\cos x \sin x - xy^2) \, dx + y(1 - x^2) \, dy = 0, \quad y(0) = 2.$$

SOLUTION   The equation is exact because

$$\frac{\partial M}{\partial y} = -2xy = \frac{\partial N}{\partial x}.$$

Now

$$\frac{\partial f}{\partial y} = y(1 - x^2)$$

$$f(x, y) = \frac{y^2}{2}(1 - x^2) + h(x)$$

$$\frac{\partial f}{\partial x} = -xy^2 + h'(x) = \cos x \sin x - xy^2.$$

The last equation implies that $h'(x) = \cos x \sin x$. Integrating gives

$$h(x) = -\int (\cos x)(-\sin x \, dx) = -\frac{1}{2} \cos^2 x.$$

Thus

$$\frac{y^2}{2}(1-x^2) - \frac{1}{2}\cos^2 x = c_1 \qquad \text{or} \qquad y^2(1-x^2) - \cos^2 x = c,$$

where $2c_1$ has been replaced by c. The initial condition $y = 2$ when $x = 0$ demands that $4(1) - \cos^2(0) = c$ or that $c = 3$. Thus a solution of the problem is $y^2(1-x^2) - \cos^2 x = 3$. ∎

Remark

When testing an equation for exactness, make sure it is of the precise form $M(x, y)\, dx + N(x, y)\, dy = 0$. Sometimes a differential equation is written $G(x, y)\, dx = H(x, y)\, dy$. In this case, first rewrite it as $G(x, y)\, dx - H(x, y)\, dy = 0$ and then identify $M(x, y) = G(x, y)$ and $N(x, y) = -H(x, y)$ before using (4).

SECTION 2.2 EXERCISES

Answers to odd-numbered problems begin on page A-1.

In Problems 1–24 determine whether the given equation is exact. If it is exact, solve it.

1. $(2x - 1)\, dx + (3y + 7)\, dy = 0$

2. $(2x + y)\, dx - (x + 6y)\, dy = 0$

3. $(5x + 4y)\, dx + (4x - 8y^3)\, dy = 0$

4. $(\sin y - y \sin x)\, dx + (\cos x + x \cos y - y)\, dy = 0$

5. $(2y^2x - 3)\, dx + (2yx^2 + 4)\, dy = 0$

6. $\left(2y - \frac{1}{x} + \cos 3x\right)\frac{dy}{dx} + \frac{y}{x^2} - 4x^3 + 3y \sin 3x = 0$

7. $(x + y)(x - y)\, dx + x(x - 2y)\, dy = 0$

8. $\left(1 + \ln x + \frac{y}{x}\right) dx = (1 - \ln x)\, dy$

9. $(y^3 - y^2 \sin x - x)\, dx + (3xy^2 + 2y \cos x)\, dy = 0$

10. $(x^3 + y^3)\, dx + 3xy^2\, dy = 0$

11. $(y \ln y - e^{-xy})\, dx + \left(\frac{1}{y} + x \ln y\right) dy = 0$

12. $\frac{2x}{y}\, dx - \frac{x^2}{y^2}\, dy = 0$

13. $x\frac{dy}{dx} = 2xe^x - y + 6x^2$

14. $(3x^2y + e^y)\, dx + (x^3 + xe^y - 2y)\, dy = 0$

15. $\left(1 - \frac{3}{x} + y\right) dx + \left(1 - \frac{3}{y} + x\right) dy = 0$

16. $(e^y + 2xy \cosh x)y' + xy^2 \sinh x + y^2 \cosh x = 0$

17. $\left(x^2y^3 - \dfrac{1}{1 + 9x^2}\right)\dfrac{dx}{dy} + x^3y^2 = 0$

18. $(5y - 2x)y' - 2y = 0$

19. $(\tan x - \sin x \sin y)\,dx + \cos x \cos y\,dy = 0$

20. $(3x \cos 3x + \sin 3x - 3)\,dx + (2y + 5)\,dy = 0$

21. $(1 - 2x^2 - 2y)\dfrac{dy}{dx} = 4x^3 + 4xy$

22. $(2y \sin x \cos x - y + 2y^2 e^{xy^2})\,dx = (x - \sin^2 x - 4xye^{xy^2})\,dy$

23. $(4x^3y - 15x^2 - y)\,dx + (x^4 + 3y^2 - x)\,dy = 0$

24. $\left(\dfrac{1}{x} + \dfrac{1}{x^2} - \dfrac{y}{x^2 + y^2}\right)dx + \left(ye^y + \dfrac{x}{x^2 + y^2}\right)dy = 0$

In Problems 25–30 solve the given differential equation subject to the indicated initial condition.

25. $(x + y)^2\,dx + (2xy + x^2 - 1)\,dy = 0, \quad y(1) = 1$

26. $(e^x + y)\,dx + (2 + x + ye^y)\,dy = 0, \quad y(0) = 1$

27. $(4y + 2x - 5)\,dx + (6y + 4x - 1)\,dy = 0, \quad y(-1) = 2$

28. $\left(\dfrac{3y^2 - x^2}{y^5}\right)\dfrac{dy}{dx} + \dfrac{x}{2y^4} = 0, \quad y(1) = 1$

29. $(y^2 \cos x - 3x^2y - 2x)\,dx + (2y \sin x - x^3 + \ln y)\,dy = 0, \quad y(0) = e$

30. $\left(\dfrac{1}{1 + y^2} + \cos x - 2xy\right)\dfrac{dy}{dx} = y(y + \sin x), \quad y(0) = 1$

In Problems 31–34 find the value of k so that the given differential equation is exact.

31. $(y^3 + kxy^4 - 2x)\,dx + (3xy^2 + 20x^2y^3)\,dy = 0$

32. $(2x - y \sin xy + ky^4)\,dx - (20xy^3 + x \sin xy)\,dy = 0$

33. $(2xy^2 + ye^x)\,dx + (2x^2y + ke^x - 1)\,dy = 0$

34. $(6xy^3 + \cos y)\,dx + (kx^2y^2 - x \sin y)\,dy = 0$

35. Determine a function $M(x, y)$ so that the following differential equation is exact:

$$M(x, y)\,dx + \left(xe^{xy} + 2xy + \dfrac{1}{x}\right)dy = 0.$$

36. Determine a function $N(x, y)$ so that the following differential equation is exact:

$$\left(y^{1/2}x^{-1/2} + \dfrac{x}{x^2 + y}\right)dx + N(x, y)\,dy = 0.$$

It is sometimes possible to transform a nonexact differential equation $M(x, y)\,dx + N(x, y)\,dy = 0$ into an exact equation by multiplying it by an integrating factor $\mu(x, y)$. In Problems 37–42 solve the given equation by verifying that the indicated function $\mu(x, y)$ is an integrating factor.

37. $6xy\,dx + (4y + 9x^2)\,dy = 0. \quad \mu(x, y) = y^2$

38. $-y^2\,dx + (x^2 + xy)\,dy = 0, \quad \mu(x, y) = \dfrac{1}{x^2y}$

39. $(-xy \sin x + 2y \cos x) \, dx + 2x \cos x \, dy = 0, \quad \mu(x, y) = xy$

40. $y(x + y + 1) \, dx + (x + 2y) \, dy = 0, \quad \mu(x, y) = e^x$

41. $(2y^2 + 3x) \, dx + 2xy \, dy = 0, \quad \mu(x, y) = x$

42. $(x^2 + 2xy - y^2) \, dx + (y^2 + 2xy - x^2) \, dy = 0, \quad \mu(x, y) = (x + y)^{-2}$

Discussion Problem

43. Consider the concept of an integrating factor introduced in Problems 37–42. Are the two equations $M \, dx + N \, dy = 0$ and $\mu M \, dx + \mu N \, dy = 0$ necessarily equivalent in the sense that a solution of one is also a solution of the other?

2.3 LINEAR EQUATIONS

■ *Definition of a linear DE* ■ *Standard form of a linear equation* ■ *Variation of parameters*
■ *Method of solution* ■ *Integrating factor* ■ *General solution* ■ *Function defined by an integral*

In Chapter 1 we defined the general form of a linear differential equation of order n to be

$$a_n(x) \frac{d^n y}{dx^n} + a_{n-1}(x) \frac{d^{n-1} y}{dx^{n-1}} + \cdots + a_1(x) \frac{dy}{dx} + a_0(x)y = g(x).$$

Remember that linearity means that all coefficients are functions of x only, and that y and all its derivatives are raised to the first power. Now when $n = 1$, we obtain a linear first-order equation.

> **DEFINITION 2.3**   **Linear Equation**
>
> A first-order differential equation of the form
>
> $$a_1(x) \frac{dy}{dx} + a_0(x)y = g(x) \tag{1}$$
>
> is said to be a **linear equation.**

Dividing both sides of (1) by the lead coefficient $a_1(x)$ yields a more useful form, the **standard form,** of a linear equation:

$$\frac{dy}{dx} + P(x)y = f(x). \tag{2}$$

We seek a solution of (2) on an interval I for which both functions P and f are continuous.

In the discussion that follows we illustrate a property and a procedure and end up with a formula representing a solution of (2). The property and the procedure are more important than the formula since both of these concepts carry over to linear equations of higher order.

The Property   You should be able to verify by direct substitution that the differential equation (2) has the property that its solution is the **sum** of the two solutions, $y = y_c + y_p$, where y_c is a solution of

$$\frac{dy}{dx} + P(x)y = 0 \tag{3}$$

and y_p is a particular solution of (2). We can find y_c by separation of variables. Writing (3) as

$$\frac{dy}{y} + P(x)\, dx = 0,$$

integrating, and solving for y gives $y_c = ce^{-\int P(x)dx}$. For convenience let us write $y_c = cy_1(x)$, where $y_1 = e^{-\int P(x)dx}$. The fact that $dy_1/dx + P(x)y_1 = 0$ will be used immediately in determining y_p.

The Procedure   We can now find a particular solution of equation (2) by a procedure known as **variation of parameters.** The basic idea here is to find a function u so that $y_p = u(x)y_1(x)$, where y_1 as defined in the preceding paragraph is a solution of (2). In other words, our assumption for y_p is the same as $y_c = cy_1(x)$ except that c is replaced by the "variable parameter" u. Substituting $y_p = uy_1$ into (2) gives

$$\frac{d}{dx}[uy_1] + P(x)uy_1 = f(x)$$

$$u\frac{dy_1}{dx} + y_1\frac{du}{dx} + P(x)uy_1 = f(x)$$

$$\overset{\text{zero}}{u\left[\frac{dy_1}{dx} + P(x)y_1\right]} + y_1\frac{du}{dx} = f(x)$$

so that

$$y_1\frac{du}{dx} = f(x).$$

Separating variables and integrating then gives

$$du = \frac{f(x)}{y_1(x)}\, dx \qquad \text{and} \qquad u = \int \frac{f(x)}{y_1(x)}\, dx.$$

From the definition of y_1 we see that

$$y_p = uy_1 = e^{-\int P(x)dx}\int e^{\int P(x)dx}f(x)\, dx.$$

Thus

$$y = y_c + y_p = ce^{-\int P(x)dx} + e^{-\int P(x)dx}\int e^{\int P(x)dx}f(x)\, dx. \tag{4}$$

Thus if (1) has a solution, it must be of form (4). Conversely, it is a straightforward exercise in differentiation to verify that (4) constitutes a one-parameter family of solutions of equation (2).

No attempt should be made to memorize the formula given in (4). There is an equivalent but easier way of solving (2). If (4) is multiplied by

$$e^{\int P(x)dx} \tag{5}$$

and then

$$e^{\int P(x)dx}y = c + \int e^{\int P(x)dx}f(x)\, dx \tag{6}$$

is differentiated,
$$\frac{d}{dx}\left[e^{\int P(x)\,dx}y\right] = e^{\int P(x)\,dx}f(x),$$
(7)

we get
$$e^{\int P(x)\,dx}\frac{dy}{dx} + P(x)e^{\int P(x)\,dx}y = e^{\int P(x)\,dx}f(x).$$
(8)

Dividing the last result by $e^{\int P(x)\,dx}$ gives (2).

Method of Solution   The recommended method of solving (2) actually consists of working (6)–(8) in reverse order. In other words, if (2) is multiplied by (5) we get (8). The left side of (8) is recognized as the derivative of the product of $e^{\int P(x)\,dx}$ and y. This gets us to (7). We then integrate both sides of (7) to get the solution (6). Because we can solve (2) by integration after multiplication by $e^{\int P(x)\,dx}$, we call this function an **integrating factor** for the differential equation. For convenience we summarize these results.

Solving a Linear First-Order Equation

(*i*) To solve a linear first-order equation, first put it into the form (2); that is, make the coefficient of dy/dx unity.

(*ii*) Identify $P(x)$, and find the integrating factor $e^{\int P(x)\,dx}$.

(*iii*) Multiply the equation obtained in step (*i*) by the integrating factor:
$$e^{\int P(x)\,dx}\frac{dy}{dx} + P(x)e^{\int P(x)\,dx}y = e^{\int P(x)\,dx}f(x).$$

(*iv*) The left side of the equation in step (*iii*) is the derivative of the product of the integrating factor and the dependent variable y; that is,
$$\frac{d}{dx}\left[e^{\int P(x)\,dx}y\right] = e^{\int P(x)\,dx}f(x).$$

(*v*) Integrate both sides of the equation found in step (*iv*).

EXAMPLE 1   Solving a Linear DE

Solve $x\dfrac{dy}{dx} - 4y = x^{6}e^{x}$.

SOLUTION   By dividing by x we get the standard form
$$\frac{dy}{dx} - \frac{4}{x}y = x^{5}e^{x}.$$
(9)

From this form we identify $P(x) = -4/x$, and so the integrating factor is
$$e^{-4\int dx/x} = e^{-4\ln|x|} = e^{\ln x^{-4}} = x^{-4}.$$

Here we have used the basic identity $b^{\log_{b}N} = N,\, N > 0$. Now we multiply (9) by this term,
$$x^{-4}\frac{dy}{dx} - 4x^{-5}y = xe^{x},$$
(10)

and obtain
$$\frac{d}{dx}[x^{-4}y] = xe^x. \tag{11}$$

It follows from integration by parts that

$$x^{-4}y = xe^x - e^x + c \quad \text{or} \quad y = x^5e^x - x^4e^x + cx^4. \quad ■$$

EXAMPLE 2 **Solving a Linear DE**

Solve $\dfrac{dy}{dx} - 3y = 0$.

SOLUTION   This linear equation can be solved by separation of variables. Alternatively, since the equation is already in the standard form (2), we see that the integrating factor is $e^{\int(-3)dx} = e^{-3x}$. Multiplying the given equation by this factor gives $e^{-3x}dy/dx - 3e^{-3x}y = 0$. This last equation is the same as

$$\frac{d}{dx}[e^{-3x}y] = 0.$$

Integrating then gives $e^{-3x}y = c$, and therefore $y = ce^{3x}$.   ■

General Solution   If it is assumed that $P(x)$ and $f(x)$ are continuous on an interval I and that x_0 is any point in the interval, then it follows from Theorem 1.1 that there exists only one solution of the initial-value problem

$$\frac{dy}{dx} + P(x)y = f(x), \quad y(x_0) = y_0. \tag{12}$$

But we saw earlier that (2) possesses a family of solutions and that every solution of the equation on the interval I is of form (4). Thus obtaining the solution of (12) is a simple matter of finding an appropriate value of c, in (4). Consequently, we are justified in calling (4) the **general solution** of the differential equation. In retrospect, you should recall that in several instances we found singular solutions of nonlinear equations. This cannot happen in the case of a linear equation if proper attention is paid to solving the equation over a common interval on which $P(x)$ and $f(x)$ are continuous.

EXAMPLE 3 **A General Solution**

Find the general solution of $(x^2 + 9)\dfrac{dy}{dx} + xy = 0$.

SOLUTION   We write $\dfrac{dy}{dx} + \dfrac{x}{x^2 + 9}y = 0.$

The function $P(x) = x/(x^2 + 9)$ is continuous on $(-\infty, \infty)$. Now the integrating factor for the equation is

$$e^{\int xdx/(x^2+9)} = e^{\frac{1}{2}\int 2xdx/(x^2+9)} = e^{\frac{1}{2}\ln(x^2+9)} = \sqrt{x^2 + 9}$$

so that
$$\sqrt{x^2+9}\,\frac{dy}{dx}+\frac{x}{\sqrt{x^2+9}}\,y=0.$$

Integrating
$$\frac{d}{dx}[\sqrt{x^2+9}\,y]=0 \qquad \text{gives} \qquad \sqrt{x^2+9}\,y=c.$$

Hence the general solution on the interval is

$$y=\frac{c}{\sqrt{x^2+9}}.$$ ∎

Except in the case when the lead coefficient is 1, the recasting of equation (1) into the standard form (2) requires division by $a_1(x)$. If $a_1(x)$ is not a constant, then close attention should be paid to the points where $a_1(x)=0$. Specifically, in (2) the points at which $P(x)$, formed by dividing $a_0(x)$ by $a_1(x)$, is discontinuous are potentially troublesome.

EXAMPLE 4   **An Initial-Value Problem**

Solve the initial-value problem $x\dfrac{dy}{dx}+y=2x, \quad y(1)=0.$

SOLUTION   Write the given equation as

$$\frac{dy}{dx}+\frac{1}{x}y=2,$$

and observe that $P(x)=1/x$ is continuous on any interval not containing the origin. In view of the initial condition, we solve the problem on the interval $(0,\infty)$.

The integrating factor is $e^{\int dx/x}=e^{\ln x}=x$, and so

$$\frac{d}{dx}[xy]=2x$$

gives $xy=x^2+c.$ Solving for y yields the general solution

$$y=x+\frac{c}{x}. \tag{13}$$

But $y(1)=0$ implies $c=-1$. Hence we obtain

$$y=x-\frac{1}{x}, \quad 0<x<\infty. \tag{14}$$

Considered as a one-parameter family of curves, the graph of (13) is given in Figure 2.5. The solution (14) of the initial-value problem is indicated by the solid-colored portion of the graph. ∎

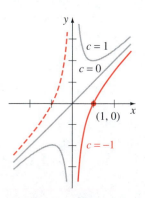

FIGURE 2.5

The next example illustrates how to solve (2) when f has a jump discontinuity.

EXAMPLE 5   **A Discontinuous $f(x)$**

Find a continuous solution satisfying

$$\frac{dy}{dx} + y = f(x), \qquad \text{where} \qquad f(x) = \begin{cases} 1, & 0 \le x \le 1 \\ 0, & x > 1 \end{cases}$$

and the initial condition $y(0) = 0$.

SOLUTION   From Figure 2.6 we see that f is piecewise continuous with a discontinuity at $x = 1$. Consequently, we solve the problem in two parts corresponding to the two intervals over which f is defined. For $0 \le x \le 1$ we have

$$\frac{dy}{dx} + y = 1 \qquad \text{or, equivalently,} \qquad \frac{d}{dx}[e^x y] = e^x.$$

Integrating the last equation and solving for y gives $y = 1 + c_1 e^{-x}$. Since $y(0) = 0$ we must have $c_1 = -1$, and therefore

$$y = 1 - e^{-x}, \quad 0 \le x \le 1.$$

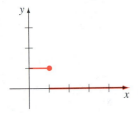

FIGURE 2.6

For $x > 1$, $$\frac{dy}{dx} + y = 0$$

leads to $y = c_2 e^{-x}$. Hence we can write

$$y = \begin{cases} 1 - e^{-x}, & 0 \le x \le 1 \\ c_2 e^{-x}, & x > 1. \end{cases}$$

Now, in order for y to be a continuous function, we certainly want $\lim_{x \to 1^+} y(x) = y(1)$. This latter requirement is equivalent to $c_2 e^{-1} = 1 - e^{-1}$, or $c_2 = e - 1$. As Figure 2.7 shows, the function

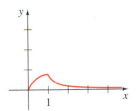

FIGURE 2.7

$$y = \begin{cases} 1 - e^{-x}, & 0 \le x \le 1 \\ (e - 1)e^{-x}, & x > 1 \end{cases}$$

is continuous but not differentiable at $x = 1$. ■

Functions Defined by Integrals   Some simple functions do not possess antiderivatives that are elementary functions, and integrals of these kinds of functions are called **nonelementary.** For example, you may have seen in calculus that $\int e^{x^2}\, dx$ and $\int \sin x^2\, dx$ are nonelementary integrals. In applied mathematics some important functions are *defined* in terms of nonelementary integrals. Two such functions are the **error function** and the **complementary error function:**

$$\text{erf}(x) = \frac{2}{\sqrt{\pi}} \int_0^x e^{-t^2}\, dt \qquad \text{and} \qquad \text{erfc}(x) = \frac{2}{\sqrt{\pi}} \int_x^\infty e^{-t^2}\, dt. \qquad \textbf{(15)}$$

Since $(2/\sqrt{\pi}) \int_0^\infty e^{-t^2}\, dt = 1$, we see from (15) that the complementary error function, erfc(x), is related to erf(x) by $\text{erf}(x) + \text{erfc}(x) = 1$. Because of its importance in areas such as probability and statistics, the error function has been extensively tabulated. Note, however, that erf(0) = 0 is one obvious functional value. Values of erf(x) can also be found using a CAS.

EXAMPLE 6 **The Error Function**

Solve the initial-value problem $\dfrac{dy}{dx} - 2xy = 2$, $y(0) = 1$.

SOLUTION   Since the equation is already in standard form we see that the integrating factor is e^{-x^2}, and so from

$$\frac{d}{dx}[e^{-x^2}y] = 2e^{-x^2} \qquad \text{we get} \qquad y = 2e^{x^2}\int_0^x e^{-t^2}\,dt + ce^{x^2}. \quad \textbf{(16)}$$

Applying $y(0) = 1$ to the last expression then gives $c = 1$. Hence the solution to the problem is

$$y = 2e^{x^2}\int_0^x e^{-t^2}\,dt + e^{x^2} = e^{x^2}[1 + \sqrt{\pi}\,\text{erf}(x)].$$

The graph of this solution, shown in color in Figure 2.8 among other members of the family defined by (16), was obtained with the aid of a CAS.   ∎

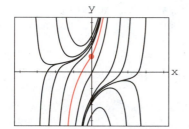

FIGURE 2.8
Some solutions of $y' - 2xy = 2$

Use of Computers

Some computer algebra systems are capable of producing explicit solutions for some kinds of differential equations. For example, to solve the equation $y' + 2y = x$, the input commands

DSolve[y′ [x] + 2 y[x] == x, y[x], x]   (in Mathematica)

and   dsolve(diff(y(x), x) + 2*y = x, y(x));   (in Maple)

yield, respectively, the output

$$\text{y[x]} -> - \left(\frac{1}{4}\right) + \frac{\text{x}}{2} + \frac{\text{C[1]}}{\text{E}^{2\,\text{x}}}$$

and   y(x) = 1/2 x − 1/4 + exp(− 2 x)_C1

Translated to standard symbols, this means $y = -\frac{1}{4} + \frac{1}{2}x + ce^{-2x}$.

Remark

Occasionally a first-order differential equation is not linear in one variable but is linear in the other. For example, because of the y^2 term, the differential equation

$$\frac{dy}{dx} = \frac{1}{x + y^2}$$

is not linear in the variable y. But its reciprocal

$$\frac{dx}{dy} = x + y^2 \qquad \text{or} \qquad \frac{dx}{dy} - x = y^2$$

is linear in x. You should verify that the integrating factor is $e^{\int(-1)dy} = e^{-y}$ and that integration by parts yields $x = -y^2 - 2y - 2 + ce^y$.

SECTION 2.3 EXERCISES

Answers to odd-numbered problems begin on page A-1.

In Problems 1–40 find the general solution of the given differential equation. State an interval on which the general solution is defined.

1. $\dfrac{dy}{dx} = 5y$

2. $\dfrac{dy}{dx} + 2y = 0$

3. $3\dfrac{dy}{dx} + 12y = 4$

4. $x\dfrac{dy}{dx} + 2y = 3$

5. $\dfrac{dy}{dx} + y = e^{3x}$

6. $\dfrac{dy}{dx} = y + e^x$

7. $y' + 3x^2 y = x^2$

8. $y' + 2xy = x^3$

9. $x^2 y' + xy = 1$

10. $y' = 2y + x^2 + 5$

11. $(x + 4y^2)\,dy + 2y\,dx = 0$

12. $\dfrac{dx}{dy} = x + y$

13. $x\,dy = (x \sin x - y)\,dx$

14. $(1 + x^2)\,dy + (xy + x^3 + x)\,dx = 0$

15. $(1 + e^x)\dfrac{dy}{dx} + e^x y = 0$

16. $(1 - x^3)\dfrac{dy}{dx} = 3x^2 y$

17. $\cos x \dfrac{dy}{dx} + y \sin x = 1$

18. $\dfrac{dy}{dx} + y \cot x = 2 \cos x$

19. $x\dfrac{dy}{dx} + 4y = x^3 - x$

20. $(1 + x)y' - xy = x + x^2$

21. $x^2 y' + x(x + 2)y = e^x$

22. $xy' + (1 + x)y = e^{-x} \sin 2x$

23. $\cos^2 x \sin x\,dy + (y \cos^3 x - 1)\,dx = 0$

24. $(1 - \cos x)\,dy + (2y \sin x - \tan x)\,dx = 0$

25. $y\,dx + (xy + 2x - ye^y)\,dy = 0$

26. $(x^2 + x)\,dy = (x^5 + 3xy + 3y)\,dx$

27. $x\dfrac{dy}{dx} + (3x + 1)y = e^{-3x}$

28. $(x + 1)\dfrac{dy}{dx} + (x + 2)y = 2xe^{-x}$

29. $y\,dx - 4(x + y^6)\,dy = 0$

30. $xy' + 2y = e^x + \ln x$

31. $\dfrac{dy}{dx} + y = \dfrac{1 - e^{-2x}}{e^x + e^{-x}}$

32. $\dfrac{dy}{dx} - y = \sinh x$

33. $y\,dx + (x + 2xy^2 - 2y)\,dy = 0$

34. $y\,dx = (ye^y - 2x)\,dy$

35. $\dfrac{dr}{d\theta} + r \sec \theta = \cos \theta$

36. $\dfrac{dP}{dt} + 2tP = P + 4t - 2$

37. $(x + 2)^2 \dfrac{dy}{dx} = 5 - 8y - 4xy$

38. $(x^2 - 1)\dfrac{dy}{dx} + 2y = (x + 1)^2$

39. $y' = (10 - y) \cosh x$

40. $dx = (3e^y - 2x)\,dy$

In Problems 41–50 solve the given differential equation subject to the indicated initial condition.

41. $\dfrac{dy}{dx} + 5y = 20, \quad y(0) = 2$

42. $y' = 2y + x(e^{3x} - e^{2x}), \quad y(0) = 2$

43. $L\dfrac{di}{dt} + Ri = E; \quad L, R,$ and E constants, $i(0) = i_0$

44. $y\dfrac{dx}{dy} - x = 2y^2, \quad y(1) = 5$

45. $y' + (\tan x)y = \cos^2 x, \quad y(0) = -1$

46. $\dfrac{dQ}{dx} = 5x^4 Q, \quad Q(0) = -7$

47. $\dfrac{dT}{dt} = k(T - 50); \quad k$ a constant, $T(0) = 200$

48. $x\,dy + (xy + 2y - 2e^{-x})\,dx = 0, \quad y(1) = 0$

49. $(x + 1)\dfrac{dy}{dx} + y = \ln x, \quad y(1) = 10$

50. $xy' + y = e^x, \quad y(1) = 2$

In Problems 51–54 find a continuous solution satisfying the given differential equation and the indicated initial condition. Use a graphing utility to graph the solution curve.

51. $\dfrac{dy}{dx} + 2y = f(x), \quad f(x) = \begin{cases} 1, & 0 \le x \le 3 \\ 0, & x > 3 \end{cases}, \quad y(0) = 0$

52. $\dfrac{dy}{dx} + y = f(x), \quad f(x) = \begin{cases} 1, & 0 \le x \le 1 \\ -1, & x > 1 \end{cases}, \quad y(0) = 1$

53. $\dfrac{dy}{dx} + 2xy = f(x), \quad f(x) = \begin{cases} x, & 0 \le x < 1 \\ 0, & x \ge 1 \end{cases}, \quad y(0) = 2$

54. $(1 + x^2)\dfrac{dy}{dx} + 2xy = f(x), \quad f(x) = \begin{cases} x, & 0 \le x < 1 \\ -x, & x \ge 1 \end{cases}, \quad y(0) = 0$

55. The **sine integral function** is defined by $\text{Si}(x) = \displaystyle\int_0^x \dfrac{\sin t}{t}\,dt$, where the integrand is defined to be 1 at $t = 0$. Express the solution of the initial-value problem

$$x^3\dfrac{dy}{dx} + 2x^2 y = 10 \sin x, \qquad y(1) = 0$$

in terms of $\text{Si}(x)$. Use tables or a CAS to calculate $y(2)$. Use an ODE solver or a CAS to graph the solution for $x > 0$.

56. Show that the solution of the initial-value problem

$$\dfrac{dy}{dx} - 2xy = -1, \qquad y(0) = \dfrac{\sqrt{\pi}}{2}$$

is $y = (\sqrt{\pi}/2)e^{x^2}\,\text{erfc}(x)$. Use tables or a CAS to compute $y(2)$. Use an ODE solver or a CAS to graph the solution curve.

57. Express the solution of the initial-value problem

$$\frac{dy}{dx} - 2xy = 1, \qquad y(1) = 1$$

in terms of $\text{erf}(x)$.

Discussion Problems

58. The analysis of nonlinear differential equations sometimes begins with either neglecting nonlinear terms in the equation or replacing these terms with linear terms. The resulting linear differential equation is called a **linearization** of the first equation. For example, the nonlinear differential equation

$$\frac{dP}{dt} = rP\left(1 - \frac{P}{K}\right) = rP - \frac{r}{K}P^2, \qquad (17)$$

where r and K are positive constants, is often used as a model for a growing but bounded population. It can be argued that when P is close to zero the nonlinear term P^2 is negligible. A linearization of the first equation is then

$$\frac{dP}{dt} = rP. \qquad (18)$$

Suppose $r = 0.02$ and $K = 300$. Use an ODE solver to compare the solution of equation (17) with the solution of equation (18) for the same initial value $P(0)$. Do this for an initial value ranging in magnitude from small, say, $P(0) = 0.5$, $P(0) = 2$, to large, say, $P(0) = 200$. Write up your observations.

59. The following system of differential equations is encountered in the study of a special type of radioactive series of elements:

$$\frac{dx}{dt} = -\lambda_1 x$$

$$\frac{dy}{dt} = \lambda_1 x - \lambda_2 y,$$

where λ_1 and λ_2 are constants. Discuss how to solve the system subject to $x(0) = x_0$, $y(0) = y_0$.

60. In the two parts of this problem assume α and β are constants; $P(x)$, $f(x)$, $f_1(x)$, and $f_2(x)$ are continuous on an interval I; and x_0 is any point in I.
   (a) Suppose that y_1 is a solution of the initial-value problem $y' + P(x)y = 0$, $y(x_0) = \alpha$ and that y_2 is a solution of $y' + P(x)y = f(x)$, $y(x_0) = 0$. Find a solution of $y' + P(x)y = f(x)$, $y(x_0) = \alpha$. Prove your assertion.
   (b) Suppose that y_1 is a solution of $y' + P(x)y = f_1(x)$, $y(x_0) = \alpha$ and that y_2 is a solution of $y' + P(x)y = f_2(x)$, $y(x_0) = \beta$. If y is a solution of $y' + P(x)y = f_1(x) + f_2(x)$, what is the value $y(x_0)$? Prove your assertion. If y is a solution of $y' + P(x)y = c_1 f_1(x) + c_2 f_2(x)$, where c_1 and c_2 are arbitrarily specified constants, what is the value $y(x_0)$? Prove your assertion.

2.4   SOLUTIONS BY SUBSTITUTIONS

■ *Substitution in a differential equation* ■ *Homogeneous function* ■ *Homogeneous DE*
■ *Bernoulli's DE*

Substitutions   We solve a differential equation by recognizing it as a certain kind of equation (say, separable) and then carrying out a procedure, consisting of equation-specific mathematical steps, that yields a sufficiently differentiable function that satisfies the equation. Often the first step in solving a given differential equation consists of transforming it into another differential equation by means of a **substitution.** For example, suppose we wish to transform the first-order equation $dy/dx = f(x, y)$ by the substitution $y = g(x, u)$, where u is regarded as a function of the variable x. If g possesses first-partial derivatives, then the Chain Rule gives

$$\frac{dy}{dx} = g_x(x, u) + g_u(x, u)\frac{du}{dx}.$$

By replacing dy/dx by $f(x, y)$ and y by $g(x, u)$ in the foregoing derivative, we get the new first-order differential equation

$$f(x, g(x, u)) = g_x(x, u) + g_u(x, u)\frac{du}{dx}$$

which, after we solve for du/dx, has the form $du/dx = F(x, u)$. If we can determine a solution $u = \phi(x)$ of this second equation, then a solution of the original differential equation is $y = g(x, \phi(x))$.

Use of Substitutions: Homogeneous Equations   If a function f possesses the property

$$f(tx, ty) = t^\alpha f(x, y)$$

for some real number α, then f is said to be a **homogeneous function** of degree α. For example, $f(x, y) = x^3 + y^3$ is homogeneous of degree 3 since

$$f(tx, ty) = (tx)^3 + (ty)^3 = t^3(x^3 + y^3) = t^3 f(x, y),$$

whereas $f(x, y) = x^3 + y^3 + 1$ is seen not to be homogeneous.
     A first-order differential equation

$$M(x, y)\,dx + N(x, y)\,dy = 0 \tag{1}$$

is said to be **homogeneous** if both coefficients M and N are homogeneous functions of the *same* degree. In other words, (1) is homogeneous if

$$M(tx, ty) = t^\alpha M(x, y) \qquad \text{and} \qquad N(tx, ty) = t^\alpha N(x, y).$$

Method of Solution   A homogeneous differential equation $M(x, y)\,dx + N(x, y)\,dy = 0$ can be solved by means of an algebraic substitution. Specifically, *either* substitution $y = ux$ or $x = vy$, where u and v are new dependent variables, *will reduce the equation to a separable first-order differential equation.* To show this, we will substitute $y = ux$ and its differential $dy = u\,dx + x\,du$ into (1):

$$M(x, ux)\,dx + N(x, ux)[u\,dx + x\,du] = 0.$$

Now, by the homogeneity property we can write

$$x^\alpha M(1, u)\, dx + x^\alpha N(1, u)[u\, dx + x\, du] = 0$$

or

$$[M(1, u) + uN(1, u)]\, dx + xN(1, u)\, du = 0,$$

which gives

$$\frac{dx}{x} + \frac{N(1, u)\, du}{M(1, u) + uN(1, u)} = 0.$$

We hasten to point out that the preceding formula should not be memorized; rather, the *procedure should be worked through each time*. The proof that the substitution $x = vy$ in (1) also leads to a separable equation follows in an analogous manner.

EXAMPLE 1   **Solving a Homogeneous DE**

Solve $(x^2 + y^2)\, dx + (x^2 - xy)\, dy = 0$.

SOLUTION   Inspection of $M(x, y) = x^2 + y^2$ and $N(x, y) = x^2 - xy$ shows that these coefficients are homogeneous functions of degree 2. If we let $y = ux$, then $dy = u\, dx + x\, du$ so that, after substituting, the given equation becomes

$$(x^2 + u^2 x^2)\, dx + (x^2 - ux^2)[u\, dx + x\, du] = 0$$

$$x^2(1 + u)\, dx + x^3(1 - u)\, du = 0$$

$$\frac{1 - u}{1 + u}\, du + \frac{dx}{x} = 0$$

$$\left[-1 + \frac{2}{1 + u}\right] du + \frac{dx}{x} = 0. \qquad \leftarrow \text{long division}$$

After integration the last line becomes

$$-u + 2\ln|1 + u| + \ln|x| = \ln|c|$$

$$-\frac{y}{x} + 2\ln\left|1 + \frac{y}{x}\right| + \ln|x| = \ln|c|. \qquad \leftarrow \text{resubstituting } u = y/x$$

Using the properties of logarithms, we can write the preceding solution as

$$\ln\left|\frac{(x + y)^2}{cx}\right| = \frac{y}{x} \qquad \text{or, equivalently,} \quad (x + y)^2 = cxe^{y/x}. \qquad \blacksquare$$

Although either of the indicated substitutions can be used for every homogeneous differential equation, in practice we try $x = vy$ whenever the function $M(x, y)$ is simpler than $N(x, y)$. Also, it could happen that after using one substitution we encounter integrals that are difficult or impossible to evaluate in closed form; switching substitutions may result in an easier problem.

Use of Substitutions: Bernoulli's Equation   The differential equation

$$\frac{dy}{dx} + P(x)\, y = f(x)\, y^n, \qquad\qquad\qquad \textbf{(2)}$$

where n is any real number, is called **Bernoulli's equation.** Note that for $n = 0$ and $n = 1$, equation (2) is linear. For $n \neq 0$ and $n \neq 1$, the substitution $u = y^{1-n}$ reduces any equation of form (2) to a linear equation.

EXAMPLE 2   **Solving a Bernoulli DE**

Solve $x\dfrac{dy}{dx} + y = x^2 y^2$.

SOLUTION   We first rewrite the equation as

$$\frac{dy}{dx} + \frac{1}{x}y = xy^2$$

by dividing by x. With $n = 2$ we next substitute

$$y = u^{-1} \qquad \text{and} \qquad \frac{dy}{dx} = -u^{-2}\frac{du}{dx} \qquad \leftarrow \text{Chain Rule}$$

into the given equation and simplify. The result is

$$\frac{du}{dx} - \frac{1}{x}u = -x.$$

The integrating factor for this linear equation on, say, $(0, \infty)$ is

$$e^{-\int dx/x} = e^{-\ln x} = e^{\ln x^{-1}} = x^{-1}.$$

Integrating $\qquad\qquad \dfrac{d}{dx}[x^{-1}u] = -1$

gives $\qquad\qquad x^{-1}u = -x + c \qquad \text{or} \qquad u = -x^2 + cx.$

Since $y = u^{-1}$ we have $y = 1/u$, and so a solution of the equation is

$$y = \frac{1}{-x^2 + cx}. \qquad\qquad\blacksquare$$

Note that we have not obtained the general solution of the original nonlinear differential equation in Example 2 since $y = 0$ is a singular solution of the equation.

Use of Substitutions: Reduction to Separation of Variables

A differential equation of the form

$$\frac{dy}{dx} = f(Ax + By + C) \qquad\qquad\qquad (3)$$

can always be reduced to an equation with separable variables by means of the substitution $u = Ax + By + C, B \neq 0$. Example 3 illustrates the technique.

EXAMPLE 3   **Using a Substitution**

Solve $\dfrac{dy}{dx} = (-5x + y)^2 - 4$.

SOLUTION   If we let $u = -5x + y$ then $du/dx = -5 + dy/dx$, and so the given equation is transformed into

$$\frac{du}{dx} + 5 = u^2 - 4 \qquad \text{or} \qquad \frac{du}{dx} = u^2 - 9.$$

Separating variables, using partial fractions, and then integrating yields

$$\frac{du}{(u-3)(u+3)} = dx$$

$$\frac{1}{6}\left[\frac{1}{u-3} - \frac{1}{u+3}\right] = dx$$

$$\frac{1}{6}\ln\left|\frac{u-3}{u+3}\right| = x + c_1$$

$$\frac{u-3}{u+3} = e^{6x+6c_1} = ce^{6x}. \qquad \leftarrow \text{replace } e^{6c_1} \text{ by } c$$

Solving the last equation for u and then resubstituting gives the solution

$$u = \frac{3(1 + ce^{6x})}{1 - ce^{6x}} \qquad \text{or} \qquad y = 5x + \frac{3(1 + ce^{6x})}{1 - ce^{6x}}. \qquad \blacksquare$$

SECTION 2.4 EXERCISES

Answers to odd-numbered problems begin on page A-2.

In Problems 1–10 solve the given homogeneous equation by using an appropriate substitution.

1. $(x - y)\,dx + x\,dy = 0$

2. $(x + y)\,dx + x\,dy = 0$

3. $x\,dx + (y - 2x)\,dy = 0$

4. $y\,dx = 2(x + y)\,dy$

5. $(y^2 + yx)\,dx - x^2\,dy = 0$

6. $(y^2 + yx)\,dx + x^2\,dy = 0$

7. $\dfrac{dy}{dx} = \dfrac{y - x}{y + x}$

8. $\dfrac{dy}{dx} = \dfrac{x + 3y}{3x + y}$

9. $-y\,dx + (x + \sqrt{xy})\,dy = 0$

10. $x\dfrac{dy}{dx} - y = \sqrt{x^2 + y^2}$

In Problems 11–14 solve the given homogeneous equation subject to the indicated initial condition.

11. $xy^2\dfrac{dy}{dx} = y^3 - x^3$, $y(1) = 2$

12. $(x^2 + 2y^2)\dfrac{dx}{dy} = xy$, $y(-1) = 1$

13. $(x + ye^{y/x})\,dx - xe^{y/x}\,dy = 0$, $y(1) = 0$

14. $y\,dx + x(\ln x - \ln y - 1)\,dy = 0$, $y(1) = e$

In Problems 15–20 solve the given Bernoulli equation by using an appropriate substitution.

15. $x\dfrac{dy}{dx} + y = \dfrac{1}{y^2}$

16. $\dfrac{dy}{dx} - y = e^x y^2$

17. $\dfrac{dy}{dx} = y(xy^3 - 1)$

18. $x\dfrac{dy}{dx} - (1 + x)y = xy^2$

19. $x^2 \dfrac{dy}{dx} + y^2 = xy$ **20.** $3(1 + x^2) \dfrac{dy}{dx} = 2xy(y^3 - 1)$

In Problems 21 and 22 solve the given Bernoulli equation subject to the indicated initial condition.

21. $x^2 \dfrac{dy}{dx} - 2xy = 3y^4, \quad y(1) = \dfrac{1}{2}$ **22.** $y^{1/2} \dfrac{dy}{dx} + y^{3/2} = 1, \quad y(0) = 4$

In Problems 23 and 28 use the procedure illustrated in Example 3 to solve the given differential equation.

23. $\dfrac{dy}{dx} = (x + y + 1)^2$ **24.** $\dfrac{dy}{dx} = \dfrac{1 - x - y}{x + y}$

25. $\dfrac{dy}{dx} = \tan^2(x + y)$ **26.** $\dfrac{dy}{dx} = \sin(x + y)$

27. $\dfrac{dy}{dx} = 2 + \sqrt{y - 2x + 3}$ **28.** $\dfrac{dy}{dx} = 1 + e^{y-x+5}$

In Problems 29 and 30 solve the given equation subject to the indicated initial condition.

29. $\dfrac{dy}{dx} = \cos(x + y), \quad y(0) = \dfrac{\pi}{4}$

30. $\dfrac{dy}{dx} = \dfrac{3x + 2y}{3x + 2y + 2}, \quad y(-1) = -1$

Discussion Problems

31. Explain why it is always possible to express any homogeneous differential equation $M(x, y)\, dx + N(x, y)\, dy = 0$ in the form

$$\frac{dy}{dx} = F\left(\frac{y}{x}\right) \qquad \text{or} \qquad \frac{dy}{dx} = G\left(\frac{x}{y}\right).$$

You might start by writing down some examples of differential equations with these forms. Do these general forms suggest why the substitutions $y = ux$ and $x = vy$ are appropriate for homogeneous first-order differential equations?

32. The differential equation

$$\frac{dy}{dx} = P(x) + Q(x)y + R(x)y^2$$

is known as **Ricatti's equation.**

(a) A Ricatti equation can be solved by a succession of two substitutions *provided* we know a particular solution y_1 of the equation. First use the substitution $y = y_1 + u$, and then discuss what to do next.

(b) Find a one-parameter family of solutions for the differential equation

$$\frac{dy}{dx} = -\frac{4}{x^2} - \frac{1}{x}y + y^2,$$

where $y_1 = 2/x$ is a known solution of the equation.

CHAPTER 2 REVIEW EXERCISES

Answers to odd-numbered problems begin on page A-2.

In Problems 1–14 classify each differential equation as separable, exact, linear, homogeneous, or Bernoulli. Some equations may be more than one kind. Do not solve the equations.

1. $\dfrac{dy}{dx} = \dfrac{x - y}{x}$

2. $\dfrac{dy}{dx} = \dfrac{1}{y - x}$

3. $(x + 1)\dfrac{dy}{dx} = -y + 10$

4. $\dfrac{dy}{dx} = \dfrac{1}{x(x - y)}$

5. $\dfrac{dy}{dx} = \dfrac{y^2 + y}{x^2 + x}$

6. $\dfrac{dy}{dx} = 5y + y^2$

7. $y\,dx = (y - xy^2)\,dy$

8. $x\dfrac{dy}{dx} = ye^{x/y} - x$

9. $xyy' + y^2 = 2x$

10. $2xyy' + y^2 = 2x^2$

11. $y\,dx + x\,dy = 0$

12. $\left(x^2 + \dfrac{2y}{x}\right)dx = (3 - \ln x^2)\,dy$

13. $\dfrac{dy}{dx} = \dfrac{x}{y} + \dfrac{y}{x} + 1$

14. $\dfrac{y}{x^2}\dfrac{dy}{dx} + e^{2x^3 + y^2} = 0$

In Problems 15–20 solve the given differential equation.

15. $(y^2 + 1)\,dx = y\sec^2 x\,dy$

16. $y(\ln x - \ln y)\,dx = (x\ln x - x\ln y - y)\,dy$

17. $(6x + 1)y^2\dfrac{dy}{dx} + 3x^2 + 2y^3 = 0$

18. $\dfrac{dx}{dy} = -\dfrac{4y^2 + 6xy}{3y^2 + 2x}$

19. $t\dfrac{dQ}{dt} + Q = t^4\ln t$

20. $(2x + y + 1)y' = 1$

In Problems 21–26 solve the given initial-value problem.

21. $\dfrac{y}{t}\dfrac{dy}{dt} = \dfrac{e^t}{\ln y}, \quad y(1) = 1$

22. $tx\dfrac{dx}{dt} = 3x^2 + t^2, \quad x(-1) = 2$

23. $(x^2 + 4)\dfrac{dy}{dx} + 8xy = 2x, \quad y(0) = -1$

24. $x\dfrac{dy}{dx} + 4y = x^4y^2, \quad y(1) = 1$

25. $y' = e^{2y - x}, \quad y(0) = 0$

26. $(2r^2\cos\theta\sin\theta + r\cos\theta)\,d\theta + (4r + \sin\theta - 2r\cos^2\theta)\,dr = 0,$

$r\left(\dfrac{\pi}{2}\right) = 2$

3

MODELING WITH FIRST-ORDER DIFFERENTIAL EQUATIONS

INTRODUCTION

We saw in Section 1.3 that mathematical models for population growth, radioactive decay, continuous compound interest, chemical reactions, cooling of bodies, liquid draining through a hole in a tank, velocity of a falling body, rate of memorization, and current in a series circuit are often first-order differential equations. We are now in a position to *solve* some of the linear and nonlinear differential equations that commonly arise in applications. The chapter concludes with the natural next step: systems of first-order differential equations as mathematical models.

3.1 LINEAR EQUATIONS

- *Exponential growth and decay* ■ *Half-life* ■ *Carbon dating*
- *Newton's law of cooling* ■ *Mixtures* ■ *Series circuits* ■ *Transient term*
- *Steady-state term*

Growth and Decay   The initial-value problem

$$\frac{dx}{dt} = kx, \qquad x(t_0) = x_0, \tag{1}$$

where k is a constant of proportionality, serves as a model for diverse phenomena involving either **growth** or **decay.** We saw in Section 1.3 that biologists have observed that, over short periods of time, the rate of growth of certain populations (such as bacteria or small animals) is proportional to the population present at any time. If we know a population at some arbitrary initial time that we can take to be $t = 0$, we can then use the solution of (1) to predict the population in the future—that is, at subsequent times $t > 0$. In physics an initial-value problem such as (1) provides a model for approximating the remaining amount of a substance that is disintegrating, or decaying, through radioactivity. The differential equation in (1) could also determine the temperature of a cooling body. In chemistry the amount of a substance remaining during certain reactions is described by (1).

The constant of proportionality k in (1) can be determined from the solution of the initial-value problem, using a subsequent measurement of x at a time $t_1 > t_0$.

EXAMPLE 1   Bacterial Growth

A culture initially has a number N_0 of bacteria. At $t = 1$ h the number of bacteria is measured to be $\frac{3}{2}N_0$. If the rate of growth is proportional to the number of bacteria present, determine the time necessary for the number of bacteria to triple.

SOLUTION   We first solve the differential equation

$$\frac{dN}{dt} = kN \tag{2}$$

subject to $N(0) = N_0$. Then we use the empirical condition $N(1) = \frac{3}{2}N_0$ to determine the constant of proportionality k.

Now (2) is both separable and linear. When it is put in the form

$$\frac{dN}{dt} - kN = 0,$$

we can see by inspection that the integrating factor is e^{-kt}. Multiplying both sides of the equation by this term immediately gives

$$\frac{d}{dt}[e^{-kt}N] = 0.$$

Integrating both sides of the last equation yields the general solution

$$e^{-kt}N = c \qquad \text{or} \qquad N(t) = ce^{kt}.$$

At $t = 0$ it follows that $N_0 = ce^0 = c$, and so $N(t) = N_0 e^{kt}$. At $t = 1$ we have $\frac{3}{2}N_0 = N_0 e^k$ or $e^k = \frac{3}{2}$. From the last equation we get $k = \ln \frac{3}{2} = 0.4055$. Thus

$$N(t) = N_0 e^{0.4055t}.$$

To find the time at which the number of bacteria has tripled, we solve $3N_0 = N_0 e^{0.4055t}$ for t. It follows that $0.4055t = \ln 3$, and so

$$t = \frac{\ln 3}{0.4055} \approx 2.71 \text{ h}.$$

See Figure 3.1.   ∎

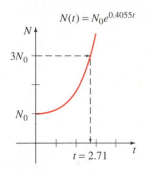

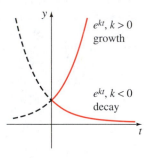

FIGURE 3.1

FIGURE 3.2

Notice in Example 1 that the actual number N_0 of bacteria present at time $t = 0$ played no part in determining the time required for the number in the culture to triple. The time necessary for an initial population of, say, 100 or 1,000,000 bacteria to triple is still approximately 2.71 hours.

As shown in Figure 3.2, the exponential function e^{kt} increases as t increases for $k > 0$ and decreases as t increases for $k < 0$. Thus problems describing growth (whether of populations, bacteria, or even capital) are characterized by a positive value of k, whereas problems involving decay (as in radioactive disintegration) yield a negative k value. Accordingly, we say that k is either a **growth constant** ($k > 0$) or a **decay constant** ($k < 0$).

Half-Life   In physics the **half-life** is a measure of the stability of a radioactive substance. The half-life is simply the time it takes for one-half of the atoms in an initial amount A_0 to disintegrate, or transmute, into the atoms of another element. The longer the half-life of a substance, the more stable it is. For example, the half-life of highly radioactive radium, Ra-226, is about 1700 years. In 1700 years one-half of a given quantity of Ra-226 is transmuted into radon, Rn-222. The most commonly occurring uranium isotope, U-238, has a half-life of approximately 4,500,000,000 years. In about 4.5 billion years, one-half of a quantity of U-238 is transmuted into lead, Pb-206.

EXAMPLE 2   **Half-Life of Plutonium**

A breeder reactor converts relatively stable uranium 238 into the isotope plutonium 239. After 15 years it is determined that 0.043% of the initial amount A_0 of the plutonium has disintegrated. Find the half-life of this isotope if the rate of disintegration is proportional to the amount remaining.

SOLUTION   Let $A(t)$ denote the amount of plutonium remaining at any time. As in Example 1, the solution of the initial-value problem

$$\frac{dA}{dt} = kA, \qquad A(0) = A_0$$

is $A(t) = A_0 e^{kt}$. If 0.043% of the atoms of A_0 have disintegrated, then 99.957% of the substance remains. To find the decay constant k we use $0.99957A_0 = A(15)$, that is, $0.99957A_0 = A_0 e^{15k}$. Solving for k then gives $k = \frac{1}{15} \ln 0.99957 = -0.00002867$. Hence

$$A(t) = A_0 e^{-0.00002867t}.$$

Now the half-life is the corresponding value of time at which $A(t) = A_0/2$. Solving for t gives $A_0/2 = A_0 e^{-0.00002867t}$ or $\frac{1}{2} = e^{-0.00002867t}$. The last equation yields

$$t = \frac{\ln 2}{0.00002867} \approx 24{,}180 \text{ years.} \qquad \blacksquare$$

Carbon Dating About 1950 the chemist Willard Libby devised a method of using radioactive carbon as a means of determining the approximate ages of fossils. The theory of **carbon dating** is based on the fact that the isotope carbon-14 is produced in the atmosphere by the action of cosmic radiation on nitrogen. The ratio of the amount of C-14 to ordinary carbon in the atmosphere appears to be a constant, and as a consequence the proportionate amount of the isotope present in all living organisms is the same as that in the atmosphere. When an organism dies, the absorption of C-14, by either breathing or eating, ceases. Thus by comparing the proportionate amount of C-14 present, say, in a fossil with the constant ratio found in the atmosphere, it is possible to obtain a reasonable estimation of its age. The method is based on the knowledge that the half-life of the radioactive C-14 is approximately 5600 years. For his work Libby won the Nobel Prize for chemistry in 1960. Libby's method has been used to date wooden furniture in Egyptian tombs and the woven flax wrappings of the Dead Sea scrolls.

EXAMPLE 3 **Age of a Fossil**

A fossilized bone is found to contain one-thousandth the original amount of C-14. Determine the age of the fossil.

SOLUTION The starting point is again $A(t) = A_0 e^{kt}$. To determine the value of the decay constant k we use the fact that $A_0/2 = A(5600)$ or $A_0/2 = A_0 e^{5600k}$. From $5600k = \ln \frac{1}{2} = -\ln 2$ we then get $k = -(\ln 2)/5600 = -0.00012378$. Therefore

$$A(t) = A_0 e^{-0.00012378t}.$$

With $A(t) = A_0/1000$ we have $A_0/1000 = A_0 e^{-0.00012378t}$ so that $-0.00012378t = \ln \frac{1}{1000} = -\ln 1000$. Thus

$$t = \frac{\ln 1000}{0.00012378} \approx 55{,}800 \text{ years.} \qquad \blacksquare$$

The date found in Example 3 is really at the border of accuracy for this method. The usual carbon-14 technique is limited to about 9 half-lives of the isotope, or about 50,000 years. One reason is that the chemical analysis needed to obtain an accurate measurement of the remaining

C-14 becomes somewhat formidable around the point of $A_0/1000$. Also, this analysis demands the destruction of a rather large sample of the specimen. If this measurement is accomplished indirectly, based on the actual radioactivity of the specimen, then it is very difficult to distinguish between the radiation from the fossil and the normal background radiation. But recently the use of a particle accelerator has enabled scientists to separate the C-14 from the stable C-12 directly. When the precise value of the ratio of C-14 to C-12 is computed, the accuracy of this method can be extended to 70,000–100,000 years. Other isotopic techniques such as using potassium 40 and argon 40 can give dates of several million years. Nonisotopic methods based on the use of amino acids are also sometimes possible.

Newton's Law of Cooling

In equation (10) of Section 1.3 we saw that the mathematical formulation of Newton's empirical law of cooling of an object is given by the linear first-order differential equation

$$\frac{dT}{dt} = k(T - T_m), \tag{3}$$

where k is a constant of proportionality, $T(t)$ is the temperature of the object for $t > 0$, and T_m is the ambient temperature—that is, the temperature of the medium around the object. In Example 4 we assume that T_m is constant.

EXAMPLE 4   **Cooling of a Cake**

When a cake is removed from an oven, its temperature is measured at 300°F. Three minutes later its temperature is 200°F. How long will it take for the cake to cool off to a room temperature of 70°F?

SOLUTION   In (3) we make the identification $T_m = 70$. We must then solve the initial-value problem

$$\frac{dT}{dt} = k(T - 70), \qquad T(0) = 300 \tag{4}$$

and determine the value of k so that $T(3) = 200$.

Equation (4) is both linear and separable. Separating variables,

$$\frac{dT}{T - 70} = k \, dt,$$

yields $\ln|T - 70| = kt + c_1$, and so $T = 70 + c_2 e^{kt}$. When $t = 0$, $T = 300$, so that $300 = 70 + c_2$ gives $c_2 = 230$ and, therefore, $T = 70 + 230 e^{kt}$. Finally, the measurement $T(3) = 200$ leads to $e^{3k} = \frac{13}{23}$ or $k = \frac{1}{3} \ln \frac{13}{23} = -0.19018$. Thus

$$T(t) = 70 + 230 e^{-0.19018t}. \tag{5}$$

We note that (5) furnishes no finite solution to $T(t) = 70$ since $\lim_{t \to \infty} T(t) = 70$. Yet intuitively we expect the cake to reach the room temperature after a reasonably long period of time. How long is "long"? Of course, we should not be disturbed by the fact that the model (4) does not quite live up to our physical intuition. Parts (a) and (b) of Figure 3.3 clearly show that the cake will be approximately at room temperature in about one-half hour.

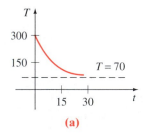

(a)

$T(t)$	t (min)
75°	20.1
74°	21.3
73°	22.8
72°	24.9
71°	28.6
70.5°	32.3

(b)

FIGURE 3.3

AZT AND SURVIVABILITY WITH AIDS

Ivan Kramer

*University of Maryland,
Baltimore County*

This essay will investigate the impact of zidovudine (formerly known as azidothymidine, or AZT) on the survivability of those who develop acquired immunodeficiency syndrome (AIDS) after being infected with the human immunodeficiency virus (HIV).

Like all other viruses, HIV is not a cell, has no metabolism, and cannot reproduce outside of a living cell. The genetic information of the virus is contained in its two identical strands of RNA. To reproduce, HIV must use the reproductive apparatus of the cell it invades to produce exact copies of the viral RNA. What HIV does is transcribe its RNA into DNA using an enzyme—reverse transcriptase—contained in the virus. The double-stranded viral DNA migrates to the nucleus of the invaded cell and is inserted into the cell's genome with the aid of another viral enzyme—integrase. The viral DNA and the invaded cell's DNA are then integrated. When the invaded cell is stimulated to reproduce, the proviral DNA is transcribed into viral RNA, and new viral particles are synthesized. Since zidovudine inhibits the HIV enzyme reverse transcriptase and stops proviral DNA chain synthesis in the laboratory, the drug held promise of being effective in slowing or stopping the progression of HIV infection in humans.

What makes HIV so dangerous is the fact that, in addition to being a rapidly mutating virus, it selectively attacks the vital T-helper cells of the host's immune system by binding to the CD4 molecule on such a cell's surface. T-helper cells (CD4 T-cells, or T4 cells) are critical to the mounting of a defense against any infection. Although the immunological parameters of the immune system of a host infected with HIV change quasi-statically after the acute infection stage, literally billions of infected T4 cells and HIV particles are destroyed—and replaced—daily during an incubation period that can last two decades or more.

In about 95% of all people infected with HIV, the immune system gradually loses its long battle with the virus. The T4 cell density in the peripheral blood of these patients begins to drop from normal levels (between 250 and 2500 cells/mm^3) toward zero at a critical point in the infection; eventually the patient develops one of the more than twenty opportunistic infections characteristic of clinical AIDS. The HIV infection has now reached its potentially fatal stage. The

T4 cell density is a popular surrogate marker of disease progression since, for some not-understood reason, its decline parallels the deterioration of the HIV-infected immune system. Remarkably, about 5% of those infected with HIV show no sign of immune system deterioration within the first ten years of the infection; these persons, termed "long-term non-progressors," may in fact be immune to developing AIDS from the HIV infection and are being intensively studied.

In order to model survivability with AIDS, the elapsed time after members of a group of HIV-infected people develop clinical AIDS will be denoted by t. Let $S(t)$ denote the *fraction* of the group that remains alive at time t. One possible survival model postulates that AIDS is not a fatal condition for a fraction of this group, denoted by S_i, to be called the "immortal fraction" here. For the remaining part of the group, the probability of dying per unit time at time t will be assumed to be a constant k. Thus the survival fraction $S(t)$ for this model is a solution of the first-order differential equation

$$\frac{dS(t)}{dt} = -k(S(t) - S_i), \qquad (1)$$

where k must be positive.

Using the separation of variables technique discussed in Chapter 2, we see that the solution of equation (1) for the survival fraction is given by

$$S(t) = S_i + (1 - S_i)e^{-kt}. \qquad (2)$$

Setting $T = k^{-1} \ln 2$, we can rewrite equation (2) in the equivalent form

$$S(t) = S_i + (1 - S_i)2^{-t/T}, \qquad (3)$$

where, in an analogy to radioactive nuclear decay, T is the time required for half of the mortal part of the cohort to die—that is, the survival half-life. See Problem 8 in Exercises 3.1.

Using a least-squares program to fit the survival fraction function in (3) to the actual survival data for the 159 Marylanders who developed AIDS in 1985 produces an immortal fraction value of $S_i = 0.0665$ and a survival half-life value of $T = 0.666$ year. Thus only about 10% of Marylanders who developed AIDS in 1985 survived three years with this condition [1]. The 1985 Maryland AIDS survival curve is virtually identical to those of 1983 and 1984. Since zidovudine was not known to have an impact on the HIV infection in 1985 and was not common therapy before 1987, it is reasonable to conclude that the survival of the 1985 Maryland AIDS patients was not significantly influenced by zidovudine therapy.

Particles of HIV (blue spheres), the virus that causes AIDS, bud from an infected white blood cell before moving on to infect other cells. The immune system controls such spread at first but is eventually outmaneuvered by the virus.
Copyright Boehringer Ingelheim, International GmbH/Photo by Lennart Nilsson

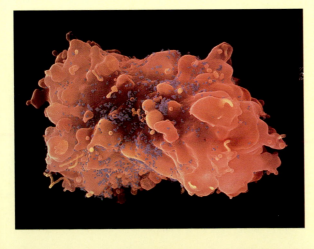

The small but nonzero value for the immortal fraction S_i obtained from the Maryland data is probably an artifact of the method that Maryland and other states use to determine the survivability of their citizens. Residents with AIDS who changed their name and then died or who died abroad would still be counted as alive by the Maryland Department of Health and Mental Hygiene. Thus the immortal fraction value of $S_i = 0.0665$ (6.65%) obtained from the Maryland data is clearly an *upper* limit to its true value.

Detailed data on the survivability of 1,415 *zidovudine-treated* HIV-infected people whose T4 cell densities dropped below normal values were published by Easterbrook et al. in 1993 [2]. As their T4 cell densities drop toward zero, these people develop clinical AIDS and begin to die. The longest survivors of this disease live to see their T4 densities fall below 10 cells/mm^3. If the time $t = 0$ is redefined to mean the moment the T4 cell density of an HIV-infected person falls below 10 cells/mm^3, then the survivability $S(t)$ of such people was determined by Easterbrook to be 0.470, 0.316, and 0.178 at elapsed times of 1 year, 1.5 years, and 2 years, respectively.

A least-squares fit of the survival fraction function in (3) to the Easterbrook data for HIV infecteds with T4 cell densities in the 0–10 cells/mm^3 range yields a value for the immortal fraction of $S_i = 0$ and a survival half-life of $T = 0.878$ year [3]. These results clearly show that zidovudine is not effective in halting replication in all strains of HIV, since those who receive this drug eventually die at nearly the same rate as those who do not. In fact, the small difference of 2.5 months between the survival half-life for 1993 infecteds with T4 cell densities below 10 cells/mm^3 on zidovudine therapy ($T = 0.878$ year) and that of 1985 infected Marylanders not taking zidovudine ($T = 0.666$ year) may be due entirely to improved hospitalization and improvements in the treatment of the opportunistic infections associated with AIDS over the years. Thus the initial ability of zidovudine to prolong survivability with HIV disease ultimately wears off, and the infection resumes its progression. Zidovudine therapy has been estimated to extend the survivability of an HIV-infected patient by perhaps five or six months on the average [3]. However, as the drug eventually loses its effectiveness, is expensive to the patient, and produces adverse side effects, *prolonged* therapy solely with zidovudine is difficult to justify.

Finally, putting the modeling results for both sets of data together, we find that the value for the immortal fraction falls somewhere within the range $0 \leq S_i \leq 0.0665$. The percentage of people for whom AIDS is not a fatal disease is less than 6.65% and may be 0.

References

1. Kramer, Ivan. Is AIDS an invariably fatal disease?: A model analysis of AIDS survival cures. *Mathematical and Computer Modelling* 15, no. 9 (1991): 1–19.
2. Easterbrook, Philippa J., Javad Emami, Graham Moyle, and Brian G. Gazzard. Progressive CD4 cell depletion and death in zidovudine-treated patients. *JAIDS* 6, no. 8 (1993).
3. Kramer, Ivan. The impact of zidovudine (AZT) therapy on the survivability of those with the progressive HIV infection. *Mathematical and Computer Modelling,* forthcoming.

About the Author

Ivan Kramer earned a B.S. in Physics and Mathematics from The City College of New York in 1961 and a Ph.D. from the University of California, Berkeley in 1969. He is currently Associate Professor of Physics at the University of Maryland, Baltimore County. Dr. Kramer was Project Director for AIDS/HIV Case Projections for Maryland, for which he received a grant from the AIDS Administration of the Maryland Department of Health and Hygiene in 1990. Since 1987 he has published numerous articles on the subject of AIDS and has been an invited speaker on the subject of mathematical modeling of the AIDS epidemic at several conferences and universities.

C. J. Knickerbocker

St. Lawrence University

A Modeling Application

WOLF POPULATION DYNAMICS

Early in 1995, after much controversy, public debate, and a 70-year absence, gray wolves were reintroduced into Yellowstone National Park and central Idaho. Since the extermination of the wolf in the 1920s, significant changes had been recorded in the populations of other animals residing in the park. For instance, the coyote population (as well as other predator populations) had grown in the absence of competition from the larger gray wolf. Therefore, with the reintroduction of the gray wolf, we anticipate changes in both the predator and the prey animal populations in the Yellowstone Park ecosystem, and the success of the wolf will depend on how it influences and is influenced by the other species in the ecosystem.

Consider the following simplified model for the interaction among the elk, coyotes, and wolves within the Yellowstone ecosystem:

$$\frac{dE}{dt} = 0.04E - 0.003EC - 0.85EW$$

$$\frac{dC}{dt} = -0.06C + 0.001EC$$

$$\frac{dW}{dt} = -0.12W + 0.005EW$$

$$E(0) = 60.0, \quad C(0) = 2.0, \quad W(t) = 0.015,$$

where $E(t)$ is the elk population, $C(t)$ is the coyote population, and $W(t)$ is the wolf population. All populations are measured in thousands of animals. The variable t represents time measured in years from 1995. So from the initial conditions we have 60,000 elk, 2,000 coyotes, and 15 wolves in the year 1995.

Before a solution is found, a qualitative analysis of the system can yield a number of interesting properties of the solutions. For example, from the equation $dC/dt = -0.06C + 0.001EC$ we see that the coyote population has a negative effect on its own growth, as more coyotes means more competition for food. But the interaction between the elk and the coyotes has a positive impact, since the coyotes find more food.

Since an explicit solution to this initial-value problem cannot be found, we need to rely on technology to find approximate solutions. For example, following is a set of instructions for finding numerical solutions to the initial-value problem using the MAPLE computer algebra system.

```
e1:=diff(e(t),t)–0.04*e(t)+0.003*e(t)*c(t)+0.85*e(t)*w(t);
e2:=diff(c(t),t)+0.06*c(t)–0.001*e(t)*c(t);
e3:=diff(w(t),t)+0.12*w(t)–0.005*e(t)*w(t);
sys:={e1,e2,e3};
ic:={e(0)=60.0,c(0)=2.0,w(0)=0.015};
ivp:=sys union ic;
H:=dsolve(ivp,{e(t),c(t),w(t)},numeric);
```

A tranquilized female wolf is released to Yellowstone by members of the U.S. Fish & Wildlife Service.
Courtesy U.S. Fish & Wildlife Service/ Dept. of Interior

From this example we can understand how animal populations are influenced by the wolves. But is the impact always negative? Let's consider a more detailed analysis of the changes in the elk population.

Between 1985 and 1995 the Yellowstone elk population increased by 40%, and numerous studies indicate that with the introduction of wolves the elk population could decrease by as much as 25%. But those animals hunted by wolves will be the very young, the very old, and the unhealthy members of the herd, which potentially leaves more food for the stronger members of the community, thereby strengthening the prey population.

The classic predator-prey model discussed in Section 3.3 is given by

$$\frac{dW}{dt} = -a_0W + a_1EW$$

$$\frac{dE}{dt} = b_0E - b_1EW,$$

where $W(t)$ is the population of wolves and $E(t)$ is the population of elk. All of the constants are positive, with a_0 measuring the death rate of the wolves, b_0 the birth rate of the elk, and a_1, b_1 the interactions between the two species.

This model suggests that the likelihood of capture by wolves is the same for every elk. Therefore, given the assumption that the weaker animals are those most often hunted by the wolves, the classic predator-prey system is not an appropriate choice of model.

To improve the model, define the total elk population by $E(t) = E_w(t) + E_s(t)$, where $E_w(t)$ and $E_s(t)$ represent the weak and strong elk, respectively. The new equation governing the change in the wolf population will be very close to the classic equation. The only change needed is to replace $E(t)$ with $E_w(t)$, since the wolves hunt only the weaker animals. This yields

$$\frac{dW}{dt} = -\alpha_0W + \alpha_1E_wW.$$

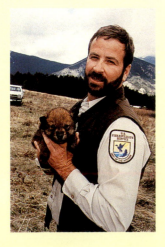

A member of the Fish & Wildlife Service holds one of the first pups born to relocated wolves in Yellowstone.
Courtesy U.S. Fish & Wildlife Service/ Dept. of Interior

The new equation governing the changes in the weak elk population is determined by noting that $E_w(t)$ is dependent not only on itself but also on the strong elk population, $E_s(t)$, since the two populations interbreed and compete for food. The interaction between the weak elk and the wolves must also be accounted for, thus yielding

$$\frac{dE_w}{dt} = \beta_0E_w - \beta_1E_wW + \beta_2E_s.$$

The third equation, for the changes in the strong elk population, is similar to the equation for the changes in the weak elk population, except there is no contribution from $W(t)$ since this group is not hunted by the wolves:

$$\frac{dE_s}{dt} = \gamma_0 E_s + \gamma_1 E_w.$$

Therefore an autonomous system of ordinary differential equations for this application can easily be constructed:

$$\frac{dW}{dt} = -\alpha_0 W + \alpha_1 E_w W$$

$$\frac{dE_w}{dt} = \beta_0 E_w - \beta_1 E_w W + \beta_2 E_s$$

$$\frac{dE_s}{dt} = \gamma_0 E_s + \gamma_1 E_w.$$

Information on the reintroduction of wolves into Yellowstone Park and central Idaho can be found on the Internet. For example, read the U.S. Fish and Wildlife Service news release of November 23, 1994. This report can be found using any search engine on the World Wide Web.

References

1. Ferris, Robert M. Return of a native. *Defenders* (Winter 1994/95).
2. Fischer, Hank. Wolves for Yellowstone. *Defenders* (Summer 1993).
3. U.S. Fish and Wildlife Service. Final rules clear the way for wolf reintroduction in Yellowstone National Park and central Idaho. News release, November 23, 1994.

"Little John" lives in a protected environment and helps educate humans about wolf behavior through the efforts of Wolf Haven International.
Photo by Pat Colton/Wolf Haven International, Tenino, WA

About the Author

C. J. Knickerbocker received his Ph.D. in Mathematics from Clarkson University in 1984. He is currently Professor of Mathematics at St. Lawrence University, where he also serves as Associate Dean for Faculty Affairs. Dr. Knickerbocker has authored numerous articles, including (with T. Greene) *Computer Analysis of Aesthetic Districts* for the American Psychological Association in 1990. He contributed material to *A First Course in Differential Equations 5/e* and *Differential Equations with Boundary-Value Problems 3/e*, both by Dennis G. Zill, and has also served as a consultant for publishers, software companies, and government agencies.

Mixtures The mixing of two fluids sometimes gives rise to a linear first-order differential equation. When we discussed the mixing of two brine solutions in Section 1.3, we assumed that the rate $A'(t)$ at which the amount of salt in the mixing tank changes was a net rate:

$$\frac{dA}{dt} = \left(\begin{array}{c}\text{rate of}\\ \text{substance entering}\end{array}\right) - \left(\begin{array}{c}\text{rate of}\\ \text{substance leaving}\end{array}\right) = R_1 - R_2. \quad \textbf{(6)}$$

In Example 5 we solve equation (12) of Section 1.3.

EXAMPLE 5   Mixture of Two Salt Solutions

Recall that the large tank considered in Section 1.3 initially held 300 gallons of a brine solution. Salt was entering and leaving the tank; a brine solution was being pumped into the tank at the rate of 3 gal/min, mixed with the solution there, and then pumped out of the tank at the rate of 3 gal/min. The concentration of the entering solution was 2 lb/gal, and so salt was entering the tank at the rate $R_1 = (2 \text{ lb/gal}) \cdot (3 \text{ gal/min}) = 6 \text{ lb/min}$ and leaving the tank at the rate $R_2 = (3 \text{ gal/min}) \cdot (A/300 \text{ lb/gal}) = A/100 \text{ lb/min}$. From these data and (6) we get equation (12) of Section 1.3. Let us pose this question: If 50 pounds of salt was dissolved in the initial 300 gallons, how much salt would be in the tank after a long time?

SOLUTION   To find $A(t)$ we solve the initial-value problem

$$\frac{dA}{dt} = 6 - \frac{A}{100}, \qquad A(0) = 50.$$

Note here that the side condition is the initial amount of salt, $A(0) = 50$, and not the initial amount of liquid. Now since the integrating factor of the linear differential equation is $e^{t/100}$, we can write the equation as

$$\frac{d}{dt}[e^{t/100}A] = 6e^{t/100}.$$

Integrating the last equation and solving for A gives the general solution $A = 600 + ce^{-t/100}$. When $t = 0$, $A = 50$, so we find that $c = -550$. Thus the amount of salt in the tank at any time t is given by

$$A(t) = 600 - 550e^{-t/100}. \qquad \textbf{(7)}$$

The solution (7) was used to construct the table in Figure 3.4(b). Also, it can be seen from (7) and Figure 3.4 that $A \to 600$ as $t \to \infty$. Of course, this is what we would expect in this case; over a long time the number of pounds of salt in the solution must be $(300 \text{ gal})(2 \text{ lb/gal}) = 600 \text{ lb}$. ∎

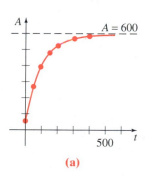

(a)

t (min)	A (lb)
50	266.41
100	397.67
150	477.27
200	525.57
300	572.62
400	589.93

(b)

FIGURE 3.4

In Example 5 we assumed that the rate at which the solution was pumped in was the same as the rate at which the solution was pumped out. However, this need not be the situation; the mixed brine solution could be pumped out at a rate faster or slower than the rate at which the other solution was pumped in. For example, if the well-stirred solution in Example 5 is pumped out at a slower rate of, say, 2 gal/min, liquid will accumulate in the tank at the rate of $(3 - 2)$ gal/min = 1 gal/min. After t minutes the tank will contain $300 + t$ gallons of brine. The rate at which the salt is leaving is then

$$R_2 = (2 \text{ gal/min}) \left(\frac{A}{300 + t} \text{ lb/gal} \right).$$

Hence equation (6) becomes

$$\frac{dA}{dt} = 6 - \frac{2A}{300 + t} \quad \text{or} \quad \frac{dA}{dt} + \frac{2}{300 + t} A = 6.$$

You should verify that the solution of the last equation subject to $A(0) = 50$ is

$$A(t) = 600 + 2t - (4.95 \times 10^7)(300 + t)^{-2}.$$

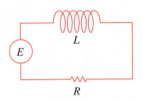

FIGURE 3.5
LR Series Circuit

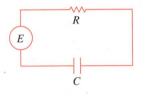

FIGURE 3.6
RC Series Circuit

Series Circuits   For a series circuit containing only a resistor and an inductor, Kirchhoff's second law states that the sum of the voltage drop across the inductor ($L(di/dt)$) and the voltage drop across the resistor (iR) is the same as the impressed voltage ($E(t)$) on the circuit. See Figure 3.5.

Thus we obtain the linear differential equation for the current $i(t)$,

$$L \frac{di}{dt} + Ri = E(t), \tag{8}$$

where L and R are constants known as the inductance and the resistance, respectively. The current $i(t)$ is also called the **response** of the system.

The voltage drop across a capacitor with capacitance C is given by $q(t)/C$, where q is the charge on the capacitor. Hence, for the series circuit shown in Figure 3.6, Kirchhoff's second law gives

$$Ri + \frac{1}{C} q = E(t). \tag{9}$$

But current i and charge q are related by $i = dq/dt$, so (9) becomes the linear differential equation

$$R \frac{dq}{dt} + \frac{1}{C} q = E(t). \tag{10}$$

EXAMPLE 6   **Series Circuit**

A 12-volt battery is connected to a series circuit in which the inductance is $\frac{1}{2}$ henry and the resistance is 10 ohms. Determine the current i if the initial current is zero.

SOLUTION   From (8) we see that we must solve

$$\frac{1}{2} \frac{di}{dt} + 10i = 12$$

subject to $i(0) = 0$. First, we multiply the differential equation by 2 and read off the integrating factor e^{20t}. We then obtain

$$\frac{d}{dt}[e^{20t}i] = 24e^{20t}.$$

Integrating each side of the last equation and solving for i gives $i = \frac{6}{5} + ce^{-20t}$. Now $i(0) = 0$ implies $0 = \frac{6}{5} + c$ or $c = -\frac{6}{5}$. Therefore the response is

$$i(t) = \frac{6}{5} - \frac{6}{5}e^{-20t}. \qquad \blacksquare$$

From (4) of Section 2.3 we can write a general solution of (8):

$$i(t) = \frac{e^{-(R/L)t}}{L}\int e^{(R/L)t}E(t)\,dt + ce^{-(R/L)t}. \tag{11}$$

In particular, when $E(t) = E_0$ is a constant, (11) becomes

$$i(t) = \frac{E_0}{R} + ce^{-(R/L)t}. \tag{12}$$

Note that as $t \to \infty$, the second term in equation (12) approaches zero. Such a term is usually called a **transient term;** any remaining terms are called the **steady-state** part of the solution. In this case E_0/R is also called the **steady-state current;** for large values of time it then appears that the current in the circuit is simply governed by Ohm's law ($E = iR$).

--- **Remark** ---

Consider the differential equation in Example 1 that describes the growth of bacteria. The solution $N(t) = N_0 e^{0.4055t}$ of the initial-value problem $dN/dt = kN$, $N(t_0) = N_0$ is of course a continuous function. But in the example we are talking about a population of bacteria, and so common sense dictates that N take on only positive integer values. Moreover, the population does not necessarily grow continuously—that is, every second, every microsecond, and so on—as predicted by the function $N(t) = N_0 e^{0.4055t}$; there may be time intervals $[t_1, t_2]$ over which there is no growth at all. Perhaps, then, the graph shown in Figure 3.7(a) is a more realistic description of N than is the graph of an exponential function. Using a continuous

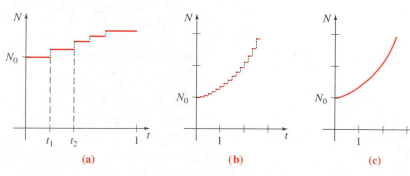

(a)　　　　　　(b)　　　　　　(c)

FIGURE 3.7

function to describe a discrete phenomenon is often more convenient than accurate. However, for some purposes we may be satisfied if our model fairly accurately describes the system when viewed macroscopically in time, as in Figures 3.7(b) and (c), rather than microscopically.

SECTION 3.1 EXERCISES

Answers to odd-numbered problems begin on page A-2.

1. The population of a certain community is known to increase at a rate proportional to the number of people present at any time. If the population has doubled in 5 years, how long will it take to triple? to quadruple?

2. Suppose it is known that the population of the community in Problem 1 is 10,000 after 3 years. What was the initial population? What will be the population in 10 years?

3. The population of a town grows at a rate proportional to the population at any time. Its initial population of 500 increases by 15% in 10 years. What will be the population in 30 years?

4. The population of bacteria in a culture grows at a rate proportional to the number of bacteria present at any time. After 3 hours it is observed that there are 400 bacteria present. After 10 hours there are 2000 bacteria present. What was the initial number of bacteria?

5. The radioactive isotope of lead, Pb-209, decays at a rate proportional to the amount present at any time and has a half-life of 3.3 hours. If 1 gram of lead is present initially, how long will it take for 90% of the lead to decay?

6. Initially 100 milligrams of a radioactive substance was present. After 6 hours the mass had decreased by 3%. If the rate of decay is proportional to the amount of the substance present at any time, find the amount remaining after 24 hours.

7. Determine the half-life of the radioactive substance described in Problem 6.

8. (a) Consider the initial-value problem $dA/dt = kA$, $A(0) = A_0$ as the model for the decay of a radioactive substance. Show that in general the half-life T of the substance is $T = -(\ln 2)/k$.
   (b) Show that the solution of the initial-value problem in part (a) can be written $A(t) = A_0 2^{-t/T}$.

9. When a vertical beam of light passes through a transparent substance, the rate at which its intensity I decreases is proportional to $I(t)$, where t represents the thickness of the medium (in feet). In clear seawater the intensity 3 feet below the surface is 25% of the initial intensity I_0 of the incident beam. What is the intensity of the beam 15 feet below the surface?

10. When interest is compounded continuously, the amount of money S increases at a rate proportional to the amount present at any time: $dS/dt = rS$, where r is the annual rate of interest (see (6) of Section 1.3).

(a) Find the amount of money accrued at the end of 5 years when $5000 is deposited in a savings account drawing $5\frac{3}{4}\%$ annual interest compounded continuously.

(b) In how many years will the initial sum deposited have doubled?

(c) Use a calculator to compare the number obtained in part (a) with the value

$$S = 5000 \left(1 + \frac{0.0575}{4}\right)^{5(4)}.$$

This value represents the amount accrued when interest is compounded quarterly.

11. In a piece of burned wood, or charcoal, it was found that 85.5% of the C-14 had decayed. Use the information in Example 3 to determine the approximate age of the wood. (It is precisely these data that archaeologists used to date prehistoric paintings in a cave in Lascaux, France.)

12. A thermometer is taken from an inside room to the outside, where the air temperature is 5°F. After 1 minute the thermometer reads 55°F, and after 5 minutes the reading is 30°F. What is the initial temperature of the room?

13. A thermometer is removed from a room where the air temperature is 70°F to the outside, where the temperature is 10°F. After $\frac{1}{2}$ minute the thermometer reads 50°F. What is the reading at $t = 1$ min? How long will it take for the thermometer to reach 15°F?

14. Formula (3) also holds when an object absorbs heat from the surrounding medium. If a small metal bar whose initial temperature is 20°C is dropped into a container of boiling water, how long will it take for the bar to reach 90°C if it is known that its temperature increased 2° in 1 second? How long will it take the bar to reach 98°C?

15. A 30-volt electromotive force is applied to an LR series circuit in which the inductance is 0.1 henry and the resistance is 50 ohms. Find the current $i(t)$ if $i(0) = 0$. Determine the current as $t \to \infty$.

16. Solve equation (8) under the assumption that $E(t) = E_0 \sin \omega t$ and $i(0) = i_0$.

17. A 100-volt electromotive force is applied to an RC series circuit in which the resistance is 200 ohms and the capacitance is 10^{-4} farad. Find the charge $q(t)$ on the capacitor if $q(0) = 0$. Find the current $i(t)$.

18. A 200-volt electromotive force is applied to an RC series circuit in which the resistance is 1000 ohms and the capacitance is 5×10^{-6} farad. Find the charge $q(t)$ on the capacitor if $i(0) = 0.4$. Determine the charge and current at $t = 0.005$ s. Determine the charge as $t \to \infty$.

19. An electromotive force

$$E(t) = \begin{cases} 120, & 0 \le t \le 20 \\ 0, & t > 20 \end{cases}$$

is applied to an LR series circuit in which the inductance is 20 henries and the resistance is 2 ohms. Find the current $i(t)$ if $i(0) = 0$.

20. Suppose an *RC* series circuit has a variable resistor. If the resistance at any time t is given by $R = k_1 + k_2t$, where $k_1 > 0$ and $k_2 > 0$ are known constants, then (10) becomes

$$(k_1 + k_2t)\frac{dq}{dt} + \frac{1}{C}q = E(t).$$

Show that if $E(t) = E_0$ and $q(0) = q_0$, then

$$q(t) = E_0C + (q_0 - E_0C)\left(\frac{k_1}{k_1 + k_2t}\right)^{1/Ck_2}.$$

21. A tank contains 200 liters of fluid in which 30 grams of salt is dissolved. Brine containing 1 gram of salt per liter is then pumped into the tank at a rate of 4 L/min; the well-mixed solution is pumped out at the same rate. Find the number $A(t)$ of grams of salt in the tank at any time.

22. Solve Problem 21 assuming pure water is pumped into the tank.

23. A large tank is filled with 500 gallons of pure water. Brine containing 2 pounds of salt per gallon is pumped into the tank at a rate of 5 gal/min. The well-mixed solution is pumped out at the same rate. Find the number $A(t)$ of pounds of salt in the tank at any time.

24. Solve Problem 23 under the assumption that the solution is pumped out at a faster rate of 10 gal/min. When is the tank empty?

25. A large tank is partially filled with 100 gallons of fluid in which 10 pounds of salt is dissolved. Brine containing $\frac{1}{2}$ pound of salt per gallon is pumped into the tank at a rate of 6 gal/min. The well-mixed solution is then pumped out at a slower rate of 4 gal/min. Find the number of pounds of salt in the tank after 30 minutes.

26. In Example 5 the size of the tank containing the salt mixture was not given. Suppose, as discussed on page 66, that the rate at which brine is pumped into the tank is the same but the mixed solution is pumped out at a rate of 2 gal/min. It stands to reason that since brine is accumulating in the tank at the rate of 1 gal/min, any finite tank must eventually overflow. Now suppose that the tank has an open top and has a total capacity of 400 gallons.
   (a) When will the tank overflow?
   (b) How many pounds of salt will be in the tank at the instant it overflows?
   (c) Assume that the tank is overflowing and that brine continues to be pumped in at a rate of 3 gal/min and the well-stirred solution continues to be pumped out at a rate of 2 gal/min. Devise a method for determining the number of pounds of salt in the tank at $t = 150$ min.
   (d) Determine the number of pounds of salt in the tank as $t \to \infty$. Does your answer agree with your intuition?
   (e) Use a graphing utility to obtain the graph $A(t)$ on the interval $[0, \infty)$.

27. A differential equation governing the velocity v of a falling mass m subjected to air resistance proportional to the instantaneous velocity is

$$m \frac{dv}{dt} = mg - kv,$$

where k is a positive constant of proportionality.
(a) Solve the equation subject to the initial condition $v(0) = v_0$.
(b) Determine the limiting, or terminal, velocity of the mass.
(c) If distance s is related to velocity $ds/dt = v$, find an explicit expression for s if it is further known that $s(0) = s_0$.

28. The rate at which a drug disseminates into the bloodstream is governed by the differential equation

$$\frac{dx}{dt} = r - kx,$$

where r and k are positive constants. The function $x(t)$ describes the concentration of the drug in the bloodstream at any time t. Find the limiting value of $x(t)$ as $t \to \infty$. At what time is the concentration one-half this limiting value? Assume that $x(0) = 0$.

29. In one model of the changing population $P(t)$ of a community, it is assumed that

$$\frac{dP}{dt} = \frac{dB}{dt} - \frac{dD}{dt},$$

where dB/dt and dD/dt are the birth and death rates, respectively.
(a) Solve for $P(t)$ if

$$\frac{dB}{dt} = k_1 P \qquad \text{and} \qquad \frac{dD}{dt} = k_2 P.$$

(b) Analyze the cases $k_1 > k_2$, $k_1 = k_2$, and $k_1 < k_2$.

30. The differential equation

$$\frac{dP}{dt} = (k \cos t)P,$$

where k is a positive constant, is often used as a model of a population that undergoes yearly seasonal fluctuations. Solve for $P(t)$, and graph the solution. Assume $P(0) = P_0$.

31. When forgetfulness is taken into account, the rate of memorization of a subject is given by

$$\frac{dA}{dt} = k_1(M - A) - k_2 A,$$

where $k_1 > 0$, $k_2 > 0$, $A(t)$ is the amount of material memorized in time t, M is the total amount to be memorized, and $M - A$ is the amount remaining to be memorized. Solve for $A(t)$, and graph the solution. Assume $A(0) = 0$. Find the limiting value of A as $t \to \infty$, and interpret the result.

32. When all the curves in a family $G(x, y, c_1) = 0$ intersect orthogonally all the curves in another family $H(x, y, c_2) = 0$, the families are said to be **orthogonal trajectories** of each other. See Figure 3.8. If

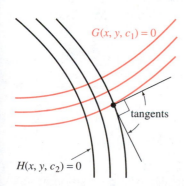

$G(x, y, c_1) = 0$

tangents

$H(x, y, c_2) = 0$

FIGURE 3.8

Year	Population
1790	3.929
1800	5.308
1810	7.240
1820	9.638
1830	12.866
1840	17.069
1850	23.192
1860	31.433
1870	38.558
1880	50.156
1890	62.948
1900	75.996
1910	91.972
1920	105.711
1930	122.775
1940	131.669
1950	150.697

$dy/dx = f(x, y)$ is the differential equation of one family, then the differential equation for the orthogonal trajectories of this family is $dy/dx = -1/f(x, y)$.

(a) Find a differential equation for the family $y = -x - 1 + c_1 e^x$.

(b) Find the orthogonal trajectories for the family in part (a).

33. The U.S. census population figures (in millions) for the years 1790, 1800, 1810, . . . , 1950 are given in the table on the left.

(a) Use these data to construct a model of the form

$$\frac{dP}{dt} = kP, \qquad P(0) = P_0.$$

(b) Construct a table comparing the population predicted by the model in part (a) with the census population. Compute the error and the percentage error for each entry pair.

Discussion Problems

34. Suppose a medical examiner, arriving at the scene of a homicide, finds that the temperature of the corpse is 82°F. Make up additional, but plausible, data necessary to determine an approximate time of death of the victim, using Newton's law of cooling (3).

35. Mr. Jones puts two cups of coffee on the breakfast table at the same time. He immediately pours cream into his coffee from a pitcher that was sitting on the table for a long time. He then reads the morning paper for 5 minutes before taking his first sip. Mrs. Jones arrives at the table 5 minutes after the cups were set down, adds cream to her coffee, and takes a sip. Assume that Mr. and Mrs. Jones add exactly the same amount of cream to their cups. Discuss who drinks the hotter cup of coffee. Defend your assertion with sound mathematics.

36. A linear model for the spread of an epidemic in a community of n persons is given by the initial-value problem

$$\frac{dx}{dt} = r(n - x), \qquad x(0) = x_0,$$

where $x(t)$ denotes the population at time t, $r > 0$ is a constant rate, and x_0 is some small positive integer (such as 1). Explain why, according to this model, everyone in the community will contract the disease. Determine how long it will take for the epidemic to run its course.

3.2 NONLINEAR EQUATIONS

■ *Population models* ■ *Relative growth rate* ■ *Logistic differential equation*
■ *Logistic function* ■ *Second-order chemical reaction*

Population Models If $P(t)$ denotes the size of a population at any time t, the model for exponential growth begins with the assumption that $dP/dt = kP$ for some $k > 0$. In this model the **specific,** or **relative, growth**

rate defined by

$$\frac{dP/dt}{P} \qquad (1)$$

is assumed to be a constant k. True cases of exponential growth over long periods of time are hard to find, because the limited resources of the environment will at some time exert restrictions on the growth of a population. Thus (1) can be expected to decrease as P increases in size.

The assumption that the rate at which a population grows (or declines) is dependent only on the number present and not on any time-dependent mechanisms such as seasonal phenomena (see Problem 18 in Exercises 1.3) can be stated as

$$\frac{dP/dt}{P} = f(P) \qquad \text{or} \qquad \frac{dP}{dt} = Pf(P). \qquad (2)$$

The differential equation in (2), which is widely assumed in models of animal populations, is called the **density-dependent hypothesis.**

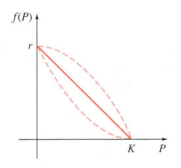

FIGURE 3.9

Logistic Equation

Suppose an environment is capable of sustaining no more than a fixed number K of individuals in its population. The quantity K is called the **carrying capacity** of the environment. Hence for the function f in (2) we have $f(K) = 0$, and we simply let $f(0) = r$. Figure 3.9 shows three functions f that satisfy these two conditions. The simplest assumption we can make is that $f(P)$ is linear—that is, $f(P) = c_1 P + c_2$. If we use the conditions $f(0) = r$ and $f(K) = 0$, we find $c_2 = r$ and $c_1 = -r/K$, respectively, and f takes on the form $f(P) = r - (r/K)P$. Equation (2) becomes

$$\frac{dP}{dt} = P\left(r - \frac{r}{K}P\right). \qquad (3)$$

Relabeling constants, we find that the nonlinear equation (3) is the same as

$$\frac{dP}{dt} = P(a - bP). \qquad (4)$$

About 1840 the Belgian mathematician-biologist P. F. Verhulst was concerned with mathematical models for predicting the human population of various countries. One of the equations he studied was (4), where $a > 0$, $b > 0$. Equation (4) came to be known as the **logistic equation,** and its solution is called the **logistic function.** The graph of a logistic function is called a **logistic curve.**

The differential equation $dP/dt = kP$ does not provide a very accurate model for population when the population itself is very large. Overcrowded conditions with the resulting detrimental effects on the environment, such as pollution and excessive and competitive demands for food and fuel, can have an inhibiting effect on population growth. As we shall now see, the solution of (4) is bounded as $t \to \infty$. If we rewrite (4) as $dP/dt = aP - bP^2$, the nonlinear term $-bP^2$, $b > 0$ can be interpreted as an "inhibition" or "competition" term. Also, in most applications the positive constant a is much larger than the constant b.

Logistic curves have proved to be quite accurate in predicting the growth patterns, in a limited space, of certain types of bacteria, protozoa, water fleas (*Daphnia*), and fruit flies (*Drosophila*).

Solution of the Logistic Equation   One method of solving (4) is by separation of variables. Decomposing the left side of $dP/P(a - bP) = dt$ into partial fractions and integrating gives

$$\left(\frac{1/a}{P} + \frac{b/a}{a - bP}\right) dP = dt$$

$$\frac{1}{a}\ln|P| - \frac{1}{a}\ln|a - bP| = t + c$$

$$\ln\left|\frac{P}{a - bP}\right| = at + ac$$

$$\frac{P}{a - bP} = c_1 e^{at}.$$

It follows from the last equation that

$$P(t) = \frac{ac_1 e^{at}}{1 + bc_1 e^{at}} = \frac{ac_1}{bc_1 + e^{-at}}.$$

If $P(0) = P_0$, $P_0 \neq a/b$, we find $c_1 = P_0/(a - bP_0)$, and so, after substituting and simplifying, the solution becomes

$$P(t) = \frac{aP_0}{bP_0 + (a - bP_0)e^{-at}}. \tag{5}$$

Graphs of $P(t)$ The basic shape of the graph of the logistic function $P(t)$ can be obtained without too much effort. Although the variable t usually represents time and we are seldom concerned with applications in which $t < 0$, it is nonetheless of some interest to include this interval when displaying the various graphs of P. From (5) we see that

$$P(t) \to \frac{aP_0}{bP_0} = \frac{a}{b} \text{ as } t \to \infty \qquad \text{and} \qquad P(t) \to 0 \text{ as } t \to -\infty.$$

The dashed line $P = a/2b$ shown in Figure 3.10 corresponds to the ordinate of a point of inflection of the logistic curve. To show this, we differentiate (4) by the product rule:

$$\frac{d^2P}{dt^2} = P\left(-b\frac{dP}{dt}\right) + (a - bP)\frac{dP}{dt} = \frac{dP}{dt}(a - 2bP)$$

$$= P(a - bP)(a - 2bP)$$

$$= 2b^2 P\left(P - \frac{a}{b}\right)\left(P - \frac{a}{2b}\right).$$

From calculus recall that the points where $d^2P/dt^2 = 0$ are possible points of inflection, but $P = 0$ and $P = a/b$ can obviously be ruled out. Hence $P = a/2b$ is the only possible ordinate value at which the concavity of the graph can change. For $0 < P < a/2b$ it follows that $P'' > 0$, and $a/2b < P < a/b$ implies $P'' < 0$. Thus, as we read from left to right, the graph changes from concave up to concave down at the point corresponding to

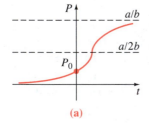

(a)

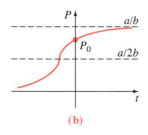

(b)

FIGURE 3.10

$P = a/2b$. When the initial value satisfies $0 < P_0 < a/2b$, the graph of $P(t)$ assumes the shape of an S, as we see in Figure 3.10(a). For $a/2b < P_0 < a/b$ the graph is still S-shaped, but the point of inflection occurs at a negative value of t, as shown in Figure 3.10(b).

We have already seen equation (4) in (9) of Section 1.3 in the form $dx/dt = kx(n + 1 - x)$, $k > 0$. This differential equation provides a reasonable model for describing the spread of an epidemic brought about initially by introducing an infected individual into a static population. The solution $x(t)$ represents the number of individuals infected with the disease at any time.

EXAMPLE 1   **Logistic Growth**

Suppose a student carrying a flu virus returns to an isolated college campus of 1000 students. If it is assumed that the rate at which the virus spreads is proportional not only to the number x of infected students but also to the number of students not infected, determine the number of infected students after 6 days if it is further observed that after 4 days $x(4) = 50$.

SOLUTION   Assuming that no one leaves the campus throughout the duration of the disease, we must solve the initial-value problem

$$\frac{dx}{dt} = kx(1000 - x), \qquad x(0) = 1.$$

By making the identifications $a = 1000k$ and $b = k$, we have immediately from (5) that

$$x(t) = \frac{1000k}{k + 999ke^{-1000kt}} = \frac{1000}{1 + 999e^{-1000kt}}.$$

Now, using the information $x(4) = 50$, we determine k from

$$50 = \frac{1000}{1 + 999e^{-4000k}}.$$

We find $-1000k = \frac{1}{4}\ln\frac{19}{999} = -0.9906$. Thus

$$x(t) = \frac{1000}{1 + 999e^{-0.9906t}}.$$

Finally $x(6) = \dfrac{1000}{1 + 999e^{-5.9436}} = 276$ students.

Additional calculated values of $x(t)$ are given in the table in Figure 3.11(b).   ■

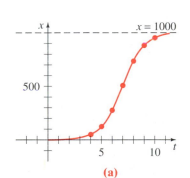

(a)

t (days)	x (number infected)
4	50 (observed)
5	124
6	276
7	507
8	735
9	882
10	953

(b)

FIGURE 3.11

Gompertz Curves   Another equation of the form given in (2) is a modification of the logistic equation

$$\frac{dP}{dt} = P(a - b \ln P), \tag{6}$$

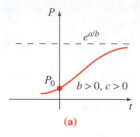

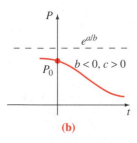

FIGURE 3.12

where a and b are constants. It is readily shown by separation of variables (see Problem 5 in Exercises 3.2) that a solution of (6) is

$$P(t) = e^{a/b}e^{-ce^{-bt}}, \tag{7}$$

where c is an arbitrary constant. We note that when $b > 0$, $P \to e^{a/b}$ as $t \to \infty$; whereas for $b < 0$, $c > 0$, $P \to 0$ as $t \to \infty$. The graph of the function (7), called a **Gompertz curve,** is quite similar to the graph of the logistic function. Figure 3.12 shows two possibilities for the graph of $P(t)$.

Functions such as (7) are encountered, for example, in the growth or decline of certain populations, in the growth of solid tumors, in actuarial predictions, and in the growth of revenue in the sale of a commercial product.

Chemical Reactions　Suppose that a grams of chemical A are combined with b grams of chemical B. If there are M parts of A and N parts of B formed in the compound and $X(t)$ is the number of grams of chemical C formed, then the numbers of grams of chemicals A and B remaining at any time are, respectively,

$$a - \frac{M}{M + N}X \quad \text{and} \quad b - \frac{N}{M + N}X.$$

By the law of mass action, the rate of the reaction satisfies

$$\frac{dX}{dt} \propto \left(a - \frac{M}{M + N}X\right)\left(b - \frac{N}{M + N}X\right). \tag{8}$$

If we factor out $M/(M + N)$ from the first factor and $N/(M + N)$ from the second and introduce a constant $k > 0$ of proportionality, (8) has the form

$$\frac{dX}{dt} = k(\alpha - X)(\beta - X), \tag{9}$$

where $\alpha = a(M + N)/M$ and $\beta = b(M + N)/N$. Recall from (7) of Section 1.3 that a chemical reaction governed by the nonlinear differential equation (9) is said to be a **second-order reaction.**

EXAMPLE 2　**Second-Order Chemical Reaction**

A compound C is formed when two chemicals A and B are combined. The resulting reaction between the two chemicals is such that for each gram of A, 4 grams of B is used. It is observed that 30 grams of the compound C is formed in 10 minutes. Determine the amount of C at any time if the rate of the reaction is proportional to the amounts of A and B remaining and if initially there are 50 grams of A and 32 grams of B. How much of the compound C is present at 15 minutes? Interpret the solution as $t \to \infty$.

SOLUTION　Let $X(t)$ denote the number of grams of the compound C present at any time t. Clearly $X(0) = 0$ g and $X(10) = 30$ g.

If, for example, 2 grams of compound C is present, we must have used, say, a grams of A and b grams of B so that $a + b = 2$ and $b = 4a$. Thus we must use $a = \frac{2}{5} = 2(\frac{1}{5})$ g of chemical A and $b = \frac{8}{5} = 2(\frac{4}{5})$ g of B. In general, for X grams of C we must use

$$\frac{X}{5} \text{ grams of } A \qquad \text{and} \qquad \frac{4}{5}X \text{ grams of } B.$$

The amounts of A and B remaining at any time are then

$$50 - \frac{X}{5} \qquad \text{and} \qquad 32 - \frac{4}{5}X,$$

respectively.

Now we know that the rate at which compound C is formed satisfies

$$\frac{dX}{dt} \propto \left(50 - \frac{X}{5}\right)\left(32 - \frac{4}{5}X\right).$$

To simplify the subsequent algebra, we factor $\frac{1}{5}$ from the first term and $\frac{4}{5}$ from the second and then introduce the constant of proportionality:

$$\frac{dX}{dt} = k(250 - X)(40 - X).$$

By separation of variables and partial fractions we can write

$$-\frac{1/210}{250 - X}\,dX + \frac{1/210}{40 - X}\,dX = k\,dt.$$

Integrating gives

$$\ln\left|\frac{250 - X}{40 - X}\right| = 210kt + c_1 \qquad \text{or} \qquad \frac{250 - X}{40 - X} = c_2 e^{210kt}. \qquad \textbf{(10)}$$

When $t = 0$, $X = 0$, so it follows at this point that $c_2 = \frac{25}{4}$. Using $X = 30$ g at $t = 10$, we find $210k = \frac{1}{10}\ln\frac{88}{25} = 0.1258$. With this information we solve the last equation in (10) for X:

$$X(t) = 1000\frac{1 - e^{-0.1258t}}{25 - 4e^{-0.1258t}}. \qquad \textbf{(11)}$$

The behavior of X as a function of time is displayed in Figure 3.13. It is clear from the accompanying table and (11) that $X \to 40$ as $t \to \infty$. This means that 40 grams of compound C is formed, leaving

$$50 - \frac{1}{5}(40) = 42 \text{ g of } A \qquad \text{and} \qquad 32 - \frac{4}{5}(40) = 0 \text{ g of } B. \qquad \blacksquare$$

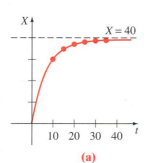

(a)

t (min)	x (g)
10	30 (measured)
15	34.78
20	37.25
25	38.54
30	39.22
35	39.59

(b)

FIGURE 3.13

Note

Integration result number 20 in the Table of Integrals on the endpapers of this text notwithstanding, the alternative form in terms of the inverse hyperbolic tangent,

$$\int \frac{du}{a^2 - u^2} = \frac{1}{a}\tanh^{-1}\frac{u}{a} + c,$$

might be more useful in several problems in Exercises 3.2.

SECTION 3.2 EXERCISES

Answers to odd-numbered problems begin on page A-3.

1. The number $C(t)$ of supermarkets throughout the country that are using a computerized checkout system is described by the initial-value problem

$$\frac{dC}{dt} = C(1 - 0.0005C), \qquad C(0) = 1,$$

where $t > 0$. How many supermarkets are using the computerized method when $t = 10$? How many companies are expected to adopt the new procedure over a long period of time?

2. The number $N(t)$ of people in a community who are exposed to a particular advertisement is governed by the logistic equation. Initially $N(0) = 500$, and it is observed that $N(1) = 1000$. If it is predicted that the limiting number of people in the community who will see the advertisement is 50,000, determine $N(t)$ at any time.

3. A model for the population $P(t)$ at any time in a suburb of a large city is given by the initial-value problem

$$\frac{dP}{dt} = P(10^{-1} - 10^{-7} P), \qquad P(0) = 5000,$$

where t is measured in months. What is the limiting value of the population? At what time will the population be equal to one-half of this limiting value?

4. Find a solution of the **modified logistic equation**

$$\frac{dP}{dt} = P(a - bP)(1 - cP^{-1}), \quad a, b, c > 0.$$

5. (a) Solve equation (6):

$$\frac{dP}{dt} = P(a - b \ln P).$$

   (b) Determine the value of c in (7) if $P(0) = P_0$.

6. Assuming $0 < P_0 < e^{a/b}$ and $a > 0$, use (6) to find the ordinate of the point of inflection for a Gompertz curve.

7. Two chemicals A and B are combined to form a chemical C. The rate, or velocity, of the reaction is proportional to the product of the instantaneous amounts of A and B not converted to chemical C. Initially there are 40 grams of A and 50 grams of B, and for each gram of B, 2 grams of A is used. It is observed that 10 grams of C is formed in 5 minutes. How much is formed in 20 minutes? What is the limiting amount of C after a long time? How much of chemicals A and B remains after a long time?

8. Solve Problem 7 if 100 grams of chemical A is present initially. At what time is chemical C half-formed?

9. Obtain a solution of the equation

$$\frac{dX}{dt} = k(\alpha - X)(\beta - X)$$

governing second-order reactions in the two cases $\alpha \neq \beta$ and $\alpha = \beta$.

10. In a third-order chemical reaction the number X of grams of a compound obtained by combining three chemicals is governed by

$$\frac{dX}{dt} = k(\alpha - X)(\beta - X)(\gamma - X).$$

Solve the equation under the assumption $\alpha \neq \beta \neq \gamma$.

11. The height h of water that is draining through a hole in a tank in the form of a right circular cylinder is given by

$$\frac{dh}{dt} = -\frac{A_o}{A_w}\sqrt{2gh}, \qquad g = 32 \text{ ft/s}^2,$$

where A_w and A_o are the cross-sectional areas of the water and the hole, respectively (see (14) in Section 1.3). Solve the equation if the initial height of the water is 20 feet and $A_w = 50$ ft² and $A_o = \frac{1}{4}$ ft². At what time is the tank empty?

12. How long will it take the tank in Problem 11 to empty if the friction/contraction factor at the hole is $c = 0.6$ (see Problem 7 in Exercises 1.3)?

13. Solve the differential equation of the **tractrix**

$$\frac{dy}{dx} = -\frac{y}{\sqrt{s^2 - y^2}}$$

(see Problem 13 in Exercises 1.3). Assume that the initial point on the y-axis is $(0, 10)$ and the length of rope is $s = 10$ ft.

14. According to **Stefan's law** of radiation, the rate of change of temperature from a body at absolute temperature T is

$$\frac{dT}{dt} = k(T^4 - T_m{}^4),$$

where T_m is the absolute temperature of the surrounding medium. Find a solution of this differential equation. It can be shown that when $T - T_m$ is small compared to T_m, this particular equation is closely approximated by Newton's law of cooling (see (10) in Section 1.3).

15. A differential equation governing the velocity v of a falling mass m subjected to air resistance proportional to the square of the instantaneous velocity is

$$m\frac{dv}{dt} = mg - kv^2,$$

where k is a positive constant of proportionality.
   (a) Solve this equation subject to the initial condition $v(0) = v_0$.
   (b) Determine the limiting, or terminal, velocity of the mass.

(c) If distance s is related to velocity by $ds/dt = v$, find an explicit expression for s if it is further known that $s(0) = s_0$.

16. (a) Determine a differential equation for the velocity $v(t)$ of a mass m sinking in water that imparts a resistance proportional to the square of the instantaneous velocity and also exerts an upward buoyant force whose magnitude is given by Archimedes' principle. Assume the positive direction is downward.

(b) Solve the differential equation in part (a).

(c) Determine the limiting, or terminal, velocity of the sinking mass.

17. (a) If a constant number h of animals is removed or harvested per unit time, then a model for the population $P(t)$ of animals at any time t is given by

$$\frac{dP}{dt} = P(a - bP) - h, \qquad P(0) = P_0,$$

where a, b, h, and P_0 are positive constants. Solve the problem when $a = 5$, $b = 1$, and $h = 4$.

(b) Use an ODE solver to determine the long-term behavior of the population in part (a) in the cases $P_0 > 4$, $1 < P_0 < 4$, and $0 < P_0 < 1$.

(c) If the population becomes extinct in a finite time, find that time.

18. (a) Use the census data from 1790, 1850, and 1910 (given in the table accompanying Problem 33 in Exercises 3.1) to construct a population model of the form

$$\frac{dP}{dt} = P(a - bP), \qquad P(0) = P_0.$$

(b) Construct a table comparing the population predicted by the model in part (a) with the census population. Compute the error and the percentage error for each entry pair.

19. Find the orthogonal trajectories for the family $y = 1/(x + c_1)$ (see Problem 32 in Exercises 3.1). Use a graphing utility to graph both families on the same set of coordinate axes.

20. If it is assumed that a snowball melts in such a manner that its shape is always spherical, then a mathematical model for its volume V is given by

$$\frac{dV}{dt} = kS,$$

where S is the surface area of a sphere of radius r and $k < 0$ is a constant of proportionality. (See Problem 19 in Exercises 1.3.)

(a) Rewrite the differential equation in terms of $V(t)$.

(b) Solve the equation in part (a) subject to the initial condition $V(0) = V_0$.

(c) If $r(0) = r_0$, determine the radius of the snowball as a function of time t. When does the snowball disappear?

21. The differential equation

$$\frac{dy}{dx} = \frac{-x + \sqrt{x^2 + y^2}}{y}$$

describes the shape of a plane curve C that will reflect all incoming light beams to the same point. (See Problem 17 in Exercises 1.3.) There are several ways of solving this equation.

(a) First, verify that the differential equation is homogeneous (see Section 2.4). Show that the substitution $y = ux$ yields

$$\frac{u\,du}{\sqrt{1 + u^2}\,(1 - \sqrt{1 + u^2})} = \frac{dx}{x}.$$

Use a CAS (or another judicious substitution) to integrate the left side of the equation. Show that the curve C must be a parabola with focus at the origin and symmetric with respect to the x-axis.

(b) Next, show that the first differential equation can be written in the alternative form $y = 2xy' + y(y')^2$. Let $w = y^2$, and then use Problem 54 in Exercises 1.1 to solve the resulting differential equation. Reconcile any difference between this answer and that obtained in part (a).

(c) Finally, show that the first differential equation can also be solved by means of the substitution $u = x^2 + y^2$.

22. A simple model for the shape of a tsunami, or tidal wave, is given by

$$\frac{1}{2}\left(\frac{dW}{dx}\right)^2 = 2W^2 - W^3,$$

where $W(x)$ is the height of the wave expressed as a function of its position relative to a fixed point offshore.

(a) By inspection, find all constant solutions of the differential equation.

(b) Use a CAS to find a nonconstant solution of the differential equation.

(c) Use a graphing utility to graph all solutions that satisfy the initial condition $W(0) = 2$.

Discussion Problem

23. A sky diver weighing 160 pounds steps out the door of an airplane that is flying at an altitude of 12,000 feet, After freely falling for 15 seconds, the sky diver opens the parachute. Assume that air resistance is proportional to v^2 while the parachute is unopened and is proportional to the velocity v after the parachute is opened. (See Figure 3.14.) For a person of this weight, typical values for the constant k in the models given in Problem 27 of Exercises 3.1 and Problem 15 above are $k = 7.857$ and $k = 0.0053$, respectively. Determine the time it takes the sky diver to reach the ground. What is the sky diver's impact velocity?

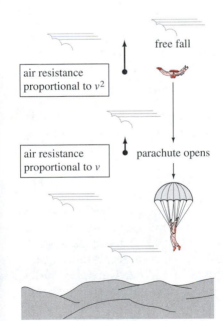

free fall

air resistance proportional to v^2

air resistance proportional to v

parachute opens

FIGURE 3.14

3.3 SYSTEMS OF LINEAR AND NONLINEAR EQUATIONS

- *System of DEs as a mathematical model* ■ *Linear and nonlinear systems*
- *Radioactive decay* ■ *Mixtures* ■ *Lotka-Volterra predator-prey model*
- *Competition models* ■ *Electrical networks*

Up to now all the mathematical models that we have considered have been single differential equations. A single differential equation can describe a single population in an environment; but if there are, say, two interacting and perhaps competing species living in the same environment (for example, rabbits and foxes), then a model for their populations $x(t)$ and $y(t)$ *might* be a system of two first-order differential equations such as

$$\frac{dx}{dt} = g_1(t, x, y)$$

$$\frac{dy}{dt} = g_2(t, x, y). \tag{1}$$

When g_1 and g_2 are linear in the variables x and y, that is, $g_1(x, y) = c_1 x + c_2 y + f_1(t)$ and $g_2(x, y) = c_3 x + c_4 y + f_2(t)$, then (1) is said to be a **linear system.** A system of differential equations that is not linear is said to be **nonlinear.**

In this section we are going to discuss mathematical models based on some of the topics that we have already discussed in the preceding two sections. This section will be similar to Section 1.3 in that we are just going to discuss certain mathematical models that are systems of first-order differential equations; we are not going to develop any methods for solving these systems. There are reasons for not solving systems at this point: First, we do not as yet possess the necessary mathematical tools for solving systems, and second, some of the systems that we will discuss simply cannot be solved.

We shall examine solution methods for systems of linear first-order equations in Chapter 8 and for systems of linear higher-order differential equations in Chapters 4 and 7.

Radioactive Series   In the discussion of radioactive decay in Sections 1.3 and 3.1 we assumed that the rate of decay was proportional to the number $A(t)$ of nuclei of the substance present at time t. When a substance decays by radioactivity, it usually doesn't just transmute into one stable substance and the process stops; rather, the first substance decays into another radioactive substance, this substance in turn decays into a third substance, and so on. This process, called a **radioactive decay series,** continues until a stable element is reached. For example, the uranium decay series is U-238 $\rightarrow$ Th-234 $\rightarrow$ $\cdots$ $\rightarrow$ Pb-206, where Pb-206 is a stable isotope of lead. The half-lives of the various elements in a radioactive series can range from billions of years (4.5×10^9 years for U-238) to a fraction of a second. Suppose a radioactive series is described schematically by $X \xrightarrow{-\lambda_1} Y \xrightarrow{-\lambda_2} Z$, where $k_1 = -\lambda_1 < 0$ and $k_2 = -\lambda_2 < 0$ are the decay constants for substances X and Y, respectively, and Z is a stable element. Suppose, too, that $x(t)$, $y(t)$, and $z(t)$ denote amounts of substances X, Y, and Z, respectively, remaining at any time. The decay of element X is described by

$$\frac{dx}{dt} = -\lambda_1 x,$$

whereas the rate at which the second element Y decays is the net rate

$$\frac{dy}{dt} = \lambda_1 x - \lambda_2 y$$

since it is *gaining* atoms from the decay of X and at the same time *losing* atoms because of its own decay. Since Z is a stable element, it is simply gaining atoms from the decay of element Y:

$$\frac{dz}{dt} = \lambda_2 y.$$

In other words, a model of the radioactive decay series for three elements is the linear system of three first-order differential equations

$$\frac{dx}{dt} = -\lambda_1 x$$
$$\frac{dy}{dt} = \lambda_1 x - \lambda_2 y \qquad\qquad \textbf{(2)}$$
$$\frac{dz}{dt} = \lambda_2 y.$$

Mixtures Consider the two tanks shown in Figure 3.15. Let us suppose for the sake of discussion that tank A contains 50 gallons of water in which 25 pounds of salt is dissolved. Suppose tank B contains 50 gallons of pure water. Liquid is pumped in and out of the tanks as indicated in the figure; the mixture exchanged between the two tanks and the liquid pumped out of tank B are assumed to be well stirred. We wish to construct a mathematical model that describes the numbers $x_1(t)$ and $x_2(t)$ of pounds of salt in tanks A and B, respectively, at any time t.

By an analysis similar to that on page 22 in Section 1.3 and Example 5 in Section 3.1, we see for tank A that the net rate of change of $x_1(t)$ is

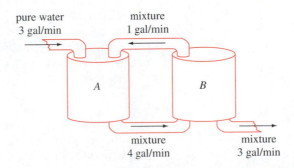

pure water
3 gal/min

mixture
1 gal/min

A

B

mixture
4 gal/min

mixture
3 gal/min

FIGURE 3.15

$$\overbrace{\frac{dx_1}{dt} = (3\ \text{gal/min}) \cdot (0\ \text{lb/gal}) + (1\ \text{gal/min}) \cdot \left(\frac{x_2}{50}\ \text{lb/gal}\right)}^{\text{input rate of salt}} - \overbrace{(4\ \text{gal/min}) \cdot \left(\frac{x_1}{50}\ \text{lb/gal}\right)}^{\text{output rate of salt}}$$

$$= -\frac{2}{25}x_1 + \frac{1}{50}x_2.$$

Similarly, for tank B the net rate of change of $x_2(t)$ is

$$\frac{dx_2}{dt} = 4 \cdot \frac{x_1}{50} - 3 \cdot \frac{x_2}{50} - 1 \cdot \frac{x_2}{50}$$

$$= \frac{2}{25}x_1 - \frac{2}{25}x_2.$$

Thus we obtain the linear system

$$\frac{dx_1}{dt} = -\frac{2}{25}x_1 + \frac{1}{50}x_2$$

$$\frac{dx_2}{dt} = \frac{2}{25}x_1 - \frac{2}{25}x_2. \tag{3}$$

Observe that the foregoing system is accompanied by the initial conditions $x_1(0) = 25$, $x_2(0) = 0$.

A Predator-Prey Model

Suppose that two different species of animals interact within the same environment or ecosystem, and suppose further that the first species eats only vegetation and the second eats only the first species. In other words, one species is a predator and the other is a prey. For example, wolves hunt grass-eating caribou, sharks devour little fish, and the snowy owl pursues an arctic rodent called the lemming. For the sake of discussion, let us imagine that the predators are foxes and the prey are rabbits.

Let $x(t)$ and $y(t)$ denote, respectively, the fox and rabbit populations at any time t. If there were no rabbits, then one might expect that the foxes, lacking an adequate food supply, would decline in number according to

$$\frac{dx}{dt} = -ax, \quad a > 0. \tag{4}$$

When rabbits are present in the environment, however, it seems reasonable that the number of encounters or interactions between these two species per unit time is jointly proportional to their populations x and y, that is, proportional to the product xy. Thus when rabbits are present there is a supply of food, and so foxes are added to the system at a rate bxy, $b > 0$. Adding this last rate to (4) gives a model for the fox population:

$$\frac{dx}{dt} = -ax + bxy. \tag{5}$$

On the other hand, were there no foxes, then the rabbits would, with an added assumption of unlimited food supply, grow at a rate that is proportional to the number of rabbits present at any time:

$$\frac{dy}{dt} = dy, \quad d > 0. \tag{6}$$

But when foxes are present, a model for the rabbit population is (6) decreased by cxy, $c > 0$, that is, decreased by the rate at which the rabbits are eaten during their encounters with the foxes:

$$\frac{dy}{dt} = dy - cxy. \tag{7}$$

Equations (5) and (7) constitute a system of nonlinear differential equations

$$\frac{dx}{dt} = -ax + bxy = x(-a + by)$$

$$\frac{dy}{dt} = dy - cxy = y(d - cx), \tag{8}$$

where a, b, c, and d are positive constants. This famous system of equations is known as the **Lotka-Volterra predator-prey model.**

Except for two constant solutions, $x(t) = 0$, $y(t) = 0$ and $x(t) = d/c$, $y(t) = a/b$, the nonlinear system (8) cannot be solved in terms of elementary functions. However, we can analyze such systems quantitatively and qualitatively. See Chapter 9, Numerical Methods for Ordinary Differential Equations, and Chapter 10, Plane Autonomous Systems and Stability.*

EXAMPLE 1   **Predator-Prey Model**

Suppose

$$\frac{dx}{dt} = -0.16x + 0.08xy$$

$$\frac{dy}{dt} = 4.5y - 0.9xy$$

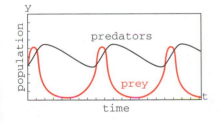

FIGURE 3.16

represents a predator-prey model. Since we are dealing with populations, we have $x(t) \geq 0$, $y(t) \geq 0$. Figure 3.16, obtained with the aid of an ODE solver, shows typical population curves of the predators and prey for this model superimposed on the same coordinate axes. The initial conditions used were $x(0) = 4$, $y(0) = 4$. The curve in black represents the population $x(t)$ of the predator (foxes), and the colored curve is the population $y(t)$ of the prey (rabbits). Observe that the model seems to predict that both populations $x(t)$ and $y(t)$ are periodic in time. This makes intuitive sense since, as the number of prey decreases, the predator population eventually decreases because of a diminished food supply; but attendant to a decrease in the number of predators is an increase in the number of prey; this in turn gives rise to an increased number of predators, which ultimately brings about another decrease in the number of prey. ∎

Competition Models   Now suppose two different species of animals occupy the same ecosystem, not as predator and prey but rather as compet-

---

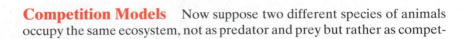

*Chapters 10–15 are in the expanded version of this text, *Differential Equations with Boundary-Value Problems.*

itors for the same resources (such as food and living space) in the system. In the absence of the other, let us assume that the rate at which each population grows is given by

$$\frac{dx}{dt} = ax \qquad \text{and} \qquad \frac{dy}{dt} = cy, \qquad (9)$$

respectively.

Since the two species compete, another assumption might be that each of these rates is diminished simply by the influence, or existence, of the other population. Thus a model for the two populations is given by the linear system

$$\frac{dx}{dt} = ax - by$$

$$\frac{dy}{dt} = cy - dx, \qquad (10)$$

where a, b, c, and d are positive constants.

On the other hand, we might assume, as we did in (5), that each growth rate in (9) should be reduced by a rate proportional to the number of interactions between the two species:

$$\frac{dx}{dt} = ax - bxy$$

$$\frac{dy}{dt} = cy - dxy. \qquad (11)$$

Inspection shows that this nonlinear system is similar to the Lotka-Volterra predator-prey model. Lastly, it might be more realistic to replace the rates in (9), which indicate that the population of each species in isolation grows exponentially, with rates indicating that each population grows logistically (that is, over a long time the population is bounded):

$$\frac{dx}{dt} = a_1x - b_1x^2 \qquad \text{and} \qquad \frac{dy}{dt} = a_2y - b_2y^2. \qquad (12)$$

When these new rates are decreased by rates proportional to the number of interactions, we obtain another nonlinear model

$$\frac{dx}{dt} = a_1x - b_1x^2 - c_1xy = x(a_1 - b_1x - c_1y)$$

$$\frac{dy}{dt} = a_2y - b_2y^2 - c_2xy = y(a_2 - b_2y - c_2x), \qquad (13)$$

where all coefficients are positive. The linear system (10) and the nonlinear systems (11) and (13) are, of course, called **competition models.**

Networks   An electrical network having more than one loop also gives rise to simultaneous differential equations. As shown in Figure 3.17, the current $i_1(t)$ splits in the directions shown at point B_1, called a *branch point* of the network. By Kirchhoff's first law we can write

$$i_1(t) = i_2(t) + i_3(t). \qquad (14)$$

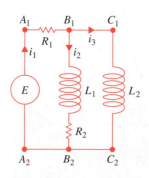

FIGURE 3.17

In addition, we can also apply **Kirchhoff's second law** to each loop. For loop $A_1B_1B_2A_2A_1$, summing the voltage drops across each part of the loop gives

$$E(t) = i_1R_1 + L_1\frac{di_2}{dt} + i_2R_2. \tag{15}$$

Similarly, for loop $A_1B_1C_1C_2B_2A_2A_1$ we find

$$E(t) = i_1R_1 + L_2\frac{di_3}{dt}. \tag{16}$$

Using (14) to eliminate i_1 in (15) and (16) yields two linear first-order equations for the currents $i_2(t)$ and $i_3(t)$:

$$L_1\frac{di_2}{dt} + (R_1 + R_2)i_2 + R_1i_3 = E(t)$$

$$L_2\frac{di_3}{dt} + \qquad R_1i_2 + R_1i_3 = E(t). \tag{17}$$

We leave it as an exercise (see Problem 14) to show that the system of differential equations describing the currents $i_1(t)$ and $i_2(t)$ in the network containing a resistor, an inductor, and a capacitor shown in Figure 3.18 is

$$L\frac{dt_1}{dt} + Ri_2 \quad = E(t)$$

$$RC\frac{di_2}{dt} + i_2 - i_1 = 0. \tag{18}$$

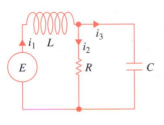

FIGURE 3.18

SECTION 3.3 EXERCISES

Answers to odd-numbered problems begin on page A-3.

1. We have not discussed methods by which systems of first-order differential equations can be solved. Nevertheless, systems such as (2) can be solved with no knowledge other than how to solve a single linear first-order equation. Find a solution of (2) subject to the initial conditions $x(0) = x_0$, $y(0) = 0$, and $z(0) = 0$.

2. In Problem 1, suppose that time is measured in days, that the decay constants are $k_1 = -0.138629$ and $k_2 = -0.004951$, and that $x_0 = 20$. Use a graphing utility to obtain the graphs of the solutions $x(t)$, $y(t)$, and $z(t)$ on the same set of coordinate axes. Use the graphs to approximate the half-lives of substances X and Y.

3. Use the graphs in Problem 2 to approximate the times when the amounts $x(t)$ and $y(t)$ are the same, the times when the amounts $x(t)$ and $z(t)$ are the same, and the times when the amounts $y(t)$ and $z(t)$ are the same. Why does the time determined when the amounts $y(t)$ and $z(t)$ are the same make intuitive sense?

4. Construct a mathematical model for a radioactive series of four elements W, X, Y, and Z, where Z is a stable element.

5. Consider two tanks A and B with liquid being pumped in and out at the same rates, as described by the system of equations (3). What is

the system of differential equations if, instead of pure water, a brine solution containing 2 pounds of salt per gallon is pumped into tank A?

6. Use the information given in Figure 3.19 to construct a mathematical model for the numbers $x_1(t)$, $x_2(t)$, and $x_3(t)$ of pounds of salt at any time in tanks A, B, and C, respectively.

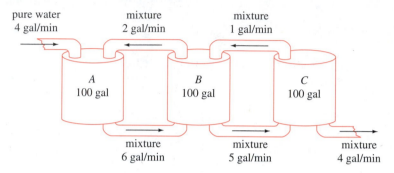

FIGURE 3.19

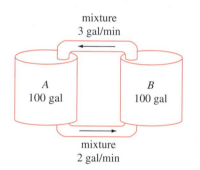

FIGURE 3.20

7. Initially, two large tanks A and B each hold 100 gallons of brine; 100 pounds of salt is dissolved in the solution in tank A and 50 pounds of salt in tank B. The system is closed in that well-stirred liquid is pumped only between the tanks, as shown in Figure 3.20. Use the information given in the figure to construct a mathematical model for the numbers $x_1(t)$ and $x_2(t)$ of pounds of salt at any time in tanks A and B, respectively.

8. Consider the closed two-tank system in Problem 7. There is a relationship between the variables $x_1(t)$ and $x_2(t)$ that holds at any time. What is it? Use this relationship to help find the amount of salt in tank B at $t = 30$ min.

9. Consider the Lotka-Volterra predator-prey model defined by

$$\frac{dx}{dt} = -0.1x + 0.02xy$$

$$\frac{dy}{dt} = 0.2y - 0.025xy,$$

where the populations $x(t)$ (predators) and $y(t)$ (prey) are measured in the thousands. Use an ODE solver with $x(0) = 6$, $y(0) = 6$ to approximate the time $t > 0$ when the two populations are first equal. Use the graphs to approximate the period of each population.

10. Consider the competition model defined by

$$\frac{dx}{dt} = x(2 - 0.4x - 0.3y)$$

$$\frac{dy}{dt} = y(1 - 0.1y - 0.3x),$$

where the populations $x(t)$ and $y(t)$ are measured in thousands and t in years. Use an ODE solver to analyze the populations over a long period of time for each of the following cases:

(a) $x(0) = 1.5$, $y(0) = 3.5$ **(b)** $x(0) = 1$, $y(0) = 1$
(c) $x(0) = 2$, $y(0) = 7$ **(d)** $x(0) = 4.5$, $y(0) = 0.5$

11. Consider the competition model defined by

$$\frac{dx}{dt} = x(1 - 0.1x - 0.05y)$$

$$\frac{dy}{dt} = y(1.7 - 0.1y - 0.15x),$$

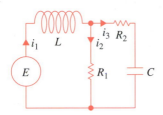

FIGURE 3.21

where the populations $x(t)$ and $y(t)$ are measured in thousands and t in years. Use an ODE solver to analyze the populations over a long period of time for each of the following cases:

(a) $x(0) = 1$, $y(0) = 1$ **(b)** $x(0) = 4$, $y(0) = 10$
(c) $x(0) = 9$, $y(0) = 4$ **(d)** $x(0) = 5.5$, $y(0) = 3.5$

12. Show that a system of differential equations describing the currents $i_2(t)$ and $i_3(t)$ in the electrical network shown in Figure 3.21 is

$$L\frac{di_2}{dt} + L\frac{di_3}{dt} + R_1 i_2 = E(t)$$

$$-R_1\frac{di_2}{dt} + R_2\frac{di_3}{dt} + \frac{1}{C}i_3 = 0.$$

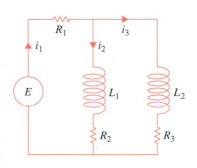

FIGURE 3.22

13. Determine a system of first-order differential equations describing the currents $i_2(t)$ and $i_3(t)$ in the electrical network shown in Figure 3.22.

14. Show that the linear system given in (18) describes the currents $i_1(t)$ and $i_2(t)$ in the network shown in Figure 3.18. [*Hint: dq/dt = i_3.*]

15. A communicable disease is spread throughout a small community, with a fixed population of n people, by contact between infected persons and persons who are susceptible to the disease. Suppose initially that everyone is susceptible to the disease and that no one leaves the community while the epidemic is spreading. At time t, let $s(t)$, $i(t)$, and $r(t)$ denote, respectively, the number of people in the community (measured in hundreds) who are *susceptible* to the disease but not yet infected with it, the number of people who are *infected* with the disease, and the number of people who have *recovered* from the disease. Explain why the system of differential equations

$$\frac{ds}{dt} = -k_1 si$$

$$\frac{di}{dt} = -k_2 i + k_1 si$$

$$\frac{dr}{dt} = k_2 i,$$

where k_1 (called the *infection rate*) and k_2 (called the *removal rate*) are positive constants, is a reasonable mathematical model for the spread of the epidemic throughout the community. Give plausible initial conditions associated with this system of equations.

16. **(a)** Explain why, in Problem 15, it is sufficient to analyze only

$$\frac{ds}{dt} = -k_1 si$$

$$\frac{di}{dt} = -k_2 i + k_1 si.$$

(b) Suppose $k_1 = 0.2$, $k_2 = 0.7$, and $n = 10$. Choose various values of $i(0) = i_0$, $0 < i_0 < 10$. Use an ODE solver to determine what the model predicts about the epidemic in the cases $s_0 > k_2/k_1$ and $s_0 \le k_2/k_1$. In the case of an epidemic, estimate the number of people eventually infected.

Discussion Problems

fluid at   fluid at
concentration   concentration
$x(t)$ $y(t)$

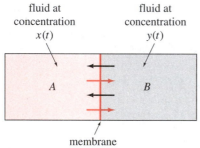

membrane

FIGURE 3.23

17. Suppose compartments A and B shown in Figure 3.23 are filled with fluids and are separated by a permeable membrane. The figure is a compartmental representation of the exterior and interior of a cell. Suppose, too, that a nutrient necessary for cell growth passes through the membrane. A model for the concentrations $x(t)$ and $y(t)$ of the nutrient in compartments A and B, respectively, at any time t is given by the linear system of differential equations

$$\frac{dx}{dt} = \frac{\kappa}{V_A}(y - x)$$

$$\frac{dy}{dt} = \frac{\kappa}{V_B}(x - y),$$

where V_A and V_B are the volumes of the compartments and $\kappa > 0$ is a permeability factor. Let $x(0) = x_0$ and $y(0) = y_0$ denote the initial concentrations of the nutrient. Based solely on the equations in the system and the assumption $x_0 > y_0 > 0$, sketch, on the same set of coordinate axes, possible solution curves of the system. Explain your reasoning. Discuss the behavior of the solutions over a long period of time.

18. The system in Problem 17, like the system in (2), can be solved with no advanced knowledge. Solve for $x(t)$ and $y(t)$, and compare their graphs with your conjecture in Problem 17. [*Hint:* Subtract the two equations, and let $z(t) = x(t) - y(t)$.] Determine the limiting values of $x(t)$ and $y(t)$ as $t \to \infty$. Explain why these values make intuitive sense.

19. Based solely on the physical description of the mixture problem on pages 83–84 and in Figure 3.15, discuss the nature of the functions $x_1(t)$ and $x_2(t)$. What is the behavior of each function over a long period of time? Sketch possible graphs of $x_1(t)$ and $x_2(t)$. Check your conjectures by using an ODE solver to obtain the solution curves of (3) subject to $x_1(0) = 25$, $x_2(0) = 0$.

CHAPTER 3 REVIEW EXERCISES

Answers to odd-numbered problems begin on page A-4.

1. In March 1976 the world population reached 4 billion. A popular news magazine predicted that with an average yearly growth rate of 1.8%, the world population would be 8 billion in 45 years. How does this value compare with that predicted by the model that says the rate of increase is proportional to the population at any time?

2. Air containing 0.06% carbon dioxide is pumped into a room whose volume is 8000 ft^3. The air is pumped in at a rate of 2000 ft^3/min, and the circulated air is then pumped out at the same rate. If there is an initial concentration of 0.2% carbon dioxide, determine the subsequent amount in the room at any time. What is the concentration at 10 minutes? What is the steady-state, or equilibrium, concentration of carbon dioxide?

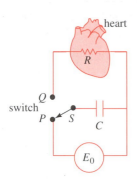

FIGURE 3.24

3. A heart pacemaker, as shown in Figure 3.24, consists of a battery, a capacitor, and the heart as a resistor. When the switch S is at P the capacitor charges; when S is at Q the capacitor discharges, sending an electrical stimulus to the heart. During this time the voltage E applied to the heart is given by

$$\frac{dE}{dt} = -\frac{1}{RC}E, \quad t_1 < t < t_2,$$

where R and C are constants. Determine $E(t)$ if $E(t_1) = E_0$. (Of course, the opening and closing of the switch are periodic in time, to simulate the natural heartbeat.)

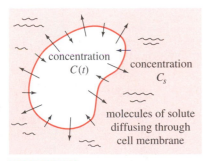

FIGURE 3.25

4. Suppose a cell is suspended in a solution containing a solute of constant concentration C_s. Suppose further that the cell has constant volume V and that the area of its permeable membrane is the constant A. By Fick's law the rate of change of its mass m is directly proportional to the area A and the difference $C_s - C(t)$, where $C(t)$ is the concentration of the solute inside the cell at any time t. Find $C(t)$ if $m = VC(t)$ and $C(0) = C_0$. See Figure 3.25.

5. Consider Newton's law of cooling $dT/dt = k(T - T_m)$, $k < 0$, where the temperature of the surrounding medium T_m changes with time. Suppose the initial temperature of a body is T_1 and the initial temperature of the surrounding medium is T_2 and $T_m = T_2 + B(T_1 - T)$, where $B > 0$ is a constant.
   (a) Find the temperature of the body at any time t.
   (b) What is the limiting value of the temperature as $t \to \infty$?
   (c) What is the limiting value of T_m as $t \to \infty$?

6. An LR series circuit has a variable inductor with the inductance defined by

$$L = \begin{cases} 1 - \dfrac{t}{10}, & 0 \le t < 10 \\ 0, & t \ge 10 \end{cases}.$$

Find the current $i(t)$ if the resistance is 0.2 ohm, the impressed voltage is $E(t) = 4$, and $i(0) = 0$. Graph $i(t)$.

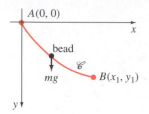

FIGURE 3.26

7. A classical problem in the calculus of variations is to find the shape of a curve $\mathscr{C}$ such that a bead, under the influence of gravity, will slide from point $A(0, 0)$ to point $B(x_1, y_1)$ in the least time. See Figure 3.26. It can be shown that a nonlinear differential equation for the shape $y(x)$ of the path is $y[1 + (y')^2] = k$, where k is a constant. First solve for dx in terms of y and dy, and then use the substitution $y = k \sin^2\theta$ to obtain the parametric form of the solution. The curve $\mathscr{C}$ turns out to be a cycloid.

8. A projectile is shot vertically into the air with an initial velocity of v_0 ft/s. Assuming that air resistance is proportional to the square of the instantaneous velocity, the motion is described by the pair of differential equations

$$m\frac{dv}{dt} = -mg - kv^2, \quad k > 0,$$

positive y-axis up, origin at ground level so that $v = v_0$ at $y = 0$;

and

$$m\frac{dv}{dt} = mg - kv^2, \quad k > 0,$$

positive y-axis down, origin at the maximum height so that $v = 0$ at $y = h$. These equations describe the motion of the projectile when rising and falling, respectively. Prove that the impact velocity v_i of the projectile is less than the initial velocity v_0. It can also be shown that the time t_1 needed for the projectile to attain its maximum height h is less than the time t_2 that it takes to fall from this height. See Figure 3.27.

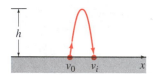

FIGURE 3.27

9. The populations of two species of animals are described by the nonlinear system of first-order differential equations

$$\frac{dx}{dt} = k_1 x(\alpha - x)$$

$$\frac{dy}{dt} = k_2 xy.$$

Solve for x and y in terms of t.

10. Initially, two large tanks A and B each hold 100 gallons of brine. The well-stirred liquid is pumped between the tanks as shown in Figure 3.28. Use the information given in the figure to construct a mathematical model for the numbers $x_1(t)$ and $x_2(t)$ of pounds of salt at any time in tanks A and B, respectively.

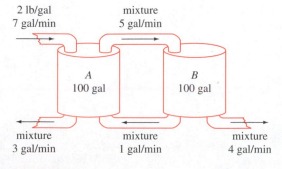

FIGURE 3.28

CHAPTER

4

DIFFERENTIAL EQUATIONS OF HIGHER ORDER

INTRODUCTION

We turn now to the solution of differential equations of order two or higher. In the first seven sections of this chapter we examine some of the underlying theory and methods for solving certain kinds of *linear* equations. The elimination method for solving systems of linear ordinary differential equations is introduced in Section 4.8 since this basic method simply uncouples a system into individual linear higher-order equations in each dependent variable. The chapter concludes with a brief examination of *nonlinear* higher-order equations.

4.1 PRELIMINARY THEORY: LINEAR EQUATIONS

- *Linear higher-order DEs* ■ *Initial-value problem* ■ *Existence and uniqueness*
- *Boundary-value problem* ■ *Homogeneous and nonhomogeneous DEs*
- *Linear differential operator* ■ *Linear dependence* ■ *Linear independence*
- *Wronskian* ■ *Fundamental set of solutions* ■ *Superposition principles*
- *General solution* ■ *Complementary function* ■ *Particular solution*

4.1.1   Initial-Value and Boundary-Value Problems

Initial-Value Problem   In Section 1.2 we defined an initial-value problem for a general nth-order differential equation. For a linear differential equation, an ***n*th-order initial-value problem** is

Solve: $$a_n(x)\frac{d^n y}{dx^n} + a_{n-1}(x)\frac{d^{n-1}y}{dx^{n-1}} + \cdots + a_1(x)\frac{dy}{dx} + a_0(x)y = g(x)$$

Subject to: $y(x_0) = y_0, \quad y'(x_0) = y_1, \quad \ldots, \quad y^{(n-1)}(x_0) = y_{n-1}.$ **(1)**

Recall that for a problem such as this we seek a function defined on some interval I containing x_0 that satisfies the differential equation and the n initial conditions specified at x_0: $y(x_0) = y_0, y'(x_0) = y_1, \ldots, y^{(n-1)}(x_0) = y_{n-1}$. We have already seen that in the case of a second-order initial-value problem a solution curve must pass through the point (x_0, y_0) and have slope y_1 at this point.

Existence and Uniqueness   In Section 1.2 we stated a theorem that gave conditions under which the existence and uniqueness of a solution of a first-order initial-value problem were guaranteed. The theorem that follows gives sufficient conditions for the existence of a unique solution of the problem in (1).

THEOREM 4.1   **Existence of a Unique Solution**

Let $a_n(x), a_{n-1}(x), \ldots, a_1(x), a_0(x)$ and $g(x)$ be continuous on an interval I, and let $a_n(x) \neq 0$ for every x in this interval. If $x = x_0$ is any point in this interval, then a solution $y(x)$ of the initial-value problem (1) exists on the interval and is unique.

EXAMPLE 1   **Unique Solution of an IVP**

The initial-value problem

$$3y''' + 5y'' - y' + 7y = 0, \qquad y(1) = 0, \quad y'(1) = 0, \quad y''(1) = 0$$

possesses the trivial solution $y = 0$. Since the third-order equation is linear with constant coefficients, it follows that all the conditions of Theorem 4.1 are fulfilled. Hence $y = 0$ is the *only* solution on any interval containing $x = 1$.   ∎

EXAMPLE 2 Unique Solution of an IVP

You should verify that the function $y = 3e^{2x} + e^{-2x} - 3x$ is a solution of the initial-value problem

$$y'' - 4y = 12x, \qquad y(0) = 4, \quad y'(0) = 1.$$

Now the differential equation is linear, the coefficients as well as $g(x) = 12x$ are continuous, and $a_2(x) = 1 \neq 0$ on any interval I containing $x = 0$. We conclude from Theorem 4.1 that the given function is the unique solution on I. ∎

The requirements in Theorem 4.1 that $a_i(x)$, $i = 0, 1, 2, \ldots, n$ be continuous and $a_n(x) \neq 0$ for every x in I are both important. Specifically, if $a_n(x) = 0$ for some x in the interval, then the solution of a linear initial-value problem may not be unique or even exist. For example, you should verify that the function $y = cx^2 + x + 3$ is a solution of the initial-value problem

$$x^2 y'' - 2xy' + 2y = 6, \qquad y(0) = 3, \quad y'(0) = 1$$

on the interval $(-\infty, \infty)$ for any choice of the parameter c. In other words, there is no unique solution of the problem. Although most of the conditions of Theorem 4.1 are satisfied, the obvious difficulties are that $a_2(x) = x^2$ is zero at $x = 0$ and that the initial conditions are also imposed at $x = 0$.

Boundary-Value Problem Another type of problem consists of solving a linear differential equation of order two or greater in which the dependent variable y or its derivatives are specified at *different points*. A problem such as

$$\textit{Solve:} \qquad a_2(x)\frac{d^2y}{dx^2} + a_1(x)\frac{dy}{dx} + a_0(x)y = g(x)$$

$$\textit{Subject to:} \quad y(a) = y_0, \quad y(b) = y_1$$

is called a **boundary-value problem (BVP).** The prescribed values $y(a) = y_0$ and $y(b) = y_1$ are called **boundary conditions.** A solution of the foregoing problem is a function satisfying the differential equation on some interval I, containing a and b, whose graph passes through the two points (a, y_0) and (b, y_1). See Figure 4.1.

For a second-order differential equation, other pairs of boundary conditions could be

$$y'(a) = y_0, \quad y(b) = y_1$$
$$y(a) = y_0, \quad y'(b) = y_1$$
$$y'(a) = y_0, \quad y'(b) = y_1,$$

where y_0 and y_1 denote arbitrary constants. These three pairs of conditions are just special cases of the general boundary conditions

$$\alpha_1 y(a) + \beta_1 y'(a) = \gamma_1$$
$$\alpha_2 y(b) + \beta_2 y'(b) = \gamma_2.$$

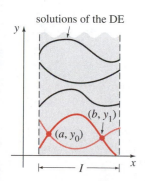

solutions of the DE

FIGURE 4.1

The next examples show that even when the conditions of Theorem 4.1 are fulfilled, a boundary-value problem may have (i) several solutions (as suggested in Figure 4.1), (ii) a unique solution, or (iii) no solution at all.

EXAMPLE 3   A BVP Can Have Many, One, or No Solutions

In Example 5 of Section 1.1 we saw that the two-parameter family of solutions of the differential equation $x'' + 16x = 0$ is

$$x = c_1 \cos 4t + c_2 \sin 4t. \qquad (2)$$

(a) Suppose we now wish to determine that solution of the equation that further satisfies the boundary conditions $x(0) = 0$, $x(\pi/2) = 0$. Observe that the first condition $0 = c_1 \cos 0 + c_2 \sin 0$ implies $c_1 = 0$, so that $x = c_2 \sin 4t$. But when $t = \pi/2$, $0 = c_2 \sin 2\pi$ is satisfied for any choice of c_2 since $\sin 2\pi = 0$. Hence the boundary-value problem

$$x'' + 16x = 0, \qquad x(0) = 0, \quad x\left(\frac{\pi}{2}\right) = 0 \qquad (3)$$

has infinitely many solutions. Figure 4.2 shows the graphs of some of the members of the one-parameter family $x = c_2 \sin 4t$ that pass through the two points $(0, 0)$ and $(\pi/2, 0)$.

(b) If the boundary-value problem in (3) is changed to

$$x'' + 16x = 0, \qquad x(0) = 0, \quad x\left(\frac{\pi}{8}\right) = 0 \qquad (4)$$

then $x(0) = 0$ still requires $c_1 = 0$ in the solution (2). But applying $x(\pi/8) = 0$ to $x = c_2 \sin 4t$ demands that $0 = c_2 \sin(\pi/2) = c_2 \cdot 1$. Hence $x = 0$ is a solution of this new boundary-value problem. Indeed, it can be proved that $x = 0$ is the *only* solution of (4).

(c) Finally, if we change the problem to

$$x'' + 16x = 0, \qquad x(0) = 0, \quad x\left(\frac{\pi}{2}\right) = 1 \qquad (5)$$

we find again that $c_1 = 0$ from $x(0) = 0$, but that applying $x(\pi/2) = 1$ to $x = c_2 \sin 4t$ leads to the contradiction $1 = c_2 \sin 2\pi = c_2 \cdot 0 = 0$. Hence the boundary-value problem (5) has no solution. ∎

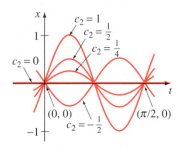

FIGURE 4.2

4.1.2   Homogeneous Equations

A linear nth-order differential equation of the form

$$a_n(x)\frac{d^n y}{dx^n} + a_{n-1}(x)\frac{d^{n-1}y}{dx^{n-1}} + \cdots + a_1(x)\frac{dy}{dx} + a_0(x)y = 0 \qquad (6)$$

is said to be **homogeneous,** whereas an equation

$$a_n(x)\frac{d^n y}{dx^n} + a_{n-1}(x)\frac{d^{n-1}y}{dx^{n-1}} + \cdots + a_1(x)\frac{dy}{dx} + a_0(x)y = g(x), \qquad (7)$$

with $g(x)$ not identically zero, is said to be **nonhomogeneous.** For example, $2y'' + 3y' - 5y = 0$ is a homogeneous linear second-order differential equation, whereas $x^3 y''' + 6y' + 10y = e^x$ is a nonhomogeneous linear third-order differential equation. The word *homogeneous* in this context does not refer to coefficients that are homogeneous functions, as in Section 2.4.

We shall see that in order to solve a nonhomogeneous linear equation (7), we must first be able to solve the associated homogeneous equation (6).

Note

To avoid needless repetition throughout the remainder of this text we shall, as a matter of course, make the following important assumptions when stating definitions and theorems about the linear equations (6) and (7): On some common interval I

- the coefficients $a_i(x)$, $i = 0, 1, 2, \ldots, n$ are continuous;
- the right-hand member $g(x)$ is continuous; and
- $a_n(x) \neq 0$ for every x in the interval.

Differential Operators

In calculus, differentiation is often denoted by the capital letter D, that is, $dy/dx = Dy$. The symbol D is called a **differential operator** since it transforms a differentiable function into another function. For example, $D(\cos 4x) = -4 \sin 4x$ and $D(5x^3 - 6x^2) = 15x^2 - 12x$. Higher-order derivatives can be expressed in terms of D in a natural manner:

$$\frac{d}{dx}\left(\frac{dy}{dx}\right) = \frac{d^2y}{dx^2} = D(Dy) = D^2y \qquad \text{and in general} \qquad \frac{d^ny}{dx^n} = D^ny,$$

where y represents a sufficiently differentiable function. Polynomial expressions involving D, such as $D + 3$, $D^2 + 3D - 4$, and $5x^3D^3 - 6x^2D^2 + 4xD + 9$, are also differential operators. In general, we define an ***n*th-order differential operator** to be

$$L = a_n(x)D^n + a_{n-1}(x)D^{n-1} + \cdots + a_1(x)D + a_0(x). \tag{8}$$

As a consequence of two basic properties of differentiation, $D(cf(x)) = cDf(x)$, c is a constant, and $D\{f(x) + g(x)\} = Df(x) + Dg(x)$, the differential operator L possesses a linearity property; that is, L operating on a linear combination of two differentiable functions is the same as the linear combination of L operating on the individual functions. In symbols this means that

$$L\{\alpha f(x) + \beta g(x)\} = \alpha L(f(x)) + \beta L(g(x)), \tag{9}$$

where α and β are constants. Because of (9) we say that the nth-order differential operator L is a **linear operator.**

Differential Equations

Any linear differential equation can be expressed in terms of the D notation. For example, the differential equation $y'' + 5y' + 6y = 5x - 3$ can be written as $D^2y + 5Dy + 6y = 5x - 3$ or $(D^2 + 5D + 6)y = 5x - 3$. Using (8), the linear nth-order differential equations (6) and (7) can be written compactly as

$$L(y) = 0 \qquad \text{and} \qquad L(y) = g(x),$$

respectively.

Superposition Principle   In the next theorem we see that the sum, or **superposition,** of two or more solutions of a homogeneous linear differential equation is also a solution.

THEOREM 4.2   **Superposition Principle–Homogeneous Equations**

Let $y_1, y_2, \ldots, y_k$ be solutions of the homogeneous nth-order differential equation (6) on an interval I. Then the linear combination

$$y = c_1 y_1(x) + c_2 y_2(x) + \cdots + c_k y_k(x),$$

where the $c_i, i = 1, 2, \ldots, k$ are arbitrary constants, is also a solution on the interval.

PROOF   We prove the case $k = 2$. Let L be the differential operator defined in (8), and let $y_1(x)$ and $y_2(x)$ be solutions of the homogeneous equation $L(y) = 0$. If we define $y = c_1 y_1(x) + c_2 y_2(x)$, then by linearity of L we have

$$L(y) = L\{c_1 y_1(x) + c_2 y_2(x)\} = c_1 L(y_1) + c_2 L(y_2) = c_1 \cdot 0 + c_2 \cdot 0 = 0.$$ ∎

Corollaries to Theorem 4.2

(A) A constant multiple $y = c_1 y_1(x)$ of a solution $y_1(x)$ of a homogeneous linear differential equation is also a solution.
(B) A homogeneous linear differential equation always possesses the trivial solution $y = 0$.

EXAMPLE 4   **Superposition–Homogeneous DE**

The functions $y_1 = x^2$ and $y_2 = x^2 \ln x$ are both solutions of the homogeneous linear equation $x^3 y''' - 2xy' + 4y = 0$ on the interval $(0, \infty)$. By the superposition principle, the linear combination

$$y = c_1 x^2 + c_2 x^2 \ln x$$

is also a solution of the equation on the interval.   ∎

The function $y = e^{7x}$ is a solution of $y'' - 9y' + 14y = 0$. Since the differential equation is linear and homogeneous, the constant multiple $y = ce^{7x}$ is also a solution. For various values of c we see that $y = 9e^{7x}$, $y = 0$, $y = -\sqrt{5}\, e^{7x}, \ldots$ are all solutions of the equation.

Linear Dependence and Linear Independence   The next two concepts are basic to the study of linear differential equations.

DEFINITION 4.1 **Linear Dependence/Independence**

A set of functions $f_1(x), f_2(x), \ldots, f_n(x)$ is said to be **linearly dependent** on an interval I if there exist constants $c_1, c_2, \ldots, c_n$, not all zero, such that

$$c_1 f_1(x) + c_2 f_2(x) + \cdots + c_n f_n(x) = 0$$

for every x in the interval. If the set of functions is not linearly dependent on the interval, it is said to be **linearly independent.**

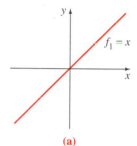

(a)

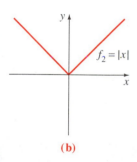

(b)

FIGURE 4.3

In other words, a set of functions is linearly independent on an interval if the only constants for which

$$c_1 f_1(x) + c_2 f_2(x) + \cdots + c_n f_n(x) = 0$$

for every x in the interval are $c_1 = c_2 = \cdots = c_n = 0$.

It is easy to understand these definitions in the case of two functions $f_1(x)$ and $f_2(x)$. If the functions are linearly dependent on an interval, then there exist constants c_1 and c_2 that are not both zero such that for every x in the interval $c_1 f_1(x) + c_2 f_2(x) = 0$. Therefore, if we assume that $c_1 \neq 0$, it follows that

$$f_1(x) = -\frac{c_2}{c_1} f_2(x);$$

that is, *if two functions are linearly dependent, then one is simply a constant multiple of the other.* Conversely, if $f_1(x) = c_2 f_2(x)$ for some constant c_2, then

$$(-1) \cdot f_1(x) + c_2 f_2(x) = 0$$

for every x on some interval. Hence the functions are linearly dependent since at least one of the constants (namely, $c_1 = -1$) is not zero. We conclude that *two functions are linearly independent when neither is a constant multiple of the other* on an interval. For example, the functions $f_1(x) = \sin 2x$ and $f_2(x) = \sin x \cos x$ are linearly dependent on $(-\infty, \infty)$ because $f_1(x)$ is a constant multiple of $f_2(x)$. Recall from the double angle formula for the sine that $\sin 2x = 2 \sin x \cos x$. On the other hand, the functions $f_1(x) = x$ and $f_2(x) = |x|$ are linearly independent on $(-\infty, \infty)$. Inspection of Figure 4.3 should convince you that neither function is a constant multiple of the other on the interval.

It follows from the preceding discussion that the ratio $f_2(x)/f_1(x)$ is not a constant on an interval on which $f_1(x)$ and $f_2(x)$ are linearly independent. This little fact will be used in the next section.

EXAMPLE 5 **Linearly Dependent Functions**

The functions $f_1(x) = \cos^2 x$, $f_2(x) = \sin^2 x$, $f_3(x) = \sec^2 x$, $f_4(x) = \tan^2 x$ are linearly dependent on the interval $(-\pi/2, \pi/2)$ since

$$c_1 \cos^2 x + c_2 \sin^2 x + c_3 \sec^2 x + c_4 \tan^2 x = 0,$$

when $c_1 = c_2 = 1$, $c_3 = -1$, $c_4 = 1$. We used here $\cos^2 x + \sin^2 x = 1$ and $1 + \tan^2 x = \sec^2 x$. ∎

A set of functions $f_1(x), f_2(x), \ldots, f_n(x)$ is linearly dependent on an interval if at least one function can be expressed as a linear combination of the remaining functions.

EXAMPLE 6   Linearly Dependent Functions

The functions $f_1(x) = \sqrt{x} + 5$, $f_2(x) = \sqrt{x} + 5x$, $f_3(x) = x - 1$, $f_4(x) = x^2$ are linearly dependent on the interval $(0, \infty)$ since f_2 can be written as a linear combination of f_1, f_3, and f_4. Observe that

$$f_2(x) = 1 \cdot f_1(x) + 5 \cdot f_3(x) + 0 \cdot f_4(x)$$

for every x in the interval $(0, \infty)$. ∎

Solutions of Differential Equations   We are primarily interested in linearly independent functions or, more to the point, linearly independent solutions of a linear differential equation. Although we could always appeal directly to Definition 4.1, it turns out that the question of whether n solutions $y_1, y_2, \ldots, y_n$ of a homogeneous linear nth-order differential equation (6) are linearly independent can be settled somewhat mechanically using a determinant.

DEFINITION 4.2   Wronskian

Suppose each of the functions $f_1(x), f_2(x), \ldots, f_n(x)$ possesses at least $n - 1$ derivatives. The determinant

$$W(f_1, f_2, \ldots, f_n) = \begin{vmatrix} f_1 & f_2 & \cdots & f_n \\ f_1' & f_2' & \cdots & f_n' \\ \vdots & \vdots & & \vdots \\ f_1^{(n-1)} & f_2^{(n-1)} & \cdots & f_n^{(n-1)} \end{vmatrix},$$

where the primes denote derivatives, is called the **Wronskian** of the functions.

THEOREM 4.3   Criterion for Linearly Independent Solutions

Let $y_1, y_2, \ldots, y_n$ be n solutions of the homogeneous linear nth-order differential equation (6) on an interval I. Then the set of solutions is **linearly independent** on I if and only if

$$W(y_1, y_2, \ldots, y_n) \neq 0$$

for every x in the interval.

It follows from Theorem 4.3 that when $y_1, y_2, \ldots, y_n$ are n solutions of (6) on an interval I, the Wronskian $W(y_1, y_2, \ldots, y_n)$ is either identically zero or never zero on the interval.

A set of n linearly independent solutions of a homogeneous linear nth-order differential equation is given a special name.

DEFINITION 4.3 **Fundamental Set of Solutions**

Any set $y_1, y_2, \ldots, y_n$ of n linearly independent solutions of the homogeneous linear nth-order differential equation (6) on an interval I is said to be a **fundamental set of solutions** on the interval.

The basic question of whether a fundamental set of solutions exists for a linear equation is answered in the next theorem.

THEOREM 4.4 **Existence of a Fundamental Set**

There exists a fundamental set of solutions for the homogeneous linear nth-order differential equation (6) on an interval I.

Analogous to the fact that any vector in three dimensions can be expressed as a linear combination of the *linearly independent* vectors **i**, **j**, **k**, any solution of an nth-order homogeneous linear differential equation on an interval I can be expressed as a linear combination of n linearly independent solutions on I. In other words, n linearly independent solutions $y_1, y_2, \ldots, y_n$ are the basic building blocks for the general solution of the equation.

THEOREM 4.5 **General Solution–Homogeneous Equations**

Let $y_1, y_2, \ldots, y_n$ be a fundamental set of solutions of the homogeneous linear nth-order differential equation (6) on an interval I. Then the **general solution** of the equation on the interval is

$$y = c_1 y_1(x) + c_2 y_2(x) + \cdots + c_n y_n(x),$$

where $c_i, i = 1, 2, \ldots, n$ are arbitrary constants.

Theorem 4.5 states that if $Y(x)$ is any solution of (6) on the interval, then constants $C_1, C_2, \ldots, C_n$ can always be found so that

$$Y(x) = C_1 y_1(x) + C_2 y_2(x) + \cdots + C_n y_n(x).$$

We will prove the case when $n = 2$.

PROOF   Let Y be a solution and y_1 and y_2 be linearly independent solutions of $a_2 y'' + a_1 y' + a_0 y = 0$ on an interval I. Suppose $x = t$ is a point in I for which $W(y_1(t), y_2(t)) \neq 0$. Suppose also that $Y(t) = k_1$ and $Y'(t) = k_2$. If we now examine the equations

$$C_1 y_1(t) + C_2 y_2(t) = k_1$$
$$C_1 y_1'(t) + C_2 y_2'(t) = k_2,$$

it follows that we can determine C_1 and C_2 uniquely, provided that the determinant of the coefficients satisfies

$$\begin{vmatrix} y_1(t) & y_2(t) \\ y_1'(t) & y_2'(t) \end{vmatrix} \neq 0.$$

But this determinant is simply the Wronskian evaluated at $x = t$, and, by assumption, $W \neq 0$. If we define $G(x) = C_1 y_1(x) + C_2 y_2(x)$, we observe that **(i)** $G(x)$ satisfies the differential equation since it is a superposition of two known solutions; **(ii)** $G(x)$ satisfies the initial conditions

$$G(t) = C_1 y_1(t) + C_2 y_2(t) = k_1$$
$$G'(t) = C_1 y_1'(t) + C_2 y_2'(t) = k_2;$$

(iii) $Y(x)$ satisfies the *same* linear equation and the *same* initial conditions. Since the solution of this linear initial-value problem is unique (Theorem 4.1), we have $Y(x) = G(x)$ or $Y(x) = C_1 y_1(x) + C_2 y_2(x)$. ∎

EXAMPLE 7   **General Solution of a Homogeneous DE**

The functions $y_1 = e^{3x}$ and $y_2 = e^{-3x}$ are both solutions of the homogeneous linear equation $y'' - 9y = 0$ on the interval $(-\infty, \infty)$. By inspection, the solutions are linearly independent on the x-axis. This fact can be corroborated by observing that the Wronskian

$$W(e^{3x}, e^{-3x}) = \begin{vmatrix} e^{3x} & e^{-3x} \\ 3e^{3x} & -3e^{-3x} \end{vmatrix} = -6 \neq 0$$

for every x. We conclude that y_1 and y_2 form a fundamental set of solutions, and consequently

$$y = c_1 e^{3x} + c_2 e^{-3x}$$

is the general solution of the equation on the interval. ∎

EXAMPLE 8   **A Solution Obtained from a General Solution**

The function $y = 4 \sinh 3x - 5e^{3x}$ is a solution of the differential equation in Example 7. (Verify this.) In view of Theorem 4.5, we must be able to obtain this solution from the general solution $y = c_1 e^{3x} + c_2 e^{-3x}$. Observe that if we choose $c_1 = 2$ and $c_2 = -7$, then $y = 2e^{3x} - 7e^{-3x}$ can be rewritten as

$$y = 2e^{3x} - 2e^{-3x} - 5e^{-3x} = 4 \left(\frac{e^{3x} - e^{-3x}}{2} \right) - 5e^{-3x}.$$

The last expression is recognized as $y = 4 \sinh 3x - 5e^{-3x}$. ∎

| EXAMPLE 9 | **General Solution of a Homogeneous DE** |

The functions $y_1 = e^x$, $y_2 = e^{2x}$, and $y_3 = e^{3x}$ satisfy the third-order equation

$$\frac{d^3y}{dx^3} - 6\frac{d^2y}{dx^2} + 11\frac{dy}{dx} - 6y = 0.$$

Since
$$W(e^x, e^{2x}, e^{3x}) = \begin{vmatrix} e^x & e^{2x} & e^{3x} \\ e^x & 2e^{2x} & 3e^{3x} \\ e^x & 4e^{2x} & 9e^{3x} \end{vmatrix} = 2e^{6x} \neq 0$$

for every real value of x, the functions y_1, y_2, and y_3 form a fundamental set of solutions on $(-\infty, \infty)$. We conclude that

$$y = c_1e^x + c_2e^{2x} + c_3e^{3x}$$

is the general solution of the differential equation on the interval. ∎

4.1.3 Nonhomogeneous Equations

Any function y_p, free of arbitrary parameters, that satisfies (7) is said to be a **particular solution** or **particular integral** of the equation. For example, it is a straightforward task to show that the constant function $y_p = 3$ is a particular solution of the nonhomogeneous equation $y'' + 9y = 27$.

Now if $y_1, y_2, \ldots, y_k$ are solutions of (6) on an interval I and y_p is any particular solution of (7) on I, then the linear combination

$$y = c_1y_1(x) + c_2y_2(x) + \cdots + c_ky_k(x) + y_p \tag{10}$$

is also a solution of the nonhomogeneous equation (7). If you think about it, this makes sense, because the linear combination $c_1y_1(x) + c_2y_2(x) + \cdots + c_ky_k(x)$ is mapped into 0 by the operator $L = a_nD^n + a_{n-1}D^{n-1} + \cdots + a_1D + a_0$, whereas y_p is mapped into $g(x)$. If we use $k = n$ linearly independent solutions of the nth-order equation (6), then the expression in (10) becomes the general solution of (7).

| THEOREM 4.6 | **General Solution–Nonhomogeneous Equations** |

Let y_p be any particular solution of the nonhomogeneous linear nth-order differential equation (7) on an interval I, and let $y_1, y_2, \ldots, y_n$ be a fundamental set of solutions of the associated homogeneous differential equation (6) on I. Then the **general solution** of the equation on the interval is

$$y = c_1y_1(x) + c_2y_2(x) + \cdots + c_ny_n(x) + y_p,$$

where the c_i, $i = 1, 2, \ldots, n$ are arbitrary constants.

PROOF Let L be the differential operator defined in (8), and let $Y(x)$ and $y_p(x)$ be particular solutions of the nonhomogeneous equation $L(y) = g(x)$. If we define $u(x) = Y(x) - y_p(x)$, then by linearity of L

we have

$$L(u) = L\{Y(x) - y_p(x)\} = L(Y(x)) - L(y_p(x)) = g(x) - g(x) = 0.$$

This shows that $u(x)$ is a solution of the homogeneous equation $L(y) = 0$. Hence, by Theorem 4.5, $u(x) = c_1 y_1(x) + c_2 y_2(x) + \cdots + c_n y_n(x)$, and so

$$Y(x) - y_p(x) = c_1 y_1(x) + c_2 y_2(x) + \cdots + c_n y_n(x)$$

or

$$Y(x) = c_1 y_1(x) + c_2 y_2(x) + \cdots + c_n y_n(x) + y_p(x). \qquad \blacksquare$$

Complementary Function   We see in Theorem 4.6 that the general solution of a nonhomogeneous linear equation consists of the sum of two functions:

$$y = c_1 y_1(x) + c_2 y_2(x) + \cdots + c_n y_n(x) + y_p(x) = y_c(x) + y_p(x).$$

The linear combination $y_c(x) = c_1 y_1(x) + c_2 y_2(x) + \cdots + c_n y_n(x)$, which is the general solution of (6), is called the **complementary function** for equation (7). In other words, to solve a nonhomogeneous linear differential equation we first solve the associated homogeneous equation and then find any particular solution of the nonhomogeneous equation. The general solution of the nonhomogeneous equation is then

y = complementary function + any particular solution.

EXAMPLE 10   **General Solution of a Nonhomogeneous DE**

By substitution, the function $y_p = -\frac{11}{12} - \frac{1}{2}x$ is readily shown to be a particular solution of the nonhomogeneous equation

$$\frac{d^3y}{dx^3} - 6\frac{d^2y}{dx^2} + 11\frac{dy}{dx} - 6y = 3x. \qquad (11)$$

In order to write the general solution of (11), we must also be able to solve the associated homogeneous equation

$$\frac{d^3y}{dx^3} - 6\frac{d^2y}{dx^2} + 11\frac{dy}{dx} - 6y = 0.$$

But in Example 9 we saw that the general solution of this latter equation on the interval $(-\infty, \infty)$ was $y_c = c_1 e^x + c_2 e^{2x} + c_3 e^{3x}$. Hence the general solution of (11) on the interval is

$$y = y_c + y_p = c_1 e^x + c_2 e^{2x} + c_3 e^{3x} - \frac{11}{12} - \frac{1}{2}x. \qquad \blacksquare$$

Another Superposition Principle   The last theorem of this discussion will be useful in Section 4.4 when we consider a method for finding particular solutions of nonhomogeneous equations.

THEOREM 4.7 **Superposition Principle—Nonhomogeneous Equations**

Let $y_{p_1}, y_{p_2}, \ldots, y_{p_k}$ be k particular solutions of the nonhomogeneous linear nth-order differential equation (7) on an interval I corresponding, in turn, to k distinct functions $g_1, g_2, \ldots, g_k$. That is, suppose y_{p_i} denotes a particular solution of the corresponding differential equation

$$a_n(x)y^{(n)} + a_{n-1}(x)y^{(n-1)} + \cdots + a_1(x)y' + a_0(x)y = g_i(x), \qquad (12)$$

where $i = 1, 2, \ldots, k$. Then

$$y_p = y_{p_1}(x) + y_{p_2}(x) + \cdots + y_{p_k}(x) \qquad (13)$$

is a particular solution of

$$a_n(x)y^{(n)} + a_{n-1}(x)y^{(n-1)} + \cdots + a_1(x)y' + a_0(x)y$$
$$= g_1(x) + g_2(x) + \cdots + g_k(x). \qquad (14)$$

PROOF We prove the case $k = 2$. Let L be the differential operator defined in (8), and let $y_{p_1}(x)$ and $y_{p_2}(x)$ be particular solutions of the nonhomogeneous equations $L(y) = g_1(x)$ and $L(y) = g_2(x)$, respectively. If we define $y_p = y_{p_1}(x) + y_{p_2}(x)$, we want to show that y_p is a particular solution of $L(y) = g_1(x) + g_2(x)$. The result follows again by the linearity of the operator L:

$$L(y_p) = L\{y_{p_1}(x) + y_{p_2}(x)\} = L(y_{p_1}(x)) + L(y_{p_2}(x)) = g_1(x) + g_2(x).$$

∎

EXAMPLE 11 **Superposition—Nonhomogeneous DE**

You should verify that

$y_{p_1} = -4x^2$ is a particular solution of $y'' - 3y' + 4y = -16x^2 + 24x - 8$,

$y_{p_2} = e^{2x}$ is a particular solution of $y'' - 3y' + 4y = 2e^{2x}$,

$y_{p_3} = xe^x$ is a particular solution of $y'' - 3y' + 4y = 2xe^x - e^x$.

It follows from Theorem 4.7 that the superposition of y_{p_1}, y_{p_2}, and y_{p_3},

$$y = y_{p_1} + y_{p_2} + y_{p_3} = -4x^2 + e^{2x} + xe^x,$$

is a solution of

$$y'' - 3y' + 4y = \underbrace{-16x^2 + 24x - 8}_{g_1(x)} + \underbrace{2e^{2x}}_{g_2(x)} + \underbrace{2xe^x - e^x}_{g_3(x)}.$$

∎

Note

If the y_{p_i} are particular solutions of (12) for $i = 1, 2, \ldots, k$, then the linear combination

$$y_p = c_1 y_{p_1} + c_2 y_{p_2} + \cdots + c_k y_{p_k},$$

where the c_i are constants, is also a particular solution of (14) when the right-hand member of the equation is the linear combination

$$c_1 g_1(x) + c_2 g_2(x) + \cdots + c_k g_k(x).$$

Before we actually start solving homogeneous and nonhomogeneous linear differential equations, we need one additional bit of theory presented in the next section.

Remark

This remark is a continuation of the brief discussion of dynamical systems given at the end of Section 1.3.

A dynamical system whose rule or mathematical model is a linear nth-order differential equation

$$a_n(t)y^{(n)} + a_{n-1}(t)y^{(n-1)} + \cdots + a_1(t)y' + a_0(t)y = g(t)$$

is said to be an nth-order **linear system.** The n time-dependent functions $y(t), y'(t), \ldots, y^{(n-1)}(t)$ are the **state variables** of the system. Recall that their values at some time t give the **state of the system.** The function g is variously called the **input function, forcing function,** or **excitation function.** A solution $y(t)$ of the differential equation is said to be the **output** or **response of the system.** Under the conditions stated in Theorem 4.1, the output or response $y(t)$ is uniquely determined by the input and the state of the system prescribed at a time t_0, that is, by the initial conditions $y(t_0), y'(t_0), \ldots, y^{(n-1)}(t_0)$. The dependence of the output on the input is illustrated in Figure 4.4.

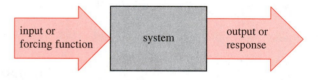

FIGURE 4.4

In order for a dynamical system to be a linear system, it is necessary that the superposition principle (Theorem 4.7) hold in the system; that is, the response of the system to a superposition of inputs is a superposition of outputs. We have already examined some simple linear systems in Section 3.1 (linear first-order equations); in Section 5.1 we examine linear systems in which the mathematical models are second-order differential equations.

SECTION 4.1 EXERCISES

Answers to odd-numbered problems begin on page A-4.

4.1.1

1. Given that $y = c_1 e^x + c_2 e^{-x}$ is a two-parameter family of solutions of $y'' - y = 0$ on the interval $(-\infty, \infty)$, find a member of the family satisfying the initial conditions $y(0) = 0$, $y'(0) = 1$.

2. Find a solution of the differential equation in Problem 1 satisfying the boundary conditions $y(0) = 0$, $y(1) = 1$.

3. Given that $y = c_1 e^{4x} + c_2 e^{-x}$ is a two-parameter family of solutions of $y'' - 3y' - 4y = 0$ on the interval $(-\infty, \infty)$, find a member of the family satisfying the initial conditions $y(0) = 1$, $y'(0) = 2$.

4. Given that $y = c_1 + c_2 \cos x + c_3 \sin x$ is a three-parameter family of solutions of $y''' + y' = 0$ on the interval $(-\infty, \infty)$, find a member of the family satisfying the initial conditions $y(\pi) = 0$, $y'(\pi) = 2$, $y''(\pi) = -1$.

5. Given that $y = c_1 x + c_2 x \ln x$ is a two-parameter family of solutions of $x^2 y'' - xy' + y = 0$ on the interval $(-\infty, \infty)$, find a member of the family satisfying the initial conditions $y(1) = 3$, $y'(1) = -1$.

6. Given that $y = c_1 + c_2 x^2$ is a two-parameter family of solutions of $xy'' - y' = 0$ on the interval $(-\infty, \infty)$, show that constants c_1 and c_2 cannot be found so that a member of the family satisfies the initial conditions $y(0) = 0$, $y'(0) = 1$. Explain why this does not violate Theorem 4.1.

7. Find two members of the family of solutions of $xy'' - y' = 0$ given in Problem 6 satisfying the initial conditions $y(0) = 0$, $y'(0) = 0$.

8. Find a member of the family of solutions of $xy'' - y' = 0$ given in Problem 6 satisfying the boundary conditions $y(0) = 1$, $y'(1) = 6$. Does Theorem 4.1 guarantee that this solution is unique?

9. Given that $y = c_1 e^x \cos x + c_2 e^x \sin x$ is a two-parameter family of solutions of $y'' - 2y' + 2y = 0$ on the interval $(-\infty, \infty)$, determine whether a member of the family can be found that satisfies the boundary conditions

   (a) $y(0) = 1$, $y'(0) = 0$ **(b)** $y(0) = 1$, $y(\pi) = -1$

   (c) $y(0) = 1$, $y\left(\dfrac{\pi}{2}\right) = 1$ **(d)** $y(0) = 0$, $y(\pi) = 0$.

10. Given that $y = c_1 x^2 + c_2 x^4 + 3$ is a two-parameter family of solutions of $x^2 y'' - 5xy' + 8y = 24$ on the interval $(-\infty, \infty)$, determine whether a member of the family can be found that satisfies the boundary conditions

   (a) $y(-1) = 0$, $y(1) = 4$ **(b)** $y(0) = 1$, $y(1) = 2$
   (c) $y(0) = 3$, $y(1) = 0$ **(d)** $y(1) = 3$, $y(2) = 15$.

In Problems 11 and 12 find an interval around $x = 0$ for which the given initial-value problem has a unique solution.

11. $(x - 2)y'' + 3y = x$, $y(0) = 0$, $y'(0) = 1$

12. $y'' + (\tan x)y = e^x$, $y(0) = 1$, $y'(0) = 0$

13. Given that $x = c_1 \cos \omega t + c_2 \sin \omega t$ is a two-parameter family of solutions of $x'' + \omega^2 x = 0$ on the interval $(-\infty, \infty)$, show that a solution satisfying the initial conditions $x(0) = x_0$, $x'(0) = x_1$ is given by

$$x(t) = x_0 \cos \omega t + \frac{x_1}{\omega} \sin \omega t.$$

14. Use the two-parameter family $x = c_1 \cos \omega t + c_2 \sin \omega t$ to show that a solution of the differential equation satisfying $x(t_0) = x_0$, $x'(t_0) = x_1$ is the solution of the initial-value problem in Problem 13 shifted by an amount t_0:

$$x(t) = x_0 \cos \omega(t - t_0) + \frac{x_1}{\omega} \sin \omega(t - t_0).$$

4.1.2

In Problems 15–22 determine whether the given functions are linearly independent or dependent on $(-\infty, \infty)$.

15. $f_1(x) = x$, $f_2(x) = x^2$, $f_3(x) = 4x - 3x^2$

16. $f_1(x) = 0$, $f_2(x) = x$, $f_3(x) = e^x$

17. $f_1(x) = 5$, $f_2(x) = \cos^2 x$, $f_3(x) = \sin^2 x$

18. $f_1(x) = \cos 2x$, $f_2(x) = 1$, $f_3(x) = \cos^2 x$

19. $f_1(x) = x$, $f_2(x) = x - 1$, $f_3(x) = x + 3$

20. $f_1(x) = 2 + x$, $f_2(x) = 2 + |x|$

21. $f_1(x) = 1 + x$, $f_2(x) = x$, $f_3(x) = x^2$

22. $f_1(x) = e^x$, $f_2(x) = e^{-x}$, $f_3(x) = \sinh x$

In Problems 23–30 verify that the given functions form a fundamental set of solutions of the differential equation on the indicated interval. Form the general solution.

23. $y'' - y' - 12y = 0$; $e^{-3x}, e^{4x}, (-\infty, \infty)$

24. $y'' - 4y = 0$; $\cosh 2x, \sinh 2x, (-\infty, \infty)$

25. $y'' - 2y' + 5y = 0$; $e^x \cos 2x, e^x \sin 2x, (-\infty, \infty)$

26. $4y'' - 4y' + y = 0$; $e^{x/2}, xe^{x/2}, (-\infty, \infty)$

27. $x^2 y'' - 6xy' + 12y = 0$; $x^3, x^4, (0, \infty)$

28. $x^2 y'' + xy' + y = 0$; $\cos(\ln x), \sin(\ln x), (0, \infty)$

29. $x^3 y''' + 6x^2 y'' + 4xy' - 4y = 0$; $x, x^{-2}, x^{-2} \ln x, (0, \infty)$

30. $y^{(4)} + y'' = 0$; $1, x, \cos x, \sin x, (-\infty, \infty)$

Discussion Problems

31. (a) Verify that $y_1 = x^3$ and $y_2 = |x|^3$ are linearly independent solutions of the differential equation $x^2 y'' - 4xy' + 6y = 0$ on the interval $(-\infty, \infty)$.

    (b) Show that $W(y_1, y_2) = 0$ for every real number x. Does this result violate Theorem 4.3? Explain.

    (c) Verify that $Y_1 = x^3$ and $Y_2 = x^2$ are also linearly independent solutions of the differential equation in part (a) on the interval $(-\infty, \infty)$.

    (d) Find a solution of the differential equation satisfying $y(0) = 0$, $y'(0) = 0$.

    (e) By the superposition principle, Theorem 4.2, both linear combinations

$$y = c_1 y_1 + c_2 y_2 \quad \text{and} \quad Y = c_1 Y_1 + c_2 Y_2$$

    are solutions of the differential equation. Discuss whether one, both, or neither of the linear combinations is a general solution of the differential equation on the interval $(-\infty, \infty)$.

32. Suppose that $y_1 = e^x$ and $y_2 = e^{-x}$ are two solutions of a homogeneous linear differential equation. Explain why $y_3 = \cosh x$ and $y_4 = \sinh x$ are also solutions of the equation.

4.1.3

In Problems 33–36 verify that the given two-parameter family of functions is the general solution of the nonhomogeneous differential equation on the indicated interval.

33. $y'' - 7y' + 10y = 24e^x$
$y = c_1 e^{2x} + c_2 e^{5x} + 6e^x, \quad (-\infty, \infty)$

34. $y'' + y = \sec x$

$y = c_1 \cos x + c_2 \sin x + x \sin x + (\cos x) \ln(\cos x), \quad \left(-\dfrac{\pi}{2}, \dfrac{\pi}{2} \right)$

35. $y'' - 4y' + 4y = 2e^{2x} + 4x - 12$
$y = c_1 e^{2x} + c_2 x e^{2x} + x^2 e^{2x} + x - 2, \quad (-\infty, \infty)$

36. $2x^2 y'' + 5xy' + y = x^2 - x$

$y = c_1 x^{-1/2} + c_2 x^{-1} + \dfrac{1}{15} x^2 - \dfrac{1}{6} x, \quad (0, \infty)$

37. Given that $y_{p_1} = 3e^{2x}$ and $y_{p_2} = x^2 + 3x$ are particular solutions of

$$y'' - 6y' + 5y = -9e^{2x}$$

and $\qquad\qquad y'' - 6y' + 5y = 5x^2 + 3x - 16,$

respectively, find particular solutions of

$$y'' - 6y' + 5y = 5x^2 + 3x - 16 - 9e^{2x}$$

and $\qquad\qquad y'' - 6y' + 5y = -10x^2 - 6x + 32 + e^{2x}.$

38. (a) By inspection determine a particular solution of $y'' + 2y = 10$.
(b) By inspection determine a particular solution of $y'' + 2y = -4x$.
(c) Find a particular solution of $y'' + 2y = -4x + 10$.
(d) Find a particular solution of $y'' + 2y = 8x + 5$.

4.2 REDUCTION OF ORDER

- *Reducing a linear second-order DE to a first-order DE*
- *Standard form of a homogeneous linear second-order DE*

It is one of the more interesting facts of mathematical life in the study of linear *second-order* differential equations that we can construct a second solution y_2 of

$$a_2(x)y'' + a_1(x)y' + a_0(x)y = 0 \qquad\qquad (1)$$

on an interval I from a known nontrivial solution y_1. We seek a second solution $y_2(x)$ of (1) so that y_1 and y_2 are linearly independent on I. Recall that if y_1 and y_2 are linearly independent, then their ratio y_2/y_1 is nonconstant on I; that is, $y_2/y_1 = u(x)$ or $y_2(x) = u(x)y_1(x)$. The idea is to find the function $u(x)$ by substituting $y_2(x) = u(x)y_1(x)$ into the given differential equation. The method is called **reduction of order** since we must solve a linear *first-order* equation to find u.

EXAMPLE 1   **A Second Solution by Reduction of Order**

Given that $y_1 = e^x$ is a solution of $y'' - y = 0$ on the interval $(-\infty, \infty)$, use reduction of order to find a second solution y_2.

SOLUTION   If $y = u(x)y_1(x) = u(x)e^x$, then the product rule gives

$$y' = ue^x + e^x u', \qquad y'' = ue^x + 2e^x u' + e^x u'',$$

and so $$y'' - y = e^x(u'' + 2u') = 0.$$

Since $e^x \neq 0$, the last equation requires $u'' + 2u' = 0$. If we make the substitution $w = u'$, this linear second-order equation in u becomes $w' + 2w = 0$, which is a linear first-order equation in w. Using the integrating factor e^{2x}, we can write

$$\frac{d}{dx}[e^{2x}w] = 0.$$

After integrating we get $w = c_1 e^{-2x}$ or $u' = c_1 e^{-2x}$. Integrating again then yields

$$u = -\frac{c_1}{2}e^{-2x} + c_2.$$

Thus $$y = u(x)e^x = -\frac{c_1}{2}e^{-x} + c_2 e^x. \qquad \textbf{(2)}$$

By picking $c_2 = 0$ and $c_1 = -2$ we obtain the desired second solution, $y_2 = e^{-x}$. Because $W(e^x, e^{-x}) \neq 0$ for every x, the solutions are linearly independent on $(-\infty, \infty)$.   ∎

Since we have shown that $y_1 = e^x$ and $y_2 = e^{-x}$ are linearly independent solutions of a linear second-order equation, the expression in (2) is actually the general solution of $y'' - y = 0$ on $(-\infty, \infty)$.

General Case   Suppose we divide by $a_2(x)$ in order to put equation (1) in the **standard form**

$$y'' + P(x)y' + Q(x)y = 0, \qquad \textbf{(3)}$$

where $P(x)$ and $Q(x)$ are continuous on some interval I. Let us suppose further that $y_1(x)$ is a known solution of (3) on I and that $y_1(x) \neq 0$ for every x in the interval. If we define $y = u(x)y_1(x)$, it follows that

$$y' = uy_1' + y_1 u', \qquad y'' = uy_1'' + 2y_1' u' + y_1 u''$$

$$y'' + Py' + Qy = u[\underbrace{y_1'' + Py_1' + Qy_1}_{\text{zero}}] + y_1 u'' + (2y_1' + Py_1)u' = 0.$$

This implies that we must have

$$y_1 u'' + (2y_1' + Py_1)u' = 0 \qquad \text{or} \qquad y_1 w' + (2y_1' + Py_1)w = 0, \qquad \textbf{(4)}$$

where we have let $w = u'$. Observe that the last equation in (4) is both linear and separable. Separating variables and integrating, we obtain

$$\frac{dw}{w} + 2\frac{y_1'}{y_1}\, dx + P\, dx = 0$$

$$\ln|wy_1^2| = -\int P\, dx + c \qquad \text{or} \qquad wy_1^2 = c_1 e^{-\int P\, dx}.$$

We solve the last equation for w, use $w = u'$, and integrate again:

$$u = c_1 \int \frac{e^{-\int P\, dx}}{y_1^2}\, dx + c_2.$$

By choosing $c_1 = 1$ and $c_2 = 0$ we find from $y = u(x)y_1(x)$ that a second solution of equation (3) is

$$y_2 = y_1(x) \int \frac{e^{-\int P(x)\, dx}}{y_1^2(x)}\, dx. \tag{5}$$

It makes a good review of differentiation to verify that the function $y_2(x)$ defined in (5) satisfies equation (3) and that y_1 and y_2 are linearly independent on any interval on which $y_1(x)$ is not zero. See Problem 29 in Exercises 4.2.

EXAMPLE 2   **A Second Solution by Formula (5)**

The function $y_1 = x^2$ is a solution of $x^2 y'' - 3xy' + 4y = 0$. Find the general solution on the interval $(0, \infty)$.

SOLUTION   From the standard form of the equation,

$$y'' - \frac{3}{x}y' + \frac{4}{x^2}y = 0,$$

we find from (5) $y_2 = x^2 \int \dfrac{e^{3\int dx/x}}{x^4}\, dx \qquad \leftarrow e^{3\int dx/x} = e^{\ln x^3} = x^3$

$$= x^2 \int \frac{dx}{x} = x^2 \ln x.$$

The general solution on $(0, \infty)$ is given by $y = c_1 y_1 + c_2 y_2$; that is,

$$y = c_1 x^2 + c_2 x^2 \ln x. \qquad \blacksquare$$

Remark

We have derived and illustrated how to use (5) because this formula appears again in the next section and in Section 6.1. We use (5) simply to save time in obtaining a desired result. Your instructor will tell you whether you should memorize (5) or whether you should know the first principles of reduction of order.

SECTION 4.2 EXERCISES

Answers to odd-numbered problems begin on page A-4.

In Problems 1–24 find a second solution of each differential equation. Use reduction of order or formula (5) as instructed. Assume an appropriate interval of validity.

1. $y'' + 5y' = 0$; $y_1 = 1$

2. $y'' - y' = 0$; $y_1 = 1$

3. $y'' - 4y' + 4y = 0$; $y_1 = e^{2x}$

4. $y'' + 2y' + y = 0$; $y_1 = xe^{-x}$

5. $y'' + 16y = 0$; $y_1 = \cos 4x$

6. $y'' + 9y = 0$; $y_1 = \sin 3x$

7. $y'' - y = 0$; $y_1 = \cosh x$

8. $y'' - 25y = 0$; $y_1 = e^{5x}$

9. $9y'' - 12y' + 4y = 0$; $y_1 = e^{2x/3}$

10. $6y'' + y' - y = 0$; $y_1 = e^{x/3}$

11. $x^2 y'' - 7xy' + 16y = 0$; $y_1 = x^4$

12. $x^2 y'' + 2xy' - 6y = 0$; $y_1 = x^2$

13. $xy'' + y' = 0$; $y_1 = \ln x$

14. $4x^2 y'' + y = 0$; $y_1 = x^{1/2} \ln x$

15. $(1 - 2x - x^2)y'' + 2(1 + x)y' - 2y = 0$; $y_1 = x + 1$

16. $(1 - x^2)y'' - 2xy' = 0$; $y_1 = 1$

17. $x^2 y'' - xy' + 2y = 0$; $y_1 = x \sin(\ln x)$

18. $x^2 y'' - 3xy' + 5y = 0$; $y_1 = x^2 \cos(\ln x)$

19. $(1 + 2x)y'' + 4xy' - 4y = 0$; $y_1 = e^{-2x}$

20. $(1 + x)y'' + xy' - y = 0$; $y_1 = x$

21. $x^2 y'' - xy' + y = 0$; $y_1 = x$

22. $x^2 y'' - 20y = 0$; $y_1 = x^{-4}$

23. $x^2 y'' - 5xy' + 9y = 0$; $y_1 = x^3 \ln x$

24. $x^2 y'' + xy' + y = 0$; $y_1 = \cos(\ln x)$

In Problems 25–28 use the method of reduction of order to find a solution of the given nonhomogeneous equation. The indicated function $y_1(x)$ is a solution of the associated homogeneous equation. Determine a second solution of the homogeneous equation and a particular solution of the nonhomogeneous equation.

25. $y'' - 4y = 2$; $y_1 = e^{-2x}$

26. $y'' + y' = 1$; $y_1 = 1$

27. $y'' - 3y' + 2y = 5e^{3x}$; $y_1 = e^x$

28. $y'' - 4y' + 3y = x$; $y_1 = e^x$

29. **(a)** Verify by direct substitution that (5) satisfies (3).
   (b) Show that $W(y_1(x), y_2(x)) = u'y_1^2 = e^{-\int P(x)\,dx}$.

Discussion Problem

30. (a) Give a convincing demonstration that the second-order equation $ay'' + by' + cy = 0$, a, b, and c constants, always possesses at least one solution of the form $y_1 = e^{m_1 x}$, m_1 a constant.

   (b) Explain why the differential equation in part (a) must then have a second solution either of the form $y_2 = e^{m_2 x}$ or of the form $y_2 = xe^{m_1 x}$, m_1 and m_2 constants.

   (c) Reexamine Problems 1–10. Can you explain why the statements in parts (a) and (b) are not contradicted by the answers to Problems 5–7?

4.3 HOMOGENEOUS LINEAR EQUATIONS WITH CONSTANT COEFFICIENTS

■ *Auxiliary equation* ■ *Roots of a quadratic auxiliary equation* ■ *Euler's formula*
■ *Forms of the general solution of a homogeneous linear second-order DE with constant coefficients* ■ *Higher-order DEs* ■ *Roots of auxiliary equations of degree greater than two*

We have seen that the linear first-order equation $dy/dx + ay = 0$, where a is a constant, has the exponential solution $y = c_1 e^{-ax}$ on the interval $(-\infty, \infty)$. Therefore it is natural to seek to determine whether exponential solutions exist on $(-\infty, \infty)$ for homogeneous linear higher-order equations

$$a_n y^{(n)} + a_{n-1} y^{(n-1)} + \cdots + a_2 y'' + a_1 y' + a_0 y = 0, \tag{1}$$

where the coefficients a_i, $i = 0, 1, \ldots, n$ are real constants and $a_n \neq 0$. The surprising fact is that all solutions of (1) are exponential functions or are constructed out of exponential functions.

Method of Solution   We begin by considering the special case of the second-order equation

$$ay'' + by' + cy = 0. \tag{2}$$

If we try a solution of the form $y = e^{mx}$, then $y' = me^{mx}$ and $y'' = m^2 e^{mx}$ so that equation (2) becomes

$$am^2 e^{mx} + bme^{mx} + ce^{mx} = 0 \quad \text{or} \quad e^{mx}(am^2 + bm + c) = 0.$$

Because e^{mx} is never zero for real values of x, it is apparent that the only way this exponential function can satisfy the differential equation is if we choose m so that it is a root of the quadratic equation

$$am^2 + bm + c = 0. \tag{3}$$

This latter equation is called the **auxiliary equation** or **characteristic equation** of the differential equation (2). We consider three cases—the solutions of the auxiliary equation corresponding to distinct real roots, real but equal roots, and a conjugate pair of complex roots.

CASE I: Distinct Real Roots   Under the assumption that the auxiliary equation (3) has two unequal real roots m_1 and m_2, we find two solutions, $y_1 = e^{m_1 x}$ and $y_2 = e^{m_2 x}$. We see that these functions are linearly independent on $(-\infty, \infty)$ and hence form a fundamental set. It follows that the general solution of (2) on this interval is

$$y = c_1 e^{m_1 x} + c_2 e^{m_2 x}. \tag{4}$$

CASE II: Repeated Real Roots   When $m_1 = m_2$ we necessarily obtain only one exponential solution, $y_1 = e^{m_1 x}$. From the quadratic formula we find that $m_1 = -b/2a$ since the only way to have $m_1 = m_2$ is to have $b^2 - 4ac = 0$. It follows from the discussion in Section 4.2 that a second solution of the equation is

$$y_2 = e^{m_1 x} \int \frac{e^{2m_1 x}}{e^{2m_1 x}}\, dx = e^{m_1 x} \int dx = xe^{m_1 x}. \tag{5}$$

In (5) we have used the fact that $-b/a = 2m_1$. The general solution is then

$$y = c_1 e^{m_1 x} + c_2 x e^{m_1 x}. \tag{6}$$

CASE III: Conjugate Complex Roots   If m_1 and m_2 are complex, then we can write $m_1 = \alpha + i\beta$ and $m_2 = \alpha - i\beta$, where α and $\beta > 0$ are real and $i^2 = -1$. Formally, there is no difference between this case and Case I, and hence

$$y = C_1 e^{(\alpha + i\beta)x} + C_2 e^{(\alpha - i\beta)x}.$$

However, in practice we prefer to work with real functions instead of complex exponentials. To this end we use Euler's formula:

$$e^{i\theta} = \cos\theta + i\sin\theta,$$

where θ is any real number.* It follows from this formula that

$$e^{i\beta x} = \cos\beta x + i\sin\beta x \qquad \text{and} \qquad e^{-i\beta x} = \cos\beta x - i\sin\beta x, \tag{7}$$

where we have used $\cos(-\beta x) = \cos\beta x$ and $\sin(-\beta x) = -\sin\beta x$. Note that by first adding and then subtracting the two equations in (7), we obtain, respectively,

$$e^{i\beta x} + e^{-i\beta x} = 2\cos\beta x \qquad \text{and} \qquad e^{i\beta x} - e^{-i\beta x} = 2i\sin\beta x.$$

Since $y = C_1 e^{(\alpha + i\beta)x} + C_2 e^{(\alpha - i\beta)x}$ is a solution of (2) for any choice of the constants C_1 and C_2, the choices $C_1 = C_2 = 1$ and $C_1 = 1, C_2 = -1$ give, in turn, two solutions:

$$y_1 = e^{(\alpha + i\beta)x} + e^{(\alpha - i\beta)x} \qquad \text{and} \qquad y_2 = e^{(\alpha + i\beta)x} - e^{(\alpha - i\beta)x}.$$

But $$y_1 = e^{\alpha x}(e^{i\beta x} + e^{-i\beta x}) = 2e^{\alpha x}\cos\beta x$$

and $$y_2 = e^{\alpha x}(e^{i\beta x} - e^{-i\beta x}) = 2ie^{\alpha x}\sin\beta x.$$

---

*A formal derivation of Euler's formula can be obtained from the Maclaurin series $e^x = \sum_{n=0}^{\infty} \frac{x^n}{n!}$ by substituting $x = i\theta$, using $i^2 = -1$, $i^3 = -i, \ldots$, and then separating the series into real and imaginary parts. The plausibility thus established, we can adopt $\cos\theta + i\sin\theta$ as the *definition* of $e^{i\theta}$.

Hence from Corollary (A) of Theorem 4.2 the last two results show that the *real* functions $e^{\alpha x} \cos \beta x$ and $e^{\alpha x} \sin \beta x$ are solutions of (2). Moreover, these solutions form a fundamental set on $(-\infty, \infty)$. Consequently, the general solution is

$$y = c_1 e^{\alpha x} \cos \beta x + c_2 e^{\alpha x} \sin \beta x$$
$$= e^{\alpha x}(c_1 \cos \beta x + c_2 \sin \beta x). \tag{8}$$

EXAMPLE 1   **Second-Order DEs**

Solve the following differential equations:

(a) $2y'' - 5y' - 3y = 0$ (b) $y'' - 10y' + 25y = 0$ (c) $y'' + y' + y = 0$

SOLUTION   We give the auxiliary equations, the roots, and the corresponding general solutions.

(a) $2m^2 - 5m - 3 = (2m + 1)(m - 3) = 0$, $m_1 = -\dfrac{1}{2}$, $m_2 = 3$,

$\quad y = c_1 e^{-x/2} + c_2 e^{3x}$

(b) $m^2 - 10m + 25 = (m - 5)^2 = 0$, $m_1 = m_2 = 5$,

$\quad y = c_1 e^{5x} + c_2 x e^{5x}$

(c) $m^2 + m + 1 = 0$, $m_1 = -\dfrac{1}{2} + \dfrac{\sqrt{3}}{2} i$, $m_2 = -\dfrac{1}{2} - \dfrac{\sqrt{3}}{2} i$,

$$y = e^{-x/2}\left(c_1 \cos \frac{\sqrt{3}}{2} x + c_2 \sin \frac{\sqrt{3}}{2} x \right) \qquad \blacksquare$$

EXAMPLE 2   **An Initial-Value Problem**

Solve the initial-value problem

$$y'' - 4y' + 13y = 0, \qquad y(0) = -1, \quad y'(0) = 2.$$

SOLUTION   The roots of the auxiliary equation $m^2 - 4m + 13 = 0$ are $m_1 = 2 + 3i$ and $m_2 = 2 - 3i$ so that

$$y = e^{2x}(c_1 \cos 3x + c_2 \sin 3x).$$

Applying condition $y(0) = -1$, we see from $-1 = e^0(c_1 \cos 0 + c_2 \sin 0)$ that $c_1 = -1$. Differentiating $y = e^{2x}(-\cos 3x + c_2 \sin 3x)$ and then using $y'(0) = 2$ gives $2 = 3c_2 - 2$ or $c_2 = \frac{4}{3}$. Hence the solution is

$$y = e^{2x}\left(-\cos 3x + \frac{4}{3} \sin 3x \right). \qquad \blacksquare$$

The two differential equations $y'' + k^2 y = 0$ and $y'' - k^2 y = 0$, k real, are important in applied mathematics. For the former equation, the auxiliary equation $m^2 + k^2 = 0$ has imaginary roots $m_1 = ki$ and $m_2 = -ki$. It follows from (8) with $\alpha = 0$ and $\beta = k$ that its general solution is

$$y = c_1 \cos kx + c_2 \sin kx. \tag{9}$$

The auxiliary equation of the second equation $m^2 - k^2 = 0$ has distinct real roots $m_1 = k$ and $m_2 = -k$. Hence its general solution is

$$y = c_1 e^{kx} + c_2 e^{-kx}. \tag{10}$$

Notice that if we choose $c_1 = c_2 = \frac{1}{2}$ and then $c_1 = \frac{1}{2}$, $c_2 = -\frac{1}{2}$ in (10), we get the particular solutions $y = (e^{kx} + e^{-kx})/2 = \cosh kx$ and $y = (e^{kx} - e^{-kx})/2 = \sinh kx$. Because $\cosh kx$ and $\sinh kx$ are linearly independent on any interval of the x-axis, an alternative form for the general solution of $y'' - k^2 y = 0$ is

$$y = c_1 \cosh kx + c_2 \sinh kx.$$

Higher-Order Equations   In general, to solve an nth-order differential equation

$$a_n y^{(n)} + a_{n-1} y^{(n-1)} + \cdots + a_2 y'' + a_1 y' + a_0 y = 0, \tag{11}$$

where the a_i, $i = 0, 1, \ldots, n$ are real constants, we must solve an nth-degree polynomial equation

$$a_n m^n + a_{n-1} m^{n-1} + \cdots + a_2 m^2 + a_1 m + a_0 = 0. \tag{12}$$

If all the roots of (12) are real and distinct, then the general solution of (11) is

$$y = c_1 e^{m_1 x} + c_2 e^{m_2 x} + \cdots + c_n e^{m_n x}.$$

It is somewhat harder to summarize the analogues of Cases II and III because the roots of an auxiliary equation of degree greater than two can occur in many combinations. For example, a fifth-degree equation could have five distinct real roots, or three distinct real and two complex roots, or one real and four complex roots, or five real but equal roots, or five real roots but two of them equal, and so on. When m_1 is a root of multiplicity k of an nth-degree auxiliary equation (that is, k roots are equal to m_1), it can be shown that the linearly independent solutions are

$$e^{m_1 x}, x e^{m_1 x}, x^2 e^{m_1 x}, \ldots, x^{k-1} e^{m_1 x}$$

and the general solution must contain the linear combination

$$c_1 e^{m_1 x} + c_2 x e^{m_1 x} + c_3 x^2 e^{m_1 x} + \cdots + c_k x^{k-1} e^{m_1 x}.$$

Lastly, it should be remembered that when the coefficients are real, complex roots of an auxiliary equation always appear in conjugate pairs. Thus, for example, a cubic polynomial equation can have at most two complex roots.

EXAMPLE 3   **Third-Order DE**

Solve $y''' + 3y'' - 4y = 0$.

SOLUTION   It should be apparent from inspection of $m^3 + 3m^2 - 4 = 0$ that one root is $m_1 = 1$. Now if we divide $m^3 + 3m^2 - 4$ by $m - 1$, we find

$$m^3 + 3m^2 - 4 = (m - 1)(m^2 + 4m + 4) = (m - 1)(m + 2)^2,$$

and so the other roots are $m_2 = m_3 = -2$. Thus the general solution is

$$y = c_1 e^x + c_2 e^{-2x} + c_3 x e^{-2x}.$$ ∎

EXAMPLE 4 **Fourth-Order DE**

Solve $\dfrac{d^4 y}{dx^4} + 2\dfrac{d^2 y}{dx^2} + y = 0$.

SOLUTION The auxiliary equation $m^4 + 2m^2 + 1 = (m^2 + 1)^2 = 0$ has roots $m_1 = m_3 = i$ and $m_2 = m_4 = -i$. Thus from Case II the solution is

$$y = C_1 e^{ix} + C_2 e^{-ix} + C_3 x e^{ix} + C_4 x e^{-ix}.$$

By Euler's formula the grouping $C_1 e^{ix} + C_2 e^{-ix}$ can be rewritten as

$$c_1 \cos x + c_2 \sin x$$

after a relabeling of constants. Similarly, $x(C_3 e^{ix} + C_4 e^{-ix})$ can be expressed as $x(c_3 \cos x + c_4 \sin x)$. Hence the general solution is

$$y = c_1 \cos x + c_2 \sin x + c_3 x \cos x + c_4 x \sin x.$$ ∎

Example 4 illustrates a special case when the auxiliary equation has repeated complex roots. In general, if $m_1 = \alpha + i\beta$ is a complex root of multiplicity k of an auxiliary equation with real coefficients, then its conjugate $m_2 = \alpha - i\beta$ is also a root of multiplicity k. From the $2k$ complex-valued solutions

$$e^{(\alpha+i\beta)x}, \quad xe^{(\alpha+i\beta)x}, \quad x^2 e^{(\alpha+i\beta)x}, \quad \ldots, \quad x^{k-1} e^{(\alpha+i\beta)x}$$
$$e^{(\alpha-i\beta)x}, \quad xe^{(\alpha-i\beta)x}, \quad x^2 e^{(\alpha-i\beta)x}, \quad \ldots, \quad x^{k-1} e^{(\alpha-i\beta)x}$$

we conclude, with the aid of Euler's formula, that the general solution of the corresponding differential equation must then contain a linear combination of the $2k$ real linearly independent solutions

$$e^{\alpha x} \cos \beta x, \quad xe^{\alpha x} \cos \beta x, \quad x^2 e^{\alpha x} \cos \beta x, \quad \ldots, \quad x^{k-1} e^{\alpha x} \cos \beta x$$
$$e^{\alpha x} \sin \beta x, \quad xe^{\alpha x} \sin \beta x, \quad x^2 e^{\alpha x} \sin \beta x, \quad \ldots, \quad x^{k-1} e^{\alpha x} \sin \beta x.$$

In Example 4 we identify $k = 2$, $\alpha = 0$, and $\beta = 1$.

Of course the most difficult aspect of solving constant-coefficient differential equations is finding roots of auxiliary equations of degree greater than two. For example, to solve $3y''' + 5y'' + 10y' - 4y = 0$ we must solve $3m^3 + 5m^2 + 10m - 4 = 0$. Something we can try is to test the auxiliary equation for rational roots. Recall, if $m_1 = p/q$ is a rational root (expressed in lowest terms) of an auxiliary equation $a_n m^n + \cdots + a_1 m + a_0 = 0$ with integer coefficients, then p is a factor of a_0 and q is a factor of a_n. For our specific cubic auxiliary equation, all the factors of $a_0 = -4$ and $a_n = 3$ are p: $\pm 1, \pm 2, \pm 4$ and q: $\pm 1, \pm 3$, so the possible rational roots are p/q: $\pm 1, \pm 2, \pm 4, \pm\frac{1}{3}, \pm\frac{2}{3}, \pm\frac{4}{3}$. Each of these numbers can then be tested, say, by synthetic division. In this way we discover both the root $m_1 = \frac{1}{3}$ and the factorization

$$3m^3 + 5m^2 + 10m - 4 = \left(m - \frac{1}{3}\right)(3m^2 + 6m + 12).$$

The quadratic formula then yields the remaining roots $m_2 = -1 + \sqrt{3}\, i$ and $m_3 = -1 - \sqrt{3}\, i$. Therefore the general solution of $3y''' + 5y'' + 10y' - 4y = 0$ is

$$y = c_1 e^{x/3} + e^{-x}(c_2 \cos \sqrt{3}x + c_3 \sin \sqrt{3}x).$$

Use of Computers   Finding roots or approximations of roots of polynomial equations is a routine problem with an appropriate calculator or computer software. The computer algebra systems Mathematica and Maple can solve polynomial equations (in one variable) of degree less than five in terms of algebraic formulas. For the auxiliary equation in the preceding paragraph, the commands

> Solve[3 m^3 + 5 m^2 + 10 m − 4 == 0, m]   (in Mathematica)
>
> solve(3*m^3 + 5*m^2 + 10*m − 4, m);   (in Maple)

yield immediately their representations of the roots $\frac{1}{3}$, $-1 + \sqrt{3}\, i$, $-1 - \sqrt{3}\, i$. For auxiliary equations of higher degree it may be necessary to resort to numerical commands such as NSolve and FindRoot in Mathematica. Because of their capability of solving polynomial equations, it is not surprising that some computer algebra systems are also able to give explicit solutions of homogeneous linear constant-coefficient differential equations. For example, the inputs

> DSolve [y''[x] + 2 y'[x] + 2 y[x] == 0, y[x], x]   (in Mathematica)
>
> dsolve(diff(y(x),x$2) + 2*diff(y(x),x) +2*y(x) = 0, y(x));   (in Maple)

give, respectively,

$$y[x] -> \frac{C[2]\ \text{Cos}\ [x] - C[1]\ \text{Sin}\ [x]}{E^x}$$

and $y(x) = \_C1 \exp(-x) \sin(x) + \_C2 \exp(-x) \cos(x)$

Translated, this means $y = c_2 e^{-x} \cos x + c_1 e^{-x} \sin x$ is a solution of $y'' + 2y' + 2y = 0$. Note that the minus sign in front of C[1] in the first output is superfluous. (Why?)

In the classic text *Differential Equations,* by Ralph Palmer Agnew* (used by the author as a student), the following statement is made:

> It is not reasonable to expect students in this course to have computing skill and equipment necessary for efficient solving of equations such as
>
> $$4.317 \frac{d^4y}{dx^4} + 2.179 \frac{d^3y}{dx^3} + 1.416 \frac{d^2y}{dx^2} + 1.295 \frac{dy}{dx} + 3.169y = 0. \quad \textbf{(13)}$$

Although it is debatable whether computing skills have improved in the intervening years, it is a certainty that technology has. If one has access to a computer algebra system, equation (13) could be considered reasonable. After simplification and some relabeling of the output, Mathematica yields the (approximate) general solution

$$y = c_1 e^{-0.728852x} \cos(0.618605x) + c_2 e^{-0.728852x} \sin(0.618605x)$$
$$+ c_3 e^{0.476478x} \cos(0.759081x) + c_4 e^{0.476478x} \sin(0.759081x).$$

---

*McGraw-Hill, New York, 1960.

We note in passing that the DSolve and dsolve commands in Mathematica and Maple, like most aspects of any CAS, have their limitations.

SECTION 4.3 EXERCISES

Answers to odd-numbered problems begin on page A-4.

In Problems 1–36 find the general solution of the given differential equation.

1. $4y'' + y' = 0$ **2.** $2y'' - 5y' = 0$

3. $y'' - 36y = 0$ **4.** $y'' - 8y = 0$

5. $y'' + 9y = 0$ **6.** $3y'' + y = 0$

7. $y'' - y' - 6y = 0$ **8.** $y'' - 3y' + 2y = 0$

9. $\dfrac{d^2y}{dx^2} + 8\dfrac{dy}{dx} + 16y = 0$ **10.** $\dfrac{d^2y}{dx^2} - 10\dfrac{dy}{dx} + 25y = 0$

11. $y'' + 3y' - 5y = 0$ **12.** $y'' + 4y' - y = 0$

13. $12y'' - 5y' - 2y = 0$ **14.** $8y'' + 2y' - y = 0$

15. $y'' - 4y' + 5y = 0$ **16.** $2y'' - 3y' + 4y = 0$

17. $3y'' + 2y' + y = 0$ **18.** $2y'' + 2y' + y = 0$

19. $y''' - 4y'' - 5y' = 0$ **20.** $4y''' + 4y'' + y' = 0$

21. $y''' - y = 0$ **22.** $y''' + 5y'' = 0$

23. $y''' - 5y'' + 3y' + 9y = 0$ **24.** $y''' + 3y'' - 4y' - 12y = 0$

25. $y''' + y'' - 2y = 0$ **26.** $y''' - y'' - 4y = 0$

27. $y''' + 3y'' + 3y' + y = 0$ **28.** $y''' - 6y'' + 12y' - 8y = 0$

29. $\dfrac{d^4y}{dx^4} + \dfrac{d^3y}{dx^3} + \dfrac{d^2y}{dx^2} = 0$ **30.** $\dfrac{d^4y}{dx^4} - 2\dfrac{d^2y}{dx^2} + y = 0$

31. $16\dfrac{d^4y}{dx^4} + 24\dfrac{d^2y}{dx^2} + 9y = 0$ **32.** $\dfrac{d^4y}{dx^4} - 7\dfrac{d^2y}{dx^2} - 18y = 0$

33. $\dfrac{d^5y}{dx^5} - 16\dfrac{dy}{dx} = 0$ **34.** $\dfrac{d^5y}{dx^5} - 2\dfrac{d^4y}{dx^4} + 17\dfrac{d^3y}{dx^3} = 0$

35. $\dfrac{d^5y}{dx^5} + 5\dfrac{d^4y}{dx^4} - 2\dfrac{d^3y}{dx^3} - 10\dfrac{d^2y}{dx^2} + \dfrac{dy}{dx} + 5y = 0$

36. $2\dfrac{d^5y}{dx^5} - 7\dfrac{d^4y}{dx^4} + 12\dfrac{d^3y}{dx^3} + 8\dfrac{d^2y}{dx^2} = 0$

In Problems 37–52 solve the given differential equation subject to the indicated initial conditions.

37. $y'' + 16y = 0$, $y(0) = 2, y'(0) = -2$

38. $y'' - y = 0$, $y(0) = y'(0) = 1$

39. $y'' + 6y' + 5y = 0$, $y(0) = 0, y'(0) = 3$

40. $y'' - 8y' + 17y = 0$, $y(0) = 4, y'(0) = -1$

41. $2y'' - 2y' + y = 0$, $y(0) = -1, y'(0) = 0$

42. $y'' - 2y' + y = 0$, $y(0) = 5, y'(0) = 10$

43. $y'' + y' + 2y = 0$, $y(0) = y'(0) = 0$

44. $4y'' - 4y' - 3y = 0$, $y(0) = 1, y'(0) = 5$

45. $y'' - 3y' + 2y = 0$, $y(1) = 0, y'(1) = 1$

46. $y'' + y = 0$, $y\left(\dfrac{\pi}{3}\right) = 0, y'\left(\dfrac{\pi}{3}\right) = 2$

47. $y''' + 12y'' + 36y' = 0$, $y(0) = 0, y'(0) = 1, y''(0) = -7$

48. $y''' + 2y'' - 5y' - 6y = 0$, $y(0) = y'(0) = 0, y''(0) = 1$

49. $y''' - 8y = 0$, $y(0) = 0, y'(0) = -1, y''(0) = 0$

50. $\dfrac{d^4y}{dx^4} = 0$, $y(0) = 2, y'(0) = 3, y''(0) = 4, y'''(0) = 5$

51. $\dfrac{d^4y}{dx^4} - 3\dfrac{d^3y}{dx^3} + 3\dfrac{d^2y}{dx^2} - \dfrac{dy}{dx} = 0$, $y(0) = y'(0) = 0, y''(0) = y'''(0) = 1$

52. $\dfrac{d^4y}{dx^4} - y = 0$, $y(0) = y'(0) = y''(0) = 0, y'''(0) = 1$

In Problems 53–56 solve the given differential equation subject to the indicated boundary conditions.

53. $y'' - 10y' + 25y = 0$, $y(0) = 1, y(1) = 0$

54. $y'' + 4y = 0$, $y(0) = 0, y(\pi) = 0$

55. $y'' + y = 0$, $y'(0) = 0, y'\left(\dfrac{\pi}{2}\right) = 2$

56. $y'' - y = 0$, $y(0) = 1, y'(1) = 0$

In Problems 57–60 use a computer either as an aid in solving the auxiliary equation or as a means of directly obtaining the general solution of the given differential equation. If you use a CAS to obtain the general solution, simplify the output and write the solution in terms of real functions.

57. $y''' - 6y'' + 2y' + y = 0$

58. $6.11y''' + 8.59y'' + 7.93y' + 0.778y = 0$

59. $3.15y^{(4)} - 5.34y'' + 6.33y' - 2.03y = 0$

60. $y^{(4)} + 2y'' - y' + 2y = 0$

Discussion Problems

61. (a) The roots of a quadratic auxiliary equation are $m_1 = 4$ and $m_2 = -5$. What is the corresponding homogeneous linear differential equation?

(b) Two roots of a cubic auxiliary equation with real coefficients are $m_1 = -\frac{1}{2}$ and $m_2 = 3 + i$. What is the corresponding homogeneous linear differential equation?

(c) $y_1 = e^{-4x} \cos x$ is a solution of $y''' + 6y'' + y' - 34y = 0$. What is the general solution of the differential equation?

62. What conditions should be imposed on the constant coefficients a, b, and c in order to guarantee that all solutions of the second-order differential equation $ay'' + by' + cy = 0$ are bounded on the interval $[0, \infty)$?

63. Consider the differential equation $xy'' + y' + xy = 0$ or $y'' + (1/x)y' + y = 0$ for $x > 0$. Discuss how this last equation enables us to discern the qualitative behavior of the solutions as $x \to \infty$. Verify your conjectures using an ODE solver.

4.4 UNDETERMINED COEFFICIENTS—SUPERPOSITION APPROACH

■ *General solution of a nonhomogeneous linear DE* ■ *The form of a particular solution*
■ *Superposition principle for nonhomogeneous DEs*
■ *Cases for using undetermined coefficients*

Note to the Instructor   In this section the method of undetermined coefficients is developed from the viewpoint of the superposition principle for nonhomogeneous differential equations (Theorem 4.7). In Section 4.5 an entirely different approach to this method will be presented, one utilizing the concept of differential annihilator operators. Take your pick.

To solve a nonhomogeneous linear differential equation

$$a_n y^{(n)} + a_{n-1} y^{(n-1)} + \cdots + a_1 y' + a_0 y = g(x) \tag{1}$$

we must do two things:

- (i) Find the complementary function y_c.
- (ii) Find *any* particular solution y_p of the nonhomogeneous equation.

Then, as discussed in Section 4.1, the general solution of (1) on an interval is $y = y_c + y_p$.

The complementary function y_c is the general solution of the associated homogeneous equation $a_n y^{(n)} + a_{n-1} y^{(n-1)} + \cdots + a_1 y' + a_0 y = 0$. In the last section we saw how to solve these kinds of equations when the coefficients were constants. The first of two ways we shall consider for obtaining a particular solution y_p is called the **method of undetermined coefficients.** The underlying idea in this method is a conjecture, or educated guess, about the form of y_p, motivated by the kinds of functions that comprise the input $g(x)$. Basically straightforward, the method regrettably is limited to nonhomogeneous linear equations such as (1) where

- the coefficients a_i, $i = 0, 1, \ldots, n$ are constants and
- $g(x)$ is a constant k, a polynomial function, an exponential function $e^{\alpha x}$, sine or cosine functions $\sin \beta x$, $\cos \beta x$, or finite sums and products of these functions.

Note

Strictly speaking, $g(x) = k$ (a constant) is a polynomial function. Since a constant function is probably not the first thing that comes to mind when you think of polynomial functions, for emphasis we shall continue to use the redundancy "constant functions, polynomials, ..."

The following are some examples of the types of functions $g(x)$ that are appropriate for this discussion:

$$g(x) = 10, \quad g(x) = x^2 - 5x, \quad g(x) = 15x - 6 + 8e^{-x},$$

$$g(x) = \sin 3x - 5x \cos 2x, \quad g(x) = e^x \cos x + (3x^2 - 1)e^{-x},$$

and so on. That is, $g(x)$ is a linear combination of functions of the type

$$k \text{ (constant)}, \quad x^n, \quad x^n e^{\alpha x}, \quad x^n e^{\alpha x} \cos \beta x, \quad \text{and} \quad x^n e^{\alpha x} \sin \beta x,$$

where n is a nonnegative integer and α and β are real numbers. The method of undetermined coefficients is not applicable to equations of form (1) when

$$g(x) = \ln x, \quad g(x) = \frac{1}{x}, \quad g(x) = \tan x, \quad g(x) = \sin^{-1}x,$$

and so on. Differential equations in which the input $g(x)$ is a function of this last kind will be considered in Section 4.6.

The set of functions that consists of constants, polynomials, exponentials $e^{\alpha x}$, sines, and cosines has the remarkable property that derivatives of their sums and products are again sums and products of constants, polynomials, exponentials $e^{\alpha x}$, sines, and cosines. Since the linear combination of derivatives $a_n y_p^{(n)} + a_{n-1} y_p^{(n-1)} + \cdots + a_1 y_p' + a_0 y_p$ must be identical to $g(x)$, it seems reasonable to assume that y_p has the same form as $g(x)$.

The next two examples illustrate the basic method.

EXAMPLE 1 **General Solution Using Undetermined Coefficients**

Solve $y'' + 4y' - 2y = 2x^2 - 3x + 6$. (2)

SOLUTION Step 1. We first solve the associated homogeneous equation $y'' + 4y' - 2y = 0$. From the quadratic formula we find the roots of the auxiliary equation $m^2 + 4m - 2 = 0$ are $m_1 = -2 - \sqrt{6}$ and $m_2 = -2 + \sqrt{6}$. Hence the complementary function is

$$y_c = c_1 e^{-(2+\sqrt{6})x} + c_2 e^{(-2+\sqrt{6})x}.$$

Step 2. Now, since the function $g(x)$ is a quadratic polynomial, let us assume a particular solution that is also in the form of a quadratic polynomial:

$$y_p = Ax^2 + Bx + C.$$

We seek to determine *specific* coefficients A, B, and C for which y_p is a solution of (2). Substituting y_p and the derivatives

$$y_p' = 2Ax + B \qquad \text{and} \qquad y_p'' = 2A$$

into the given differential equation (2), we get

$$y_p'' + 4y_p' - 2y_p = 2A + 8Ax + 4B - 2Ax^2 - 2Bx - 2C$$
$$= 2x^2 - 3x + 6.$$

Since the last equation is supposed to be an identity, the coefficients of like powers of x must be equal:

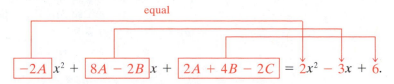

That is,

$$-2A = 2, \qquad 8A - 2B = -3, \qquad 2A + 4B - 2C = 6.$$

Solving this system of equations leads to the values $A = -1$, $B = -\frac{5}{2}$, and $C = -9$. Thus a particular solution is

$$y_p = -x^2 - \frac{5}{2}x - 9.$$

Step 3. The general solution of the given equation is

$$y = y_c + y_p = c_1 e^{-(2+\sqrt{6})x} + c_2 e^{(-2+\sqrt{6})x} - x^2 - \frac{5}{2}x - 9. \qquad \blacksquare$$

EXAMPLE 2 **Particular Solution Using Undetermined Coefficients**

Find a particular solution of $y'' - y' + y = 2 \sin 3x$.

SOLUTION A natural first guess for a particular solution would be $A \sin 3x$. But since successive differentiations of $\sin 3x$ produce $\sin 3x$ *and* $\cos 3x$, we are prompted instead to assume a particular solution that includes both of these terms:

$$y_p = A \cos 3x + B \sin 3x.$$

Differentiating y_p and substituting the results into the differential equation gives, after regrouping,

$$y_p'' - y_p' + y_p = (-8A - 3B) \cos 3x + (3A - 8B) \sin 3x = 2 \sin 3x$$

or

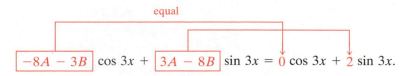

From the resulting system of equations,

$$-8A - 3B = 0, \qquad 3A - 8B = 2,$$

we get $A = \frac{6}{73}$ and $B = -\frac{16}{73}$. A particular solution of the equation is

$$y_p = \frac{6}{73} \cos 3x - \frac{16}{73} \sin 3x. \qquad \blacksquare$$

As we mentioned, the form that we assume for the particular solution y_p is an educated guess; it is not a blind guess. This educated guess must take into consideration not only the types of functions that make up $g(x)$ but also, as we shall see in Example 4, the functions that make up the complementary function y_c.

EXAMPLE 3   **Forming y_p by Superposition**

Solve $y'' - 2y' - 3y = 4x - 5 + 6xe^{2x}$.   **(3)**

SOLUTION   **Step 1.** First, the solution of the associated homogeneous equation $y'' - 2y' - 3y = 0$ is found to be $y_c = c_1 e^{-x} + c_2 e^{3x}$.

Step 2. Next, the presence of $4x - 5$ in $g(x)$ suggests that the particular solution includes a linear polynomial. Furthermore, since the derivative of the product xe^{2x} produces $2xe^{2x}$ and e^{2x}, we also assume that the particular solution includes both xe^{2x} and e^{2x}. In other words, g is the sum of two basic kinds of functions:

$$g(x) = g_1(x) + g_2(x) = \textit{polynomial} + \textit{exponentials}.$$

Correspondingly, the superposition principle for nonhomogeneous equations (Theorem 4.7) suggests that we seek a particular solution

$$y_p = y_{p_1} + y_{p_2},$$

where $y_{p_1} = Ax + B$ and $y_{p_2} = Cxe^{2x} + Ee^{2x}$. Substituting

$$y_p = Ax + B + Cxe^{2x} + Ee^{2x}$$

into the given equation (3) and grouping like terms gives

$$y_p'' - 2y_p' - 3y_p = -3Ax - 2A - 3B - 3Cxe^{2x} + (2C - 3E)e^{2x} = 4x - 5 + 6xe^{2x}. \quad \textbf{(4)}$$

From this identity we obtain the four equations

$$-3A = 4, \quad -2A - 3B = -5, \quad -3C = 6, \quad 2C - 3E = 0.$$

The last equation in this system results from the interpretation that the coefficient of e^{2x} in the right member of (4) is zero. Solving, we find $A = -\frac{4}{3}$, $B = \frac{23}{9}$, $C = -2$, and $E = -\frac{4}{3}$. Consequently,

$$y_p = -\frac{4}{3}x + \frac{23}{9} - 2xe^{2x} - \frac{4}{3}e^{2x}.$$

Step 3. The general solution of the equation is

$$y = c_1 e^{-x} + c_2 e^{3x} - \frac{4}{3}x + \frac{23}{9} - \left(2x + \frac{4}{3}\right)e^{2x}. \quad ■$$

In light of the superposition principle (Theorem 4.7) we can also approach Example 3 from the viewpoint of solving two simpler problems. You should verify that substituting

$$y_{p_1} = Ax + B \qquad \text{into} \quad y'' - 2y' - 3y = 4x - 5$$

and $y_{p_2} = Cxe^{2x} + Ee^{2x}$ into $y'' - 2y' - 3y = 6xe^{2x}$

yields, in turn, $y_{p_1} = -\frac{4}{3}x + \frac{23}{9}$ and $y_{p_2} = -(2x + \frac{4}{3})e^{2x}$. A particular solution of (3) is then $y_p = y_{p_1} + y_{p_2}$.

The next example illustrates that sometimes the "obvious" assumption for the form of y_p is not a correct assumption.

EXAMPLE 4   **A Glitch in the Method**

Find a particular solution of $y'' - 5y' + 4y = 8e^x$.

SOLUTION   Differentiation of e^x produces no new functions. Thus, proceeding as we did in the earlier examples, we can reasonably assume a particular solution of the form $y_p = Ae^x$. But substitution of this expression into the differential equation yields the contradictory statement

$$0 = 8e^x,$$

and so we have clearly made the wrong guess for y_p.

The difficulty here is apparent upon examining the complementary function $y_c = c_1 e^x + c_2 e^{4x}$. Observe that our assumption Ae^x is already present in y_c. This means that e^x is a solution of the associated homogeneous differential equation, and a constant multiple Ae^x when substituted into the differential equation necessarily produces zero.

What then should be the form of y_p? Inspired by Case II of Section 4.3, let's see whether we can find a particular solution of the form

$$y_p = Axe^x.$$

Substituting $y_p' = Axe^x + Ae^x$ and $y_p'' = Axe^x + 2Ae^x$ into the differential equation and simplifying gives

$$y_p'' - 5y_p' + 4y_p = -3Ae^x = 8e^x.$$

From the last equality we see that the value of A is now determined as $A = -\frac{8}{3}$. Therefore a particular solution of the given equation is

$$y_p = -\frac{8}{3}xe^x. \qquad \blacksquare$$

The difference in the procedures used in Examples 1–3 and in Example 4 suggests that we consider two cases. The first case reflects the situation in Examples 1–3.

CASE I: No function in the assumed particular solution is a solution of the associated homogeneous differential equation.

In Table 4.1 we illustrate some specific examples of $g(x)$ in (1) along with the corresponding form of the particular solution. We are, of course, taking for granted that no function in the assumed particular solution y_p is duplicated by a function in the complementary function y_c.

TABLE 4.1   Trial Particular Solutions

$g(x)$	Form of y_p
1. 1 (any constant)	A
2. $5x + 7$	$Ax + B$
3. $3x^2 - 2$	$Ax^2 + Bx + C$
4. $x^3 - x + 1$	$Ax^3 + Bx^2 + Cx + E$
5. $\sin 4x$	$A \cos 4x + B \sin 4x$
6. $\cos 4x$	$A \cos 4x + B \sin 4x$
7. e^{5x}	Ae^{5x}
8. $(9x - 2)e^{5x}$	$(Ax + B)e^{5x}$
9. $x^2 e^{5x}$	$(Ax^2 + Bx + C)e^{5x}$
10. $e^{3x} \sin 4x$	$Ae^{3x} \cos 4x + Be^{3x} \sin 4x$
11. $5x^2 \sin 4x$	$(Ax^2 + Bx + C) \cos 4x + (Ex^2 + Fx + G) \sin 4x$
12. $xe^{3x} \cos 4x$	$(Ax + B)e^{3x} \cos 4x + (Cx + E)e^{3x} \sin 4x$

EXAMPLE 5   **Forms of Particular Solutions—Case I**

Determine the form of a particular solution of

(a) $y'' - 8y' + 25y = 5x^3 e^{-x} - 7e^{-x}$ (b) $y'' + 4y = x \cos x$

SOLUTION   (a) We can write $g(x) = (5x^3 - 7)e^{-x}$. Using entry 9 in Table 4.1 as a model, we assume a particular solution of the form

$$y_p = (Ax^3 + Bx^2 + Cx + E)e^{-x}.$$

Note that there is no duplication between the terms in y_p and the terms in the complementary function $y_c = e^{4x}(c_1 \cos 3x + c_2 \sin 3x)$.

(b) The function $g(x) = x \cos x$ is similar to entry 11 in Table 4.1 except, of course, that we use a linear rather than a quadratic polynomial and $\cos x$ and $\sin x$ instead of $\cos 4x$ and $\sin 4x$ in the form of y_p:

$$y_p = (Ax + B) \cos x + (Cx + E) \sin x.$$

Again observe that there is no duplication of terms between y_p and $y_c = c_1 \cos 2x + c_2 \sin 2x$.   ∎

If $g(x)$ consists of a sum of, say, m terms of the kind listed in the table, then (as in Example 3) the assumption for a particular solution y_p consists of the sum of the trial forms $y_{p_1}, y_{p_2}, \ldots, y_{p_m}$ corresponding to these terms:

$$y_p = y_{p_1} + y_{p_2} + \cdots + y_{p_m}.$$

The foregoing sentence can be put another way.

Form Rule for Case I The form of y_p is a linear combination of all linearly independent functions that are generated by repeated differentiations of $g(x)$.

EXAMPLE 6   **Forming y_p by Superposition—Case I**

Determine the form of a particular solution of

$$y'' - 9y' + 14y = 3x^2 - 5 \sin 2x + 7xe^{6x}.$$

SOLUTION

Corresponding to $3x^2$ we assume $y_{p_1} = Ax^2 + Bx + C.$

Corresponding to $-5\sin 2x$ we assume $y_{p_2} = E\cos 2x + F\sin 2x.$

Corresponding to $7xe^{6x}$ we assume $y_{p_3} = (Gx + H)e^{6x}.$

The assumption for the particular solution is then

$$y_p = y_{p_1} + y_{p_2} + y_{p_3} = Ax^2 + Bx + C + E\cos 2x + F\sin 2x + (Gx + H)e^{6x}.$$

No term in this assumption duplicates a term in $y_c = c_1 e^{2x} + c_2 e^{7x}.$ ■

CASE II   A function in the assumed particular solution is also a solution of the associated homogeneous differential equation.

The next example is similar to Example 4.

EXAMPLE 7   **Particular Solution—Case II**

Find a particular solution of $y'' - 2y' + y = e^x.$

SOLUTION   The complementary function is $y_c = c_1 e^x + c_2 x e^x.$ As in Example 4, the assumption $y_p = Ae^x$ will fail since it is apparent from y_c that e^x is a solution of the associated homogeneous equation $y'' - 2y' + y = 0.$ Moreover, we will not be able to find a particular solution of the form $y_p = Axe^x$ since the term xe^x is also duplicated in $y_c.$ We next try

$$y_p = Ax^2 e^x.$$

Substituting into the given differential equation yields

$$2Ae^x = e^x \qquad \text{and so} \qquad A = \frac{1}{2}.$$

Thus a particular solution is $y_p = \frac{1}{2}x^2 e^x.$ ■

Suppose again that $g(x)$ consists of m terms of the kind given in Table 4.1, and suppose further that the usual assumption for a particular solution is

$$y_p = y_{p_1} + y_{p_2} + \cdots + y_{p_m},$$

where the y_{p_i}, $i = 1, 2, \ldots, m$ are the trial particular solution forms corresponding to these terms. Under the circumstances described in Case II, we can make up the following general rule.

Multiplication Rule for Case II If any y_{p_i} contains terms that duplicate terms in y_c, then that y_{p_i} must be multiplied by x^n, where n is the smallest positive integer that eliminates that duplication.

EXAMPLE 8   **An Initial-Value Problem**

Solve the initial-value problem

$$y'' + y = 4x + 10 \sin x, \qquad y(\pi) = 0, \quad y'(\pi) = 2.$$

SOLUTION   The solution of the associated homogeneous equation $y'' + y = 0$ is $y_c = c_1 \cos x + c_2 \sin x$. Since $g(x) = 4x + 10 \sin x$ is the sum of a linear polynomial and a sine function, our normal assumption for y_p, from entries 2 and 5 of Table 4.1, would be the sum of $y_{p_1} = Ax + B$ and $y_{p_2} = C \cos x + E \sin x$:

$$y_p = Ax + B + C \cos x + E \sin x. \tag{5}$$

But there is an obvious duplication of the terms $\cos x$ and $\sin x$ in this assumed form and two terms in the complementary function. This duplication can be eliminated by simply multiplying y_{p_2} by x. Instead of (5) we now use

$$y_p = Ax + B + Cx \cos x + Ex \sin x. \tag{6}$$

Differentiating this expression and substituting the results into the differential equation gives

$$y_p'' + y_p = Ax + B - 2C \sin x + 2E \cos x = 4x + 10 \sin x,$$

and so $A = 4, \quad B = 0, \quad -2C = 10, \quad 2E = 0.$

The solutions of the system are immediate: $A = 4$, $B = 0$, $C = -5$, and $E = 0$. Therefore from (6) we obtain $y_p = 4x - 5x \cos x$. The general solution of the given equation is

$$y = y_c + y_p = c_1 \cos x + c_2 \sin x + 4x - 5x \cos x.$$

We now apply the prescribed initial conditions to the general solution of the equation. First, $y(\pi) = c_1 \cos \pi + c_2 \sin \pi + 4\pi - 5\pi \cos \pi = 0$ yields $c_1 = 9\pi$ since $\cos \pi = -1$ and $\sin \pi = 0$. Next, from the derivative

$$y' = -9\pi \sin x + c_2 \cos x + 4 + 5x \sin x - 5 \cos x$$

and $y'(\pi) = -9\pi \sin \pi + c_2 \cos \pi + 4 + 5\pi \sin \pi - 5 \cos \pi = 2$

we find $c_2 = 7$. The solution of the initial value is then

$$y = 9\pi \cos x + 7 \sin x + 4x - 5x \cos x. \qquad \blacksquare$$

EXAMPLE 9   **Using the Multiplication Rule**

Solve $y'' - 6y' + 9y = 6x^2 + 2 - 12e^{3x}$.

SOLUTION   The complementary function is $y_c = c_1 e^{3x} + c_2 x e^{3x}$. And so, based on entries 3 and 7 of Table 4.1, the usual assumption for a particular solution would be

$$y_p = \underbrace{Ax^2 + Bx + C}_{y_{p_1}} + \underbrace{Ee^{3x}}_{y_{p_2}}.$$

Inspection of these functions shows that the one term in y_{p_2} is duplicated in y_c. If we multiply y_{p_2} by x, we note that the term xe^{3x} is still part of y_c. But multiplying y_{p_2} by x^2 eliminates all duplications. Thus the operative form of a particular solution is

$$y_p = Ax^2 + Bx + C + Ex^2e^{3x}.$$

Differentiating this last form, substituting into the differential equation, and collecting like terms gives

$$y_p'' - 6y_p' + 9y_p = 9Ax^2 + (-12A + 9B)x + 2A - 6B + 9C + 2Ee^{3x} = 6x^2 + 2 - 12e^{3x}.$$

It follows from this identity that $A = \frac{2}{3}$, $B = \frac{8}{9}$, $C = \frac{2}{3}$, and $E = -6$. Hence the general solution $y = y_c + y_p$ is

$$y = c_1e^{3x} + c_2xe^{3x} + \frac{2}{3}x^2 + \frac{8}{9}x + \frac{2}{3} - 6x^2e^{3x}.$$ ∎

EXAMPLE 10 Third-Order DE—Case I

Solve $y''' + y'' = e^x \cos x$.

SOLUTION From the characteristic equation $m^3 + m^2 = 0$ we find $m_1 = m_2 = 0$ and $m_3 = -1$. Hence the complementary function of the equation is $y_c = c_1 + c_2x + c_3e^{-x}$. With $g(x) = e^x \cos x$, we see from entry 10 of Table 4.1 that we should assume

$$y_p = Ae^x \cos x + Be^x \sin x.$$

Since there are no functions in y_p that duplicate functions in the complementary solution, we proceed in the usual manner. From

$$y_p''' + y_p'' = (-2A + 4B)e^x \cos x + (-4A - 2B)e^x \sin x = e^x \cos x$$

we get $-2A + 4B = 1$, $-4A - 2B = 0$.

This system gives $A = -\frac{1}{10}$ and $B = \frac{1}{5}$, so that a particular solution is $y_p = -\frac{1}{10}e^x \cos x + \frac{1}{5}e^x \sin x$. The general solution of the equation is

$$y = y_c + y_p = c_1 + c_2x + c_3e^{-x} - \frac{1}{10}e^x \cos x + \frac{1}{5}e^x \sin x.$$ ∎

EXAMPLE 11 Fourth-Order DE—Case II

Determine the form of a particular solution of $y^{(4)} + y''' = 1 - x^2e^{-x}$.

SOLUTION Comparing $y_c = c_1 + c_2x + c_3x^2 + c_4e^{-x}$ with our normal assumption for a particular solution

$$y_p = \underbrace{A}_{y_{p_1}} + \underbrace{Bx^2e^{-x} + Cxe^{-x} + Ee^{-x}}_{y_{p_2}},$$

we see that the duplications between y_c and y_p are eliminated when y_{p_1} is multiplied by x^3 and y_{p_2} is multiplied by x. Thus the correct assumption for a particular solution is

$$y_p = Ax^3 + Bx^3e^{-x} + Cx^2e^{-x} + Exe^{-x}.$$

     ■

Remark

In Problems 27–36 of Exercises 4.4 you are asked to solve initial-value problems, and in Problems 37 and 38, boundary-value problems. As illustrated in Example 8, be sure to apply the initial conditions or the boundary conditions to the general solution $y = y_c + y_p$. Students often make the mistake of applying these conditions to only the complementary function y_c since it is that part of the solution that contains the constants.

SECTION 4.4 EXERCISES

Answers to odd-numbered problems begin on page A-4.

In Problems 1–26 solve the given differential equation by undetermined coefficients.

1. $y'' + 3y' + 2y = 6$ **2.** $4y'' + 9y = 15$

3. $y'' - 10y' + 25y = 30x + 3$ **4.** $y'' + y' - 6y = 2x$

5. $\frac{1}{4}y'' + y' + y = x^2 - 2x$

6. $y'' - 8y' + 20y = 100x^2 - 26xe^x$

7. $y'' + 3y = -48x^2e^{3x}$ **8.** $4y'' - 4y' - 3y = \cos 2x$

9. $y'' - y' = -3$ **10.** $y'' + 2y' = 2x + 5 - e^{-2x}$

11. $y'' - y' + \frac{1}{4}y = 3 + e^{x/2}$ **12.** $y'' - 16y = 2e^{4x}$

13. $y'' + 4y = 3 \sin 2x$ **14.** $y'' + 4y = (x^2 - 3) \sin 2x$

15. $y'' + y = 2x \sin x$ **16.** $y'' - 5y' = 2x^3 - 4x^2 - x + 6$

17. $y'' - 2y' + 5y = e^x \cos 2x$

18. $y'' - 2y' + 2y = e^{2x}(\cos x - 3 \sin x)$

19. $y'' + 2y' + y = \sin x + 3 \cos 2x$

20. $y'' + 2y' - 24y = 16 - (x + 2)e^{4x}$

21. $y''' - 6y'' = 3 - \cos x$

22. $y''' - 2y'' - 4y' + 8y = 6xe^{2x}$

23. $y''' - 3y'' + 3y' - y = x - 4e^x$

24. $y''' - y'' - 4y' + 4y = 5 - e^x + e^{2x}$

25. $y^{(4)} + 2y'' + y = (x - 1)^2$

26. $y^{(4)} - y'' = 4x + 2xe^{-x}$

In Problems 27–36 solve the given differential equation subject to the indicated initial conditions.

27. $y'' + 4y = -2$, $\quad y\left(\dfrac{\pi}{8}\right) = \dfrac{1}{2}, y'\left(\dfrac{\pi}{8}\right) = 2$

28. $2y'' + 3y' - 2y = 14x^2 - 4x - 11$, $\quad y(0) = 0, y'(0) = 0$

29. $5y'' + y' = -6x$, $\quad y(0) = 0, y'(0) = -10$

30. $y'' + 4y' + 4y = (3 + x)e^{-2x}$, $\quad y(0) = 2, y'(0) = 5$

31. $y'' + 4y' + 5y = 35e^{-4x}$, $\quad y(0) = -3, y'(0) = 1$

32. $y'' - y = \cosh x$, $\quad y(0) = 2, y'(0) = 12$

33. $\dfrac{d^2x}{dt^2} + \omega^2 x = F_0 \sin \omega t$, $\quad x(0) = 0, x'(0) = 0$

34. $\dfrac{d^2x}{dt^2} + \omega^2 x = F_0 \cos \gamma t$, $\quad x(0) = 0, x'(0) = 0$

35. $y''' - 2y'' + y' = 2 - 24e^x + 40e^{5x}$, $\quad y(0) = \dfrac{1}{2}, y'(0) = \dfrac{5}{2}, y''(0) = -\dfrac{9}{2}$

36. $y''' + 8y = 2x - 5 + 8e^{-2x}$, $\quad y(0) = -5, y'(0) = 3, y''(0) = -4$

In Problems 37 and 38 solve the given differential equation subject to the indicated boundary conditions.

37. $y'' + y = x^2 + 1$, $\quad y(0) = 5, y(1) = 0$

38. $y'' - 2y' + 2y = 2x - 2$, $\quad y(0) = 0, y(\pi) = \pi$

39. In applications the function $g(x)$ is often discontinuous. Solve the initial-value problem

$$y'' + 4y = g(x), \quad y(0) = 1, y'(0) = 2,$$

where
$$g(x) = \begin{cases} \sin x, & 0 \le x \le \dfrac{\pi}{2} \\[2mm] 0, & x > \dfrac{\pi}{2} \end{cases}.$$

[*Hint:* Solve the problem on the two intervals, and then find a solution so that y and y' are continuous at $x = \pi/2$.]

Discussion Problems

40. (a) Discuss how to solve the second-order equation $ay'' + by' = g(x)$ *without* the aid of undetermined coefficients. Suppose that $g(x)$ is continuous. Also keep in mind that

$$ay'' + by' = \dfrac{d}{dx}(ay' + by).$$

(b) Illustrate your method by solving $y'' + y' = 2x - e^{-x}$.

(c) Discuss when the method obtained in part (a) is applicable to nonhomogeneous linear differential equations of order higher than two.

41. Discuss how the method of this section can be used to find a particular solution of $y'' + y = \sin x \cos 2x$. Carry out your idea.

4.5 UNDETERMINED COEFFICIENTS—ANNIHILATOR APPROACH

■ *Factoring a differential operator* ■ *Annihilator operator*
■ *Determining the form of a particular solution* ■ *Undetermined coefficients*

We saw in Section 4.1 that a linear nth-order differential equation can be written

$$a_n D^n y + a_{n-1} D^{n-1} y + \cdots + a_1 D y + a_0 y = g(x), \qquad \textbf{(1)}$$

where $D^k y = d^k y / dx^k$, $k = 0, 1, \ldots, n$. When it suits our purpose, (1) is also written as $L(y) = g(x)$, where L denotes the linear nth-order differential operator

$$L = a_n D^n + a_{n-1} D^{n-1} + \cdots + a_1 D + a_0. \qquad \textbf{(2)}$$

The operator notation is more than a helpful shorthand—on a very practical level the *application* of differential operators enables us to obtain a particular solution of certain kinds of nonhomogeneous linear differential equations. Before investigating this, we need to examine two concepts.

Factoring Operators   When the a_i, $i = 0, 1, \ldots, n$ are real constants, a linear differential operator (2) can be *factored* whenever the characteristic polynomial $a_n m^n + a_{n-1} m^{n-1} + \cdots + a_1 m + a_0$ factors. In other words, if r_1 is a root of the auxiliary equation

$$a_n m^n + a_{n-1} m^{n-1} + \cdots + a_1 m + a_0 = 0,$$

then $L = (D - r_1) P(D)$, where the polynomial expression $P(D)$ is a linear differential operator of order $n - 1$. For example, if we treat D as an algebraic quantity, then the operator $D^2 + 5D + 6$ can be factored as $(D + 2)(D + 3)$ or as $(D + 3)(D + 2)$. Thus if a function $y = f(x)$ possesses a second derivative,

$$(D^2 + 5D + 6)y = (D + 2)(D + 3)y = (D + 3)(D + 2)y.$$

This illustrates a general property:

Factors of a linear differential operator with constant coefficients commute.

A differential equation such as $y'' + 4y' + 4y = 0$ can be written as

$$(D^2 + 4D + 4)y = 0 \quad \text{or} \quad (D + 2)(D + 2)y = 0 \quad \text{or} \quad (D + 2)^2 y = 0.$$

Annihilator Operator   If L is a linear differential operator with constant coefficients and f is a sufficiently differentiable function such that

$$L(f(x)) = 0,$$

then L is said to be an **annihilator** of the function. For example, a constant function $y = k$ is annihilated by D since $Dk = 0$. The function $y = x$ is annihilated by the differential operator D^2 since the first and second derivatives of x are 1 and 0, respectively. Similarly, $D^3 x^2 = 0$, and so on.

> The differential operator D^n annihilates each of the functions **(3)**
>
> $$1, \quad x, \quad x^2, \quad \ldots, \quad x^{n-1}.$$

As an immediate consequence of (3) and the fact that differentiation can be done term by term, a polynomial

$$c_0 + c_1x + c_2x^2 + \cdots + c_{n-1}x^{n-1} \tag{4}$$

can be annihilated by finding an operator that annihilates the highest power of x.

The functions that are annihilated by a linear nth-order differential operator L are simply those functions that can be obtained from the general solution of the homogeneous differential equation $L(y) = 0$.

> The differential operator $(D - \alpha)^n$ annihilates each of the functions **(5)**
>
> $$e^{\alpha x}, \quad xe^{\alpha x}, \quad x^2e^{\alpha x}, \quad \ldots, \quad x^{n-1}e^{\alpha x}.$$

To see this, note that the auxiliary equation of the homogeneous equation $(D - \alpha)^n y = 0$ is $(m - \alpha)^n = 0$. Since α is a root of multiplicity n, the general solution is

$$y = c_1e^{\alpha x} + c_2xe^{\alpha x} + \cdots + c_nx^{n-1}e^{\alpha x}. \tag{6}$$

EXAMPLE 1　Annihilator Operators

Find a differential operator that annihilates the given function.

(a) $1 - 5x^2 + 8x^3$ (b) e^{-3x} (c) $4e^{2x} - 10xe^{2x}$

SOLUTION　(a) From (3) we know that $D^4x^3 = 0$, and so it follows from (4) that

$$D^4(1 - 5x^2 + 8x^3) = 0.$$

(b) From (5), with $\alpha = -3$ and $n = 1$, we see that

$$(D + 3)e^{-3x} = 0.$$

(c) From (5) and (6), with $\alpha = 2$ and $n = 2$, we have

$$(D - 2)^2(4e^{2x} - 10xe^{2x}) = 0. \qquad \blacksquare$$

When α and β are real numbers, the quadratic formula reveals that $[m^2 - 2\alpha m + (\alpha^2 + \beta^2)]^n = 0$ has complex roots $\alpha + i\beta$, $\alpha - i\beta$, both of multiplicity n. From the discussion at the end of Section 4.3 we have the next result.

The differential operator $[D^2 - 2\alpha D + (\alpha^2 + \beta^2)]^n$ annihilates each of the functions

$$e^{\alpha x} \cos \beta x, \quad xe^{\alpha x} \cos \beta x, \quad x^2 e^{\alpha x} \cos \beta x, \quad \ldots, \quad x^{n-1} e^{\alpha x} \cos \beta x,$$
$$e^{\alpha x} \sin \beta x, \quad xe^{\alpha x} \sin \beta x, \quad x^2 e^{\alpha x} \sin \beta x, \quad \ldots, \quad x^{n-1} e^{\alpha x} \sin \beta x. \tag{7}$$

EXAMPLE 2   Annihilator Operator

Find a differential operator that annihilates $5e^{-x} \cos 2x - 9e^{-x} \sin 2x$.

SOLUTION   Inspection of the functions $e^{-x} \cos 2x$ and $e^{-x} \sin 2x$ shows that $\alpha = -1$ and $\beta = 2$. Hence, from (7) we conclude that $D^2 + 2D + 5$ will annihilate each function. Since $D^2 + 2D + 5$ is a linear operator, it will annihilate *any* linear combination of these functions such as $5e^{-x} \cos 2x - 9e^{-x} \sin 2x$.   ∎

When $\alpha = 0$ and $n = 1$, a special case of (7) is

$$(D^2 + \beta^2) \begin{cases} \cos \beta x \\ \sin \beta x \end{cases} = 0. \tag{8}$$

For example, $D^2 + 16$ will annihilate any linear combination of $\sin 4x$ and $\cos 4x$.

We are often interested in annihilating the sum of two or more functions. As just seen in Examples 1 and 2, if L is a linear differential operator such that $L(y_1) = 0$ and $L(y_2) = 0$, then L will annihilate the linear combination $c_1 y_1(x) + c_2 y_2(x)$. This is a direct consequence of Theorem 4.2. Let us now suppose that L_1 and L_2 are linear differential operators with constant coefficients such that L_1 annihilates $y_1(x)$ and L_2 annihilates $y_2(x)$ but $L_1(y_2) \neq 0$ and $L_2(y_1) \neq 0$. Then the *product* of differential operators $L_1 L_2$ annihilates the sum $c_1 y_1(x) + c_2 y_2(x)$. We can easily demonstrate this, using linearity and the fact that $L_1 L_2 = L_2 L_1$:

$$L_1 L_2 (y_1 + y_2) = L_1 L_2 (y_1) + L_1 L_2 (y_2)$$
$$= L_2 L_1 (y_1) + L_1 L_2 (y_2)$$
$$= L_2 [\underbrace{L_1(y_1)}_{\text{zero}}] + L_1 [\underbrace{L_2(y_2)}_{\text{zero}}] = 0.$$

For example, we know from (3) that D^2 annihilates $7 - x$ and from (8) that $D^2 + 16$ annihilates $\sin 4x$. Therefore the product of operators $D^2(D^2 + 16)$ will annihilate the linear combination $7 - x + 6 \sin 4x$.

Note

The differential operator that annihilates a function is not unique. We saw in part (b) of Example 1 that $D + 3$ will annihilate e^{-3x}, but so will differential operators of higher order as long as $D + 3$ is one of the factors of the operator. For example, $(D + 3)(D + 1)$, $(D + 3)^2$, and $D^3(D + 3)$ all annihilate e^{-3x}. (Verify this.) As a matter of course, when we seek a differential annihilator for a function $y = f(x)$ we want the operator of *lowest possible order* that does the job.

Undetermined Coefficients This brings us to the point of the preceding discussion. Suppose that $L(y) = g(x)$ is a linear differential equation with constant coefficients and that the input $g(x)$ consists of finite sums and products of the functions listed in (3), (5), and (7)—that is, $g(x)$ is a linear combination of functions of the form

$$k \text{ (constant)}, \quad x^m, \quad x^m e^{\alpha x}, \quad x^m e^{\alpha x} \cos \beta x, \quad \text{and} \quad x^m e^{\alpha x} \sin \beta x,$$

where m is a nonnegative integer and α and β are real numbers. We now know that such a function $g(x)$ can be annihilated by a differential operator L_1, of lowest order, consisting of a product of the operators D^n, $(D - \alpha)^n$, and $(D^2 - 2\alpha D + \alpha^2 + \beta^2)^n$. Applying L_1 to both sides of the equation $L(y) = g(x)$ yields $L_1 L(y) = L_1(g(x)) = 0$. By solving the *homogeneous higher-order* equation $L_1 L(y) = 0$, we can discover the *form* of a particular solution y_p for the original *nonhomogeneous* equation $L(y) = g(x)$. We then substitute this assumed form into $L(y) = g(x)$ to find an explicit particular solution. This procedure for determining y_p, called the **method of undetermined coefficients,** is illustrated in the next several examples.

Before proceeding, recall that the general solution of a nonhomogeneous linear differential equation $L(y) = g(x)$ is $y = y_c + y_p$, where y_c is the complementary function—that is, the general solution of the associated homogeneous equation $L(y) = 0$. The general solution of each equation $L(y) = g(x)$ is defined on the interval $(-\infty, \infty)$.

EXAMPLE 3 **General Solution Using Undetermined Coefficients**

Solve $y'' + 3y' + 2y = 4x^2$. $\qquad\qquad$ **(9)**

SOLUTION Step 1. First, we solve the homogeneous equation $y'' + 3y' + 2y = 0$. Then, from the auxiliary equation $m^2 + 3m + 2 = (m + 1)(m + 2) = 0$ we find $m_1 = -1$ and $m_2 = -2$, and so the complementary function is

$$y_c = c_1 e^{-x} + c_2 e^{-2x}.$$

Step 2. Now, since $4x^2$ is annihilated by the differential operator D^3, we see that $D^3(D^2 + 3D + 2)y = 4D^3 x^2$ is the same as

$$D^3(D^2 + 3D + 2)y = 0. \qquad\qquad \textbf{(10)}$$

The auxiliary equation of the fifth-order equation in (10),

$$m^3(m^2 + 3m + 2) = 0 \qquad \text{or} \qquad m^3(m + 1)(m + 2) = 0,$$

has roots $m_1 = m_2 = m_3 = 0$, $m_4 = -1$, and $m_5 = -2$. Thus its general solution must be

$$y = c_1 + c_2 x + c_3 x^2 + \boxed{c_4 e^{-x} + c_5 e^{-2x}} . \qquad\qquad \textbf{(11)}$$

The terms in the shaded box in (11) constitute the complementary function of the original equation (9). We can then argue that a particular solution y_p of (9) should also satisfy equation (10). This means that the terms remaining in (11) must be the basic form of y_p:

$$y = A + Bx + Cx^2, \tag{12}$$

where, for convenience, we have replaced c_1, c_2, and c_3 by A, B, and C, respectively. For (12) to be a particular solution of (9) it is necessary to find *specific* coefficients A, B, and C. Differentiating (12), we have

$$y_p' = B + 2Cx, \qquad y_p'' = 2C,$$

and substitution into (9) then gives

$$y_p'' + 3y_p' + 2y_p = 2C + 3B + 6Cx + 2A + 2Bx + 2Cx^2 = 4x^2.$$

Since the last equation is supposed to be an identity, the coefficients of like powers of x must be equal:

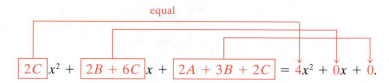

$$\boxed{2C}\,x^2 + \boxed{2B + 6C}\,x + \boxed{2A + 3B + 2C} = 4x^2 + 0x + 0.$$

That is, $\quad 2C = 4, \quad 2B + 6C = 0, \quad 2A + 3B + 2C = 0. \tag{13}$

Solving the equations in (13) gives $A = 7$, $B = -6$, and $C = 2$. Thus $y_p = 7 - 6x + 2x^2$.

Step 3. The general solution of the equation in (9) is $y = y_c + y_p$ or

$$y = c_1 e^{-x} + c_2 e^{-2x} + 7 - 6x + 2x^2. \qquad \blacksquare$$

EXAMPLE 4   **General Solution Using Undetermined Coefficients**

Solve $y'' - 3y' = 8e^{3x} + 4\sin x. \tag{14}$

SOLUTION   **Step 1.** The auxiliary equation for the associated homogeneous equation $y'' - 3y' = 0$ is $m^2 - 3m = m(m - 3) = 0$, and so $y_c = c_1 + c_2 e^{3x}$.

Step 2. Now, since $(D - 3)e^{3x} = 0$ and $(D^2 + 1)\sin x = 0$, we apply the differential operator $(D - 3)(D^2 + 1)$ to both sides of (14):

$$(D - 3)(D^2 + 1)(D^2 - 3D)y = 0. \tag{15}$$

The auxiliary equation of (15) is

$$(m - 3)(m^2 + 1)(m^2 - 3m) = 0 \qquad \text{or} \qquad m(m - 3)^2(m^2 + 1) = 0.$$

Thus $\quad y = \boxed{c_1 + c_2 e^{3x}} + c_3 x e^{3x} + c_4 \cos x + c_5 \sin x.$

After excluding the linear combination of terms in the box that corresponds to y_c, we arrive at the form of y_p:

$$y_p = Axe^{3x} + B \cos x + C \sin x.$$

Substituting y_p in (14) and simplifying yields

$$y_p'' - 3y_p' = 3Ae^{3x} + (-B - 3C) \cos x + (3B - C) \sin x = 8e^{3x} + 4 \sin x.$$

Equating coefficients gives

$$3A = 8, \quad -B - 3C = 0, \quad 3B - C = 4.$$

We find $A = \frac{8}{3}$, $B = \frac{6}{5}$, and $C = -\frac{2}{5}$, and consequently

$$y_p = \frac{8}{3} xe^{3x} + \frac{6}{5} \cos x - \frac{2}{5} \sin x.$$

Step 3. The general solution of (14) is then

$$y = c_1 + c_2 e^{3x} + \frac{8}{3} xe^{3x} + \frac{6}{5} \cos x - \frac{2}{5} \sin x.$$ ∎

EXAMPLE 5 **General Solution Using Undetermined Coefficients**

Solve $y'' + y = x \cos x - \cos x.$ **(16)**

SOLUTION The complementary function is $y_c = c_1 \cos x + c_2 \sin x$. Now by comparing $\cos x$ and $x \cos x$ with the functions in the first row of (7) we see that $\alpha = 0$ and $n = 1$, and so $(D^2 + 1)^2$ is an annihilator for the right-hand member of the equation in (16). Applying this operator to the differential equation gives

$$(D^2 + 1)^2 (D^2 + 1)y = 0 \quad \text{or} \quad (D^2 + 1)^3 y = 0.$$

Since i and $-i$ are both complex roots of multiplicity 3 of the auxiliary equation of the last differential equation, we conclude

$$y = \boxed{c_1 \cos x + c_2 \sin x} + c_3 x \cos x + c_4 x \sin x + c_5 x^2 \cos x + c_6 x^2 \sin x.$$

We substitute

$$y_p = Ax \cos x + Bx \sin x + Cx^2 \cos x + Ex^2 \sin x$$

into (16) and simplify:

$$y_p'' + y_p = 4Ex \cos x - 4Cx \sin x + (2B + 2C) \cos x + (-2A + 2E) \sin x$$
$$= x \cos x - \cos x.$$

Equating coefficients gives the equations

$$4E = 1, \quad -4C = 0, \quad 2B + 2C = -1, \quad -2A + 2E = 0,$$

from which we find $A = \frac{1}{4}$, $B = -\frac{1}{2}$, $C = 0$, and $E = \frac{1}{4}$. Hence the general solution of (16) is

$$y = c_1 \cos x + c_2 \sin x + \frac{1}{4} x \cos x - \frac{1}{2} x \sin x + \frac{1}{4} x^2 \sin x.$$ ∎

EXAMPLE 6   **Form of a Particular Solution**

Determine the form of a particular solution for

$$y'' - 2y' + y = 10e^{-2x} \cos x. \qquad \textbf{(17)}$$

SOLUTION   The complementary function for the given equation is $y_c = c_1 e^x + c_2 x e^x$.

Now from (7), with $\alpha = -2$, $\beta = 1$, and $n = 1$, we know that

$$(D^2 + 4D + 5)e^{-2x} \cos x = 0.$$

Applying the operator $D^2 + 4D + 5$ to (17) gives

$$(D^2 + 4D + 5)(D^2 - 2D + 1)y = 0. \qquad \textbf{(18)}$$

Since the roots of the auxiliary equation of (18) are $-2 - i$, $-2 + i$, 1, and 1,

$$y = \boxed{c_1 e^x + c_2 x e^x} + c_3 e^{-2x} \cos x + c_4 e^{-2x} \sin x.$$

Hence a particular solution of (17) can be found with the form

$$y_p = A e^{-2x} \cos x + B e^{-2x} \sin x. \qquad \blacksquare$$

EXAMPLE 7   **Form of a Particular Solution**

Determine the form of a particular solution for

$$y''' - 4y'' + 4y' = 5x^2 - 6x + 4x^2 e^{2x} + 3e^{5x}. \qquad \textbf{(19)}$$

SOLUTION   Observe that

$$D^3(5x^2 - 6x) = 0, \quad (D - 2)^3 x^2 e^{2x} = 0, \quad \text{and} \quad (D - 5)e^{5x} = 0.$$

Therefore $D^3(D - 2)^3(D - 5)$ applied to (19) gives

$$D^3(D - 2)^3(D - 5)(D^3 - 4D^2 + 4D)y = 0$$

or

$$D^4(D - 2)^5(D - 5)y = 0.$$

The roots of the auxiliary equation for the last differential equation are easily seen to be 0, 0, 0, 0, 2, 2, 2, 2, 2, and 5. Hence

$$y = \boxed{c_1} + c_2 x + c_3 x^2 + c_4 x^3 + \boxed{c_5 e^{2x} + c_6 x e^{2x}} + c_7 x^2 e^{2x} + c_8 x^3 e^{2x} + c_9 x^4 e^{2x} + c_{10} e^{5x}. \qquad \textbf{(20)}$$

Since the linear combination $c_1 + c_5 e^{2x} + c_6 x e^{2x}$ corresponds to the complementary function of (19), the remaining terms in (20) give the form of a particular solution of the differential equation:

$$y_p = Ax + Bx^2 + Cx^3 + Ex^2 e^{2x} + Fx^3 e^{2x} + Gx^4 e^{2x} + He^{5x}. \qquad \blacksquare$$

Summary of the Method   For your convenience the method of undetermined coefficients is summarized here.

Undetermined Coefficients–Annihilator Approach

The differential equation $L(y) = g(x)$ has constant coefficients, and the function $g(x)$ consists of finite sums and products of constants, polynomials, exponential functions $e^{\alpha x}$, sines, and cosines.

(i) Find the complementary solution y_c for the homogeneous equation $L(y) = 0$.

(ii) Operate on both sides of the nonhomogeneous equation $L(y) = g(x)$ with a differential operator L_1 that annihilates the function $g(x)$.

(iii) Find the general solution of the higher-order homogeneous differential equation $L_1 L(y) = 0$.

(iv) Delete from the solution in step (iii) all those terms that are duplicated in the complementary solution y_c found in step (i). Form a linear combination y_p of the terms that remain. This is the form of a particular solution of $L(y) = g(x)$.

(v) Substitute y_p found in step (iv) into $L(y) = g(x)$. Match coefficients of the various functions on each side of the equality, and solve the resulting system of equations for the unknown coefficients in y_p.

(vi) With the particular solution found in step (v), form the general solution $y = y_c + y_p$ of the given differential equation.

Remark

The method of undetermined coefficients is not applicable to linear differential equations with variable coefficients nor is it applicable to linear equations with constant coefficients when $g(x)$ is a function such as

$$g(x) = \ln x, \quad g(x) = \frac{1}{x}, \quad g(x) = \tan x, \quad g(x) = \sin^{-1}x,$$

and so on. Differential equations in which the input $g(x)$ is a function of this last kind will be considered in the next section.

SECTION 4.5 EXERCISES

Answers to odd-numbered problems begin on page A-5.

In Problems 1–10 write the given differential equation in the form $L(y) = g(x)$, where L is a linear differential operator with constant coefficients. If possible, factor L.

1. $9y'' - 4y = \sin x$

2. $y'' - 5y = x^2 - 2x$

3. $y'' - 4y' - 12y = x - 6$

4. $2y'' - 3y' - 2y = 1$

5. $y''' + 10y'' + 25y' = e^x$

6. $y''' + 4y' = e^x \cos 2x$

7. $y''' + 2y'' - 13y' + 10y = xe^{-x}$

8. $y''' + 4y'' + 3y' = x^2 \cos x - 3x$

9. $y^{(4)} + 8y' = 4$

10. $y^{(4)} - 8y'' + 16y = (x^3 - 2x)e^{4x}$

In Problems 11–14 verify that the given differential operator annihilates the indicated functions.

11. D^4; $y = 10x^3 - 2x$ **12.** $2D - 1$; $y = 4e^{x/2}$

13. $(D - 2)(D + 5)$; $y = e^{2x} + 3e^{-5x}$

14. $D^2 + 64$; $y = 2 \cos 8x - 5 \sin 8x$

In Problems 15–26 find a linear differential operator that annihilates the given function.

15. $1 + 6x - 2x^3$ **16.** $x^3(1 - 5x)$

17. $1 + 7e^{2x}$ **18.** $x + 3xe^{6x}$

19. $\cos 2x$ **20.** $1 + \sin x$

21. $13x + 9x^2 - \sin 4x$ **22.** $8x - \sin x + 10 \cos 5x$

23. $e^{-x} + 2xe^x - x^2e^x$ **24.** $(2 - e^x)^2$

25. $3 + e^x \cos 2x$ **26.** $e^{-x} \sin x - e^{2x} \cos x$

In Problems 27–34 find linearly independent functions that are annihilated by the given differential operator.

27. D^5 **28.** $D^2 + 4D$

29. $(D - 6)(2D + 3)$ **30.** $D^2 - 9D - 36$

31. $D^2 + 5$ **32.** $D^2 - 6D + 10$

33. $D^3 - 10D^2 + 25D$ **34.** $D^2(D - 5)(D - 7)$

In Problems 35–64 solve the given differential equation by undetermined coefficients.

35. $y'' - 9y = 54$ **36.** $2y'' - 7y' + 5y = -29$

37. $y'' + y' = 3$ **38.** $y''' + 2y'' + y' = 10$

39. $y'' + 4y' + 4y = 2x + 6$ **40.** $y'' + 3y' = 4x - 5$

41. $y''' + y'' = 8x^2$ **42.** $y'' - 2y' + y = x^3 + 4x$

43. $y'' - y' - 12y = e^{4x}$ **44.** $y'' + 2y' + 2y = 5e^{6x}$

45. $y'' - 2y' - 3y = 4e^x - 9$ **46.** $y'' + 6y' + 8y = 3e^{-2x} + 2x$

47. $y'' + 25y = 6 \sin x$ **48.** $y'' + 4y = 4 \cos x + 3 \sin x - 8$

49. $y'' + 6y' + 9y = -xe^{4x}$ **50.** $y'' + 3y' - 10y = x(e^x + 1)$

51. $y'' - y = x^2e^x + 5$ **52.** $y'' + 2y' + y = x^2e^{-x}$

53. $y'' - 2y' + 5y = e^x \sin x$

54. $y'' + y' + \dfrac{1}{4}y = e^x(\sin 3x - \cos 3x)$

55. $y'' + 25y = 20 \sin 5x$ **56.** $y'' + y = 4 \cos x - \sin x$

57. $y'' + y' + y = x \sin x$ **58.** $y'' + 4y = \cos^2 x$

59. $y''' + 8y'' = -6x^2 + 9x + 2$

60. $y''' - y'' + y' - y = xe^x - e^{-x} + 7$

61. $y''' - 3y'' + 3y' - y = e^x - x + 16$

62. $2y''' - 3y'' - 3y' + 2y = (e^x + e^{-x})^2$

63. $y^{(4)} - 2y''' + y'' = e^x + 1$ **64.** $y^{(4)} - 4y'' = 5x^2 - e^{2x}$

In Problems 65–72 solve the given differential equation subject to the indicated initial conditions.

65. $y'' - 64y = 16, \quad y(0) = 1, y'(0) = 0$

66. $y'' + y' = x, \quad y(0) = 1, y'(0) = 0$

67. $y'' - 5y' = x - 2, \quad y(0) = 0, y'(0) = 2$

68. $y'' + 5y' - 6y = 10e^{2x}, \quad y(0) = 1, y'(0) = 1$

69. $y'' + y = 8 \cos 2x - 4 \sin x, \quad y\left(\frac{\pi}{2}\right) = -1, y'\left(\frac{\pi}{2}\right) = 0$

70. $y''' - 2y'' + y' = xe^x + 5, \quad y(0) = 2, y'(0) = 2, y''(0) = -1$

71. $y'' - 4y' + 8y = x^3, \quad y(0) = 2, y'(0) = 4$

72. $y^{(4)} - y''' = x + e^x, \quad y(0) = 0, y'(0) = 0, y''(0) = 0, y'''(0) = 0$

Discussion Problem

73. Suppose L is a linear differential operator that factors but has variable coefficients. Do the factors of L commute? Defend your answer.

4.6 VARIATION OF PARAMETERS

- *Standard form of a nonhomogeneous linear second-order DE*
- *A particular solution with variable parameters* ■ *Finding variable parameters by integration*
- *Wronskian* ■ *Higher-order DEs*

The procedure that we used in Section 2.3 to find a particular solution of a linear first-order differential equation

$$\frac{dy}{dx} + P(x)y = f(x) \tag{1}$$

on an interval is applicable to linear higher-order equations as well. To adapt the method of **variation of parameters** to a linear second-order differential equation,

$$a_2(x)y'' + a_1(x)y' + a_0(x)y = g(x), \tag{2}$$

we begin as we did in Section 4.2—we put the differential equation in the standard form

$$y'' + P(x)y' + Q(x)y = f(x) \tag{3}$$

by dividing through by the lead coefficient $a_2(x)$. Here we suppose that $P(x)$, $Q(x)$, and $f(x)$ are continuous on some interval I. Equation (3) is the analogue of (1). As we have already seen in Section 4.3, there is no difficulty in obtaining the complementary function y_c of (2) when the coefficients are constants.

The Assumptions Analogous to the assumption $y_p = u(x)y_1(x)$ that we used in Section 2.3 to find a particular solution y_p of the linear first-order equation (1), for the linear second-order equation (2) we seek a solution of the form

$$y_p = u_1(x)y_1(x) + u_2(x)y_2(x), \tag{4}$$

where y_1 and y_2 form a fundamental set of solutions on I of the associated homogeneous form of (2). Using the product rule to differentiate y_p twice, we get

$$y_p' = u_1 y_1' + y_1 u_1' + u_2 y_2' + y_2 u_2'$$

$$y_p'' = u_1 y_1'' + y_1' u_1' + y_1 u_1'' + u_1' y_1' + u_2 y_2'' + y_2' u_2' + y_2 u_2'' + u_2' y_2'.$$

Substituting (4) and the foregoing derivatives into (2) and grouping terms yields

$$y_p'' + P(x)y_p' + Q(x)y_p = u_1[\overset{\text{zero}}{\overbrace{y_1'' + Py_1' + Qy_1}}] + u_2[\overset{\text{zero}}{\overbrace{y_2'' + Py_2' + Qy_2}}]$$

$$+ y_1 u_1'' + u_1' y_1' + y_2 u_2'' + u_2' y_2' + P[y_1 u_1' + y_2 u_2'] + y_1' u_1' + y_2' u_2'$$

$$= \frac{d}{dx}[y_1 u_1'] + \frac{d}{dx}[y_2 u_2'] + P[y_1 u_1' + y_2 u_2'] + y_1' u_1' + y_2' u_2'$$

$$= \frac{d}{dx}[y_1 u_1' + y_2 u_2'] + P[y_1 u_1' + y_2 u_2'] + y_1' u_1' + y_2' u_2' = f(x). \tag{5}$$

Because we seek to determine two unknown functions u_1 and u_2, reason dictates that we need two equations. We can obtain these equations by making the further assumption that the functions u_1 and u_2 satisfy $y_1 u_1' + y_2 u_2' = 0$. This assumption does not come out of the blue but is prompted by the first two terms in (5) since, if we demand that $y_1 u_1' + y_2 u_2' = 0$, then (5) reduces to $y_1' u_1' + y_2' u_2' = f(x)$. We now have our desired two equations, albeit two equations for determining the derivatives u_1' and u_2'. By Cramer's rule, the solution of the system

$$y_1 u_1' + y_2 u_2' = 0$$

$$y_1' u_1' + y_2' u_2' = f(x)$$

can be expressed in terms of determinants:

$$u_1' = \frac{W_1}{W} \qquad \text{and} \qquad u_2' = \frac{W_2}{W}, \tag{6}$$

where

$$W = \begin{vmatrix} y_1 & y_2 \\ y_1' & y_2' \end{vmatrix}, \qquad W_1 = \begin{vmatrix} 0 & y_2 \\ f(x) & y_2' \end{vmatrix}, \qquad W_2 = \begin{vmatrix} y_1 & 0 \\ y_1' & f(x) \end{vmatrix}. \tag{7}$$

The functions u_1 and u_2 are found by integrating the results in (6). The determinant W is recognized as the Wronskian of y_1 and y_2. By linear independence of y_1 and y_2 on I, we know that $W(y_1(x), y_2(x)) \neq 0$ for every x in the interval.

Summary of the Method   Usually it is not a good idea to memorize formulas in lieu of understanding a procedure. However, the foregoing procedure is too long and complicated to use each time we wish to solve a differential equation. In this case it is more efficient to simply use the formulas in (6). Thus to solve $a_2 y'' + a_1 y' + a_0 y = g(x)$, first find the complementary function $y_c = c_1 y_1 + c_2 y_2$ and then compute the Wronskian $W(y_1(x), y_2(x))$. By dividing by a_2, we put the equation into the standard form $y'' + Py' + Qy = f(x)$ to determine $f(x)$. We find u_1 and u_2 by integrating $u_1' = W_1/W$ and $u_2' = W_2/W$, where W_1 and W_2 are defined as in (7). A particular solution is $y_p = u_1 y_1 + u_2 y_2$. The general solution of the equation is then $y = y_c + y_p$.

EXAMPLE 1 **General Solution Using Variation of Parameters**

Solve $y'' - 4y' + 4y = (x + 1)e^{2x}$.

SOLUTION From auxiliary equation $m^2 - 4m + 4 = (m - 2)^2 = 0$ we have $y_c = c_1 e^{2x} + c_2 x e^{2x}$. With the identifications $y_1 = e^{2x}$ and $y_2 = xe^{2x}$, we next compute the Wronskian:

$$W(e^{2x}, xe^{2x}) = \begin{vmatrix} e^{2x} & xe^{2x} \\ 2e^{2x} & 2xe^{2x} + e^{2x} \end{vmatrix} = e^{4x}.$$

Since the given differential equation is already in form (3) (that is, the coefficient of y'' is 1), we identify $f(x) = (x + 1)e^{2x}$. From (7) we obtain

$$W_1 = \begin{vmatrix} 0 & xe^{2x} \\ (x + 1)e^{2x} & 2xe^{2x} + e^{2x} \end{vmatrix} = -(x + 1)xe^{4x}, \qquad W_2 = \begin{vmatrix} e^{2x} & 0 \\ 2e^{2x} & (x + 1)e^{2x} \end{vmatrix} = (x + 1)e^{4x},$$

and so from (6)

$$u_1' = -\frac{(x + 1)xe^{4x}}{e^{4x}} = -x^2 - x, \qquad u_2' = \frac{(x + 1)e^{4x}}{e^{4x}} = x + 1.$$

It follows that $u_1 = -\dfrac{x^3}{3} - \dfrac{x^2}{2}$ and $u_2 = \dfrac{x^2}{2} + x$.

Hence $y_p = \left(-\dfrac{x^3}{3} - \dfrac{x^2}{2}\right)e^{2x} + \left(\dfrac{x^2}{2} + x\right)xe^{2x} = \left(\dfrac{x^3}{6} + \dfrac{x^2}{2}\right)e^{2x}$

and $y = y_c + y_p = c_1 e^{2x} + c_2 xe^{2x} + \left(\dfrac{x^3}{6} + \dfrac{x^2}{2}\right)e^{2x}$. ∎

EXAMPLE 2 **General Solution Using Variation of Parameters**

Solve $4y'' + 36y = \csc 3x$.

SOLUTION We first put the equation in the standard form (3) by dividing by 4:

$$y'' + 9y = \frac{1}{4}\csc 3x.$$

Since the roots of the auxiliary equation $m^2 + 9 = 0$ are $m_1 = 3i$ and $m_2 = -3i$, the complementary function is $y_c = c_1 \cos 3x + c_2 \sin 3x$. Using $y_1 = \cos 3x$, $y_2 = \sin 3x$, and $f(x) = \frac{1}{4}\csc 3x$, we obtain

$$W(\cos 3x, \sin 3x) = \begin{vmatrix} \cos 3x & \sin 3x \\ -3\sin 3x & 3\cos 3x \end{vmatrix} = 3$$

$$W_1 = \begin{vmatrix} 0 & \sin 3x \\ \frac{1}{4}\csc 3x & 3\cos 3x \end{vmatrix} = -\frac{1}{4}, \qquad W_2 = \begin{vmatrix} \cos 3x & 0 \\ -3\sin 3x & \frac{1}{4}\csc 3x \end{vmatrix} = \frac{1}{4}\frac{\cos 3x}{\sin 3x}.$$

Integrating

$$u_1' = \frac{W_1}{W} = -\frac{1}{12} \qquad \text{and} \qquad u_2' = \frac{W_2}{W} = \frac{1}{12} \frac{\cos 3x}{\sin 3x}$$

gives

$$u_1 = -\frac{1}{12} x \qquad \text{and} \qquad u_2 = \frac{1}{36} \ln|\sin 3x|.$$

Thus a particular solution is

$$y_p = -\frac{1}{12} x \cos 3x + \frac{1}{36} (\sin 3x) \ln|\sin 3x|.$$

The general solution of the equation is

$$y = y_c + y_p = c_1 \cos 3x + c_2 \sin 3x - \frac{1}{12} x \cos 3x + \frac{1}{36} (\sin 3x) \ln|\sin 3x|. \qquad \textbf{(8)}$$

Equation (8) represents the general solution of the differential equation on, say, the interval $(0, \pi/6)$.

Constants of Integration

When computing the indefinite integrals of u_1' and u_2', we need not introduce any constants. This is because

$$\begin{aligned} y = y_c + y_p &= c_1 y_1 + c_2 y_2 + (u_1 + a_1) y_1 + (u_2 + b_1) y_2 \\ &= (c_1 + a_1) y_1 + (c_2 + b_1) y_2 + u_1 y_1 + u_2 y_2 \\ &= C_1 y_1 + C_2 y_2 + u_1 y_1 + u_2 y_2. \end{aligned}$$

EXAMPLE 3   **General Solution Using Variation of Parameters**

Solve $y'' - y = \dfrac{1}{x}$.

SOLUTION   The auxiliary equation $m^2 - 1 = 0$ yields $m_1 = -1$ and $m_2 = 1$. Therefore $y_c = c_1 e^x + c_2 e^{-x}$. Now $W(e^x, e^{-x}) = -2$ and

$$u_1' = -\frac{e^{-x}(1/x)}{-2}, \qquad u_1 = \frac{1}{2} \int_{x_0}^x \frac{e^{-t}}{t} \, dt,$$

$$u_2' = \frac{e^x(1/x)}{-2}, \qquad u_2 = -\frac{1}{2} \int_{x_0}^x \frac{e^t}{t} \, dt.$$

It is well known that the integrals defining u_1 and u_2 cannot be expressed in terms of elementary functions. Hence we write

$$y_p = \frac{1}{2} e^x \int_{x_0}^x \frac{e^{-t}}{t} \, dt - \frac{1}{2} e^{-x} \int_{x_0}^x \frac{e^t}{t} \, dt,$$

and so

$$y = y_c + y_p = c_1 e^x + c_2 e^{-x} + \frac{1}{2} e^x \int_{x_0}^x \frac{e^{-t}}{t} \, dt - \frac{1}{2} e^{-x} \int_{x_0}^x \frac{e^t}{t} \, dt.$$

In Example 3 we can integrate on any interval $x_0 \leq t \leq x$ not containing the origin.

Higher-Order Equations   The method we have just examined for nonhomogeneous second-order differential equations can be generalized to linear nth-order equations that have been put into the standard form

$$y^{(n)} + P_{n-1}(x)y^{(n-1)} + \cdots + P_1(x)y' + P_0(x)y = f(x). \qquad (9)$$

If $y_c = c_1 y_1 + c_2 y_2 + \cdots + c_n y_n$ is the complementary function for (9), then a particular solution is

$$y_p = u_1(x)y_1(x) + u_2(x)y_2(x) + \cdots + u_n(x)y_n(x),$$

where the u_k', $k = 1, 2, \ldots, n$ are determined by the n equations

$$
\begin{aligned}
y_1 u_1' + \quad y_2 u_2' + \cdots + \quad y_n u_n' &= 0 \\
y_1' u_1' + \quad y_2' u_2' + \cdots + \quad y_n' u_n' &= 0 \\
\vdots \qquad\qquad\qquad\qquad \vdots \\
y_1^{(n-1)} u_1' + y_2^{(n-1)} u_2' + \cdots + y_n^{(n-1)} u_n' &= f(x).
\end{aligned}
$$

The first $n - 1$ equations in this system, like $y_1 u_1' + y_2 u_2' = 0$ in (5), are assumptions made to simplify the resulting equation after $y_p = u_1(x)y_1(x) + \cdots + u_n(x)y_n(x)$ is substituted in (9). In this case Cramer's rule gives

$$u_k' = \frac{W_k}{W}, \quad k = 1, 2, \ldots, n,$$

where W is the Wronskian of $y_1, y_2, \ldots, y_n$ and W_k is the determinant obtained by replacing the kth column of the Wronskian by the column

$$
\begin{matrix}
0 \\
0 \\
\vdots \\
0 \\
f(x).
\end{matrix}
$$

When $n = 2$ we get (6).

Remarks

(*i*) Variation of parameters has a distinct advantage over the method of undetermined coefficients in that it will *always* yield a particular solution y_p provided the related homogeneous equation can be solved. The present method is not limited to a function $f(x)$, which is a combination of the four types of functions listed on page 121. Also, variation of parameters, unlike undetermined coefficients, is applicable to differential equations with variable coefficients.

(*ii*) In the problems that follow do not hesitate to simplify the form of y_p. Depending on how the antiderivatives of u_1' and u_2' are found, you may not obtain the same y_p as given in the answer section. For example, in Problem 3 both $y_p = \frac{1}{2}\sin x - \frac{1}{2}x\cos x$ and $y_p = \frac{1}{4}\sin x - \frac{1}{2}x\cos x$ are valid answers. In either case the general solution $y = y_c + y_p$ simplifies to $y = c_1\cos x + c_2\sin x - \frac{1}{2}x\cos x$. Why?

SECTION 4.6 EXERCISES

Answers to odd-numbered problems begin on page A-5.

In Problems 1–24 solve each differential equation by variation of parameters. State an interval on which the general solution is defined.

1. $y'' + y = \sec x$
2. $y'' + y = \tan x$
3. $y'' + y = \sin x$
4. $y'' + y = \sec x \tan x$
5. $y'' + y = \cos^2 x$
6. $y'' + y = \sec^2 x$
7. $y'' - y = \cosh x$
8. $y'' - y = \sinh 2x$
9. $y'' - 4y = \dfrac{e^{2x}}{x}$
10. $y'' - 9y = \dfrac{9x}{e^{3x}}$
11. $y'' + 3y' + 2y = \dfrac{1}{1 + e^x}$
12. $y'' - 3y' + 2y = \dfrac{e^{3x}}{1 + e^x}$
13. $y'' + 3y' + 2y = \sin e^x$
14. $y'' - 2y' + y = e^x \arctan x$
15. $y'' - 2y' + y = \dfrac{e^x}{1 + x^2}$
16. $y'' - 2y' + 2y = e^x \sec x$
17. $y'' + 2y' + y = e^{-x} \ln x$
18. $y'' + 10y' + 25y = \dfrac{e^{-10x}}{x^2}$
19. $3y'' - 6y' + 30y = e^x \tan 3x$
20. $4y'' - 4y' + y = e^{x/2} \sqrt{1 - x^2}$
21. $y''' + y' = \tan x$
22. $y''' + 4y' = \sec 2x$
23. $y''' - 2y'' - y' + 2y = e^{3x}$
24. $2y''' - 6y'' = x^2$

In Problems 25–28 solve each differential equation by variation of parameters subject to the initial conditions $y(0) = 1$, $y'(0) = 0$.

25. $4y'' - y = xe^{x/2}$
26. $2y'' + y' - y = x + 1$
27. $y'' + 2y' - 8y = 2e^{-2x} - e^{-x}$
28. $y'' - 4y' + 4y = (12x^2 - 6x)e^{2x}$

29. Given that $y_1 = x^{-1/2} \cos x$ and $y_2 = x^{-1/2} \sin x$, form a fundamental set of solutions of $x^2 y'' + xy' + (x^2 - \frac{1}{4})y = 0$ on $(0, \infty)$. Find the general solution of

$$x^2 y'' + xy' + \left(x^2 - \frac{1}{4}\right) y = x^{3/2}.$$

30. Given that $y_1 = \cos(\ln x)$ and $y_2 = \sin(\ln x)$ are known linearly independent solutions of $x^2 y'' + xy' + y = 0$ on $(0, \infty)$, find a particular solution of

$$x^2 y'' + xy' + y = \sec(\ln x).$$

Discussion Problems

31. For Problem 30 give the general solution of the differential equation. Discuss why the interval of validity of the general solution is *not* $(0, \infty)$.

32. Discuss how the methods of undetermined coefficients and variation of parameters can be combined to solve the differential equation

$$y'' + y' = 4x^2 - 3 + \frac{e^x}{x}.$$

4.7 CAUCHY-EULER EQUATION

■ *A linear DE with special variable coefficients* ■ *Auxiliary equation*
■ *Roots of a quadratic auxiliary equation* ■ *Forms of the general solution of a homogeneous linear second-order Cauchy-Euler DE* ■ *Using variation of parameters*
■ *Higher-order DEs* ■ *Reduction to constant-coefficient equations*

The relative ease with which we were able to find explicit solutions of linear higher-order differential equations with constant coefficients in the preceding sections does not, in general, carry over to linear equations with variable coefficients. We shall see in Chapter 6 that when a linear differential equation has variable coefficients, the best that we can *usually* expect is to find a solution in the form of an infinite series. However, the type of differential equation considered in this section is an exception to this rule; it is an equation with variable coefficients whose general solution can always be expressed in terms of powers of *x*, sines, cosines, and logarithmic and exponential functions. Moreover, its method of solution is quite similar to that for constant-coefficient equations.

Cauchy-Euler, or Equidimensional, Equation   Any linear differential equation of the form

$$a_n x^n \frac{d^n y}{dx^n} + a_{n-1} x^{n-1} \frac{d^{n-1} y}{dx^{n-1}} + \cdots + a_1 x \frac{dy}{dx} + a_0 y = g(x),$$

where the coefficients a_n, a_{n-1}, ..., a_0 are constants, is known diversely as a **Cauchy-Euler equation,** an **Euler-Cauchy equation,** an **Euler equation,** or an **equidimensional equation.** The observable characteristic of this type of equation is that the degree $k = n, n - 1, \ldots, 1, 0$ of the monomial coefficients x^k matches the order k of differentiation $d^k y/dx^k$:

$$a_n \overset{\text{same}}{x^n} \frac{d^n y}{dx^n} + a_{n-1} \overset{\text{same}}{x^{n-1}} \frac{d^{n-1} y}{dx^{n-1}} + \cdots.$$

As in Section 4.3, we start the discussion with a detailed examination of the forms of the general solutions of the homogeneous second-order equation

$$ax^2 \frac{d^2 y}{dx^2} + bx \frac{dy}{dx} + cy = 0.$$

The solution of higher-order equations follows analogously. Also, we can solve the nonhomogeneous equation $ax^2 y'' + bxy' + cy = g(x)$ by variation of parameters, once we have determined the complementary function $y_c(x)$.

Note

The coefficient of $d^2 y/dx^2$ is zero at $x = 0$. Hence in order to guarantee that the fundamental results of Theorem 4.1 are applicable to the Cauchy-Euler equation, we confine our attention to finding the general solution on the interval $(0, \infty)$. Solutions on the interval $(-\infty, 0)$ can be obtained by substituting $t = -x$ into the differential equation.

Method of Solution   We try a solution of the form $y = x^m$, where m is to be determined. The first and second derivatives are, respectively,

$$\frac{dy}{dx} = mx^{m-1} \quad \text{and} \quad \frac{d^2y}{dx^2} = m(m-1)x^{m-2}.$$

Consequently

$$ax^2 \frac{d^2y}{dx^2} + bx \frac{dy}{dx} + cy = ax^2 \cdot m(m-1)x^{m-2} + bx \cdot mx^{m-1} + cx^m$$

$$= am(m-1)x^m + bmx^m + cx^m = x^m(am(m-1) + bm + c).$$

Thus $y = x^m$ is a solution of the differential equation whenever m is a solution of the **auxiliary equation**

$$am(m-1) + bm + c = 0 \quad \text{or} \quad am^2 + (b-a)m + c = 0. \quad \textbf{(1)}$$

There are three different cases to be considered, depending on whether the roots of this quadratic equation are real and distinct, real and equal, or complex. In the last case the roots appear as a conjugate pair.

CASE I: Distinct Real Roots   Let m_1 and m_2 denote the real roots of (1) such that $m_1 \neq m_2$. Then $y_1 = x^{m_1}$ and $y_2 = x^{m_2}$ form a fundamental set of solutions. Hence the general solution is

$$y = c_1 x^{m_1} + c_2 x^{m_2}. \quad \textbf{(2)}$$

EXAMPLE 1   **Cauchy-Euler Equation: Distinct Roots**

Solve $x^2 \dfrac{d^2y}{dx^2} - 2x \dfrac{dy}{dx} - 4y = 0$.

SOLUTION   Rather than just memorizing equation (1), it is preferable to assume $y = x^m$ as the solution a few times in order to understand the origin and the difference between this new form of the auxiliary equation and that obtained in Section 4.3. Differentiate twice,

$$\frac{dy}{dx} = mx^{m-1}, \qquad \frac{d^2y}{dx^2} = m(m-1)x^{m-2},$$

and substitute back into the differential equation:

$$x^2 \frac{d^2y}{dx^2} - 2x \frac{dy}{dx} - 4y = x^2 \cdot m(m-1)x^{m-2} - 2x \cdot mx^{m-1} - 4x^m$$

$$= x^m(m(m-1) - 2m - 4) = x^m(m^2 - 3m - 4) = 0$$

if $m^2 - 3m - 4 = 0$. Now $(m+1)(m-4) = 0$ implies $m_1 = -1$, $m_2 = 4$ so that

$$y = c_1 x^{-1} + c_2 x^4. \qquad \blacksquare$$

CASE II: Repeated Real Roots   If the roots of (1) are repeated (that is, $m_1 = m_2$), then we obtain only one solution, namely, $y = x^{m_1}$. When the roots of the quadratic equation $am^2 + (b-a)m + c = 0$ are equal, the discriminant of the coefficients is necessarily zero. It follows from the quadratic formula that the root must be $m_1 = -(b - a)/2a$.

Now we can construct a second solution y_2, using (5) of Section 4.2. We first write the Cauchy-Euler equation in the form

$$\frac{d^2y}{dx^2} + \frac{b}{ax}\frac{dy}{dx} + \frac{c}{ax^2}y = 0$$

and make the identifications $P(x) = b/ax$ and $\int (b/ax)\, dx = (b/a)\ln x$. Thus

$$y_2 = x^{m_1} \int \frac{e^{-(b/a)\ln x}}{x^{2m_1}}\, dx$$

$$= x^{m_1} \int x^{-b/a} \cdot x^{-2m_1}\, dx \qquad \leftarrow e^{-(b/a)\ln x} = e^{\ln x^{-b/a}} = x^{-b/a}$$

$$= x^{m_1} \int x^{-b/a} \cdot x^{(b-a)/a}\, dx \qquad \leftarrow -2m_1 = (b-a)/a$$

$$= x^{m_1} \int \frac{dx}{x} = x^{m_1} \ln x.$$

The general solution is then

$$y = c_1 x^{m_1} + c_2 x^{m_1} \ln x. \tag{3}$$

EXAMPLE 2   **Cauchy-Euler Equation: Repeated Roots**

Solve $4x^2 \dfrac{d^2y}{dx^2} + 8x \dfrac{dy}{dx} + y = 0$.

SOLUTION   The substitution $y = x^m$ yields

$$4x^2\frac{d^2y}{dx^2} + 8x\frac{dy}{dx} + y = x^m(4m(m-1) + 8m + 1) = x^m(4m^2 + 4m + 1) = 0$$

when $4m^2 + 4m + 1 = 0$ or $(2m + 1)^2 = 0$. Since $m_1 = -\frac{1}{2}$, the general solution is

$$y = c_1 x^{-1/2} + c_2 x^{-1/2} \ln x. \qquad \blacksquare$$

For higher-order equations, if m_1 is a root of multiplicity k, then it can be shown that

$$x^{m_1}, \quad x^{m_1}\ln x, \quad x^{m_1}(\ln x)^2, \quad \ldots, \quad x^{m_1}(\ln x)^{k-1}$$

are k linearly independent solutions. Correspondingly, the general solution of the differential equation must then contain a linear combination of these k solutions.

CASE III: Conjugate Complex Roots   If the roots of (1) are the conjugate pair $m_1 = \alpha + i\beta$, $m_2 = \alpha - i\beta$, where α and $\beta > 0$ are real, then a solution is

$$y = C_1 x^{\alpha + i\beta} + C_2 x^{\alpha - i\beta}.$$

But when the roots of the auxiliary equation are complex, as in the case of equations with constant coefficients, we wish to write the solution in terms of real functions only. We note the identity

$$x^{i\beta} = (e^{\ln x})^{i\beta} = e^{i\beta \ln x},$$

which, by Euler's formula, is the same as

$$x^{i\beta} = \cos(\beta \ln x) + i \sin(\beta \ln x).$$

Similarly, $$x^{-i\beta} = \cos(\beta \ln x) - i \sin(\beta \ln x).$$

Adding and subtracting the last two results yields

$$x^{i\beta} + x^{-i\beta} = 2 \cos(\beta \ln x) \qquad \text{and} \qquad x^{i\beta} - x^{-i\beta} = 2i \sin(\beta \ln x),$$

respectively. From the fact that $y = C_1 x^{\alpha+i\beta} + C_2 x^{\alpha-i\beta}$ is a solution for any values of the constants, we see, in turn, for $C_1 = C_2 = 1$ and $C_1 = 1$, $C_2 = -1$ that

$$y_1 = x^{\alpha}(x^{i\beta} + x^{-i\beta}) \qquad \text{and} \qquad y_2 = x^{\alpha}(x^{i\beta} - x^{-i\beta})$$

or $$y_1 = 2x^{\alpha} \cos(\beta \ln x) \qquad \text{and} \qquad y_2 = 2ix^{\alpha} \sin(\beta \ln x)$$

are also solutions. Since $W(x^{\alpha} \cos(\beta \ln x), x^{\alpha} \sin(\beta \ln x)) = \beta x^{2\alpha-1} \neq 0$, $\beta > 0$, on the interval $(0, \infty)$, we conclude that

$$y_1 = x^{\alpha} \cos(\beta \ln x) \qquad \text{and} \qquad y_2 = x^{\alpha} \sin(\beta \ln x)$$

constitute a fundamental set of real solutions of the differential equation. Hence the general solution is

$$y = x^{\alpha}[c_1 \cos(\beta \ln x) + c_2 \sin(\beta \ln x)]. \tag{4}$$

EXAMPLE 3   An Initial-Value Problem

Solve the initial-value problem

$$x^2 \frac{d^2y}{dx^2} + 3x \frac{dy}{dx} + 3y = 0, \quad y(1) = 1, y'(1) = -5.$$

SOLUTION   We have

$$x^2 \frac{d^2y}{dx^2} + 3x \frac{dy}{dx} + 3y = x^m(m(m-1) + 3m + 3) = x^m(m^2 + 2m + 3) = 0$$

when $m^2 + 2m + 3 = 0$. From the quadratic formula we find $m_1 = -1 + \sqrt{2}i$ and $m_2 = -1 - \sqrt{2}i$. If we make the identifications $\alpha = -1$ and $\beta = \sqrt{2}$, we see from (4) that the general solution of the differential equation is

$$y = x^{-1}[c_1 \cos(\sqrt{2} \ln x) + c_2 \sin(\sqrt{2} \ln x)].$$

By applying the conditions $y(1) = 1$, $y'(1) = -5$ to the foregoing solution, we find, in turn, $c_1 = 1$ and $c_2 = -2\sqrt{2}$. Thus the solution to the initial-value problem is

$$y = x^{-1}[\cos(\sqrt{2} \ln x) - 2\sqrt{2} \sin(\sqrt{2} \ln x)].$$

The graph of this solution, obtained with the aid of computer software, is given in Figure 4.5.   ∎

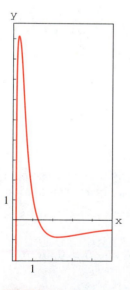

FIGURE 4.5

The next example illustrates the solution of a third-order Cauchy-Euler equation.

EXAMPLE 4 **Third-Order Cauchy-Euler Equation**

Solve $x^3 \dfrac{d^3y}{dx^3} + 5x^2 \dfrac{d^2y}{dx^2} + 7x \dfrac{dy}{dx} + 8y = 0$.

SOLUTION   The first three derivatives of $y = x^m$ are

$$\frac{dy}{dx} = mx^{m-1}, \quad \frac{d^2y}{dx^2} = m(m-1)x^{m-2}, \quad \frac{d^3y}{dx^3} = m(m-1)(m-2)x^{m-3}$$

so that the given differential equation becomes

$$x^3 \frac{d^3y}{dx^3} + 5x^2 \frac{d^2y}{dx^2} + 7x \frac{dy}{dx} + 8y = x^3 m(m-1)(m-2)x^{m-3} + 5x^2 m(m-1)x^{m-2} + 7xmx^{m-1} + 8x^m$$

$$= x^m(m(m-1)(m-2) + 5m(m-1) + 7m + 8)$$
$$= x^m(m^3 + 2m^2 + 4m + 8) = x^m(m+2)(m^2+4) = 0.$$

In this case we see that $y = x^m$ will be a solution of the differential equation for $m_1 = -2$, $m_2 = 2i$, and $m_3 = -2i$. Hence the general solution is

$$y = c_1 x^{-2} + c_2 \cos(2 \ln x) + c_3 \sin(2 \ln x).$$

■

Because the method of undetermined coefficients is applicable only to differential equations with constant coefficients, it cannot be applied directly to a nonhomogeneous Cauchy-Euler equation.

In our last example the method of variation of parameters is employed.

EXAMPLE 5 **Using Variation of Parameters**

Solve the nonhomogeneous equation

$$x^2 y'' - 3xy' + 3y = 2x^4 e^x.$$

SOLUTION   The substitution $y = x^m$ leads to the auxiliary equation

$$m(m-1) - 3m + 3 = 0 \quad \text{or} \quad (m-1)(m-3) = 0.$$

Thus $y_c = c_1 x + c_2 x^3$.

Before using variation of parameters to find a particular solution $y_p = u_1 y_1 + u_2 y_2$, recall that the formulas $u_1' = W_1/W$ and $u_2' = W_2/W$, where W_1, W_2, and W are the determinants defined on page 142, were derived under the assumption that the differential equation has been put into the standard form $y'' + P(x)y' + Q(x)y = f(x)$. Therefore we divide the given equation by x^2, and from

$$y'' - \frac{3}{x}y' + \frac{3}{x^2}y = 2x^2 e^x$$

we make the identification $f(x) = 2x^2 e^x$. Now with $y_1 = x$, $y_2 = x^3$ and

$$W = \begin{vmatrix} x & x^3 \\ 1 & 3x^2 \end{vmatrix} = 2x^3, \quad W_1 = \begin{vmatrix} 0 & x^3 \\ 2x^2 e^x & 3x^2 \end{vmatrix} = -2x^5 e^x, \quad W_2 = \begin{vmatrix} x & 0 \\ 1 & 2x^2 e^x \end{vmatrix} = 2x^3 e^x$$

we find $u_1' = -\dfrac{2x^5 e^x}{2x^3} = -x^2 e^x$ and $u_2' = \dfrac{2x^3 e^x}{2x^3} = e^x$.

The integral of the latter function is immediate, but in the case of u_1' we integrate by parts twice. The results are $u_1 = -x^2 e^x + 2xe^x - 2e^x$ and $u_2 = e^x$. Hence

$$y_p = u_1 y_1 + u_2 y_2 = (-x^2 e^x + 2xe^x - 2e^x)x + e^x x^3 = 2x^2 e^x - 2xe^x.$$

Finally we have $y = y_c + y_p = c_1 x + c_2 x^3 + 2x^2 e^x - 2xe^x$. ∎

Remark

The similarity between the forms of solutions of Cauchy-Euler equations and solutions of linear equations with constant coefficients is not just a coincidence. For example, when the roots of the auxiliary equations for $ay'' + by' + cy = 0$ and $ax^2 y'' + bxy' + cy = 0$ are distinct and real, the respective general solutions are

$$y = c_1 e^{m_1 x} + c_2 e^{m_2 x} \quad \text{and} \quad y = c_1 x^{m_1} + c_2 x^{m_2}, \quad x > 0. \tag{5}$$

In view of the identity $e^{\ln x} = x$, $x > 0$, the second solution given in (5) can be expressed in the same form as the first solution:

$$y = c_1 e^{m_1 \ln x} + c_2 e^{m_2 \ln x} = c_1 e^{m_1 t} + c_2 e^{m_2 t},$$

where $t = \ln x$. This last result illustrates another fact of mathematical life: Any Cauchy-Euler equation can *always* be rewritten as a linear differential equation with constant coefficients by means of the substitution $x = e^t$. The idea is to solve the new differential equation in terms of the variable t, using the methods of the previous sections, and once the general solution is obtained, resubstitute $t = \ln x$. Since this procedure provides a good review of the Chain Rule of differentiation, you are urged to work Problems 35–40 in Exercises 4.7.

SECTION 4.7 EXERCISES

Answers to odd-numbered problems begin on page A-5.

In Problems 1–22 solve the given differential equation.

1. $x^2 y'' - 2y = 0$ **2.** $4x^2 y'' + y = 0$

3. $xy'' + y' = 0$ **4.** $xy'' - y' = 0$

5. $x^2 y'' + xy' + 4y = 0$ **6.** $x^2 y'' + 5xy' + 3y = 0$

7. $x^2 y'' - 3xy' - 2y = 0$ **8.** $x^2 y'' + 3xy' - 4y = 0$

9. $25x^2 y'' + 25xy' + y = 0$ **10.** $4x^2 y'' + 4xy' - y = 0$

11. $x^2 y'' + 5xy' + 4y = 0$ **12.** $x^2 y'' + 8xy' + 6y = 0$

13. $x^2 y'' - xy' + 2y = 0$ **14.** $x^2 y'' - 7xy' + 41y = 0$

15. $3x^2 y'' + 6xy' + y = 0$ **16.** $2x^2 y'' + xy' + y = 0$

17. $x^3 y''' - 6y = 0$ **18.** $x^3 y''' + xy' - y = 0$

19. $x^3 \dfrac{d^3 y}{dx^3} - 2x^2 \dfrac{d^2 y}{dx^2} - 2x \dfrac{dy}{dx} + 8y = 0$

20. $x^3 \dfrac{d^3 y}{dx^3} - 2x^2 \dfrac{d^2 y}{dx^2} + 4x \dfrac{dy}{dx} - 4y = 0$

21. $x\dfrac{d^4y}{dx^4} + 6\dfrac{d^3y}{dx^3} = 0$

22. $x^4\dfrac{d^4y}{dx^4} + 6x^3\dfrac{d^3y}{dx^3} + 9x^2\dfrac{d^2y}{dx^2} + 3x\dfrac{dy}{dx} + y = 0$

In Problems 23–26 solve the given differential equation subject to the indicated initial conditions.

23. $x^2y'' + 3xy' = 0$, $\quad y(1) = 0$, $y'(1) = 4$

24. $x^2y'' - 5xy' + 8y = 0$, $\quad y(2) = 32$, $y'(2) = 0$

25. $x^2y'' + xy' + y = 0$, $\quad y(1) = 1$, $y'(1) = 2$

26. $x^2y'' - 3xy' + 4y = 0$, $\quad y(1) = 5$, $y'(1) = 3$

In Problems 27 and 28 solve the given differential equation subject to the indicated initial conditions. [*Hint:* Let $t = -x$.]

27. $4x^2y'' + y = 0$, $\quad y(-1) = 2$, $y'(-1) = 4$

28. $x^2y'' - 4xy' + 6y = 0$, $\quad y(-2) = 8$, $y'(-2) = 0$

Solve Problems 29–34 by variation of parameters.

29. $xy'' + y' = x$

30. $xy'' - 4y' = x^4$

31. $2x^2y'' + 5xy' + y = x^2 - x$

32. $x^2y'' - 2xy' + 2y = x^4e^x$

33. $x^2y'' - xy' + y = 2x$

34. $x^2y'' - 2xy' + 2y = x^3 \ln x$

In Problems 35–40 use the substitution $x = e^t$ to transform the given Cauchy-Euler equation to a differential equation with constant coefficients. Solve the original equation by solving the new equation, using the procedures in Sections 4.4 and 4.5.

35. $x^2\dfrac{d^2y}{dx^2} + 10x\dfrac{dy}{dx} + 8y = x^2$

36. $x^2y'' - 4xy' + 6y = \ln x^2$

37. $x^2y'' - 3xy' + 13y = 4 + 3x$

38. $2x^2y'' - 3xy' - 3y = 1 + 2x + x^2$

39. $x^2y'' + 9xy' - 20y = \dfrac{5}{x^3}$

40. $x^3\dfrac{d^3y}{dx^3} - 3x^2\dfrac{d^2y}{dx^2} + 6x\dfrac{dy}{dx} - 6y = 3 + \ln x^3$

Discussion Problem

41. The value of the lead coefficient a_nx^n of any Cauchy-Euler equation is zero at $x = 0$. We say that 0 is a **singular point** of the differential equation (see Section 6.2). A singular point is potentially a troublesome point in that the solutions of the differential equation *may* become unbounded or exhibit other peculiar behavior near the point. Discuss the nature of the pairs of roots m_1 and m_2 of the auxiliary equation of (1) in each of these three cases: distinct real (for example, m_1 positive and m_2 positive), repeated real, and conjugate complex. Make up corresponding solutions, and use a graphics calculator or graphing software to graph these solutions. Discuss the behavior of these solutions as $x \to 0^+$.

4.8 SYSTEMS OF LINEAR EQUATIONS

- *Solution of a system of DEs* ■ *Linear differential operators* ■ *Systematic elimination*
- *Solution by determinants*

Simultaneous ordinary differential equations involve two or more equations that contain derivatives of two or more unknown functions of a single independent variable. If x, y, and z are functions of the variable t, then

$$4\frac{d^2x}{dt^2} = -5x + y \qquad\qquad x' - 3x + y' + \;z' = 5$$

$$\text{and} \qquad x' \qquad - y' + 2z' = t^2$$

$$2\frac{d^2y}{dt^2} = 3x - y \qquad\qquad x + y' - 6z' = t - 1$$

are two examples of systems of simultaneous differential equations.

Solution of a System   A **solution** of a system of differential equations is a set of sufficiently differentiable functions $x = \phi_1(t)$, $y = \phi_2(t)$, $z = \phi_3(t)$, and so on, that satisfies each equation in the system on some common interval I.

Systematic Elimination   The first technique that we consider for solving systems of linear differential equations with constant coefficients is based on the algebraic principle of **systematic elimination** of variables. We shall see that the analogue of *multiplying* an algebraic equation by a constant is *operating* on a differential equation with some combination of derivatives. To accomplish this we rewrite each equation in a system in terms of the differential operator D. Recall from Section 4.1 that a single linear equation

$$a_n y^{(n)} + a_{n-1} y^{(n-1)} + \cdots + a_1 y' + a_0 y = g(t),$$

where the a_i, $i = 0, 1, \ldots, n$ are constants, can be written as

$$(a_n D^n + a_{n-1} D^{n-1} + \cdots + a_1 D + a_0)y = g(t).$$

The linear nth-order differential operator $a_n D^n + a_{n-1} D^{n-1} + \cdots + a_1 D + a_0$ is often abbreviated as $P(D)$. Since $P(D)$ is a polynomial in the symbol D, we may be able to factor it into differential operators of lower order. Moreover, the factors of $P(D)$ commute.

EXAMPLE 1   **System Written in Operator Notation**

Write the system of differential equations

$$x'' + 2x' + y'' = x + 3y + \sin t$$
$$x' + y' = -4x + 2y + e^{-t}$$

in operator notation.

SOLUTION   Rewrite the given system as

$$x'' + 2x' - x + y'' - 3y = \sin t$$
$$x' + 4x + y' - 2y = e^{-t}$$

so that

$$(D^2 + 2D - 1)x + (D^2 - 3)y = \sin t$$
$$(D + 4)x + (D - 2)y = e^{-t}.$$

Method of Solution   Consider the simple system of linear first-order equations

$$Dy = 2x$$
$$Dx = 3y \tag{1}$$

or, equivalently,

$$2x - Dy = 0$$
$$Dx - 3y = 0. \tag{2}$$

Operating on the first equation in (2) by D while multiplying the second by 2 and then subtracting eliminates x from the system. It follows that

$$-D^2y + 6y = 0 \quad \text{or} \quad D^2y - 6y = 0.$$

Since the roots of the auxiliary equation are $m_1 = \sqrt{6}$ and $m_2 = -\sqrt{6}$, we obtain

$$y(t) = c_1 e^{\sqrt{6}t} + c_2 e^{-\sqrt{6}t}. \tag{3}$$

Multiplying the first equation by -3 while operating on the second by D and then adding gives the differential equation for x, $D^2x - 6x = 0$. It follows immediately that

$$x(t) = c_3 e^{\sqrt{6}t} + c_4 e^{-\sqrt{6}t}. \tag{4}$$

Now (3) and (4) do not satisfy the system (1) for every choice of c_1, c_2, c_3, and c_4. Substituting $x(t)$ and $y(t)$ into the first equation of the original system (1) gives, after we simplify,

$$(\sqrt{6}c_1 - 2c_3)e^{\sqrt{6}t} + (-\sqrt{6}c_2 - 2c_4)e^{-\sqrt{6}t} = 0.$$

Since the latter expression is to be zero for all values of t, we must have

$$\sqrt{6}c_1 - 2c_3 = 0 \quad \text{and} \quad -\sqrt{6}c_2 - 2c_4 = 0$$

or

$$c_3 = \frac{\sqrt{6}}{2}c_1, \quad c_4 = -\frac{\sqrt{6}}{2}c_2. \tag{5}$$

Hence we conclude that a solution of the system must be

$$x(t) = \frac{\sqrt{6}}{2}c_1 e^{\sqrt{6}t} - \frac{\sqrt{6}}{2}c_2 e^{-\sqrt{6}t}$$
$$y(t) = c_1 e^{\sqrt{6}t} + c_2 e^{-\sqrt{6}t}.$$

You are urged to substitute (3) and (4) into the second equation of (1) and verify that the same relationship (5) holds between the constants.

EXAMPLE 2 Solution by Elimination

Solve
$$Dx + (D + 2)y = 0$$
$$(D - 3)x - \qquad 2y = 0. \tag{6}$$

SOLUTION Operating on the first equation by $D - 3$ and on the second by D and then subtracting eliminates x from the system. It follows that the differential equation for y is

$$[(D - 3)(D + 2) + 2D]y = 0 \quad \text{or} \quad (D^2 + D - 6)y = 0.$$

Since the characteristic equation of this last differential equation is $m^2 + m - 6 = (m - 2)(m + 3) = 0$, we obtain the solution

$$y(t) = c_1 e^{2t} + c_2 e^{-3t}. \tag{7}$$

Eliminating y in a similar manner yields $(D^2 + D - 6)x = 0$, from which we find

$$x(t) = c_3 e^{2t} + c_4 e^{-3t}. \tag{8}$$

As we noted in the foregoing discussion, a solution of (6) does not contain four independent constants since the system itself puts a constraint on the actual number that can be chosen arbitrarily. Substituting (7) and (8) into the first equation of (6) gives

$$(4c_1 + 2c_3)e^{2t} + (-c_2 - 3c_4)e^{-3t} = 0.$$

From $4c_1 + 2c_3 = 0$ and $-c_2 - 3c_4 = 0$ we get $c_3 = -2c_1$ and $c_4 = -\frac{1}{3}c_2$. Accordingly, a solution of the system is

$$x(t) = -2c_1 e^{2t} - \frac{1}{3} c_2 e^{-3t}$$
$$y(t) = c_1 e^{2t} + c_2 e^{-3t}. \qquad \blacksquare$$

Since we could just as easily solve for c_3 and c_4 in terms of c_1 and c_2, the solution in Example 2 can be written in the alternative form

$$x(t) = c_3 e^{2t} + c_4 e^{-3t}$$
$$y(t) = -\frac{1}{2} c_3 e^{2t} - 3c_4 e^{-3t}.$$

Also, it sometimes pays to keep one's eyes open when solving systems. Had we solved for x first, then y could be found, along with the relationship between the constants, by using the last equation in (6). You should verify that substituting $x(t)$ into $y = \frac{1}{2}(Dx - 3x)$ yields $y = -\frac{1}{2}c_3 e^{2t} - 3c_4 e^{-3t}$.

EXAMPLE 3 Solution by Elimination

Solve
$$x' - 4x + y'' = t^2$$
$$x' + \quad x + y' = 0. \tag{9}$$

SOLUTION First we write the system in differential operator notation:

$$(D - 4)x + D^2y = t^2$$
$$(D + 1)x + Dy = 0. \tag{10}$$

Then, by eliminating x, we obtain

$$[(D + 1)D^2 - (D - 4)D]y = (D + 1)t^2 - (D - 4)0$$

or

$$(D^3 + 4D)y = t^2 + 2t.$$

Since the roots of the auxiliary equation $m(m^2 + 4) = 0$ are $m_1 = 0$, $m_2 = 2i$, and $m_3 = -2i$, the complementary function is

$$y_c = c_1 + c_2 \cos 2t + c_3 \sin 2t.$$

To determine the particular solution y_p we use undetermined coefficients by assuming $y_p = At^3 + Bt^2 + Ct$. Therefore

$$y_p' = 3At^2 + 2Bt + C, \quad y_p'' = 6At + 2B, \quad y_p''' = 6A,$$
$$y_p''' + 4y_p' = 12At^2 + 8Bt + 6A + 4C = t^2 + 2t.$$

The last equality implies

$$12A = 1, \quad 8B = 2, \quad 6A + 4C = 0,$$

and hence $A = \frac{1}{12}$, $B = \frac{1}{4}$, $C = -\frac{1}{8}$. Thus

$$y = y_c + y_p = c_1 + c_2 \cos 2t + c_3 \sin 2t + \frac{1}{12}t^3 + \frac{1}{4}t^2 - \frac{1}{8}t. \tag{11}$$

Eliminating y from the system (10) leads to

$$[(D - 4) - D(D + 1)]x = t^2 \qquad \text{or} \qquad (D^2 + 4)x = -t^2.$$

It should be obvious that

$$x_c = c_4 \cos 2t + c_5 \sin 2t$$

and that undetermined coefficients can be applied to obtain a particular solution of the form $x_p = At^2 + Bt + C$. In this case the usual differentiations and algebra yield $x_p = -\frac{1}{4}t^2 + \frac{1}{8}$, and so

$$x = x_c + x_p = c_4 \cos 2t + c_5 \sin 2t - \frac{1}{4}t^2 + \frac{1}{8}. \tag{12}$$

Now c_4 and c_5 can be expressed in terms of c_2 and c_3 by substituting (11) and (12) into either equation of (9). By using the second equation, we find, after combining terms,

$$(c_5 - 2c_4 - 2c_2) \sin 2t + (2c_5 + c_4 + 2c_3) \cos 2t = 0$$

so that $c_5 - 2c_4 - 2c_2 = 0$ and $2c_5 + c_4 + 2c_3 = 0.$

Solving for c_4 and c_5 in terms of c_2 and c_3 gives

$$c_4 = -\frac{1}{5}(4c_2 + 2c_3) \qquad \text{and} \qquad c_5 = \frac{1}{5}(2c_2 - 4c_3).$$

Finally, a solution of (9) is found to be

$$x(t) = -\frac{1}{5}(4c_2 + 2c_3) \cos 2t + \frac{1}{5}(2c_2 - 4c_3) \sin 2t - \frac{1}{4}t^2 + \frac{1}{8}$$

$$y(t) = c_1 + c_2 \cos 2t + c_3 \sin 2t + \frac{1}{12}t^3 + \frac{1}{4}t^2 - \frac{1}{8}t.$$

■

EXAMPLE 4 **A Mathematical Model Revisited**

In Section 3.3 we saw that the system of linear first-order differential equations (3) describes the numbers $x_1(t)$ and $x_2(t)$ of pounds of salt in a brine mixture that flows between two tanks. At that time we were not able to solve the system. But now, in terms of differential operators, the system is

$$\left(D + \frac{2}{25}\right)x_1 - \frac{1}{50}x_2 = 0$$

$$-\frac{2}{25}x_1 + \left(D + \frac{2}{25}\right)x_2 = 0.$$

Operating on the first equation by $D + \frac{2}{25}$, multiplying the second equation by $\frac{1}{50}$, adding, and then simplifying gives

$$(625D^2 + 100D + 3)x_1 = 0.$$

From auxiliary equation $625m^2 + 100m + 3 = (25m + 1)(25m + 3) = 0$ we see immediately that

$$x_1(t) = c_1 e^{-t/25} + c_2 e^{-3t/25}.$$

In like manner we find $(625D^2 + 100D + 3)x_2 = 0$, and so

$$x_2(t) = c_3 e^{-t/25} + c_4 e^{-3t/25}.$$

Substituting $x_1(t)$ and $x_2(t)$ into, say, the first equation of the system then gives

$$(2c_1 - c_3)e^{-t/25} + (-2c_2 - c_4)e^{-3t/25} = 0.$$

From this last equation we find $c_3 = 2c_1$ and $c_4 = -2c_2$. Thus a solution of the system is

$$x_1(t) = c_1 e^{-t/25} + c_2 e^{-3t/25}$$
$$x_2(t) = 2c_1 e^{-t/25} - 2c_2 e^{-3t/25}.$$

In the original discussion we assumed that initial conditions were $x_1(0) = 25$ and $x_2(0) = 0$. Applying these conditions to the solution yields $c_1 + c_2 = 25$ and $2c_1 - 2c_2 = 0$. Solving these equations simultaneously gives $c_1 = c_2 = \frac{25}{2}$. Finally, a solution of the initial-value problem is

$$x_1(t) = \frac{25}{2}e^{-t/25} + \frac{25}{2}e^{-3t/25}$$

$$x_2(t) = 25e^{-t/25} - 25e^{-3t/25}.$$ ∎

Use of Determinants Symbolically, if L_1, L_2, L_3, and L_4 denote linear differential operators with constant coefficients, then a system of linear differential equations in two variables x and y can be written as

$$L_1 x + L_2 y = g_1(t)$$
$$L_3 x + L_4 y = g_2(t). \tag{13}$$

Eliminating variables, as we would for algebraic equations, leads to

$$(L_1 L_4 - L_2 L_3)x = f_1(t) \qquad \text{and} \qquad (L_1 L_4 - L_2 L_3)y = f_2(t), \tag{14}$$

where

$$f_1(t) = L_4 g_1(t) - L_2 g_2(t) \quad \text{and} \quad f_2(t) = L_1 g_2(t) - L_3 g_1(t).$$

Formally the results in (14) can be written in terms of determinants similar to those used in Cramer's rule:

$$\begin{vmatrix} L_1 & L_2 \\ L_3 & L_4 \end{vmatrix} x = \begin{vmatrix} g_1 & L_2 \\ g_2 & L_4 \end{vmatrix} \quad \text{and} \quad \begin{vmatrix} L_1 & L_2 \\ L_3 & L_4 \end{vmatrix} y = \begin{vmatrix} L_1 & g_1 \\ L_3 & g_2 \end{vmatrix}. \quad (15)$$

The left-hand determinant in each equation in (15) can be expanded in the usual algebraic sense, with the result then operating on the functions $x(t)$ and $y(t)$. However, some care should be exercised in the expansion of the right-hand determinants in (15). We must expand these determinants in the sense of the internal differential operators actually operating on the functions $g_1(t)$ and $g_2(t)$.

$$\text{If} \qquad \begin{vmatrix} L_1 & L_2 \\ L_3 & L_4 \end{vmatrix} \neq 0$$

in (15) and is a differential operator of order n, then

- The system (13) can be decoupled into two nth-order differential equations in x and y.
- The characteristic equations and hence the complementary functions of these differential equations are the same.
- Since x and y both contain n constants, a total of $2n$ constants appear.
- The total number of *independent* constants in the solution of the system is n.

$$\text{If} \qquad \begin{vmatrix} L_1 & L_2 \\ L_3 & L_4 \end{vmatrix} = 0$$

in (13), then the system may have a solution containing any number of independent constants or may have no solution at all. Similar remarks hold for systems larger than indicated in (13).

EXAMPLE 5 Solution by Determinants

Solve
$$x' = 3x - y - 12$$
$$y' = x + y + 4e^t. \quad (16)$$

SOLUTION Write the system in terms of differential operators:

$$(D-3)x + y = -12$$
$$-x + (D-1)y = 4e^t.$$

Then use determinants:

$$\begin{vmatrix} D-3 & 1 \\ -1 & D-1 \end{vmatrix} x = \begin{vmatrix} -12 & 1 \\ 4e^t & D-1 \end{vmatrix}$$

$$\begin{vmatrix} D-3 & 1 \\ -1 & D-1 \end{vmatrix} y = \begin{vmatrix} D-3 & -12 \\ -1 & 4e^t \end{vmatrix}.$$

After expanding, we find that

$$(D - 2)^2 x = 12 - 4e^t \quad \text{and} \quad (D - 2)^2 y = -12 - 8e^t.$$

By the usual methods it follows that

$$x = x_c + x_p = c_1 e^{2t} + c_2 te^{2t} + 3 - 4e^t \tag{17}$$
$$y = y_c + y_p = c_3 e^{2t} + c_4 te^{2t} - 3 - 8e^t. \tag{18}$$

Substituting (17) and (18) into the second equation of (16) gives

$$(c_3 - c_1 + c_4)e^{2t} + (c_4 - c_2)te^{2t} = 0,$$

which then implies $c_4 = c_2$ and $c_3 = c_1 - c_4 = c_1 - c_2$. Thus a solution of (16) is

$$x(t) = c_1 e^{2t} + c_2 te^{2t} + 3 - 4e^t$$
$$y(t) = (c_1 - c_2)e^{2t} + c_2 te^{2t} - 3 - 8e^t.$$

SECTION 4.8 EXERCISES

Answers to odd-numbered problems begin on page A-5.

In Problems 1–22 solve, if possible, the given system of differential equations by either systematic elimination or determinants.

1. $\dfrac{dx}{dt} = 2x - y$

$\dfrac{dy}{dt} = x$

2. $\dfrac{dx}{dt} = 4x + 7y$

$\dfrac{dy}{dt} = x - 2y$

3. $\dfrac{dx}{dt} = -y + t$

$\dfrac{dy}{dt} = x - t$

4. $\dfrac{dx}{dt} - 4y = 1$

$x + \dfrac{dy}{dt} = 2$

5. $(D^2 + 5)x - 2y = 0$
$-2x + (D^2 + 2)y = 0$

6. $(D + 1)x + (D - 1)y = 2$
$3x + (D + 2)y = -1$

7. $\dfrac{d^2x}{dt^2} = 4y + e^t$

$\dfrac{d^2y}{dt^2} = 4x - e^t$

8. $\dfrac{d^2x}{dt^2} + \dfrac{dy}{dt} = -5x$

$\dfrac{dx}{dt} + \dfrac{dy}{dt} = -x + 4y$

9. $Dx + D^2y = e^{3t}$
$(D + 1)x + (D - 1)y = 4e^{3t}$

10. $D^2x - Dy = t$
$(D + 3)x + (D + 3)y = 2$

11. $(D^2 - 1)x - y = 0$
$(D - 1)x + Dy = 0$

12. $(2D^2 - D - 1)x - (2D + 1)y = 1$
$(D - 1)x + Dy = -1$

13. $2\dfrac{dx}{dt} - 5x + \dfrac{dy}{dt} = e^t$

$\dfrac{dx}{dt} - x + \dfrac{dy}{dt} = 5e^t$

14. $\dfrac{dx}{dt} + \dfrac{dy}{dt} = e^t$

$-\dfrac{d^2x}{dt^2} + \dfrac{dx}{dt} + x + y = 0$

15. $(D-1)x + (D^2+1)y = 1$
$(D^2-1)x + (D+1)y = 2$

16. $D^2x - 2(D^2+D)y = \sin t$
$x + \qquad Dy = 0$

17. $Dx = y$
$Dy = z$
$Dz = x$

18. $\qquad Dx + \qquad z = e^t$
$(D-1)x + Dy + Dz = 0$
$x + 2y + Dz = e^t$

19. $\dfrac{dx}{dt} - 6y \qquad = 0$

$x - \dfrac{dy}{dt} + z = 0$

$x + y - \dfrac{dz}{dt} = 0$

20. $\dfrac{dx}{dt} = -x + z$

$\dfrac{dy}{dt} = -y + z$

$\dfrac{dz}{dt} = -x + y$

21. $2Dx + (D-1)y = t$
$Dx + \qquad Dy = t^2$

22. $\qquad Dx - \qquad 2Dy = t^2$
$(D+1)x - 2(D+1)y = 1$

In Problems 23 and 24 solve the given system subject to the indicated initial conditions.

23. $\dfrac{dx}{dt} = -5x - y$

$\dfrac{dy}{dt} = 4x - y; \quad x(1) = 0, y(1) = 1$

24. $\dfrac{dx}{dt} = y - 1$

$\dfrac{dy}{dt} = -3x + 2y; \quad x(0) = 0, y(0) = 0$

25. A projectile shot from a gun has weight $w = mg$ and velocity **v** tangent to its path of motion. Ignoring air resistance and all other forces except its weight, determine a system of differential equations that describes the motion. See Figure 4.6. Solve the system. [*Hint:* Use Newton's second law of motion in the x and y directions.]

26. Determine a system of differential equations that describes the motion in Problem 25 if the projectile encounters a retarding force **k** (of magnitude k) acting tangent to the path but opposite to the motion. See Figure 4.7. Solve the system. [*Hint:* **k** is a multiple of velocity, say $c\mathbf{v}$.]

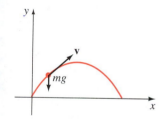

FIGURE 4.6

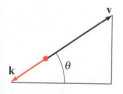

FIGURE 4.7

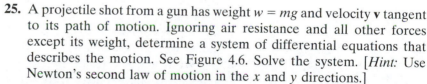

4.9 NONLINEAR EQUATIONS

■ *Some differences between linear and nonlinear DEs* ■ *Solution by substitution*
■ *Use of Taylor series* ■ *Use of ODE solvers* ■ *Autonomous equations*

There are several significant differences between linear and nonlinear differential equations. We saw in Section 4.1 that homogeneous linear equations of order two or higher have the property that a linear combination of solutions is also a solution (Theorem 4.2). Nonlinear equations do not possess this property of superposability. For example, on the interval $(-\infty, \infty)$, $y_1 = e^x$, $y_2 = e^{-x}$, $y_3 = \cos x$, and $y_4 = \sin x$ are four linearly independent solutions of the nonlinear second-order differential equation $(y'')^2 - y^2 = 0$. But linear combinations such as $y = c_1 e^x + c_3 \cos x$, $y = c_2 e^{-x} + c_4 \sin x$, and $y = c_1 e^x + c_2 e^{-x} + c_3 \cos x + c_4 \sin x$ are not solutions of the equation for arbitrary nonzero constants c_i. (See Problem 1 in Exercises 4.9.)

In Chapter 2 we saw that we could solve a few nonlinear first-order differential equations by recognizing them as separable, exact, homogeneous, or perhaps Bernoulli equations. Even though the solutions of these equations were in the form of a one-parameter family, this family did not as a rule represent the general solution of the differential equation. On the other hand, by paying attention to certain continuity conditions we obtained general solutions of linear first-order equations. Stated another way, nonlinear first-order differential equations can possess singular solutions, whereas linear equations cannot. But the major difference between linear and nonlinear equations of order two or higher lies in the realm of solvability. Given a linear equation, there is a chance that we can find some form of a solution that we can look at—an explicit solution or a solution in the form of an infinite series. On the other hand, nonlinear higher-order differential equations virtually defy solution. This does not mean that a nonlinear higher-order differential equation has no solution, but rather that there are no general methods whereby either an explicit or an implicit solution can be found. Although this sounds disheartening, there are still things that can be done. We can always analyze a nonlinear equation quantitatively (approximate a solution using a numerical procedure, graph a solution using an ODE solver) or qualitatively.

Let us make clear at the outset that nonlinear higher-order differential equations are important—dare we say even more important than linear equations?—because as we fine-tune the mathematical model of, say, a physical system, we also increase the likelihood that this higher-resolution model will be nonlinear.

We begin by illustrating a substitution method that occasionally enables us to find explicit or implicit solutions of special kinds of nonlinear equations.

Use of Substitutions   Nonlinear second-order differential equations $F(x, y', y'') = 0$, where the dependent variable y is missing, and $F(y, y', y'') = 0$, where the independent variable x is missing, can be reduced to first-order equations by means of the substitution $u = y'$.

Example 1 illustrates the substitution technique for an equation of the form $F(x, y', y'') = 0$. If $u = y'$, then the differential equation becomes $F(x, u, u') = 0$. If we can solve this last equation for u, we can find y by integration. Note that since we are solving a second-order equation, its solution will contain two arbitrary constants.

EXAMPLE 1   **Dependent Variable y Is Missing**

Solve $y'' = 2x(y')^2$.

SOLUTION   If we let $u = y'$, then $du/dx = y''$. After substituting, the second-order equation reduces to a first-order equation with separable variables; the independent variable is x, and the dependent variable is u:

$$\frac{du}{dx} = 2xu^2 \quad \text{or} \quad \frac{du}{u^2} = 2x\,dx$$

$$\int u^{-2}\,du = \int 2x\,dx$$

$$-u^{-1} = x^2 + c_1^2.$$

The constant of integration is written as c_1^2 for convenience. The reason should be obvious in the next few steps. Since $u^{-1} = 1/y'$, it follows that

$$\frac{dy}{dx} = -\frac{1}{x^2 + c_1^2},$$

and so $\quad y = -\int \dfrac{dx}{x^2 + c_1^2} \quad$ or $\quad y = -\dfrac{1}{c_1}\tan^{-1}\dfrac{x}{c_1} + c_2.$ ■

Next we show how to solve an equation of the form $F(y, y', y'') = 0$. Once more we let $u = y'$, but since the independent variable x is missing we use this substitution to transform the differential equation into one in which the independent variable is y and the dependent variable is u. To this end we use the Chain Rule to compute the second derivative of y:

$$y'' = \frac{du}{dx} = \frac{du}{dy}\frac{dy}{dx} = u\frac{du}{dy}.$$

The first-order equation that we must now solve is $F(y, u, u\,du/dy) = 0$.

EXAMPLE 2 **Independent Variable x Is Missing**

Solve $yy'' = (y')^2$.

SOLUTION With the aid of $u = y'$ and the Chain Rule shown above, the differential equation becomes

$$y\left(u\frac{du}{dy}\right) = u^2 \quad \text{or} \quad \frac{du}{u} = \frac{dy}{y}.$$

From $\quad \displaystyle\int \frac{du}{u} = \int \frac{dy}{y} \quad$ we get $\quad \ln|u| = \ln|y| + c_1.$

Solving the last equation for u in terms of y gives $u = c_2 y$, where the constant $\pm e^{c_1}$ has been relabeled as c_2. We then resubstitute $u = dy/dx$, separate variables, integrate, and once again relabel constants:

$$\int \frac{dy}{y} = c_2 \int dx \quad \text{or} \quad \ln|y| = c_2 x + c_3 \quad \text{or} \quad y = c_4 e^{c_2 x}. \qquad ■$$

Use of Taylor Series In some instances a solution of a nonlinear initial-value problem, in which the initial conditions are specified at x_0, can be approximated by a Taylor series centered at x_0.

EXAMPLE 3 Taylor Series Solution of an IVP

Let us assume that a solution of the initial-value problem

$$y'' = x + y - y^2, \qquad y(0) = -1, \quad y'(0) = 1 \tag{1}$$

exists. If we further assume that the solution $y(x)$ of the problem is analytic at 0, then $y(x)$ possesses a Taylor series expansion centered at 0:

$$y(x) = y(0) + \frac{y'(0)}{1!}x + \frac{y''(0)}{2!}x^2 + \frac{y'''(0)}{3!}x^3 + \frac{y^{(iv)}(0)}{4!}x^4 + \frac{y^{(v)}(0)}{5!}x^5 + \cdots. \tag{2}$$

Note that the values of the first and second terms in the series (2) are known since those values are the specified initial conditions $y(0) = -1$, $y'(0) = 1$. Moreover, the differential equation itself defines the value of the second derivative at 0: $y''(0) = 0 + y(0) - y(0)^2 = 0 + (-1) - (-1)^2 = -2$. We can then find expressions for the higher derivatives y''', $y^{(iv)}$, ... by calculating the successive derivatives of the differential equation:

$$y'''(x) = \frac{d}{dx}(x + y - y^2) = 1 + y' - 2yy' \tag{3}$$

$$y^{(iv)}(x) = \frac{d}{dx}(1 + y' - 2yy') = y'' - 2yy'' - 2(y')^2 \tag{4}$$

$$y^{(v)}(x) = \frac{d}{dx}(y'' - 2yy'' - 2(y')^2) = y''' - 2yy''' - 6y'y'' \tag{5}$$

and so on. Now, using $y(0) = -1$ and $y'(0) = 1$, we find from (3) that $y'''(0) = 4$. From the values $y(0) = -1$, $y'(0) = 1$, and $y''(0) = -2$ we find $y^{(iv)}(0) = -8$ from (4). With the additional information that $y'''(0) = 4$, we then see from (5) that $y^{(v)}(0) = 24$. Hence from (2) the first six terms of a series solution of the initial-value problem (1) are

$$y(x) = -1 + x - x^2 + \frac{2}{3}x^3 - \frac{1}{3}x^4 + \frac{1}{5}x^5 + \cdots. \qquad \blacksquare$$

Use of an ODE Solver   It is possible to examine the equation in Example 3 by means of an ODE solver. In order to examine a higher-order differential equation numerically it is necessary, in most software packages, to express the differential equation as a system of differential equations. To approximate the solution curve of a second-order initial-value problem

$$\frac{d^2y}{dx^2} = f(x, y, y'), \qquad y(x_0) = y_0, \quad y'(x_0) = y_1$$

we let $dy/dx = u$, then $d^2y/dx^2 = du/dx$. The second-order equation becomes a system of two first-order differential equations in the dependent variables y and u:

$$\frac{dy}{dx} = u$$

$$\frac{du}{dx} = f(x, y, u)$$

with initial conditions $y(x_0) = y_0$, $u(x_0) = y_1$.

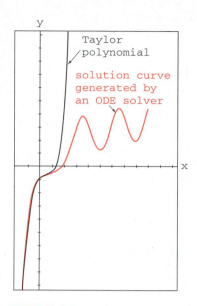

FIGURE 4.8
Comparison of two approximate solutions

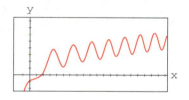

FIGURE 4.9

EXAMPLE 4 **Graphical Analysis of Example 3**

Following the foregoing procedure, the second-order initial-value problem in Example 3 is equivalent to

$$\frac{dy}{dx} = u$$

$$\frac{du}{dx} = x + y - y^2$$

with initial conditions $y(0) = -1$, $u(0) = 1$. With the aid of an ODE solver we get the solution curve shown in color in Figure 4.8. For comparison, the curve shown in black is the graph of the fifth-degree Taylor polynomial $T_5(x) = -1 + x - x^2 + \frac{2}{3}x^3 - \frac{1}{3}x^4 + \frac{1}{5}x^5$. Although we do not know the interval of convergence of the Taylor series obtained in Example 3, the closeness of the two curves in a neighborhood of the origin suggests that it is likely that the series converges on the interval $(-1, 1)$. ∎

The colored graph in Figure 4.8 raises some qualitative questions: Is the solution of the original initial-value problem oscillatory as $x \to \infty$? The graph generated by an ODE solver on the larger interval shown in Figure 4.9 would seem to *suggest* that the answer is yes. But this single example or even an assortment of examples does not answer the basic question of whether *all* solutions of the differential equation $y'' = x + y - y^2$ are oscillatory in nature. Also, what is happening to the solution curve in Figure 4.8 when x is near -1? What is the behavior of solutions of the differential equation as $x \to \infty$? Are solutions bounded as $x \to \infty$? Questions such as these are not easily answered, in general, for nonlinear second-order differential equations. But certain kinds of second-order equations lend themselves to a systematic qualitative analysis. Nonlinear second-order equations of the form

$$F(y, y', y'') = 0 \qquad \text{or} \qquad \frac{d^2y}{dx^2} = f(y, y')$$

(that is, differential equations that have no explicit dependence on the independent variable x) are called **autonomous.** The differential equation in Example 2 is autonomous; the equation in Example 3 is nonautonomous. For a treatment of the topic of stability of autonomous nonlinear differential equations the reader is referred to Chapter 10.

SECTION 4.9 EXERCISES

Answers to odd-numbered problems begin on page A-6.

In Problems 1 and 2 verify that y_1 and y_2 are solutions of the given differential equation but that $y = c_1 y_1 + c_2 y_2$ is, in general, not a solution.

1. $(y'')^2 = y^2$; $y_1 = e^x, y_2 = \cos x$ **2.** $yy'' = \frac{1}{2}(y')^2$; $y_1 = 1, y_2 = x^2$

In Problems 3–8 solve the given differential equation by using the substitution $u = y'$.

3. $y'' + (y')^2 + 1 = 0$ **4.** $y'' = 1 + (y')^2$

5. $x^2 y'' + (y')^2 = 0$ **6.** $(y + 1)y'' = (y')^2$

7. $y'' + 2y(y')^3 = 0$ **8.** $y^2 y'' = y'$

9. Find a solution of the initial-value problem

$$y'' + yy' = 0, \qquad y(0) = 1, \quad y'(0) = -1.$$

Use an ODE solver to graph the solution curve. Use a graphing utility to graph the explicit solution. Find an interval of validity for the solution of the problem.

10. Find two solutions of the initial-value problem

$$(y'')^2 + (y')^2 = 1, \qquad y\left(\frac{\pi}{2}\right) = \frac{1}{2}, \quad y'\left(\frac{\pi}{2}\right) = \frac{\sqrt{3}}{2}.$$

Use an ODE solver to graph the solution curves.

In Problems 11 and 12 show that the substitution $u = y'$ leads to a Bernoulli equation. Solve this equation (see Section 2.4).

11. $xy'' = y' + (y')^3$ **12.** $xy'' = y' + x(y')^2$

In Problems 13–16 proceed as in Example 3 and obtain the first six nonzero terms of a Taylor series solution centered at 0 of the given initial-value problem. Use an ODE solver and a graphing utility to compare the solution curve and the graph of the Taylor polynomial.

13. $y'' = x + y^2$, $y(0) = 1, y'(0) = 1$

14. $y'' + y^2 = 1$, $y(0) = 2, y'(0) = 3$

15. $y'' = x^2 + y^2 - 2y'$, $y(0) = 1, y'(0) = 1$

16. $y'' = e^y$, $y(0) = 0, y'(0) = -1$

17. In calculus the curvature of a curve that is defined by $y = f(x)$ is defined as

$$\kappa = \frac{y''}{[1 + (y')^2]^{3/2}}.$$

Find a function $y = f(x)$ for which $\kappa = 1$. [*Hint:* For simplicity, ignore constants of integration.]

18. A mathematical model for the position $x(t)$ of a body moving rectilinearly on the x-axis in an inverse-square force field is given by

$$\frac{d^2 x}{dt^2} = -\frac{k^2}{x^2}.$$

Suppose that at $t = 0$ the body starts from rest from the position $x = x_0, x_0 > 0$. Show that the velocity of the body at any time is given by

$$\frac{v^2}{2} = k^2 \left(\frac{1}{x} - \frac{1}{x_0}\right).$$

Use the last equation and a CAS to carry out the integration to express time t in terms of x.

Discussion Problems

19. A mathematical model for the position $x(t)$ of a moving object is

$$\frac{d^2x}{dt^2} + \sin x = 0.$$

Use an ODE solver to investigate the solutions of the equation subject to $x(0) = 0$, $x'(0) = \beta$, $\beta \geq 0$. Discuss the motion of the object for $t \geq 0$ and various choices of β. Investigate the equation

$$\frac{d^2x}{dt^2} + \frac{dx}{dt} + \sin x = 0$$

in the same manner. Discuss a possible physical interpretation of the dx/dt term.

20. We saw that $\sin x$, $\cos x$, e^x, and e^{-x} were four solutions of the nonlinear equation $(y'')^2 - y^2 = 0$. Without attempting to solve the differential equation, discuss how these explicit solutions can be found using knowledge about linear equations. Without attempting to verify, discuss why the two special linear combinations $y = c_1 e^x + c_2 e^{-x}$ and $y = c_3 \cos x + c_4 \sin x$ must satisfy the differential equation.

CHAPTER 4 REVIEW EXERCISES

Answers to odd-numbered problems begin on page A-6.

Answer Problems 1–10 without referring back to the text. Fill in the blank or answer true/false. In some cases there may be more than one correct answer.

1. The only solution of $y'' + x^2 y = 0$, $y(0) = 0$, $y'(0) = 0$ is _____.

2. If two differentiable functions $f_1(x)$ and $f_2(x)$ are linearly independent on an interval, then $W(f_1(x), f_2(x)) \neq 0$ for at least one point in the interval. _____

3. Two functions $f_1(x)$ and $f_2(x)$ are linearly independent on an interval if one is not a constant multiple of the other. _____

4. The functions $f_1(x) = x^2$, $f_2(x) = 1 - x^2$, and $f_3(x) = 2 + x^2$ are linearly _____ on the interval $(-\infty, \infty)$.

5. The functions $f_1(x) = x^2$ and $f_2(x) = x|x|$ are linearly independent on the interval _____, whereas they are linearly dependent on the interval _____.

6. Two solutions y_1 and y_2 of $y'' + y' + y = 0$ are linearly dependent if $W(y_1, y_2) = 0$ for every real value of x. _____

7. A constant multiple of a solution of a differential equation is also a solution. _____

8. A fundamental set of two solutions of $(x - 2)y'' + y = 0$ exists on any interval not containing the point _____.

9. For the method of undetermined coefficients, the assumed form of the particular solution y_p for $y'' - y = 1 + e^x$ is _____.

10. A differential operator that annihilates $e^{2x}(x + \sin x)$ is _____.

In Problems 11 and 12 find a second solution for the differential equation, given that $y_1(x)$ is a known solution.

11. $y'' + 4y = 0, \quad y_1 = \cos 2x$

12. $xy'' - 2(x + 1)y' + (x + 2)y = 0, \quad y_1 = e^x$

In Problems 13–20 find the general solution of each differential equation.

13. $y'' - 2y' - 2y = 0$ **14.** $2y'' + 2y' + 3y = 0$

15. $y''' + 10y'' + 25y' = 0$ **16.** $2y''' + 9y'' + 12y' + 5y = 0$

17. $3y''' + 10y'' + 15y' + 4y = 0$

18. $2\dfrac{d^4y}{dx^4} + 3\dfrac{d^3y}{dx^3} + 2\dfrac{d^2y}{dx^2} + 6\dfrac{dy}{dx} - 4y = 0$

19. $6x^2y'' + 5xy' - y = 0$

20. $2x^3y''' + 19x^2y'' + 39xy' + 9y = 0$

In Problems 21–24 solve each differential equation by the method of undetermined coefficients.

21. $y'' - 3y' + 5y = 4x^3 - 2x$ **22.** $y'' - 2y' + y = x^2e^x$

23. $y''' - 5y'' + 6y' = 2\sin x + 8$ **24.** $y''' - y'' = 6$

In Problems 25–28 solve the given differential equation subject to the indicated initial conditions.

25. $y'' - 2y' + 2y = 0, \quad y\left(\dfrac{\pi}{2}\right) = 0,\, y(\pi) = -1$

26. $y'' - y = x + \sin x, \quad y(0) = 2,\, y'(0) = 3$

27. $y'y'' = 4x, \quad y(1) = 5,\, y'(1) = 2$

28. $2y'' = 3y^2, \quad y(0) = 1,\, y'(0) = 1$

In Problems 29–32 solve each differential equation by the method of variation of parameters.

29. $y'' - 2y' + 2y = e^x \tan x$ **30.** $y'' - y = \dfrac{2e^x}{e^x + e^{-x}}$

31. $x^2y'' - 4xy' + 6y = 2x^4 + x^2$ **32.** $x^2y'' - xy' + y = x^3$

In Problems 33 and 34 solve the given differential equation subject to the indicated initial conditions.

33. $(2D^3 - 13D^2 + 24D - 9)y = 36, \quad y(0) = -4,\, y'(0) = 0,\, y''(0) = \dfrac{5}{2}$

34. $y'' + y = \sec^3x, \quad y(0) = 1,\, y'(0) = \dfrac{1}{2}$

In Problems 35–38 use systematic elimination or determinants to solve the given system.

35. $\begin{aligned} x' + \ y' &= 2x + 2y + 1 \\ x' + 2y' &= \quad\quad y + 3 \end{aligned}$ **36.** $\dfrac{dx}{dt} = 2x + \ y + \ t - 2$

$\qquad\qquad\qquad\qquad\qquad\qquad\qquad \dfrac{dy}{dt} = 3x + 4y - 4t$

37. $\begin{aligned} (D - 2)x - \quad\quad\quad y &= -e^t \\ -3x + (D - 4)y &= -7e^t \end{aligned}$

38. $\begin{aligned} (D + 2)x + (D + 1)y &= \sin 2t \\ 5x + (D + 3)y &= \cos 2t \end{aligned}$

5

MODELING WITH HIGHER-ORDER DIFFERENTIAL EQUATIONS

INTRODUCTION

We have seen that a single differential equation can serve as a mathematical model for different phenomena. For this reason we examine one application, the motion of a mass attached to a spring, in great detail in Section 5.1. We shall see that, except for terminology and physical interpretations of the four terms in the linear equation $ay'' + by' + cy = g(t)$, the mathematics of, say, an electrical series circuit is identical to that of a vibrating spring/mass system. Forms of this linear second-order differential equation appear in the analysis of problems in many diverse areas of science and engineering. In Section 5.1 we deal exclusively with initial-value problems, whereas in Section 5.2 we examine applications described by boundary-value problems. In Section 5.2 we also see how some boundary-value problems lead to the concepts of **eigenvalues** and **eigenfunctions.** Section 5.3 begins with a discussion of the differences between linear and nonlinear springs; we then show how the simple pendulum and a suspended wire lead to nonlinear models.

5.1 LINEAR EQUATIONS: INITIAL-VALUE PROBLEMS

■ *Linear dynamical system* ■ *Hooke's law* ■ *Newton's second law of motion*
■ *Spring/mass system* ■ *Free undamped motion* ■ *Simple harmonic motion*
■ *Equation of motion* ■ *Amplitude* ■ *Phase angle* ■ *Aging spring* ■ *Free damped motion*
■ *Driven motion* ■ *Transient and steady-state terms* ■ *Pure resonance* ■ *Series circuits*

In this section we are going to consider several linear dynamical systems (see page 106) in which each mathematical model is a second-order differential equation with constant coefficients

$$a_2 \frac{d^2 y}{dt^2} + a_1 \frac{dy}{dt} + a_0 y = g(t).$$

Recall that the function g is the **input** or **forcing function** of the system. A solution of the differential equation on an interval containing t_0 and satisfying prescribed initial conditions $y(t_0) = y_0$, $y'(t_0) = y_1$ is the **output** or **response** of the system.

5.1.1 Spring/Mass Systems: Free Undamped Motion

Hooke's Law   Suppose, as in Figure 5.1(b), that a mass m_1 is attached to a flexible spring suspended from a rigid support. When m_1 is replaced with a different mass m_2, the amount of stretch, or elongation, of the spring will of course be different.

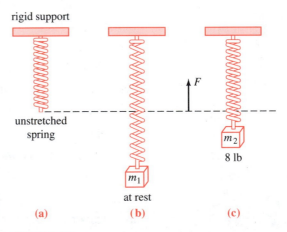

rigid support

unstretched spring

F

m_2

8 lb

m_1

at rest

(a)   (b)   (c)

FIGURE 5.1

By Hooke's law the spring itself exerts a restoring force F opposite to the direction of elongation and proportional to the amount s of elongation. Simply stated, $F = ks$, where k is a constant of proportionality called the **spring constant.** Although masses with different weights stretch a spring by different amounts, the spring is essentially characterized by the number k. For example, if a mass weighing 10 pounds stretches a spring $\frac{1}{2}$ foot, then $10 = k(\frac{1}{2})$ implies $k = 20$ lb/ft. Necessarily then, a mass weighing, say, 8 pounds stretches the same spring $\frac{2}{5}$ foot.

Newton's Second Law After a mass m is attached to a spring, it stretches the spring by an amount s and attains a position of equilibrium at which its weight W is balanced by the restoring force ks. Recall that weight is defined by $W = mg$, where mass is measured in slugs, kilograms, or grams and $g = 32$ ft/s^2, 9.8 m/s^2, or 980 cm/s^2, respectively. As indicated in Figure 5.2(b), the condition of equilibrium is $mg = ks$ or $mg - ks = 0$. If the mass is displaced by an amount x from its equilibrium position, the restoring force of the spring is then $k(x + s)$. Assuming that there are no retarding forces acting on the system and assuming that the mass vibrates free of other external forces—**free motion**—we can equate Newton's second law with the net, or resultant, force of the restoring force and the weight:

$$m\frac{d^2x}{dt^2} = -k(s + x) + mg = \underbrace{-kx + mg - ks}_{\text{zero}} = -kx. \tag{1}$$

The negative sign in (1) indicates that the restoring force of the spring acts opposite to the direction of motion. Furthermore, we can adopt the convention that displacements measured *below* the equilibrium position are positive. See Figure 5.3.

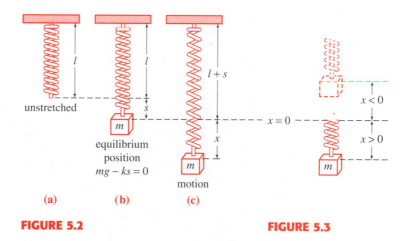

unstretched

equilibrium
position
$mg - ks = 0$

$l + s$

motion

$x < 0$

$x = 0$

$x > 0$

(a)   (b)   (c)

FIGURE 5.2   **FIGURE 5.3**

DE of Free Undamped Motion

By dividing (1) by the mass m we obtain the second-order differential equation $d^2x/dt^2 + (k/m)x = 0$ or

$$\frac{d^2x}{dt^2} + \omega^2 x = 0, \tag{2}$$

where $\omega^2 = k/m$. Equation (2) is said to describe **simple harmonic motion** or **free undamped motion.** Two obvious initial conditions associated with (2) are $x(0) = \alpha$, the amount of initial displacement, and $x'(0) = \beta$, the initial velocity of the mass. For example, if $\alpha > 0$, $\beta < 0$, the mass starts from a point *below* the equilibrium position with an imparted *upward* velocity. If $\alpha < 0$, $\beta = 0$, the mass is released from *rest* from a point $|\alpha|$ units *above* the equilibrium position, and so on.

Solution and Equation of Motion To solve equation (2) we note that the solutions of the auxiliary equation $m^2 + \omega^2 = 0$ are the complex numbers $m_1 = \omega i$, $m_2 = -\omega i$. Thus from (8) of Section 4.3 we find the general solution of (2) to be

$$x(t) = c_1 \cos \omega t + c_2 \sin \omega t. \tag{3}$$

The **period** of free vibrations described by (3) is $T = 2\pi/\omega$, and the **frequency** is $f = 1/T = \omega/2\pi$. For example, for $x(t) = 2 \cos 3t - 4 \sin 3t$ the period is $2\pi/3$ and the frequency is $3/2\pi$. The former number means that the graph of $x(t)$ repeats every $2\pi/3$ units; the latter number means that there are 3 cycles of the graph every 2π units or, equivalently, that the mass undergoes $3/2\pi$ complete vibrations per unit time. In addition, it can be shown that the period $2\pi/\omega$ is the time interval between two successive maxima of $x(t)$. Keep in mind that a maximum of $x(t)$ is a positive displacement corresponding to the mass's attaining a maximum distance *below* the equilibrium position, whereas a minimum of $x(t)$ is a negative displacement corresponding to the mass's attaining a maximum height *above* the equilibrium position. We refer to either case as an **extreme displacement** of the mass. Finally, when the initial conditions are used to determine the constants c_1 and c_2 in (3), we say that the resulting particular solution or response is the **equation of motion.**

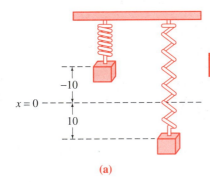

(a)

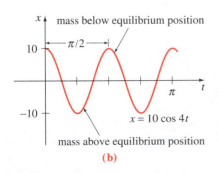

mass below equilibrium position

$x = 10 \cos 4t$

mass above equilibrium position

(b)

FIGURE 5.4

EXAMPLE 1 **Interpretation of an IVP**

Solve and interpret the initial-value problem

$$\frac{d^2x}{dt^2} + 16x = 0, \qquad x(0) = 10, \qquad x'(0) = 0.$$

SOLUTION The problem is equivalent to pulling a mass on a spring down 10 units below the equilibrium position, holding it until $t = 0$, and then releasing it from rest. Applying the initial conditions to the solution

$$x(t) = c_1 \cos 4t + c_2 \sin 4t$$

gives $x(0) = 10 = c_1 \cdot 1 + c_2 \cdot 0$ so that $c_1 = 10$. Hence

$$x(t) = 10 \cos 4t + c_2 \sin 4t.$$

From $x'(t) = -40 \sin 4t + 4c_2 \cos 4t$ we see that $x'(0) = 0 = 4c_2 \cdot 1$ and so $c_2 = 0$. Therefore the equation of motion is $x(t) = 10 \cos 4t$.

The solution clearly shows that once the system is set in motion, it stays in motion, with the mass bouncing back and forth 10 units on either side of the equilibrium position $x = 0$. As shown in Figure 5.4(b), the period of oscillation is $2\pi/4 = \pi/2$ s. ∎

EXAMPLE 2 **Free Undamped Motion**

A mass weighing 2 pounds stretches a spring 6 inches. At $t = 0$ the mass is released from a point 8 inches below the equilibrium position with an upward velocity of $\frac{4}{3}$ ft/s. Determine the equation of free motion.

SOLUTION   Because we are using the engineering system of units, the measurements given in terms of inches must be converted into feet: 6 in. $= \frac{1}{2}$ ft; 8 in. $= \frac{2}{3}$ ft. In addition, we must convert the units of weight given in pounds into units of mass. From $m = W/g$ we have $m = \frac{2}{32} = \frac{1}{16}$ slug. Also, from Hooke's law, $2 = k(\frac{1}{2})$ implies that the spring constant is $k = 4$ lb/ft. Hence (1) gives

$$\frac{1}{16}\frac{d^2x}{dt^2} = -4x \qquad \text{or} \qquad \frac{d^2x}{dt^2} + 64x = 0.$$

The initial displacement and initial velocity are $x(0) = \frac{2}{3}$, $x'(0) = -\frac{4}{3}$, where the negative sign in the last condition is a consequence of the fact that the mass is given an initial velocity in the negative, or upward, direction.

Now $\omega^2 = 64$ or $\omega = 8$, so that the general solution of the differential equation is

$$x(t) = c_1 \cos 8t + c_2 \sin 8t. \tag{4}$$

Applying the initial conditions to $x(t)$ and $x'(t)$ gives $c_1 = \frac{2}{3}$ and $c_2 = -\frac{1}{6}$. Thus the equation of motion is

$$x(t) = \frac{2}{3}\cos 8t - \frac{1}{6}\sin 8t. \tag{5} \blacksquare$$

Alternative Form of $x(t)$ When $c_1 \neq 0$ and $c_2 \neq 0$, the actual **amplitude** A of free vibrations is not obvious from inspection of equation (3). For example, although the mass in Example 2 is initially displaced $\frac{2}{3}$ foot beyond the equilibrium position, the amplitude of vibrations is a number larger than $\frac{2}{3}$. Hence it is often convenient to convert a solution of form (3) to the simpler form

$$x(t) = A\sin(\omega t + \phi), \tag{6}$$

where $A = \sqrt{c_1^2 + c_2^2}$ and ϕ is a **phase angle** defined by

$$\left.\begin{aligned} \sin\phi &= \frac{c_1}{A} \\ \cos\phi &= \frac{c_2}{A} \end{aligned}\right\} \quad \tan\phi = \frac{c_1}{c_2}. \tag{7}$$

To verify this we expand (6) by the addition formula for the sine function:

$$A\sin\omega t\cos\phi + A\cos\omega t\sin\phi = (A\sin\phi)\cos\omega t + (A\cos\phi)\sin\omega t. \tag{8}$$

It follows from Figure 5.5 that if ϕ is defined by

$$\sin\phi = \frac{c_1}{\sqrt{c_1^2 + c_2^2}} = \frac{c_1}{A}, \qquad \cos\phi = \frac{c_2}{\sqrt{c_1^2 + c_2^2}} = \frac{c_2}{A},$$

then (8) becomes

$$A\frac{c_1}{A}\cos\omega t + A\frac{c_2}{A}\sin\omega t = c_1\cos\omega t + c_2\sin\omega t = x(t).$$

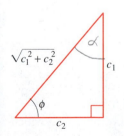

$\sqrt{c_1^2 + c_2^2}$

c_1

c_2

ϕ

α

FIGURE 5.5

EXAMPLE 3   **Alternative Form of Solution (5)**

In view of the foregoing discussion, we can write the solution (5) in Example 2 as

$$x(t) = \frac{2}{3}\cos 8t - \frac{1}{6}\sin 8t \qquad \text{or, alternatively,} \qquad x(t) = A\sin(8t + \phi).$$

The amplitude is given by

$$A = \sqrt{\left(\frac{2}{3}\right)^2 + \left(-\frac{1}{6}\right)^2} = \frac{\sqrt{17}}{6} \approx 0.69 \text{ ft.}$$

You should exercise some care when computing the phase angle ϕ defined by (7). With $c_1 = \frac{2}{3}$ and $c_2 = -\frac{1}{6}$ we find $\tan\phi = -4$, and a calculator then gives $\tan^{-1}(-4) = -1.326$ rad.* But this angle is located in the fourth quadrant and therefore contradicts the fact that $\sin\phi > 0$ and $\cos\phi < 0$ (recall that $c_1 > 0$ and $c_2 < 0$). Hence we must take ϕ to be the second-quadrant angle $\phi = \pi + (-1.326) = 1.816$ rad. Thus we have

$$x(t) = \frac{\sqrt{17}}{6}\sin(8t + 1.816). \tag{9} \quad \blacksquare$$

Form (6) is very useful since it is easy to find values of time for which the graph of $x(t)$ crosses the positive t-axis (the line $x = 0$). We observe that $\sin(\omega t + \phi) = 0$ when $\omega t + \phi = n\pi$, where n is a nonnegative integer.

Systems with Variable Spring Constants   In the model discussed above we assumed an ideal world, a world in which the physical characteristics of the spring do not change over time. In the non-ideal world, however, it seems reasonable to expect that when a spring/mass system is in motion for a long period, the spring will weaken; in other words, the "spring constant" will vary—or, more specifically, decay—with time. In one model for the **aging spring** the spring constant k in (1) is replaced by the decreasing function $K(t) = ke^{-\alpha t}$, $k > 0$, $\alpha > 0$. The differential equation $mx'' + ke^{-\alpha t}x = 0$ cannot be solved by the methods considered in Chapter 4. Nevertheless, we can obtain two linearly independent solutions using the methods in Chapter 6. See Problem 15 in Exercises 5.1, Example 3 in Section 6.4, and Problems 39 and 40 in Exercises 6.4.

When a spring/mass system is subjected to an environment in which the temperature is rapidly decreasing, it might make sense to replace the constant k with $K(t) = kt$, $k > 0$, a function that increases with time. The resulting model, $mx'' + ktx = 0$, is a form of **Airy's differential equation.** Like the equation for an aging spring, Airy's equation can be solved by the methods of Chapter 6. See Problem 16 in Exercises 5.1, Example 4 in Section 6.2, and Problems 41–43 in Exercises 6.4.

---

*The range of the inverse tangent is $-\pi/2 < \tan^{-1}x < \pi/2$.

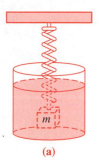

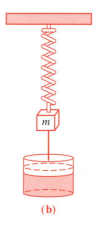

FIGURE 5.6

5.1.2 Spring/Mass Systems: Free Damped Motion

The concept of free harmonic motion is somewhat unrealistic since the motion described by equation (1) assumes that there are no retarding forces acting on the moving mass. Unless the mass is suspended in a perfect vacuum, there will be at least a resisting force due to the surrounding medium. As Figure 5.6 shows, the mass could be suspended in a viscous medium or connected to a dashpot damping device.

DE of Free Damped Motion In the study of mechanics, damping forces acting on a body are considered to be proportional to a power of the instantaneous velocity. In particular, we shall assume throughout the subsequent discussion that this force is given by a constant multiple of dx/dt. When no other external forces are impressed on the system, it follows from Newton's second law that

$$m\frac{d^2x}{dt^2} = -kx - \beta\frac{dx}{dt}, \qquad (10)$$

where β is a positive *damping constant* and the negative sign is a consequence of the fact that the damping force acts in a direction opposite to the motion.

Dividing (10) by the mass m, we find the differential equation of **free damped motion** is $d^2x/dt^2 + (\beta/m)dx/dt + (k/m)x = 0$ or

$$\frac{d^2x}{dt^2} + 2\lambda\frac{dx}{dt} + \omega^2x = 0, \qquad (11)$$

where

$$2\lambda = \frac{\beta}{m}, \quad \omega^2 = \frac{k}{m}. \qquad (12)$$

The symbol 2λ is used only for algebraic convenience since the auxiliary equation is $m^2 + 2\lambda m + \omega^2 = 0$ and the corresponding roots are then

$$m_1 = -\lambda + \sqrt{\lambda^2 - \omega^2}, \qquad m_2 = -\lambda - \sqrt{\lambda^2 - \omega^2}.$$

We can now distinguish three possible cases depending on the algebraic sign of $\lambda^2 - \omega^2$. Since each solution contains the *damping factor* $e^{-\lambda t}$, $\lambda > 0$, the displacements of the mass become negligible for large time.

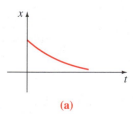

(a)

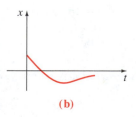

(b)

FIGURE 5.7

CASE I: $\lambda^2 - \omega^2 > 0$. In this situation the system is said to be **overdamped** since the damping coefficient β is large when compared to the spring constant k. The corresponding solution of (11) is $x(t) = c_1e^{m_1t} + c_2e^{m_2t}$ or

$$x(t) = e^{-\lambda t}(c_1e^{\sqrt{\lambda^2 - \omega^2}\,t} + c_2e^{-\sqrt{\lambda^2 - \omega^2}\,t}). \qquad (13)$$

This equation represents a smooth and nonoscillatory motion. Figure 5.7 shows two possible graphs of $x(t)$.

Case II: $\lambda^2 - \omega^2 = 0$. The system is said to be **critically damped** since any slight decrease in the damping force would result in oscillatory motion. The general solution of (11) is $x(t) = c_1e^{m_1t} + c_2te^{m_1t}$ or

$$x(t) = e^{-\lambda t}(c_1 + c_2t). \qquad (14)$$

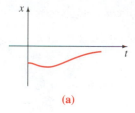

(a)

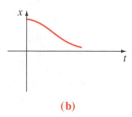

(b)

FIGURE 5.8

FIGURE 5.9

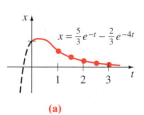

$x = \frac{5}{3}e^{-t} - \frac{2}{3}e^{-4t}$

(a)

t	$x(t)$
1	0.601
1.5	0.370
2	0.225
2.5	0.137
3	0.083

(b)

FIGURE 5.10

Some graphs of typical motion are given in Figure 5.8. Notice that the motion is quite similar to that of an overdamped system. It is also apparent from (14) that the mass can pass through the equilibrium position at most one time.

CASE III: $\lambda^2 - \omega^2 < 0$.   In this case the system is said to be **underdamped** since the damping coefficient is small compared to the spring constant. The roots m_1 and m_2 are now complex:

$$m_1 = -\lambda + \sqrt{\omega^2 - \lambda^2}\, i, \qquad m_2 = -\lambda - \sqrt{\omega^2 - \lambda^2}\, i.$$

Thus the general solution of equation (11) is

$$x(t) = e^{-\lambda t}(c_1 \cos \sqrt{\omega^2 - \lambda^2}\, t + c_2 \sin \sqrt{\omega^2 - \lambda^2}\, t). \tag{15}$$

As indicated in Figure 5.9, the motion described by (15) is oscillatory; but because of the coefficient $e^{-\lambda t}$, the amplitudes of vibration $\to 0$ as $t \to \infty$.

EXAMPLE 4   **Overdamped Motion**

It is readily verified that the solution of the initial-value problem

$$\frac{d^2x}{dt^2} + 5\frac{dx}{dt} + 4x = 0, \qquad x(0) = 1, \quad x'(0) = 1$$

is

$$x(t) = \frac{5}{3}e^{-t} - \frac{2}{3}e^{-4t}. \tag{16}$$

The problem can be interpreted as representing the overdamped motion of a mass on a spring. The mass starts from a position 1 unit *below* the equilibrium position with a *downward* velocity of 1 ft/s.

To graph $x(t)$ we find the value of t for which the function has an extremum—that is, the value of time for which the first derivative (velocity) is zero. Differentiating (16) gives $x'(t) = -\frac{5}{3}e^{-t} + \frac{8}{3}e^{-4t}$ so that $x'(t) = 0$ implies $e^{3t} = \frac{8}{5}$ or $t = \frac{1}{3}\ln\frac{8}{5} = 0.157$. It follows from the first derivative test, as well as our physical intuition, that $x(0.157) = 1.069$ ft is actually a maximum. In other words, the mass attains an extreme displacement of 1.069 feet below the equilibrium position.

We should also check to see whether the graph crosses the t-axis, that is, whether the mass passes through the equilibrium position. This cannot happen in this instance since the equation $x(t) = 0$, or $e^{3t} = \frac{2}{5}$, has the physically irrelevant solution $t = \frac{1}{3}\ln\frac{2}{5} = -0.305$.

The graph of $x(t)$, along with some other pertinent data, is given in Figure 5.10.   ∎

EXAMPLE 5   **Critically Damped Motion**

An 8-pound weight stretches a spring 2 feet. Assuming that a damping force numerically equal to 2 times the instantaneous velocity acts on the system, determine the equation of motion if the weight is released from the equilibrium position with an upward velocity of 3 ft/s.

SOLUTION From Hooke's law we see that $8 = k(2)$ gives $k = 4$ lb/ft and that $W = mg$ gives $m = \frac{8}{32} = \frac{1}{4}$ slug. The differential equation of motion is then

$$\frac{1}{4}\frac{d^2x}{dt^2} = -4x - 2\frac{dx}{dt} \quad \text{or} \quad \frac{d^2x}{dt^2} + 8\frac{dx}{dt} + 16x = 0. \quad \textbf{(17)}$$

The auxiliary equation for (17) is $m^2 + 8m + 16 = (m + 4)^2 = 0$ so that $m_1 = m_2 = -4$. Hence the system is critically damped and

$$x(t) = c_1 e^{-4t} + c_2 t e^{-4t}. \quad \textbf{(18)}$$

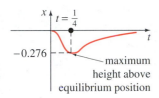

FIGURE 5.11

Applying the initial conditions $x(0) = 0$ and $x'(0) = -3$, we find, in turn, that $c_1 = 0$ and $c_2 = -3$. Thus the equation of motion is

$$x(t) = -3te^{-4t}. \quad \textbf{(19)}$$

To graph $x(t)$ we proceed as in Example 4. From $x'(t) = -3e^{-4t}(1 - 4t)$ we see that $x'(t) = 0$ when $t = \frac{1}{4}$. The corresponding extreme displacement is $x(\frac{1}{4}) = -3(\frac{1}{4})e^{-1} = -0.276$ ft. As shown in Figure 5.11, we interpret this value to mean that the weight reaches a maximum height of 0.276 foot above the equilibrium position. ∎

EXAMPLE 6   **Underdamped Motion**

A 16-pound weight is attached to a 5-foot-long spring. At equilibrium the spring measures 8.2 feet. If the weight is pushed up and released from rest at a point 2 feet above the equilibrium position, find the displacements $x(t)$ if it is further known that the surrounding medium offers a resistance numerically equal to the instantaneous velocity.

SOLUTION The elongation of the spring after the weight is attached is $8.2 - 5 = 3.2$ ft, so it follows from Hooke's law that $16 = k(3.2)$ or $k = 5$ lb/ft. In addition, $m = \frac{16}{32} = \frac{1}{2}$ slug so that the differential equation is given by

$$\frac{1}{2}\frac{d^2x}{dt^2} = -5x - \frac{dx}{dt} \quad \text{or} \quad \frac{d^2x}{dt^2} + 2\frac{dx}{dt} + 10x = 0. \quad \textbf{(20)}$$

Proceeding, we find that the roots of $m^2 + 2m + 10 = 0$ are $m_1 = -1 + 3i$ and $m_2 = -1 - 3i$, which then implies the system is underdamped and

$$x(t) = e^{-t}(c_1 \cos 3t + c_2 \sin 3t). \quad \textbf{(21)}$$

Finally, the initial conditions $x(0) = -2$ and $x'(0) = 0$ yield $c_1 = -2$ and $c_2 = -\frac{2}{3}$, so the equation of motion is

$$x(t) = e^{-t}\left(-2 \cos 3t - \frac{2}{3} \sin 3t\right). \quad \textbf{(22)} \quad ∎$$

Alternative Form of $x(t)$ In a manner identical to the procedure used on page 173, we can write any solution

$$x(t) = e^{-\lambda t}(c_1 \cos \sqrt{\omega^2 - \lambda^2}\, t + c_2 \sin \sqrt{\omega^2 - \lambda^2}\, t)$$

in the alternative form

$$x(t) = Ae^{-\lambda t}\sin(\sqrt{\omega^2 - \lambda^2}\,t + \phi), \qquad (23)$$

where $A = \sqrt{c_1{}^2 + c_2{}^2}$ and the phase angle ϕ is determined from the equations

$$\sin\phi = \frac{c_1}{A}, \quad \cos\phi = \frac{c_2}{A}, \quad \tan\phi = \frac{c_1}{c_2}.$$

The coefficient $Ae^{-\lambda t}$ is sometimes called the **damped amplitude** of vibrations. Because (23) is not a periodic function, the number $2\pi/\sqrt{\omega^2 - \lambda^2}$ is called the **quasi period** and $\sqrt{\omega^2 - \lambda^2}/2\pi$ is the **quasi frequency.** The quasi period is the time interval between two successive maxima of $x(t)$. You should verify, for the equation of motion in Example 6, that $A = 2\sqrt{10}/3$ and $\phi = 4.391$. Therefore an equivalent form of (22) is

$$x(t) = \frac{2\sqrt{10}}{3}\,e^{-t}\sin(3t + 4.391).$$

5.1.3 Spring/Mass Systems: Driven Motion

DE of Driven Motion with Damping Suppose we now take into consideration an external force $f(t)$ acting on a vibrating mass on a spring. For example, $f(t)$ could represent a driving force causing an oscillatory vertical motion of the support of the spring. See Figure 5.12. The inclusion of $f(t)$ in the formulation of Newton's second law gives the differential equation of **driven** or **forced motion:**

$$m\frac{d^2x}{dt^2} = -kx - \beta\frac{dx}{dt} + f(t). \qquad (24)$$

Dividing (24) by m gives

$$\frac{d^2x}{dt^2} + 2\lambda\frac{dx}{dt} + \omega^2 x = F(t) \qquad (25)$$

where $F(t) = f(t)/m$ and, as in the preceding section, $2\lambda = \beta/m$, $\omega^2 = k/m$. To solve the latter nonhomogeneous equation we can use either the method of undetermined coefficients or variation of parameters.

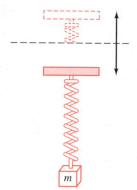

FIGURE 5.12

EXAMPLE 7   **Interpretation of an Initial-Value Problem**

Interpret and solve the initial-value problem

$$\frac{1}{5}\frac{d^2x}{dt^2} + 1.2\frac{dx}{dt} + 2x = 5\cos 4t, \qquad x(0) = \frac{1}{2}, \quad x'(0) = 0. \qquad (26)$$

SOLUTION   We can interpret the problem to represent a vibrational system consisting of a mass ($m = \frac{1}{5}$ slug or kilogram) attached to a spring ($k = 2$ lb/ft or N/m). The mass is released from rest $\frac{1}{2}$ unit (foot or meter) below the equilibrium position. The motion is damped ($\beta = 1.2$) and is being driven by an external periodic ($T = \pi/2$ s) force beginning at $t = 0$. Intuitively we would expect that even with damping the system would remain in motion until such time as the forcing function was "turned off," in which case the amplitudes would diminish. However, as the problem is given, $f(t) = 5 \cos 4t$ will remain "on" forever.

We first multiply the differential equation in (26) by 5 and solve

$$\frac{dx^2}{dt^2} + 6\frac{dx}{dt} + 10x = 0$$

by the usual methods. Since $m_1 = -3 + i$, $m_2 = -3 - i$, it follows that

$$x_c(t) = e^{-3t}(c_1 \cos t + c_2 \sin t).$$

Using the method of undetermined coefficients, we assume a particular solution of the form $x_p(t) = A \cos 4t + B \sin 4t$. Now

$$x_p' = -4A \sin 4t + 4B \cos 4t, \qquad x_p'' = -16A \cos 4t - 16B \sin 4t$$

so that

$$x_p'' + 6x_p' + 10x_p = (-6A + 24B) \cos 4t + (-24A - 6B) \sin 4t = 25 \cos 4t.$$

The resulting system of equations

$$-6A + 24B = 25, \qquad -24A - 6B = 0$$

yields $A = -\frac{25}{102}$ and $B = \frac{50}{51}$. It follows that

$$x(t) = e^{-3t}(c_1 \cos t + c_2 \sin t) - \frac{25}{102} \cos 4t + \frac{50}{51} \sin 4t. \qquad \textbf{(27)}$$

When we set $t = 0$ in the above equation, we obtain $c_1 = \frac{38}{51}$. By differentiating the expression and then setting $t = 0$, we also find that $c_2 = -\frac{86}{51}$. Therefore the equation of motion is

$$x(t) = e^{-3t}\left(\frac{38}{51} \cos t - \frac{86}{51} \sin t\right) - \frac{25}{102} \cos 4t + \frac{50}{51} \sin 4t. \qquad \textbf{(28)} \quad \blacksquare$$

Transient and Steady-State Terms   Notice that the complementary function

$$x_c(t) = e^{-3t}\left(\frac{38}{51} \cos t - \frac{86}{51} \sin t\right)$$

in (28) possesses the property that $\lim_{t\to\infty} x_c(t) = 0$. Since $x_c(t)$ becomes negligible (namely, $\to 0$) as $t \to \infty$, it is said to be a **transient term** or **transient solution.** Thus for large time the displacements of the weight in the preceding problem are closely approximated by the particular solution $x_p(t)$. This latter function is also called the **steady-state solution.** When F is a periodic function, such as $F(t) = F_0 \sin \gamma t$ or $F(t) = F_0 \cos \gamma t$, the general solution of (25) consists of

$$x(t) = \textit{transient} + \textit{steady-state.}$$

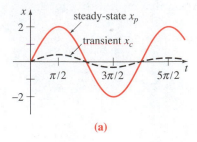

(a)

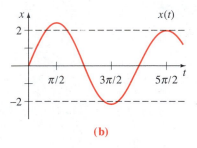

(b)

FIGURE 5.13

EXAMPLE 8   **Transient/Steady-State Solutions**

It is readily shown that the solution of the initial-value problem

$$\frac{d^2x}{dt^2} + 2\frac{dx}{dt} + 2x = 4\cos t + 2\sin t, \qquad x(0) = 0, \quad x'(0) = 3$$

is

$$x = x_c + x_p = e^{-t}\sin t + 2\sin t.$$

$$\underbrace{\phantom{e^{-t}\sin t}}_{\text{transient}}\ \underbrace{}_{\text{steady-state}}$$

Inspection of Figure 5.13 shows that the effect of the transient term on the solution is, in this case, negligible for about $t > 2\pi$. ∎

DE of Driven Motion Without Damping   With a periodic impressed force and no damping force, there is no transient term in the solution of a problem. Also, we shall see that a periodic impressed force with a frequency near or the same as the frequency of free undamped vibrations can cause a severe problem in any oscillatory mechanical system.

EXAMPLE 9   **Undamped Forced Motion**

Solve the initial-value problem

$$\frac{d^2x}{dt^2} + \omega^2 x = F_0 \sin \gamma t, \qquad x(0) = 0, \quad x'(0) = 0, \qquad \textbf{(29)}$$

where F_0 is a constant and $\gamma \neq \omega$.

SOLUTION   The complementary function is $x_c(t) = c_1 \cos \omega t + c_2 \sin \omega t$. To obtain a particular solution we assume $x_p(t) = A \cos \gamma t + B \sin \gamma t$ so that

$$x_p'' + \omega^2 x_p = A(\omega^2 - \gamma^2) \cos \gamma t + B(\omega^2 - \gamma^2) \sin \gamma t = F_0 \sin \gamma t.$$

Equating coefficients immediately gives $A = 0$ and $B = F_0/(\omega^2 - \gamma^2)$. Therefore

$$x_p(t) = \frac{F_0}{\omega^2 - \gamma^2} \sin \gamma t.$$

Applying the given initial conditions to the general solution

$$x(t) = c_1 \cos \omega t + c_2 \sin \omega t + \frac{F_0}{\omega^2 - \gamma^2} \sin \gamma t$$

yields $c_1 = 0$ and $c_2 = -\gamma F_0/\omega(\omega^2 - \gamma^2)$. Thus the solution is

$$x(t) = \frac{F_0}{\omega(\omega^2 - \gamma^2)}(-\gamma \sin \omega t + \omega \sin \gamma t), \quad \gamma \neq \omega. \qquad \textbf{(30)} \ ∎$$

Pure Resonance   Although equation (30) is not defined for $\gamma = \omega$, it is interesting to observe that its limiting value as $\gamma \to \omega$ can be obtained by applying L'Hôpital's rule. This limiting process is analogous to "tuning in" the frequency of the driving force $(\gamma/2\pi)$ to the frequency of free vibrations $(\omega/2\pi)$. Intuitively we expect that over a length of time we should be able to substantially increase the amplitudes of vibration. For $\gamma = \omega$ we define the solution to be

$$x(t) = \lim_{\gamma \to \omega} F_0 \frac{-\gamma \sin \omega t + \omega \sin \gamma t}{\omega(\omega^2 - \gamma^2)} = F_0 \lim_{\gamma \to \omega} \frac{\dfrac{d}{d\gamma}(-\gamma \sin \omega t + \omega \sin \gamma t)}{\dfrac{d}{d\gamma}(\omega^3 - \omega\gamma^2)}$$

$$= F_0 \lim_{\gamma \to \omega} \frac{-\sin \omega t + \omega t \cos \gamma t}{-2\omega\gamma}$$

$$= F_0 \frac{-\sin \omega t + \omega t \cos \omega t}{-2\omega^2}$$

$$= \frac{F_0}{2\omega^2} \sin \omega t - \frac{F_0}{2\omega} t \cos \omega t. \qquad \textbf{(31)}$$

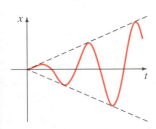

FIGURE 5.14

As suspected, when $t \to \infty$ the displacements become large; in fact, $|x(t_n)| \to \infty$ when $t_n = n\pi/\omega$, $n = 1, 2, \ldots$. The phenomenon we have just described is known as **pure resonance.** The graph given in Figure 5.14 shows typical motion in this case.

In conclusion it should be noted that there is no actual need to use a limiting process on (30) to obtain the solution for $\gamma = \omega$. Alternatively, equation (31) follows by solving the initial-value problem

$$\frac{d^2x}{dt^2} + \omega^2 x = F_0 \sin \omega t, \qquad x(0) = 0, \quad x'(0) = 0$$

directly by conventional methods.

If the displacements of a spring/mass system were actually described by a function such as (31), the system would necessarily fail. Large oscillations of the mass would eventually force the spring beyond its elastic limit. One might argue too that the resonating model presented in Figure 5.14 is completely unrealistic, because it ignores the retarding effects of ever-present damping forces. Although it is true that pure resonance cannot occur when the smallest amount of damping is taken into consideration, large and equally destructive amplitudes of vibration (although bounded as $t \to \infty$) can occur. See Problem 43 in Exercises 5.1.

5.1.4 Analogous Systems

LRC **Series Circuits** As mentioned in the introduction to this chapter, many different physical systems can be described by a linear second-order differential equation similar to the differential equation of forced motion with damping:

$$m\frac{d^2x}{dt^2} + \beta\frac{dx}{dt} + kx = f(t). \qquad \textbf{(32)}$$

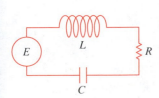

FIGURE 5.15

If $i(t)$ denotes current in the *LRC* **series electrical circuit** shown in Figure 5.15, then the voltage drops across the inductor, resistor, and capacitor are as shown in Figure 1.13. By Kirchhoff's second law, the sum of these voltages equals the voltage $E(t)$ impressed on the circuit; that is,

$$L\frac{di}{dt} + Ri + \frac{1}{C}q = E(t). \qquad \textbf{(33)}$$

But the charge $q(t)$ on the capacitor is related to the current $i(t)$ by $i = dq/dt$, and so (33) becomes the linear second-order differential equation

$$L\frac{d^2q}{dt^2} + R\frac{dq}{dt} + \frac{1}{C}q = E(t). \tag{34}$$

The nomenclature used in the analysis of circuits is similar to that used to describe spring-mass systems.

If $E(t) = 0$, the **electrical vibrations** of the circuit are said to be **free**. Since the auxiliary equation for (34) is $Lm^2 + Rm + 1/C = 0$, there will be three forms of the solution with $R \neq 0$, depending on the value of the discriminant $R^2 - 4L/C$. We say that the circuit is

overdamped if $\qquad R^2 - 4L/C > 0$,

critically damped if $\qquad R^2 - 4L/C = 0$,

and $\qquad$ underdamped if $\qquad R^2 - 4L/C < 0$.

In each of these three cases the general solution of (34) contains the factor $e^{-Rt/2L}$, and so $q(t) \to 0$ as $t \to \infty$. In the underdamped case when $q(0) = q_0$, the charge on the capacitor oscillates as it decays; in other words, the capacitor is charging and discharging as $t \to \infty$. When $E(t) = 0$ and $R = 0$, the circuit is said to be undamped and the electrical vibrations do not approach zero as t increases without bound; the response of the circuit is **simple harmonic**.

EXAMPLE 10   Underdamped Series Circuit

Find the charge $q(t)$ on the capacitor in an LRC series circuit when $L = 0.25$ henry (h), $R = 10$ ohms (Ω), $C = 0.001$ farad (f), $E(t) = 0$, $q(0) = q_0$ coulombs (C), and $i(0) = 0$.

SOLUTION   Since $1/C = 1000$, equation (34) becomes

$$\frac{1}{4}q'' + 10q' + 1000q = 0 \qquad \text{or} \qquad q'' + 40q' + 4000q = 0.$$

Solving this homogeneous equation in the usual manner, we find that the circuit is underdamped and $q(t) = e^{-20t}(c_1 \cos 60t + c_2 \sin 60t)$. Applying the initial conditions, we find $c_1 = q_0$ and $c_2 = q_0/3$. Thus

$$q(t) = q_0 e^{-20t}\left(\cos 60t + \frac{1}{3}\sin 60t\right).$$

Using (23), we can write the foregoing solution as

$$q(t) = \frac{q_0\sqrt{10}}{3}e^{-20t}\sin(60t + 1.249). \qquad \blacksquare$$

When there is an impressed voltage $E(t)$ on the circuit, the electrical vibrations are said to be **forced**. In the case when $R \neq 0$, the complementary function $q_c(t)$ of (34) is called a **transient solution**. If $E(t)$ is periodic or a constant, then the particular solution $q_p(t)$ of (34) is a **steady-state solution**.

EXAMPLE 11 **Steady-State Current**

Find the steady-state solution $q_p(t)$ and the **steady-state current** in an *LRC* series circuit when the impressed voltage is $E(t) = E_0 \sin \gamma t$.

SOLUTION The steady-state solution $q_p(t)$ is a particular solution of the differential equation

$$L \frac{d^2 q}{dt^2} + R \frac{dq}{dt} + \frac{1}{C} q = E_0 \sin \gamma t.$$

Using the method of undetermined coefficients, we assume a particular solution of the form $q_p(t) = A \sin \gamma t + B \cos \gamma t$. Substituting this expression into the differential equation, simplifying, and equating coefficients gives

$$A = \frac{E_0 \left(L\gamma - \dfrac{1}{C\gamma} \right)}{-\gamma \left(L^2 \gamma^2 - \dfrac{2L}{C} + \dfrac{1}{C^2 \gamma^2} + R^2 \right)}, \quad B = \frac{E_0 R}{-\gamma \left(L^2 \gamma^2 - \dfrac{2L}{C} + \dfrac{1}{C^2 \gamma^2} + R^2 \right)}.$$

It is convenient to express A and B in terms of some new symbols.

If $\qquad X = L\gamma - \dfrac{1}{C\gamma}, \qquad$ then $\quad X^2 = L^2 \gamma^2 - \dfrac{2L}{C} + \dfrac{1}{C^2 \gamma^2}.$

If $\qquad Z = \sqrt{X^2 + R^2}, \quad$ then $\quad Z^2 = L^2 \gamma^2 - \dfrac{2L}{C} + \dfrac{1}{C^2 \gamma^2} + R^2.$

Therefore $A = E_0 X/(-\gamma Z^2)$ and $B = E_0 R/(-\gamma Z^2)$, so the steady-state charge is

$$q_p(t) = -\frac{E_0 X}{\gamma Z^2} \sin \gamma t - \frac{E_0 R}{\gamma Z^2} \cos \gamma t.$$

Now the steady-state current is given by $i_p(t) = q_p'(t)$:

$$i_p(t) = \frac{E_0}{Z} \left(\frac{R}{Z} \sin \gamma t - \frac{X}{Z} \cos \gamma t \right). \tag{35} \quad \blacksquare$$

The quantities $X = L\gamma - 1/C\gamma$ and $Z = \sqrt{X^2 + R^2}$ defined in Example 11 are called, respectively, the **reactance** and **impedance** of the circuit. Both the reactance and the impedance are measured in ohms.

Twisted Shaft The differential equation governing the torsional motion of a weight suspended from the end of an elastic shaft is

$$I \frac{d^2 \theta}{dt^2} + c \frac{d\theta}{dt} + k\theta = T(t). \tag{36}$$

As shown in Figure 5.16, the function $\theta(t)$ represents the amount of twist of the weight at any time.

By comparing equations (25) and (34) with (36) we see that, with the exception of terminology, there is absolutely no difference between the mathematics of vibrating springs, simple series circuits, and torsional vibrations.

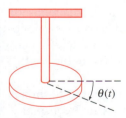

$\theta(t)$

FIGURE 5.16

Answers to odd-numbered problems begin on page A-7.

5.1.1

1. A 4-pound weight is attached to a spring whose spring constant is 16 lb/ft. What is the period of simple harmonic motion?

2. A 20-kilogram mass is attached to a spring. If the frequency of simple harmonic motion is $2/\pi$ vibrations/second, what is the spring constant k? What is the frequency of simple harmonic motion if the original mass is replaced with an 80-kilogram mass?

3. A 24-pound weight, attached to the end of a spring, stretches it 4 inches. Find the equation of motion if the weight is released from rest from a point 3 inches above the equilibrium position.

4. Determine the equation of motion if the weight in Problem 3 is released from the equilibrium position with an initial downward velocity of 2 ft/s.

5. A 20-pound weight stretches a spring 6 inches. The weight is released from rest 6 inches below the equilibrium position.
   (a) Find the position of the weight at $t = \pi/12, \pi/8, \pi/6, \pi/4, 9\pi/32$ s.
   (b) What is the velocity of the weight when $t = 3\pi/16$ s? In which direction is the weight heading at this instant?
   (c) At what times does the weight pass through the equilibrium position?

6. A force of 400 newtons stretches a spring 2 meters. A mass of 50 kilograms is attached to the end of the spring and released from the equilibrium position with an upward velocity of 10 m/s. Find the equation of motion.

7. Another spring whose constant is 20 N/m is suspended from the same rigid support but parallel to the spring/mass system in Problem 6. A mass of 20 kilograms is attached to the second spring, and both masses are released from the equilibrium position with an upward velocity of 10 m/s.
   (a) Which mass exhibits the greater amplitude of motion?
   (b) Which mass is moving faster at $t = \pi/4$ s? at $\pi/2$ s?
   (c) At what times are the two masses in the same position? Where are the masses at these times? In which directions are they moving?

8. A 32-pound weight stretches a spring 2 feet. Determine the amplitude and period of motion if the weight is released 1 foot above the equilibrium position with an initial upward velocity of 2 ft/s. How many complete vibrations will the weight have completed at the end of 4π seconds?

9. An 8-pound weight attached to a spring exhibits simple harmonic motion. Determine the equation of motion if the spring constant is 1 lb/ft and if the weight is released 6 inches below the equilibrium position with a downward velocity of $\frac{3}{2}$ ft/s. Express the solution in form (6).

10. A mass weighing 10 pounds stretches a spring $\frac{1}{4}$ foot. This mass is removed and replaced with a mass of 1.6 slugs, which is released $\frac{1}{3}$ foot above the equilibrium position with a downward velocity of $\frac{5}{4}$ ft/s. Express the solution in form (6). At what times does the mass attain a displacement below the equilibrium position numerically equal to $\frac{1}{2}$ the amplitude?

11. A 64-pound weight attached to the end of a spring stretches it 0.32 foot. From a position 8 inches above the equilibrium position the weight is given a downward velocity of 5 ft/s.
   (a) Find the equation of motion.
   (b) What are the amplitude and period of motion?
   (c) How many complete vibrations will the weight have completed at the end of 3π seconds?
   (d) At what time does the weight pass through the equilibrium position heading downward for the second time?
   (e) At what time does the weight attain its extreme displacement on either side of the equilibrium position?
   (f) What is the position of the weight at $t = 3$ s?
   (g) What is the instantaneous velocity at $t = 3$ s?
   (h) What is the acceleration at $t = 3$ s?
   (i) What is the instantaneous velocity at the times when the weight passes through the equilibrium position?
   (j) At what times is the weight 5 inches below the equilibrium position?
   (k) At what times is the weight 5 inches below the equilibrium position heading in the upward direction?

12. A mass of 1 slug is suspended from a spring whose characteristic spring constant is 9 lb/ft. Initially the mass starts from a point 1 foot above the equilibrium position with an upward velocity of $\sqrt{3}$ ft/s. Find the times for which the mass is heading downward at a velocity of 3 ft/s.

13. Under some circumstances when two parallel springs, with constants k_1 and k_2, support a single weight W, the **effective spring constant** of the system is given by $k = 4k_1k_2/(k_1 + k_2)$. A 20-pound weight stretches one spring 6 inches and another spring 2 inches. The springs are attached to a common rigid support and then to a metal plate. As shown in Figure 5.17, the 20-pound weight is attached to the center of the plate in the double-spring arrangement. Determine the effective spring constant of this system. Find the equation of motion if the weight is released from the equilibrium position with a downward velocity of 2 ft/s.

14. A certain weight stretches one spring $\frac{1}{3}$ foot and another spring $\frac{1}{2}$ foot. The two springs are attached to a common rigid support in the manner indicated in Problem 13 and Figure 5.17. The first weight is set aside, an 8-pound weight is attached to the double-spring arrangement, and the system is set in motion. If the period of motion is $\pi/15$ second, determine the numerical value of the first weight.

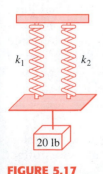

k_1 k_2

20 lb

FIGURE 5.17

Discussion Problems

15. By inspection of the differential equation only, discuss the behavior of a spring/mass system described by $4x'' + e^{-0.1t}x = 0$ over a long period of time.

16. By inspection of the differential equation only, discuss the behavior of a spring/mass system described by $4x'' + tx = 0$ over a long period of time.

5.1.2

In Problems 17–20 the given figure represents the graph of an equation of motion for a mass on a spring. The spring/mass system is damped. Use the graph to determine

(a) whether the initial displacement of the mass is above or below the equilibrium position and

(b) whether the mass is initially released from rest, heading downward, or heading upward.

17.

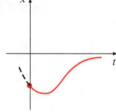

FIGURE 5.18

18.

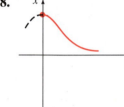

FIGURE 5.19

19.

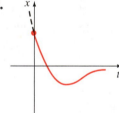

FIGURE 5.20

20.

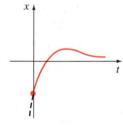

FIGURE 5.21

21. A 4-pound weight is attached to a spring whose constant is 2 lb/ft. The medium offers a resistance to the motion of the weight numerically equal to the instantaneous velocity. If the weight is released from a point 1 foot above the equilibrium position with a downward velocity of 8 ft/s, determine the time at which the weight passes through the equilibrium position. Find the time at which the weight attains its extreme displacement from the equilibrium position. What is the position of the weight at this instant?

22. A 4-foot spring measures 8 feet long after an 8-pound weight is attached to it. The medium through which the weight moves offers a resistance numerically equal to $\sqrt{2}$ times the instantaneous velocity.

Find the equation of motion if the weight is released from the equilibrium position with a downward velocity of 5 ft/s. Find the time at which the weight attains its extreme displacement from the equilibrium position. What is the position of the weight at this instant?

23. A 1-kilogram mass is attached to a spring whose constant is 16 N/m, and the entire system is then submerged in a liquid that imparts a damping force numerically equal to 10 times the instantaneous velocity. Determine the equations of motion if
    (a) the weight is released from rest 1 meter below the equilibrium position and
    (b) the weight is released 1 meter below the equilibrium position with an upward velocity of 12 m/s.

24. In parts (a) and (b) of Problem 23 determine whether the weight passes through the equilibrium position. In each case find the time at which the weight attains its extreme displacement from the equilibrium position. What is the position of the weight at this instant?

25. A force of 2 pounds stretches a spring 1 foot. A 3.2-pound weight is attached to the spring, and the system is then immersed in a medium that imparts a damping force numerically equal to 0.4 times the instantaneous velocity.
    (a) Find the equation of motion if the weight is released from rest 1 foot above the equilibrium position.
    (b) Express the equation of motion in the form given in (23).
    (c) Find the first time at which the weight passes through the equilibrium position heading upward.

26. After a 10-pound weight is attached to a 5-foot spring, the spring measures 7 feet long. The 10-pound weight is removed and replaced with an 8-pound weight, and the entire system is placed in a medium offering a resistance numerically equal to the instantaneous velocity.
    (a) Find the equation of motion if the weight is released $\frac{1}{2}$ foot below the equilibrium position with a downward velocity of 1 ft/s.
    (b) Express the equation of motion in the form given in (23).
    (c) Find the times at which the weight passes through the equilibrium position heading downward.
    (d) Graph the equation of motion.

27. A 10-pound weight attached to a spring stretches it 2 feet. The weight is attached to a dashpot damping device that offers a resistance numerically equal to β ($\beta > 0$) times the instantaneous velocity. Determine the values of the damping constant β so that the subsequent motion is (a) overdamped, (b) critically damped, and (c) underdamped.

28. A 24-pound weight stretches a spring 4 feet. The subsequent motion takes place in a medium offering a resistance numerically equal to β ($\beta > 0$) times the instantaneous velocity. If the weight starts from the equilibrium position with an upward velocity of 2 ft/s, show that if $\beta > 3\sqrt{2}$ the equation of motion is

$$x(t) = \frac{-3}{\sqrt{\beta^2 - 18}} e^{-2\beta t/3} \sinh \frac{2}{3} \sqrt{\beta^2 - 18}\, t.$$

5.1.3

29. A 16-pound weight stretches a spring $\frac{8}{3}$ feet. Initially the weight starts from rest 2 feet below the equilibrium position, and the subsequent motion takes place in a medium that offers a damping force numerically equal to $\frac{1}{2}$ the instantaneous velocity. Find the equation of motion if the weight is driven by an external force equal to $f(t) = 10 \cos 3t$.

30. A mass of 1 slug is attached to a spring whose constant is 5 lb/ft. Initially the mass is released 1 foot below the equilibrium position with a downward velocity of 5 ft/s, and the subsequent motion takes place in a medium that offers a damping force numerically equal to 2 times the instantaneous velocity.
  (a) Find the equation of motion if the mass is driven by an external force equal to $f(t) = 12 \cos 2t + 3 \sin 2t$.
  (b) Graph the transient and steady-state solutions on the same coordinate axes.
  (c) Graph the equation of motion.

31. A mass of 1 slug, when attached to a spring, stretches it 2 feet and then comes to rest in the equilibrium position. Starting at $t = 0$, an external force equal to $f(t) = 8 \sin 4t$ is applied to the system. Find the equation of motion if the surrounding medium offers a damping force numerically equal to 8 times the instantaneous velocity.

32. In Problem 31 determine the equation of motion if the external force is $f(t) = e^{-t} \sin 4t$. Analyze the displacements for $t \to \infty$.

33. When a mass of 2 kilograms is attached to a spring whose constant is 32 N/m, it comes to rest in the equilibrium position. Starting at $t = 0$, a force equal to $f(t) = 68e^{-2t} \cos 4t$ is applied to the system. Find the equation of motion in the absence of damping.

34. In Problem 33 write the equation of motion in the form $x(t) = A \sin(\omega t + \phi) + Be^{-2t} \sin(4t + \theta)$. What is the amplitude of vibrations after a very long time?

35. A mass m is attached to the end of a spring whose constant is k. After the mass reaches equilibrium, its support begins to oscillate vertically about a horizontal line L according to a formula $h(t)$. The value of h represents the distance in feet measured from L. See Figure 5.22.
  (a) Determine the differential equation of motion if the entire system moves through a medium offering a damping force numerically equal to $\beta(dx/dt)$.
  (b) Solve the differential equation in part (a) if the spring is stretched 4 feet by a weight of 16 pounds and $\beta = 2$, $h(t) = 5 \cos t$, $x(0) = x'(0) = 0$.

36. A mass of 100 grams is attached to a spring whose constant is 1600 dynes/cm. After the mass reaches equilibrium, its support oscillates according to the formula $h(t) = \sin 8t$, where h represents displacement from its original position. See Problem 35 and Figure 5.22.
  (a) In the absence of damping, determine the equation of motion if the mass starts from rest from the equilibrium position.
  (b) At what times does the mass pass through the equilibrium position?

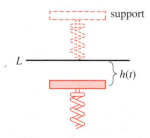

FIGURE 5.22

(c) At what times does the mass attain its extreme displacements?

(d) What are the maximum and minimum displacements?

(e) Graph the equation of motion.

In Problems 37 and 38 solve the given initial-value problem.

37. $\dfrac{d^2x}{dt^2} + 4x = -5 \sin 2t + 3 \cos 2t, \quad x(0) = -1, x'(0) = 1$

38. $\dfrac{d^2x}{dt^2} + 9x = 5 \sin 3t, \quad x(0) = 2, x'(0) = 0$

39. (a) Show that the solution of the initial-value problem

$$\frac{d^2x}{dt^2} + \omega^2 x = F_0 \cos \gamma t, \qquad x(0) = 0, \quad x'(0) = 0$$

is $\qquad\qquad x(t) = \dfrac{F_0}{\omega^2 - \gamma^2}(\cos \gamma t - \cos \omega t).$

(b) Evaluate $\displaystyle\lim_{\gamma \to \omega} \dfrac{F_0}{\omega^2 - \gamma^2}(\cos \gamma t - \cos \omega t).$

40. Compare the result obtained in part (b) of Problem 39 with the solution obtained using variation of parameters when the external force is $F_0 \cos \omega t$.

41. (a) Show that $x(t)$ given in part (a) of Problem 39 can be written in the form

$$x(t) = \frac{-2F_0}{\omega^2 - \gamma^2} \sin \frac{1}{2}(\gamma - \omega)t \sin \frac{1}{2}(\gamma + \omega)t.$$

(b) If we define $\varepsilon = \frac{1}{2}(\gamma - \omega)$, show that when ε is small an *approximate* solution is

$$x(t) = \frac{F_0}{2\varepsilon\gamma} \sin \varepsilon t \sin \gamma t.$$

When ε is small the frequency $\gamma/2\pi$ of the impressed force is close to the frequency $\omega/2\pi$ of free vibrations. When this occurs, the motion is as indicated in Figure 5.23. Oscillations of this kind are called *beats* and are due to the fact that the frequency of $\sin \varepsilon t$ is quite small in comparison to the frequency of $\sin \gamma t$. The dashed curves, or *envelope* of the graph of $x(t)$, are obtained from the graphs of $\pm(F_0/2\varepsilon\gamma) \sin \varepsilon t$. Use a graphing utility with various values of F_0, ε, and γ to verify the graph in Figure 5.23.

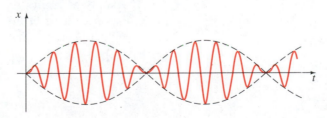

FIGURE 5.23

Discussion Problems

42. Can there be *beats* when a damping force is added to the model in part (a) of Problem 39? Defend your position with graphs obtained either from the explicit solution of the problem

$$\frac{d^2x}{dt^2} + 2\lambda\frac{dx}{dt} + \omega^2 x = F_0\cos\gamma t, \qquad x(0) = 0, \quad x'(0) = 0$$

or from solution curves obtained using an ODE solver.

43. (a) Show that the general solution of

$$\frac{d^2x}{dt^2} + 2\lambda\frac{dx}{dt} + \omega^2 x = F_0\sin\gamma t$$

is

$$x(t) = Ae^{-\lambda t}\sin\left(\sqrt{\omega^2 - \lambda^2}\,t + \phi\right) + \frac{F_0}{\sqrt{(\omega^2 - \gamma^2)^2 + 4\lambda^2\gamma^2}}\sin(\gamma t + \theta),$$

where $A = \sqrt{c_1^2 + c_2^2}$ and the phase angles ϕ and θ are, respectively, defined by $\sin\phi = c_1/A$, $\cos\phi = c_2/A$ and

$$\sin\theta = \frac{-2\lambda\gamma}{\sqrt{(\omega^2 - \gamma^2)^2 + 4\lambda^2\gamma^2}}, \qquad \cos\theta = \frac{\omega^2 - \gamma^2}{\sqrt{(\omega^2 - \gamma^2)^2 + 4\lambda^2\gamma^2}}.$$

(b) The solution in part (a) has the form $x(t) = x_c(t) + x_p(t)$. Inspection shows that $x_c(t)$ is transient, and hence, for large values of time, the solution is approximated by $x_p(t) = g(\gamma)\sin(\gamma t + \theta)$, where

$$g(\gamma) = \frac{F_0}{\sqrt{(\omega^2 - \gamma^2)^2 + 4\lambda^2\gamma^2}}.$$

Although the amplitude $g(\gamma)$ of $x_p(t)$ is bounded as $t \to \infty$, show that the maximum oscillations will occur at the value $\gamma_1 = \sqrt{\omega^2 - 2\lambda^2}$. What is the maximum value of g? The number $\sqrt{\omega^2 - 2\lambda^2}/2\pi$ is said to be the **resonance frequency** of the system.

(c) When $F_0 = 2$, $m = 1$, and $k = 4$, g becomes

$$g(\gamma) = \frac{2}{\sqrt{(4 - \gamma^2)^2 + \beta^2\gamma^2}}.$$

Construct a table of the values of γ_1 and $g(\gamma_1)$ corresponding to the damping coefficients $\beta = 2$, $\beta = 1$, $\beta = \frac{3}{4}$, $\beta = \frac{1}{2}$, and $\beta = \frac{1}{4}$. Use a graphing utility to obtain the graphs of g corresponding to these damping coefficients. Use the same coordinate axes. This family of graphs is called the **resonance curve** or **frequency response curve** of the system. What is γ_1 approaching as $\beta \to 0$? What is happening to the resonance curve as $\beta \to 0$?

44. Consider a driven undamped spring/mass system described by the initial-value problem

$$\frac{d^2x}{dt^2} + \omega^2 x = F_0\sin^n\gamma t, \qquad x(0) = 0, \quad x'(0) = 0.$$

(a) For $n = 2$, discuss why there is a single frequency $\gamma_1/2\pi$ at which the system is in pure resonance.

(b) For $n = 3$, discuss why there are two frequencies $\gamma_1/2\pi$ and $\gamma_2/2\pi$ at which the system is in pure resonance.

(c) Suppose $\omega = 1$ and $F_0 = 1$. Use an ODE solver to obtain the graph of the solution of the initial-value problem for $n = 2$ and $\gamma = \gamma_1$ in part (a). Obtain the graph of the solution of the initial-value problem for $n = 3$ corresponding, in turn, to $\gamma = \gamma_1$ and $\gamma = \gamma_2$ in part (b).

5.1.4

45. Find the charge on the capacitor in an *LRC* series circuit at $t = 0.01$ s when $L = 0.05$ h, $R = 2$ Ω, $C = 0.01$ f, $E(t) = 0$ V, $q(0) = 5$ C, and $i(0) = 0$ A. Determine the first time at which the charge on the capacitor is equal to zero.

46. Find the charge on the capacitor in an *LRC* series circuit when $L = \frac{1}{4}$ h, $R = 20$ Ω, $C = \frac{1}{300}$ f, $E(t) = 0$ V, $q(0) = 4$ C, and $i(0) = 0$ A. Is the charge on the capacitor ever equal to zero?

In Problems 47 and 48 find the charge on the capacitor and the current in the given *LRC* series circuit. Find the maximum charge on the capacitor.

47. $L = \frac{5}{3}$ h, $R = 10$ Ω, $C = \frac{1}{30}$ f, $E(t) = 300$ V, $q(0) = 0$ C, $i(0) = 0$ A.

48. $L = 1$ h, $R = 100$ Ω, $C = 0.0004$ f, $E(t) = 30$ V, $q(0) = 0$ C, $i(0) = 2$ A.

49. Find the steady-state charge and the steady-state current in an *LRC* series circuit when $L = 1$ h, $R = 2$ Ω, $C = 0.25$ f, and $E(t) = 50 \cos t$ V.

50. Show that the amplitude of the steady-state current in the *LRC* series circuit in Example 11 is given by E_0/Z, where Z is the impedance of the circuit.

51. Show that the steady-state current in an *LRC* series circuit when $L = \frac{1}{2}$ h, $R = 20$ Ω, $C = 0.001$ f, and $E(t) = 100 \sin 60t$ V is given by $i_p(t) = (4.160) \sin(60t - 0.588)$. [*Hint:* Use Problem 50.]

52. Find the steady-state current in an *LRC* series circuit when $L = \frac{1}{2}$ h, $R = 20$ Ω, $C = 0.001$ f, and $E(t) = 100 \sin 60t + 200 \cos 40t$ V.

53. Find the charge on the capacitor in an *LRC* series circuit when $L = \frac{1}{2}$ h, $R = 10$ Ω, $C = 0.01$ f, $E(t) = 150$ V, $q(0) = 1$ C, and $i(0) = 0$ A. What is the charge on the capacitor after a long time?

54. Show that if L, R, C, and E_0 are constant, then the amplitude of the steady-state current in Example 11 is a maximum when $\gamma = 1/\sqrt{LC}$. What is the maximum amplitude?

55. Show that if L, R, E_0, and γ are constant, then the amplitude of the steady-state current in Example 11 is a maximum when the capacitance is $C = 1/L\gamma^2$.

56. Find the charge on the capacitor and the current in an *LC* circuit when $L = 0.1$ h, $C = 0.1$ f, $E(t) = 100 \sin \gamma t$ V, $q(0) = 0$ C, and $i(0) = 0$ A.

57. Find the charge on the capacitor and the current in an *LC* circuit when $E(t) = E_0 \cos \gamma t$ V, $q(0) = q_0$ C, and $i(0) = i_0$ A.

58. In Problem 57 find the current when the circuit is in resonance.

5.2   LINEAR EQUATIONS: BOUNDARY-VALUE PROBLEMS

■ *DE for the deflection of a beam* ■ *Boundary conditions* ■ *Eigenvalues and eigenfunctions*
■ *Nontrivial solutions* ■ *Buckling of a thin column* ■ *Euler load* ■ *DE of a rotating string*

The preceding section was devoted to systems in which a second-order mathematical model was accompanied by prescribed initial conditions—that is, side conditions that are specified on the unknown function and its first derivative at a single point. But often the mathematical description of a physical system demands that we solve a differential equation subject to boundary conditions—that is, conditions specified on the unknown function, or on one of its derivatives, or even on a linear combination of the unknown function and one of its derivatives, at two (or more) different points.

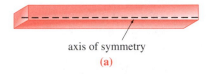

axis of symmetry
(a)

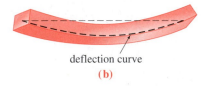

deflection curve
(b)

FIGURE 5.24

Deflection of a Beam   Many structures are constructed using girders, or beams, and these beams deflect or distort under their own weight or under the influence of some external force. As we shall now see, this deflection $y(x)$ is governed by a relatively simple linear fourth-order differential equation.

To begin, let us assume that a beam of length L is homogeneous and has uniform cross-sections along its length. In the absence of any load on the beam (including its weight), a curve joining the centroids of all its cross-sections is a straight line called the **axis of symmetry.** See Figure 5.24(a). If a load is applied to the beam in a vertical plane containing the axis of symmetry, the beam, as shown in Figure 5.24(b), undergoes a distortion, and the curve connecting the centroids of all cross-sections is called the **deflection curve** or **elastic curve.** The deflection curve approximates the shape of the beam. Now suppose that the x-axis coincides with the axis of symmetry and that the deflection $y(x)$, measured from this axis, is positive if downward. In the theory of elasticity it is shown that the bending moment $M(x)$ at a point x along the beam is related to the load per unit length $w(x)$ by the equation

$$\frac{d^2M}{dx^2} = w(x). \tag{1}$$

In addition, the bending moment $M(x)$ is proportional to the curvature κ of the elastic curve

$$M(x) = EI\kappa, \tag{2}$$

where E and I are constants; E is Young's modulus of elasticity of the material of the beam, and I is the moment of inertia of a cross-section of the beam (about an axis known as the neutral axis). The product EI is called the **flexural rigidity** of the beam.

Now, from calculus, curvature is given by $\kappa = y''/[1 + (y')^2]^{3/2}$. When the deflection $y(x)$ is small, the slope $y' \approx 0$ and so $[1 + (y')^2]^{3/2} \approx 1$. If we let $\kappa = y''$, equation (2) becomes $M = EIy''$. The second derivative of this last expression is

$$\frac{d^2M}{dx^2} = EI\frac{d^2}{dx^2}y'' = EI\frac{d^4y}{dx^4}. \tag{3}$$

DECAY OF SATELLITE ORBITS

John Ellison

Grove City College

Ever since the first Sputnik was launched into space there have been artificial satellites circling the earth. These range from small pieces of space junk to large objects like the Hubble telescope and manned space stations. For most of their lives the main force acting on these objects is the earth's gravitational field. However, drag caused by the atmosphere will cause the orbit of any satellite to decay slowly, and, left alone, the satellite will ultimately fall to earth.

Small objects burn up in the atmosphere and do not reach the earth's surface. Large objects have internal propulsion systems that can be used to maintain their orbits. However, in 1979 the internal propulsion system of one large object, Skylab, failed, and it entered the earth's atmosphere. Skylab was large enough that some small pieces survived and landed. They were very hot.

A mathematical model of the final few revolutions of the decaying orbit of a satellite is complicated and difficult to obtain. In order to get an approximation we can solve, we must make many simplifying assumptions. Two common assumptions are that the earth is a perfect sphere and that the motion of the object is essentially two dimensional. An estimation of how the drag caused by the atmosphere affects the path of the object also has to be made.

One of the most significant assumptions involves how the density of the atmosphere is modeled. This density varies considerably over the surface of the earth, depending on time of day and year, weather conditions, and even sunspot activity. The reasons for the variations in density are not particularly well understood, and for our simple model we will assume that atmospheric density depends only on altitude. Even then, finding a formula for the density is not easy. One common method is to measure the density at different altitudes experimentally and then find a curve (such as a spline) through the data. Atmospheric scientists have also devised formulas that model the density.

The equations of motion of the satellite can be derived in a manner similar

The Skylab space station photographed from Skylab 4 Command Module during the final "fly-around" before the return home.
Courtesy NASA

to that used in Chapter 5. In two dimensions, assuming the origin is at the center of the earth, they are

$$x''(t) = -\frac{m_e g x}{r^3} - kvx'$$

$$y''(t) = -\frac{m_e g y}{r^3} - kvy'$$

where $(x(t), y(t))$ is the position of the object.

In the first term on the right-hand side of each equation, m_e is the mass of the earth, g is the acceleration due to gravity, and $r = (x^2 + y^2)^{1/2}$ is the distance of the satellite from the earth's center. Note that in this term the force is inversely proportional to the square of r.

In the second term on the right-hand side of each equation, k is of the form $CA\rho/m$, where C is a proportionality constant, A is the area of the object facing the atmosphere, ρ is the density of the atmosphere, and m is the mass of the satellite; $v = [(x')^2 + (y')^2]^{1/2}$ is the velocity of the object. Hence, in this term the drag force is proportional to the square of velocity. This is a good, but not perfect, approximation.

The system of differential equations given above is highly nonlinear and cannot be solved analytically. Numerical methods and a computer must be used to get a numerical solution. Figure 1 shows the final one and a half revolutions of a decaying orbit of a satellite closely approximating Skylab. For our solution we used the numerical routines in Mathematica, which are similar to, but more sophisticated than, those presented in Section 9.5. The path of the satellite was tracked from a height of 100 kilometers with a velocity of just under 8 km/sec. We see that the satellite follows an elliptical path, spending most of the first revolution at a height greater than 100 kilometers. At the end of the first revolution the height is approximately 97 kilometers. The rate of decay then increases, and at the end of the next half orbit the height is slightly more than 80 kilometers. Then very quickly the satellite (or what remains of it) strikes the earth. This is reasonable since the density of the atmosphere increases very rapidly as the altitude decreases.

Again, it must be stressed that this analysis is based on many assumptions and simplifications. To get a more accurate path, we must improve upon these assumptions—which only makes the solution more difficult. For these reasons the point of impact of a falling satellite on the earth's surface is hard to predict. We can only hope that it is in the ocean or on uninhabited land.

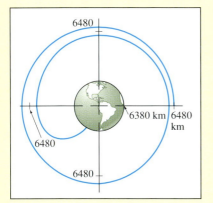

FIGURE 1

Final orbits of satellite entering earth's atmosphere

References

1. Danby, J. M. A. *Computing Applications to Differential Equations*. Reston, VA: Reston Publishing Company, 1985.
2. Heicklen, G. *Atmospheric Chemistry*. New York: Academic Press, 1976.
3. Mathematica, v 2.2.3 for Windows. Champaign, IL: Wolfram Research Inc., 1995.

About the Author

John H. Ellison received a B.A. from Whitman College, an M.A. from the University of Colorado, and a Ph.D. in Mathematics from the University of Pittsburgh. He has been a Professor of Mathematics at Grove City College for the past twenty-five years and is currently serving as Chair of the department.

A Modeling Application

THE COLLAPSE OF THE
TACOMA NARROWS SUSPENSION BRIDGE

Gilbert N. Lewis

*Michigan Technological
University*

In the summer of 1940, the Tacoma Narrows Suspension Bridge in the state of Washington was completed and opened to traffic. Almost immediately, observers noticed that the wind blowing across the roadway would sometimes set up large vertical vibrations in the roadbed. The bridge became a tourist attraction as people came to watch, and perhaps ride, the undulating bridge. Finally, on November 7, 1940, during a powerful storm, the oscillations increased beyond any previously observed, and the bridge was evacuated. Soon, the vertical oscillations became rotational, as observed by looking down the roadway. The entire span was eventually shaken apart by the large vibrations, and the bridge collapsed. See [1] for an introduction to these ideas and [2] for interesting and sometimes humorous anecdotes associated with the bridge.

The noted engineer von Karman was asked to determine the cause of the collapse. He and his co-authors [3] claimed that the wind blowing perpendicularly across the roadway separated into vortices (wind swirls) alternately above and below the roadbed, thereby setting up a vertical force acting on the bridge. It was this force that caused the oscillations. Others further hypothesized that the frequency of this periodic forcing function exactly matched the natural frequency of the bridge, thus leading to resonance, large oscillations, and destruction as described in equation (31), Section 5.1. For almost fifty years, resonance was blamed as the cause of the collapse of the bridge, although the von Karman group denied this, stating that "it is very improbable that resonance with alternating vortices plays an important role in the oscillations of suspension bridges" [3].

As we can see from equation (31), Section 5.1, resonance is a linear phenomenon. In addition, for resonance to occur there must be an exact match between the frequency of the forcing function and the natural frequency of the bridge. Furthermore, there must be absolutely no damping in the system. It should not be surprising, then, that resonance was not the culprit in the collapse.

If resonance did not cause the collapse of the bridge, what did? Recent research provides an alternative explanation for the collapse of the Tacoma Narrows Bridge. Lazer and McKenna [4] contend that nonlinear effects, and not

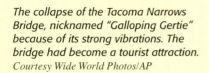

The collapse of the Tacoma Narrows Bridge, nicknamed "Galloping Gertie" because of its strong vibrations. The bridge had become a tourist attraction.
Courtesy Wide World Photos/AP

linear resonance, were the main factors leading to the large oscillations of the bridge (see [5] for a good review article). Their theory involves partial differential equations. However, a simplified model leading to a nonlinear ordinary differential equation can be constructed.

Consider a single vertical cable of the suspension bridge, acting like a linear spring, with no damping. Let the vertical deflection (positive direction downward) of the slice of the roadbed attached to this cable be denoted by $y(t)$, where t represents time and $y = 0$ represents the equilibrium position. As the roadbed is oscillating, the cable provides a linear upward restoring force (Hooke's law) as long as the deflection is downward—that is, as long as the cable is stretched. However, when the roadbed rises above its equilibrium position, the cable is no longer under tension and no longer exerts a force on the roadbed. At this point, the only forces acting on the roadbed are the vertical force due to the von Karman vortices and gravity, which is considered to be negligible. This discontinuous transition, from the linear restoring force ky for $y > 0$ to the zero restoring force for $y < 0$, leads to a nonlinearity in the governing equation. We are thus led to consider the differential equation

$$y'' + f(y) = g(t),$$

where $f(y)$ is the nonlinear function given by

$$f(y) = \begin{cases} ky, & y > 0 \\ 0, & y < 0. \end{cases}$$

Here k is the Hooke's law constant, and $g(t)$ is a (small) periodic forcing function. If we use the more general partial differential equation modeling, we are led to the slightly more general ordinary differential equation

$$y'' + f_1(y) = c + g(t),$$

where $f_1(y)$ is given by

$$f_1(y) = \begin{cases} by, & y > 0 \\ ay, & y < 0. \end{cases}$$

Here $b = EI(\pi/L)^4 + k$, $a = EI(\pi/L)^4$, EI is a constant representing certain material properties of the bridge, L is the length of the bridge, and c is a parameter related to the interactions between the bridge and the forcing function. The following are boundary conditions associated with the periodic nature of the oscillations:

$$y(0) = y(2\pi), \qquad y'(0) = y'(2\pi).$$

Note that the problem is linear on any interval on which y does not change sign, and the equation can be solved by the normal procedures outlined in the text on those intervals.

The technical details of the derivation are contained in the original paper by Lazer and McKenna [4]. They prove that multiple solutions exist when k is large enough. They also suggest the following interpretation of the solution. A large force c acting along with a small periodic function $g(t)$ produces a displacement c/b plus a small oscillation about a new equilibrium. Furthermore, if k is large, additional oscillatory solutions exist. In addition, the large-amplitude solutions can persist, even in the presence of damping. Further interesting conclusions can be inferred from the underlying partial differential equations.

Since the research underlying Lazer and McKenna's explanation has not yet been completed, it is impossible to say exactly what the final model of suspension bridges will resemble. However, it seems obvious that it will not include the phenomenon of linear resonance.

References

1. Lewis, G. N. "Tacoma Narrows Suspension Bridge Collapse." In *A First Course in Differential Equations*, Dennis G. Zill, 253–256. Boston: PWS-Kent, 1993.
2. Braun, M. *Differential Equations and Their Applications* (167–169). New York: Springer-Verlag, 1978.
3. Amann, O. H., T. von Karman, and G. B. Woodruff. *The Failure of the Tacoma Narrows Bridge*. Washington, DC: Federal Works Agency, 1941.
4. Lazer, A. C., and P. J. McKenna. Large amplitude periodic oscillations in suspension bridges: Some new connections with nonlinear analysis. *SIAM Review* 32 (December 1990): 537–578.
5. Peterson, I. Rock and roll bridge. *Science News* 137 (1991): 344–346.

The Tacoma Narrows Bridge as rebuilt following the collapse of the original in 1940
Courtesy Tacoma–Pierce County Visitor & Convention Bureau

About the Author

Gilbert N. Lewis is Associate Professor of Mathematics at Michigan Technological University, where he has taught since 1977. He received a B.S. in Applied Mathematics from Brown University in 1969 and was awarded a Ph.D., also in Applied Mathematics, from the University of Wisconsin–Milwaukee in 1976. Dr. Lewis was a visiting professor at the University of Wisconsin–Parkside and has pursued research in the areas of ordinary differential equations, asymptotic analysis, perturbation theory, and cosmology. Dr. Lewis also contributed an essay to the fifth edition of *A First Course in Differential Equations* by Dennis G. Zill.

Using the given result in (1) to replace d^2M/dx^2 in (3), we see that the deflection $y(x)$ satisfies the fourth-order differential equation

$$EI\frac{d^4y}{dx^4} = w(x). \tag{4}$$

Boundary conditions associated with equation (4) depend on how the ends of the beam are supported. A cantilever beam is **embedded** or **clamped** at one end and **free** at the other. A diving board, an outstretched arm, an airplane wing, and a balcony are common examples of such beams, but even trees, flagpoles, skyscrapers, and the George Washington monument can act as cantilever beams, because they are embedded at one end and are subject to the bending force of the wind. For a cantilever beam, the deflection $y(x)$ must satisfy the following two conditions at the embedded end $x = 0$:

- $y(0) = 0$ since there is no deflection, and
- $y'(0) = 0$ since the deflection curve is tangent to the x-axis (in other words, the slope of the deflection curve is zero at this point).

At $x = L$ the free-end conditions are

- $y''(L) = 0$ since the bending moment is zero, and
- $y'''(L) = 0$ since the shear force is zero.

Ends of the Beam	Boundary Conditions
embedded	$y = 0,\ y' = 0$
free	$y'' = 0,\ y''' = 0$
simply supported	$y = 0,\ y'' = 0$

The function $F(x) = dM/dx = EI\, d^3y/dx^3$ is called the shear force. If an end of a beam is **simply supported** (also called **pin supported, fulcrum supported,** and **hinged**), then we must have $y = 0$ and $y'' = 0$ at that end. The table on the left summarizes the boundary conditions that are associated with (4).

EXAMPLE 1 **Embedded Beam**

A beam of length L is embedded at both ends. Find the deflection of the beam if a constant load w_0 is uniformly distributed along its length—that is, $w(x) = w_0,\ 0 < x < L$.

SOLUTION From the preceding discussion we see that the deflection $y(x)$ satisfies

$$EI\frac{d^4y}{dx^4} = w_0.$$

Because the beam is embedded at both its left ($x = 0$) and its right end ($x = L$), there is no vertical deflection and the line of deflection is horizontal at these points. Thus the boundary conditions are

$$y(0) = 0, \quad y'(0) = 0, \qquad y(L) = 0, \quad y'(L) = 0.$$

We can solve the nonhomogeneous differential equation in the usual manner (find y_c by observing that $m = 0$ is a root of multiplicity four of the auxiliary equation $m^4 = 0$, and then find a particular solution

y_p by undetermined coefficients), or we can simply integrate the equation $d^4y/dx^4 = w_0/EI$ four times in succession. Either way, we find the general solution of the equation to be

$$y(x) = c_1 + c_2 x + c_3 x^2 + c_4 x^3 + \frac{w_0}{24EI} x^4.$$

Now the conditions $y(0) = 0$ and $y'(0) = 0$ give, in turn, $c_1 = 0$ and $c_2 = 0$, whereas the remaining conditions $y(L) = 0$ and $y'(L) = 0$ applied to $y(x) = c_3 x^2 + c_4 x^3 + \frac{w_0}{24EI} x^4$ yield the equations

$$c_3 L^2 + c_4 L^3 + \frac{w_0}{24EI} L^4 = 0$$

$$2c_3 L + 3c_4 L^2 + \frac{w_0}{6EI} L^3 = 0.$$

Solving this system gives $c_3 = w_0 L^2/24EI$ and $c_4 = -w_0 L/12EI$. Thus the deflection is

$$y(x) = \frac{w_0 L^2}{24EI} x^2 - \frac{w_0 L}{12EI} x^3 + \frac{w_0}{24EI} x^4 = \frac{w_0}{24EI} x^2 (x - L)^2.$$

By choosing $w_0 = 24EI$ and $L = 1$, we obtain the graph of the deflection curve in Figure 5.25. ∎

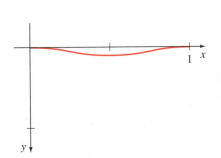

FIGURE 5.25

Eigenvalues and Eigenfunctions
Many applied problems demand that we solve a two-point boundary-value problem involving a linear differential equation that contains a parameter λ. We seek the values of λ for which the boundary-value problem has nontrivial solutions.

EXAMPLE 2   **Nontrivial Solutions of BVP**

Solve the boundary-value problem

$$y'' + \lambda y = 0, \quad y(0) = 0, \quad y(L) = 0.$$

SOLUTION   We consider three cases: $\lambda = 0$, $\lambda < 0$, and $\lambda > 0$.

Case I. For $\lambda = 0$ the solution of $y'' = 0$ is $y = c_1 x + c_2$. The conditions $y(0) = 0$ and $y(L) = 0$ imply, in turn, $c_2 = 0$ and $c_1 = 0$. Hence for $\lambda = 0$ the only solution of the boundary-value problem is the trivial solution $y = 0$.

Case II. For $\lambda < 0$ we have $y = c_1 \cosh \sqrt{-\lambda}\,x + c_2 \sinh \sqrt{-\lambda}\,x$.* Again, $y(0) = 0$ gives $c_1 = 0$, and so $y = c_2 \sinh \sqrt{-\lambda}\,x$. The second condition $y(L) = 0$ dictates that $c_2 \sinh \sqrt{-\lambda}\,L = 0$. Since $\sinh \sqrt{-\lambda}\,L \neq 0$, we must have $c_2 = 0$. Thus $y = 0$.

Case III. For $\lambda > 0$ the general solution of $y'' + \lambda y = 0$ is given by $y = c_1 \cos \sqrt{\lambda}\,x + c_2 \sin \sqrt{\lambda}\,x$. As before, $y(0) = 0$ yields $c_1 = 0$, but $y(L) = 0$ implies

$$c_2 \sin \sqrt{\lambda}\,L = 0.$$

---

*$\sqrt{-\lambda}$ looks a little strange, but bear in mind that $\lambda < 0$ is equivalent to $-\lambda > 0$.

If $c_2 = 0$, then necessarily $y = 0$. However, if $c_2 \neq 0$, then $\sin\sqrt{\lambda}\,L = 0$. The last condition implies that the argument of the sine function must be an integer multiple of π:

$$\sqrt{\lambda}\,L = n\pi \quad \text{or} \quad \lambda = \frac{n^2\pi^2}{L^2}, \quad n = 1, 2, 3, \ldots.$$

Therefore for any real nonzero c_2, $y = c_2 \sin(n\pi x/L)$ is a solution of the problem for each n. Since the differential equation is homogeneous, we may, if desired, not write c_2. In other words, for a given number in the sequence

$$\frac{\pi^2}{L^2}, \frac{4\pi^2}{L^2}, \frac{9\pi^2}{L^2}, \ldots,$$

the *corresponding* function in the sequence

$$\sin\frac{\pi}{L}x, \quad \sin\frac{2\pi}{L}x, \quad \sin\frac{3\pi}{L}x, \quad \ldots$$

is a nontrivial solution of the original problem. ∎

The numbers $\lambda_n = n^2\pi^2/L^2$, $n = 1, 2, 3, \ldots$ for which the boundary-value problem in Example 2 has a nontrivial solution are known as **characteristic values** or, more commonly, **eigenvalues.** The solutions depending on these values of λ_n, $y_n = c_2 \sin(n\pi x/L)$ or simply $y_n = \sin(n\pi x/L)$, are called **characteristic functions** or **eigenfunctions.**

Buckling of a Thin Vertical Column

In the eighteenth century Leonhard Euler was one of the first mathematicians to study an eigenvalue problem in analyzing how a thin elastic column buckles under a compressive axial force.

Consider a long slender vertical column of uniform cross-section and length L. Let $y(x)$ denote the deflection of the column when a constant vertical compressive force, or load, P is applied to its top, as shown in Figure 5.26. By comparing bending moments at any point along the column we obtain

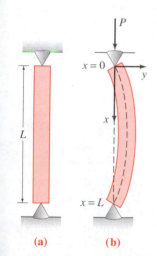

(a)   **(b)**

FIGURE 5.26

$$EI\frac{d^2y}{dx^2} = -Py \quad \text{or} \quad EI\frac{d^2y}{dx^2} + Py = 0, \tag{5}$$

where E is Young's modulus of elasticity and I is the moment of inertia of a cross-section about a vertical line through its centroid.

EXAMPLE 3   An Eigenvalue Problem

Find the deflection of a thin vertical homogeneous column of length L subjected to a constant axial load P if the column is hinged at both ends.

SOLUTION   The boundary-value problem to be solved is

$$EI\frac{d^2y}{dx^2} + Py = 0, \quad y(0) = 0, \quad y(L) = 0.$$

(a)

(b)

(c)

FIGURE 5.27

First note that $y = 0$ is a perfectly good solution of this problem. This solution has a simple intuitive interpretation: If the load P is not great enough, there is no deflection. The question then is this: For what values of P will the column bend? In mathematical terms: For what values of P does the given boundary-value problem possess nontrivial solutions?

By writing $\lambda = P/EI$ we see that

$$y'' + \lambda y = 0, \quad y(0) = 0, \quad y(L) = 0$$

is identical to the problem in Example 2. From case III of that discussion we see that the deflection curves are $y_n(x) = c_2 \sin(n\pi x/L)$, corresponding to the eigenvalues $\lambda_n = P_n/EI = n^2\pi^2/L^2$, $n = 1, 2, 3, \ldots$. Physically this means that the column will buckle or deflect only when the compressive force is one of the values $P_n = n^2\pi^2 EI/L^2$, $n = 1, 2, 3, \ldots$. These different forces are called **critical loads.** The deflection curve corresponding to the smallest critical load $P_1 = \pi^2 EI/L^2$, called the **Euler load,** is $y_1(x) = c_2 \sin(\pi x/L)$ and is known as the **first buckling mode.**

     ■

The deflection curves in Example 3 corresponding to $n = 1$, $n = 2$, and $n = 3$ are shown in Figure 5.27. Note that if the original column has some sort of physical restraint put on it at $x = L/2$, then the smallest critical load will be $P_2 = 4\pi^2 EI/L^2$ and the deflection curve will be as shown in Figure 5.27(b). If restraints are put on the column at $x = L/3$ and at $x = 2L/3$, then the column will not buckle until the critical load $P_3 = 9\pi^2 EI/L^2$ is applied and the deflection curve will be as shown in Figure 5.27(c). Where should physical restraints be placed on the column if we want the Euler load to be P_4?

Rotating String   The simple linear second-order differential equation

$$y'' + \lambda y = 0 \tag{6}$$

occurs again and again as a mathematical model. In Section 5.1 we saw (6) in the forms $d^2x/dt^2 + (k/m)x = 0$ and $d^2q/dt^2 + (1/LC)q = 0$ as models for, respectively, the simple harmonic motion of a spring/mass system and the simple harmonic response of a series circuit. It is apparent when the model for the deflection of a thin column in (5) is written as $d^2y/dx^2 + (P/EI)y = 0$ that it is the same as (6). We encounter the basic equation (6) one more time in this section: as a model that defines the deflection curve or the shape $y(x)$ assumed by a rotating string. The physical situation is analogous to when two persons hold a jump rope and twirl it in a synchronous manner. See Figure 5.28, parts (a) and (b).

Suppose a string of length L with constant linear density ρ (mass per unit length) is stretched along the x-axis and fixed at $x = 0$ and $x = L$. Suppose the string is then rotated about that axis at a constant angular speed ω. Consider a portion of the string on the inteval $[x, x + \Delta x]$, where Δx is small. If the magnitude T of the tension **T**, acting tangential to the string, is constant along the string, then the desired differential equation can be obtained by equating two different formulations of the net force

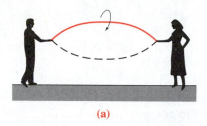

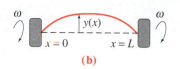

(a)

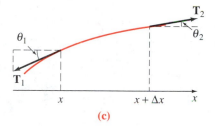

(b)

(c)

FIGURE 5.28

acting on the string on the interval $[x, x + \Delta x]$. First, we see from Figure 5.28(c) that the net vertical force is

$$F = T \sin \theta_2 - T \sin \theta_1. \tag{7}$$

When angles θ_1 and θ_2 (measured in radians) are small, we have $\sin \theta_2 \approx \tan \theta_2$ and $\sin \theta_1 \approx \tan \theta_1$. Moreover, since $\tan \theta_2$ and $\tan \theta_1$ are, in turn, slopes of the lines containing the vectors $\mathbf{T}_2$ and $\mathbf{T}_1$, we can also write

$$\tan \theta_2 = y'(x + \Delta x) \qquad \text{and} \qquad \tan \theta_1 = y'(x).$$

Thus (7) becomes

$$F \approx T[y'(x + \Delta x) - y'(x)]. \tag{8}$$

Second, we can obtain a different form of this same net force using Newton's second law, $F = ma$. Here the mass of string on the interval is $m = \rho \Delta x$; the centripetal acceleration of a body rotating with angular speed ω in a circle of radius r is $a = r\omega^2$. With Δx small we take $r = y$. Thus the net vertical force is also approximated by

$$F \approx -(\rho \Delta x) y\omega^2, \tag{9}$$

where the minus sign comes from the fact that the acceleration points in the direction opposite to the positive y-direction. Now by equating (8) and (9) we have

$$T[y'(x + \Delta x) - y'(x)] \approx -(\rho \Delta x) y\omega^2 \qquad \text{or} \qquad T\frac{y'(x + \Delta x) - y'(x)}{\Delta x} \approx -\rho\omega^2 y. \tag{10}$$

For Δx close to zero the difference quotient $[y'(x + \Delta x) - y'(x)]/\Delta x$ in (10) is approximated by the second derivative d^2y/dx^2. Finally we arrive at the model

$$T\frac{d^2y}{dx^2} = -\rho\omega^2 y \qquad \text{or} \qquad T\frac{d^2y}{dx^2} + \rho\omega^2 y = 0. \tag{11}$$

Since the string is anchored at its ends $x = 0$ and $y = L$, we expect that the solution $y(x)$ of the last equation in (11) should also satisfy the boundary conditions $y(0) = 0$ and $y(L) = 0$.

SECTION 5.2 EXERCISES

Answers to odd-numbered problems begin on page A-7.

In Problems 1–4 the beam is of length L and w_0 is constant.

1. **(a)** Solve (4) when the beam is embedded at its left end and free at its right end and $w(x) = w_0$, $0 < x < L$.
   (b) Use a graphing utility to obtain the graph of the deflection curve of the beam when $w_0 = 24EI$ and $L = 1$.

2. **(a)** Solve (4) when the beam is simply supported at both ends and $w(x) = w_0$, $0 < x < L$.
   (b) Use a graphing utility to obtain the graph of the deflection curve of the beam when $w_0 = 24EI$ and $L = 1$.

3. **(a)** Solve (4) when the beam is embedded at its left end and simply supported at its right end and $w(x) = w_0$, $0 < x < L$.
   (b) Use a graphing utility to obtain the graph of the deflection curve of the beam when $w_0 = 48EI$ and $L = 1$.

4. **(a)** Solve (4) when the beam is embedded at its left end and simply supported at its right end and $w(x) = w_0 \sin(\pi x/L)$, $0 < x < L$.
   (b) Use a graphing utility to obtain the graph of the deflection curve of the beam when $w_0 = 2\pi^3 EI$ and $L = 1$.

5. **(a)** Find the maximum deflection of the cantilever beam in Problem 1.
   (b) How does the maximum deflection of a beam that is half as long compare with the value in part (a)?

6. **(a)** Find the maximum deflection of the simply supported beam in Problem 2.
   (b) How does the maximum deflection of the simply supported beam compare with the value of maximum deflection of the embedded beam in Example 1?

7. A cantilever beam of length L is embedded at its right end, and a horizontal tensile force of P pounds is applied to its free left end. When the origin is taken at its free end, as shown in Figure 5.29, the deflection $y(x)$ of the beam can be shown to satisfy the differential equation

$$EIy'' = Py - w(x)\frac{x}{2}.$$

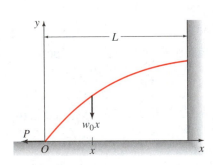

FIGURE 5.29

Find the deflection of the cantilever beam if $w(x) = w_0 x$, $0 < x < L$ and $y(0) = 0$, $y'(L) = 0$.

8. When a compressive instead of a tensile force is applied at the free end of the beam in Problem 7, the differential equation of the deflection is

$$EIy'' = -Py - w(x)\frac{x}{2}.$$

Solve this equation if $w(x) = w_0 x$, $0 < x < L$ and $y(0) = 0$, $y'(L) = 0$.

In Problems 9–22 find the eigenvalues and eigenfunctions for the given boundary-value problem.

9. $y'' + \lambda y = 0$, $\quad y(0) = 0$, $y(\pi) = 0$

10. $y'' + \lambda y = 0$, $\quad y(0) = 0$, $y\left(\dfrac{\pi}{4}\right) = 0$

11. $y'' + \lambda y = 0$, $\quad y'(0) = 0$, $y(L) = 0$

12. $y'' + \lambda y = 0$, $\quad y(0) = 0$, $y'\left(\dfrac{\pi}{2}\right) = 0$

13. $y'' + \lambda y = 0$, $\quad y'(0) = 0$, $y'(\pi) = 0$

14. $y'' + \lambda y = 0$, $\quad y(-\pi) = 0$, $y(\pi) = 0$

15. $y'' + 2y' + (\lambda + 1)y = 0$, $\quad y(0) = 0$, $y(5) = 0$

16. $y'' + (\lambda + 1)y = 0$, $\quad y'(0) = 0$, $y'(1) = 0$

17. $y'' + \lambda^2 y = 0$, $\quad y(0) = 0$, $y(L) = 0$

18. $y'' + \lambda^2 y = 0$, $\quad y(0) = 0$, $y'(3\pi) = 0$

19. $x^2y'' + xy' + \lambda y = 0, \quad y(1) = 0, y(e^\pi) = 0$

20. $x^2y'' + xy' + \lambda y = 0, \quad y'(e^{-1}) = 0, y(1) = 0$

21. $x^2y'' + xy' + \lambda y = 0, \quad y'(1) = 0, y'(e^2) = 0$

22. $x^2y'' + 2xy' + \lambda y = 0, \quad y(1) = 0, y(e^2) = 0$

23. Show that the eigenfunctions of the boundary-value problem

$$y'' + \lambda y = 0, \qquad y(0) = 0, \quad y(1) + y'(1) = 0$$

are $y_n = \sin \sqrt{\lambda_n}\, x$, where the eigenvalues λ_n of the problem are $\lambda_n = x_n^2$, with x_n, $n = 1, 2, 3, \ldots$ being the consecutive *positive* roots of the equation $\tan \sqrt{\lambda} = - \sqrt{\lambda}$.

24. (a) Use a graphing utility to convince yourself that the equation $\tan x = -x$ has an infinite number of roots. Explain why the negative roots of the equation can be ignored. Explain why $\lambda = 0$ is not an eigenvalue in Problem 23 even though $\lambda = 0$ is an obvious root of the equation $\tan \sqrt{\lambda} = - \sqrt{\lambda}$.

(b) Use a numerical procedure or a CAS to approximate the first four eigenvalues $\lambda_1, \lambda_2, \lambda_3$, and λ_4.

25. Consider the boundary-value problem introduced in the construction of the mathematical model for the shape of a rotating string:

$$T\frac{d^2y}{dx^2} + \rho\omega^2 y = 0, \qquad y(0) = 0, \quad y(L) = 0.$$

For constant T and ρ, define the critical speeds of angular rotation ω_n as the values of ω for which the boundary-value problem has nontrivial solutions. Find the critical speeds ω_n and the corresponding deflection curves $y_n(x)$.

26. When the magnitude of tension T is not constant, then a model for the deflection curve or shape $y(x)$ assumed by a rotating string is given by

$$\frac{d}{dx}\left[T(x)\frac{dy}{dx}\right] + \rho\omega^2 y = 0.$$

Suppose that $1 < x < e$ and that $T(x) = x^2$.

(a) If $y(1) = 0$, $y(e) = 0$, and $\rho\omega^2 > 0.25$, find the critical speeds ω_n and the corresponding deflection curves $y_n(x)$.

(b) In the expression for $y_n(x)$ there will be an arbitrary constant—say, c_2. Use a graphing utility to graph the deflection curves on the interval $[1, e]$ for $n = 1, 2, 3$. Choose $c_2 = 1$.

27. Consider two concentric spheres of radius $r = a$ and $r = b$, $a < b$, as shown in Figure 5.30. The temperature $u(r)$ in the region between the spheres is determined from the boundary-value problem

$$r\frac{d^2u}{dr^2} + 2\frac{du}{dr} = 0, \qquad u(a) = u_0, \quad u(b) = u_1,$$

where u_0 and u_1 are constants. Solve for $u(r)$.

28. The temperature $u(r)$ in the circular ring shown in Figure 5.31 is determined from the boundary-value problem

$$r\frac{d^2u}{dr^2} + \frac{du}{dr} = 0, \qquad u(a) = u_0, \quad u(b) = u_1,$$

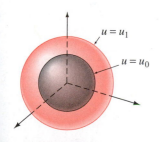

FIGURE 5.30

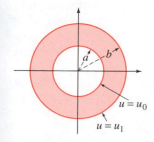

FIGURE 5.31

where u_0 and u_1 are constants. Show that

$$u(r) = \frac{u_0 \ln(r/b) - u_1 \ln(r/a)}{\ln(a/b)}.$$

Discussion Problems

29. Consider the boundary-value problem

$$y'' + 16y = 0, \qquad y(0) = y_0, \quad y\left(\frac{\pi}{2}\right) = y_1.$$

Discuss whether it is possible to determine values of y_0 and y_1 so that the problem possesses **(a)** precisely one nontrivial solution, **(b)** more than one solution, **(c)** no solution, **(d)** the trivial solution.

30. Consider the boundary-value problem

$$y'' + 16y = 0, \qquad y(0) = 1, \quad y(L) = 1.$$

Discuss whether it is possible to determine values of $L > 0$ so that the problem possesses **(a)** precisely one nontrivial solution, **(b)** more than one solution, **(c)** no solution, **(d)** the trivial solution.

5.3 NONLINEAR EQUATIONS

■ *Linear and nonlinear springs* ■ *Hard and soft springs* ■ *DE of a nonlinear pendulum*
■ *Linearization* ■ *DE of a suspended wire* ■ *Catenary* ■ *Rocket motion*

Nonlinear Springs   The mathematical model in (1) of Section 5.1 has the form

$$m\frac{d^2x}{dt^2} + F(x) = 0, \tag{1}$$

where $F(x) = kx$. Since x denotes the displacement of the mass from its equilibrium position, $F(x) = kx$ is Hooke's law—that is, the force exerted by the spring that tends to restore the mass to the equilibrium position. A spring acting under a linear restoring force $F(x) = kx$ is naturally referred to as a **linear spring.** But springs are seldom perfectly linear. Depending on how it is constructed and the material used, a spring can range from "mushy," or soft, to "stiff," or hard, so that its restorative force may vary from something below to something above that given by the linear law. In the case of free motion, if we assume that a non-aging spring possesses some nonlinear characteristics, then it might be reasonable to assume that the restorative force $F(x)$ of a spring is proportional to, say, the cube of the displacement x of the mass beyond its equilibrium position or that $F(x)$ is a linear combination of powers of the displacement such as that given by the nonlinear function $F(x) = kx + k_1x^3$. A spring whose mathematical model incorporates a nonlinear restorative force, such as

$$m\frac{d^2x}{dt^2} + kx^3 = 0 \qquad \text{or} \qquad m\frac{d^2x}{dt^2} + kx + k_1x^3 = 0, \tag{2}$$

is called a **nonlinear spring.** In addition, we examined mathematical models in which damping imparted to the motion was proportional to the instantaneous velocity dx/dt and the restoring force of a spring was given by the linear function $F(x) = kx$. But these were simply assumptions; in more realistic situations damping could be proportional to some power of the instantaneous velocity dx/dt. The nonlinear differential equation

$$m\frac{d^2x}{dt^2} + \beta \left|\frac{dx}{dt}\right| \frac{dx}{dt} + kx = 0 \qquad (3)$$

is one model of a free spring/mass system with damping proportional to the square of the velocity. One can then envision other kinds of models: linear damping and nonlinear restoring force, nonlinear damping and nonlinear restoring force, and so on. The point is, nonlinear characteristics of a physical system lead to a mathematical model that is nonlinear.

Notice in (2) that both $F(x) = kx^3$ and $F(x) = kx + k_1x^3$ are odd functions of x. To see why a polynomial function containing only odd powers of x provides a reasonable model for the restoring force, let us express F as power series centered at the equilibrium position $x = 0$:

$$F(x) = c_0 + c_1x + c_2x^2 + c_3x^3 + \cdots.$$

When the displacements x are small, the values of x^n are negligible for n sufficiently large. If we truncate the power series with, say, the fourth term, then

$$F(x) = c_0 + c_1x + c_2x^2 + c_3x^3.$$

In order for the force at $x > 0$ ($F(x) = c_0 + c_1x + c_2x^2 + c_3x^3$) and the force at $-x < 0$ ($F(-x) = c_0 - c_1x + c_2x^2 - c_3x^3$) to have the same magnitude but act in opposite directions, we must have $F(-x) = -F(x)$. Since this means that F is an odd function, we must have $c_0 = 0$ and $c_2 = 0$ and so $F(x) = c_1x + c_3x^3$. Had we used only the first two terms in the series, the same argument would yield $F(x) = c_1x$. For discussion purposes we shall write $c_1 = k$ and $c_2 = k_1$. A restoring force with mixed powers, such as $F(x) = kx + k_1x^2$, and the corresponding vibrations are said to be unsymmetrical.

Hard and Soft Springs   Let us take a closer look at the equation in (1) in the case where the restoring force is given by $F(x) = kx + k_1x^3$, $k > 0$. Graphs of three types of restoring forces are illustrated in Figure 5.32. The spring is said to be **hard** if $k_1 > 0$ and **soft** if $k_1 < 0$. Example 1 illustrates these two special cases of the differential equation $m\, d^2x/dt^2 + kx + k_1x^3 = 0$, $m > 0$, $k > 0$.

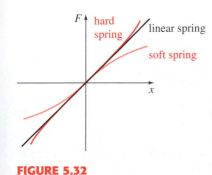

FIGURE 5.32

EXAMPLE 1   **Comparison of Hard and Soft Springs**

The differential equations

$$\frac{d^2x}{dt^2} + x + x^3 = 0 \qquad (4)$$

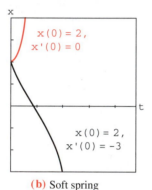

(a) Hard spring

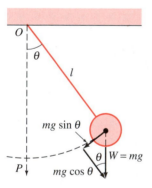

(b) Soft spring

FIGURE 5.33

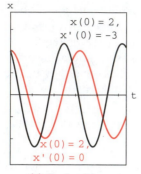

FIGURE 5.34

and

$$\frac{d^2x}{dt^2} + x - x^3 = 0 \qquad (5)$$

are special cases of (2) and are models of a hard spring and a soft spring, respectively. Figure 5.33(a) shows two solutions of (4) and Figure 5.33(b) shows two solutions of (5) obtained from an ODE solver. The curves shown in black are solutions satisfying the initial conditions $x(0) = 2$, $x'(0) = -3$; the two curves in color are solutions satisfying $x(0) = 2$, $x'(0) = 0$. These solution curves certainly suggest that the motion of a mass on the hard spring is oscillatory, whereas the motion of a mass on the soft spring is not oscillatory. But we must be careful about drawing conclusions based on a couple of solution curves. A more complete picture of the nature of the solutions of both of these equations can be obtained from the qualitative analysis discussed in Chapter 10. Also see Problem 2 in Exercises 5.3. ∎

Nonlinear Pendulum　Any object that swings back and forth is called a **physical pendulum.** The **simple pendulum** is a special case of the physical pendulum and consists of a rod of length l to which a mass m is attached at one end. In describing the motion of a simple pendulum in a vertical plane, we make the simplifying assumptions that the mass of the rod is negligible and that no external damping or driving forces act on the system. The displacement angle θ of the pendulum, measured from the vertical as shown in Figure 5.34, is considered positive when measured to the right of OP and negative when measured to the left of OP. Now recall that the arc s of a circle of radius l is related to the central angle θ by the formula $s = l\theta$. Hence angular acceleration is

$$a = \frac{d^2s}{dt^2} = l\frac{d^2\theta}{dt^2}.$$

From Newton's second law we then have

$$F = ma = ml\frac{d^2\theta}{dt^2}.$$

From Figure 5.34 we see that the tangential component of the force due to the weight W is $mg \sin \theta$. We equate the two different versions of the tangential force to obtain $ml\, d^2\theta/dt^2 = -mg \sin \theta$ or

$$\frac{d^2\theta}{dt^2} + \frac{g}{l}\sin \theta = 0. \qquad (6)$$

Linearization　Because of the presence of $\sin \theta$, the model in (6) is nonlinear. In an attempt to understand the behavior of the solutions of nonlinear higher-order differential equations, we sometimes try to simplify the problem by replacing nonlinear terms with certain approximations. For example, the Maclaurin series for $\sin \theta$ is given by

$$\sin \theta = \theta - \frac{\theta^3}{3!} + \frac{\theta^5}{5!} - \cdots,$$

and so if we use the approximation $\sin \theta \approx \theta - \theta^3/6$ equation (6) becomes $d^2\theta/dt^2 + (g/l)\theta - (g/6l)\theta^3 = 0$. Observe that this last equation is the same as the second nonlinear equation in (2) with $m = 1$, $k = g/l$, and $k_1 = -g/6l$. However, if we assume that the displacements θ are small enough to justify using the replacement $\sin\theta \approx \theta$, then (6) becomes

$$\frac{d^2\theta}{dt^2} + \frac{g}{l}\theta = 0. \tag{7}$$

If we set $\omega^2 = g/l$, we recognize (7) as the differential equation (2) of Section 5.1 governing the free undamped vibrations of a linear spring/mass system. In other words, (7) is again the basic linear equation $y'' + \lambda y = 0$ discussed on page 196 of Section 5.2. As a consequence we say that equation (7) is a **linearization** of equation (6). Since the general solution of (7) is $\theta(t) = c_1 \cos\omega t + c_2 \sin\omega t$, this linearization suggests that for initial conditions amenable to small oscillations the motion of the pendulum described by (6) will be periodic.

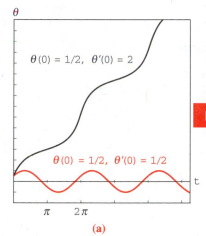

$\theta(0) = 1/2$, $\theta'(0) = 2$

$\theta(0) = 1/2$, $\theta'(0) = 1/2$

(a)

EXAMPLE 2   Nonlinear Pendulum

The graphs in Figure 5.35(a) were obtained with the aid of an ODE solver and represent solution curves of (6) when $\omega^2 = 1$. The colored curve depicts the solution of (6) that satisfies the initial conditions $\theta(0) = \frac{1}{2}$, $\theta'(0) = \frac{1}{2}$, whereas the black curve is the solution of (6) that satisfies $\theta(0) = \frac{1}{2}$, $\theta'(0) = 2$. The colored curve represents a periodic solution—the pendulum oscillating back and forth as shown in Figure 5.35(b), with an apparent amplitude $A \leq 1$. The black curve shows θ increasing without bound as time increases—the pendulum, starting from the same initial displacement, is given an initial velocity of magnitude great enough to send it over the top; in other words, the pendulum is whirling about its pivot, as shown in Figure 5.35(c). In the absence of damping, the motion in each case is continued indefinitely. ∎

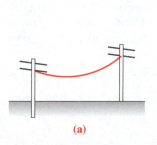

(b) $\theta(0) = \frac{1}{2}$, **(c)** $\theta(0) = \frac{1}{2}$,

$\theta'(0) = \frac{1}{2}$ $\theta'(0) = 2$

FIGURE 5.35

Suspended Wire   Suppose a suspended wire hangs under its own weight. As Figure 5.36(a) shows, a physical model for this could be a long telephone wire strung between two posts. Our goal is to construct a mathematical model that describes the shape that the hanging wire assumes.

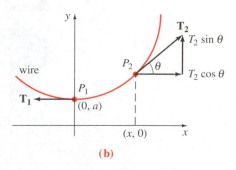

(a)   **(b)**

FIGURE 5.36

To begin, let us suppose that the y-axis in Figure 5.36(b) is chosen to pass through the lowest point P_1 on the curve and that the x-axis is a units below P_1. Let us also agree to examine only that portion of the wire between the lowest point P_1 and any arbitrary point P_2. Three forces are acting on the wire: the weight of the segment P_1P_2 and the tensions $\mathbf{T}_1$ and $\mathbf{T}_2$ in the wire at P_1 and P_2, respectively. If w is the linear density of the wire (measured, say, in lb/ft) and s is the length of the segment P_1P_2, then its weight is ws. Now the tension $\mathbf{T}_2$ resolves into horizontal and vertical components (scalar quantities) $T_2 \cos \theta$ and $T_2 \sin \theta$. Because of equilibrium we can write

$$|\mathbf{T}_1| = T_1 = T_2 \cos \theta \qquad \text{and} \qquad ws = T_2 \sin \theta.$$

Dividing the last two equations, we find $\tan \theta = ws/T_1$. That is,

$$\frac{dy}{dx} = \frac{ws}{T_1}. \tag{8}$$

Now since the arc length between points P_1 and P_2 is given by

$$s = \int_0^x \sqrt{1 + \left(\frac{dy}{dx}\right)^2}\, dx, \tag{9}$$

it follows from the fundamental theorem of calculus that the derivative of (9) is

$$\frac{ds}{dx} = \sqrt{1 + \left(\frac{dy}{dx}\right)^2}. \tag{10}$$

Differentiating (8) with respect to x and using (10) leads to

$$\frac{d^2y}{dx^2} = \frac{w}{T_1}\frac{ds}{dx} \qquad \text{or} \qquad \frac{d^2y}{dx^2} = \frac{w}{T_1}\sqrt{1 + \left(\frac{dy}{dx}\right)^2}. \tag{11}$$

One might conclude from Figure 5.36 that the shape the hanging wire assumes is parabolic. The next example shows that this is not the case—the curve assumed by the suspended cable is called a **catenary.** Before proceeding we observe that the nonlinear differential equation in (11) is one of those equations $F(y, y', y'') = 0$ discussed in Section 4.9 that we have a chance of solving by means of a substitution.

EXAMPLE 3 **Initial-Value Problem**

From the position of the y-axis in Figure 5.36(b) it is apparent that initial conditions associated with the second differential equation in (11) are $y(0) = a$ and $y'(0) = 0$. If we substitute $u = y'$,

$$\frac{d^2y}{dx^2} = \frac{w}{T_1}\sqrt{1 + \left(\frac{dy}{dx}\right)^2} \qquad \text{becomes} \qquad \frac{du}{dx} = \frac{w}{T_1}\sqrt{1 + u^2}.$$

Separating variables, we find

$$\int \frac{du}{\sqrt{1 + u^2}} = \frac{w}{T_1}\int dx \qquad \text{gives} \qquad \sinh^{-1} u = \frac{w}{T_1}x + c_1.$$

Now $y'(0) = 0$ is equivalent to $u(0) = 0$. Since $\sinh^{-1} 0 = 0$, we find that $c_1 = 0$ and so $u = \sinh(wx/T_1)$. Finally, by integrating

$$\frac{dy}{dx} = \sinh\frac{w}{T_1}x \qquad \text{we get} \qquad y = \frac{T_1}{w}\cosh\frac{w}{T_1}x + c_2.$$

If we use $y(0) = a$, $\cosh 0 = 1$, the last equation implies that $c_2 = a - T_1/w$. Thus we see that the shape of the hanging wire is defined by

$$y = \frac{T_1}{w}\cosh\frac{w}{T_1}x + a - \frac{T_1}{w}. \qquad \blacksquare$$

In Example 3, had we been clever enough at the start to choose $a = T_1/w$, the solution of the problem would have been simply the hyperbolic cosine $y = (T_1/w)\cosh(wx/T_1)$.

Rocket Motion In Section 1.3 we saw that the differential equation of a free-falling body of mass m near the surface of the earth is given by

$$m\frac{d^2s}{dt^2} = -mg \qquad \text{or simply} \qquad \frac{d^2s}{dt^2} = -g,$$

where s represents the distance from the surface of the earth to the object and the positive direction is considered to be upward. In other words, the underlying assumption here is that the distance s to the object is small when compared with the radius R of the earth; put yet another way, the distance y from the center of the earth to the object is approximately the same as R. If, on the other hand, the distance y to an object—such as a rocket or a space probe—is large compared to R, then we combine Newton's second law of motion and his universal law of gravitation to derive a differential equation in the variable y.

Suppose a rocket is shot vertically upward from the ground as shown in Figure 5.37. If the positive direction is upward and air resistance is ignored, then the differential equation of motion after fuel burnout is

$$m\frac{d^2y}{dt^2} = -k\frac{Mm}{y^2} \qquad \text{or} \qquad \frac{d^2y}{dt^2} = -k\frac{M}{y^2}, \tag{12}$$

where k is a constant of proportionality, y is the distance from the center of the earth to the rocket, M is the mass of the earth, and m is the mass of the rocket. To determine the constant k we use the fact that when $y = R$, $kMm/R^2 = mg$ or $k = gR^2/M$. Thus the last equation in (12) becomes

$$\frac{d^2y}{dt^2} = -g\frac{R^2}{y^2}. \tag{13}$$

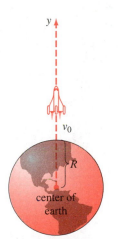

FIGURE 5.37

EXAMPLE 4 **Escape Velocity**

Since $v = dy/dt$ is velocity, we can write the acceleration of the rocket discussed above as

$$\frac{d^2y}{dt^2} = \frac{dv}{dt} = \frac{dv}{dy}\frac{dy}{dt} = v\frac{dv}{dy}.$$

Hence (13) is transformed into a first-order equation in v; that is,

$$v \frac{dv}{dy} = -g \frac{R^2}{y^2}. \tag{14}$$

This last equation can be solved by separation of variables. From

$$\int v \, dv = -gR^2 \int y^{-2} \, dy \qquad \text{we get} \qquad \frac{v^2}{2} = g \frac{R^2}{y} + c. \tag{15}$$

If we assume that the velocity of the rocket is $v = v_0$ at burnout and that $y \approx R$ at that instant, we can obtain the (approximate) value of c. From (15) we find $c = -gR + v_0^2/2$. Substituting this value into (15) and multiplying the resulting equation by 2 yields

$$v^2 = 2g \frac{R^2}{y} - 2gR + v_0^2. \tag{16}$$

You might object, correctly, that we have not really solved the original differential equation (13) for y. Actually the particular solution (16) of equation (14) gives quite a bit of information. It is this solution that can be used to determine the minimum velocity, the so-called escape velocity needed by a rocket to break free of the earth's gravitational attraction. Since we have done the hard part by getting to (16), we leave the actual determination of the escape velocity from the earth as an exercise. See Problem 14 in Exercises 5.3.   ■

SECTION 5.3 EXERCISES

Answers to odd-numbered problems begin on page A-8.

In Problems 1–4 the given differential equation is a model of an undamped spring/mass system in which the restoring force $F(x)$ in (1) is nonlinear. For each equation use an ODE solver to obtain the solution curves satisfying the given initial conditions. If the solutions are periodic, use the solution curve to estimate the period T of oscillations.

1. $\dfrac{d^2x}{dt^2} + x^3 = 0$

$x(0) = 1, \, x'(0) = 1; \quad x(0) = \dfrac{1}{2}, x'(0) = -1$

2. $\dfrac{d^2x}{dt^2} + 4x - 16x^3 = 0$

$x(0) = 1, \, x'(0) = 1; \quad x(0) = -2, \, x'(0) = 2$

3. $\dfrac{d^2x}{dt^2} + 2x - x^2 = 0$

$x(0) = 1, \, x'(0) = 1; \quad x(0) = \dfrac{3}{2}, x'(0) = -1$

4. $\dfrac{d^2x}{dt^2} + xe^{0.01x} = 0$

$x(0) = 1, \, x'(0) = 1; \quad x(0) = 3, \, x'(0) = -1$

5. In Problem 3 suppose the mass is released from the initial position $x(0) = 1$ with an initial velocity $x'(0) = x_1$. Use an ODE solver to estimate the smallest value of $|x_1|$ at which the motion of the mass is nonperiodic.

6. In Problem 3 suppose the mass is released from an initial position $x(0) = x_0$ with the initial velocity $x'(0) = 1$. Use an ODE solver to estimate an interval $a \leq x_0 \leq b$ for which the motion is oscillatory.

7. Find a linearization of the differential equation in Problem 4.

8. Consider the model of an undamped nonlinear spring/mass system given by

$$\frac{d^2x}{dt^2} + 8x - 6x^3 + x^5 = 0.$$

Use an ODE solver to discuss the nature of the oscillations of the system corresponding to the following initial conditions:

$x(0) = 1, x'(0) = 1;$ $x(0) = -2, x'(0) = 0.5;$ $x(0) = \sqrt{2}, x'(0) = 1;$
$x(0) = 2, x'(0) = 0.5;$ $x(0) = -2, x'(0) = 0;$ $x(0) = -\sqrt{2}, x'(0) = -1.$

In Problems 9 and 10 the given differential equation is a model of a damped nonlinear spring/mass system.
(a) Predict the behavior of each system as $t \to \infty$.
(b) For each equation use an ODE solver to obtain the solution curves satisfying the given initial conditions.

9. $\dfrac{d^2x}{dt^2} + \dfrac{dx}{dt} + x + x^3 = 0$

$x(0) = -3, x'(0) = 4;$ $x(0) = 0, x'(0) = -8$

10. $\dfrac{d^2x}{dt^2} + \dfrac{dx}{dt} + x - x^3 = 0$

$x(0) = 0, x'(0) = \dfrac{3}{2};$ $x(0) = -1, x'(0) = 1$

11. The model $mx'' + kx + k_1x^3 = F_0 \cos \omega t$ of an undamped periodically driven spring/mass system is called **Duffing's differential equation.** Consider the initial-value problem

$$x'' + x + k_1x^3 = 5 \cos t, \qquad x(0) = 1, \quad x'(0) = 0.$$

Use an ODE solver to investigate the behavior of the system for values of $k_1 > 0$ ranging from $k_1 = 0.01$ to $k_1 = 100$. State your conclusions.

12. **(a)** In Problem 11 find values of $k_1 < 0$ for which the system is oscillatory.
(b) Consider the initial-value problem

$$x'' + x + k_1x^3 = \cos \frac{3}{2}t, \qquad x(0) = 0, \quad x'(0) = 0.$$

Find values for $k_1 < 0$ for which the system is oscillatory.

13. Consider the model of the free damped nonlinear pendulum given by

$$\frac{d^2\theta}{dt^2} + 2\lambda\frac{d\theta}{dt} + \omega^2 \sin \theta = 0.$$

Use an ODE solver to investigate whether the motion in the two cases $\lambda^2 - \omega^2 > 0$ and $\lambda^2 - \omega^2 < 0$ corresponds, respectively, to the overdamped and underdamped cases discussed in Section 5.1 for spring/mass systems. Choose appropriate initial conditions and values of λ and ω.

14. **(a)** Use (16) to show that the escape velocity of the rocket is given by $v_0 = \sqrt{2gR}$. [*Hint:* Take $y \to \infty$ in (16) and assume that $v > 0$ for all time t.]

   (b) The result in part (a) holds for any body in the solar system. Use the values $g = 32$ ft/s and $R = 4000$ mi to show that the escape velocity from the earth is (approximately) $v_0 = 25{,}000$ mi/h.

   (c) Find the escape velocity from the moon if the acceleration of gravity is $0.165g$ and $R = 1080$ mi.

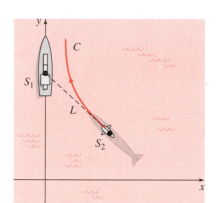

FIGURE 5.38

15. In a naval exercise a ship S_1 is pursued by a submarine S_2 as shown in Figure 5.38. Ship S_1 departs point $(0, 0)$ at $t = 0$ and proceeds along a straight-line course (the y-axis) at a constant speed v_1. The submarine S_2 keeps ship S_1 in visual contact, indicated by the straight dashed line L in the figure, while traveling at a constant speed v_2 along a curve C. Assume that S_2 starts at the point $(a, 0)$, $a > 0$, at $t = 0$ and that L is tangent to C.

   (a) Determine a mathematical model that describes the curve C. [*Hint:* $\dfrac{dt}{dx} = \dfrac{dt}{ds}\dfrac{ds}{dx}$, where s is arc length measured along C.]

   (b) Find an explicit solution of the differential equation. For convenience define $r = v_1/v_2$.

   (c) Determine whether the paths of S_1 and S_2 will ever intersect by considering the cases $r > 1$, $r < 1$, and $r = 1$.

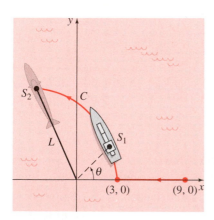

FIGURE 5.39

16. In another naval exercise a destroyer S_1 pursues a submerged submarine S_2. Suppose that S_1 at $(9, 0)$ on the x-axis detects S_2 at $(0, 0)$ and that S_2 simultaneously detects S_1. The captain of the destroyer S_1 assumes that the submarine will take immediate evasive action and conjectures that its likely new course is the straight line indicated in Figure 5.39. When S_1 is at $(3, 0)$, it changes from its straight-line course toward the origin to a pursuit curve C. Assume that the speed of the destroyer is, at all times, a constant 30 mi/h and the submarine's speed is a constant 15 mi/h.

   (a) Explain why the captain waits until S_1 reaches $(3, 0)$ before ordering a course change to C.

   (b) Using polar coordinates, find an equation $r = f(\theta)$ for the curve C.

   (c) Explain why the time, measured from the initial detection, at which the destroyer intercepts the submarine must be less than $(1 + e^{2\pi/\sqrt{3}})/5$.

Discussion Problems

17. **(a)** Consider the nonlinear pendulum whose oscillations are defined by (6). Use an ODE solver as an aid to determine whether a pendulum of length l will oscillate faster on the earth or on the moon. Use the same initial conditions, but choose these initial conditions so that the pendulum oscillates back and forth.

(b) Which pendulum in part (a) has the greater amplitude?

(c) Are the conclusions in parts (a) and (b) the same when the linear model (7) is used?

18. Consider the initial-value problem

$$\frac{d^2\theta}{dt^2} + \sin\theta = 0, \qquad \theta(0) = \frac{\pi}{12}, \quad \theta'(0) = -\frac{1}{3}$$

for the nonlinear pendulum. Since we cannot solve the differential equation, we can find no explicit solution of this problem. But suppose we wish to determine the first time $t_1 > 0$ for which the pendulum, starting from its initial position to the right, reaches the position OP in Figure 5.34; that is, find the first positive root of $\theta(t) = 0$. In this problem and the next we examine several ways to proceed.

(a) Approximate t_1 by solving the linear problem $d^2\theta/dt^2 + \theta = 0$, $\theta(0) = \frac{\pi}{12}, \theta'(0) = -\frac{1}{3}$.

(b) Use the method illustrated in Example 3 of Section 4.9 to find the first four nonzero terms of a Taylor series solution $\theta(t)$ centered at 0 for the nonlinear initial-value problem. Give the exact values of all coefficients.

(c) Use the first two terms of the Taylor series in part (b) to approximate t_1.

(d) Use the first three terms of the Taylor series in part (b) to approximate t_1.

(e) Use a calculator with root-finding capability or a CAS and the first four terms of the Taylor series in part (b) to approximate t_1.

Mathematica Programming Problem

19. In part (a) of this problem you are led through the commands in Mathematica that enable you to approximate the root t_1 of the equation $\theta(t) = 0$, where $\theta(t)$ is the solution of the nonlinear initial-value problem given in Problem 18. The procedure is easily modified so that any root of $\theta(t) = 0$ can be approximated. (*If you do not have Mathematica, adapt the given procedure by finding the corresponding syntax for the CAS you have on hand.*)

(a) Precisely reproduce and then, in turn, execute each line in the following sequence of commands.

```
sol = NDSolve[{y''[t] + Sin[y[t]] == 0, y[0] == Pi/12, y'[0] == -1/3},
    y, {t, 0, 5}]//Flattensolution = y[t] /. sol
Clear[y]
y[t_] : = Evaluate[solution]
y[t]
gr1 = Plot[y[t], {t, 0, 5}]
root = FindRoot[y[t] == 0, {t, 1}]
```

(b) Appropriately modify the syntax in part (a), and find the next two positive roots of $\theta(t) = 0$.

CHAPTER 5 REVIEW EXERCISES

Answers to odd-numbered problems begin on page A-8.

Answer Problems 1–9 without referring back to the text. Fill in the blank or answer true/false.

1. If a 10-pound weight stretches a spring 2.5 feet, a 32-pound weight will stretch it _____ feet.

2. The period of simple harmonic motion of an 8-pound weight attached to a spring whose constant is 6.25 lb/ft is _____ seconds.

3. The differential equation of a weight on a spring is $x'' + 16x = 0$. If the weight is released at $t = 0$ from 1 meter above the equilibrium position with a downward velocity of 3 m/s, the amplitude of vibrations is _____ meter.

4. Pure resonance cannot take place in the presence of a damping force. _____

5. In the presence of damping, the displacements of a weight on a spring will always approach zero as $t \to \infty$. _____

6. A weight on a spring whose motion is critically damped can possibly pass through the equilibrium position twice. _____

7. At critical damping any increase in damping will result in an _____ system.

8. If simple harmonic motion is described by $x = (\sqrt{2}/2)\sin(2t + \phi)$, the phase angle ϕ is _____ when $x(0) = -\frac{1}{2}$ and $x'(0) = 1$.

9. A 16-pound weight attached to a spring exhibits simple harmonic motion. If the frequency of oscillations is $3/2\pi$ vibrations/second, the spring constant is _____.

10. A 12-pound weight stretches a spring 2 feet. The weight is released from a point 1 foot below the equilibrium position with an upward velocity of 4 ft/s.
    (a) Find the equation describing the resulting simple harmonic motion.
    (b) What are the amplitude, period, and frequency of motion?
    (c) At what times does the weight return to the point 1 foot below the equilibrium position?
    (d) At what times does the weight pass through the equilibrium position moving upward? moving downward?
    (e) What is the velocity of the weight at $t = 3\pi/16$ s?
    (f) At what times is the velocity zero?

11. A force of 2 pounds stretches a spring 1 foot. With one end held fixed, an 8-pound weight is attached to the other end. The system lies on a table that imparts a frictional force numerically equal to $\frac{3}{2}$ times the instantaneous velocity. Initially the weight is displaced 4 inches above the equilibrium position and released from rest. Find the equation of motion if the motion takes place along a horizontal straight line that is taken as the x-axis.

12. A 32-pound weight stretches a spring 6 inches. The weight moves through a medium offering a damping force numerically equal to β times the instantaneous velocity. Determine the values of β for which the system will exhibit oscillatory motion.

13. A spring with constant $k = 2$ is suspended in a liquid that offers a damping force numerically equal to 4 times the instantaneous velocity. If a mass m is suspended from the spring, determine the values of m for which the subsequent free motion is nonoscillatory.

14. The vertical motion of a weight attached to a spring is described by the initial-value problem

$$\frac{1}{4}\frac{d^2x}{dt^2} + \frac{dx}{dt} + x = 0, \qquad x(0) = 4, \quad x'(0) = 2.$$

Determine the maximum vertical displacement.

15. A 4-pound weight stretches a spring 18 inches. A periodic force equal to $f(t) = \cos \gamma t + \sin \gamma t$ is impressed on the system starting at $t = 0$. In the absence of a damping force, for what value of γ will the system be in a state of pure resonance?

16. Find a particular solution for $\dfrac{d^2x}{dt^2} + 2\lambda\dfrac{dx}{dt} + \omega^2 x = A$, where A is a constant force.

17. A 4-pound weight is suspended from a spring whose constant is 3 lb/ft. The entire system is immersed in a fluid offering a damping force numerically equal to the instantaneous velocity. Beginning at $t = 0$, an external force equal to $f(t) = e^{-t}$ is impressed on the system. Determine the equation of motion if the weight is released from rest at a point 2 feet below the equilibrium position.

18. **(a)** Two springs are attached in series as shown in Figure 5.40. If the mass of each spring is ignored, show that the effective spring constant k is given by $1/k = 1/k_1 + 1/k_2$.
(b) A weight of W pounds stretches one spring $\frac{1}{2}$ foot and stretches a different spring $\frac{1}{4}$ foot. The two springs are attached as in the figure, and the weight W is then attached to the double spring. Assume that the motion is free and that there is no damping force present. Determine the equation of motion if the weight is released at a point 1 foot below the equilibrium position with a downward velocity of $\frac{2}{3}$ ft/s.
(c) Show that the maximum speed of the weight is $\frac{2}{3}\sqrt{3g + 1}$.

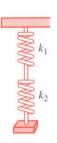

FIGURE 5.40

19. A series circuit contains an inductance of $L = 1$ h, a capacitance of $C = 10^{-4}$ f, and an electromotive force of $E(t) = 100 \sin 50t$ V. Initially the charge q and current i are zero.

(a) Find the equation for the charge at any time.
(b) Find the equation for the current at any time.
(c) Find the times for which the charge on the capacitor is zero.

20. Show that the current $i(t)$ in an LRC series circuit satisfies the differential equation

$$L\frac{d^2i}{dt^2} + R\frac{di}{dt} + \frac{1}{C}i = E'(t),$$

where $E'(t)$ denotes the derivative of $E(t)$.

21. Consider the boundary-value problem

$$y'' + \lambda y = 0, \qquad y(0) = y(2\pi), \quad y'(0) = y'(2\pi).$$

Show that except for the case $\lambda = 0$, there are two independent eigenfunctions corresponding to each eigenvalue.

CHAPTER

6

SERIES SOLUTIONS
OF LINEAR EQUATIONS

INTRODUCTION

Up to now we have primarily solved linear differential equations of order two or higher when the equation had constant coefficients. The only exception was the Cauchy-Euler equation. In applications, higher-order linear equations with variable coefficients are just as important, if not more so, than differential equations with constant coefficients. As mentioned in the introduction to Section 4.7, even a simple linear second-order equation with variable coefficients such as $y'' + xy = 0$ does not possess elementary solutions. We *can* find two linearly independent solutions of this equation, but, as we shall see in Sections 6.2 and 6.4, the solutions are defined by infinite series.

6.1 REVIEW OF POWER SERIES; POWER SERIES SOLUTIONS

■ *Power series* ■ *Radius of convergence* ■ *Interval of convergence*
■ *Analyticity at a point* ■ *Arithmetic of power series* ■ *Power series solution of a DE*

Review of Power Series Section 4.7 notwithstanding, most linear differential equations with variable coefficients cannot be solved in terms of elementary functions. A standard technique for solving higher-order linear differential equations with variable coefficients is to try to find a solution in the form of an infinite series. Often the solution can be found in the form of a power series. Because of this, it is appropriate to list some of the more important facts about power series. For an in-depth review of the infinite series concept you should consult a calculus text.

■ **Definition of a Power Series** A power series in $x - a$ is an infinite series of the form $\sum_{n=0}^{\infty} c_n (x - a)^n$. Such a series is also said to be a power series centered at a. For example, $\sum_{n=1}^{\infty} \frac{(-1)^{n+1}}{n^2} x^n$ is a power series in x; the series is centered at zero.

■ **Convergence** For a specified value of x a power series is a series of constants. If the series equals a finite real constant for the given x, then the series is said to converge at x. If the series does not converge at x, it is said to diverge at x.

■ **Interval of Convergence** Every power series has an interval of convergence. The interval of convergence is the set of all numbers for which the series converges.

■ **Radius of Convergence** Every interval of convergence has a radius of convergence R. For a power series $\sum_{n=0}^{\infty} c_n (x - a)^n$ we have just three possibilities:
   (*i*) The series converges only at its center a. In this case $R = 0$.
   (*ii*) The series converges for all x satisfying $|x - a| < R$, where $R > 0$. The series diverges for $|x - a| > R$.
   (*iii*) The series converges for all x. In this case we write $R = \infty$.

■ **Convergence at an Endpoint** Recall that the absolute-value inequality $|x - a| < R$ is equivalent to $-R < x - a < R$ or $a - R < x < a + R$. If a power series converges for $|x - a| < R$, where $R > 0$, it may or may not converge at the endpoints of the interval $a - R < x < a + R$. Figure 6.1 shows four possible intervals of convergence.

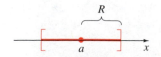

(a) $[a - R, a + R]$
Series converges
at both endpoints.

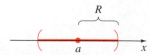

(b) $(a - R, a + R)$
Series diverges
at both endpoints.

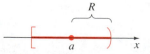

(c) $[a - R, a + R)$
Series converges at $a - R$,
diverges at $a + R$.

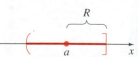

(d) $(a - R, a + R]$
Series diverges at $a - R$,
converges at $a + R$.

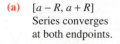

FIGURE 6.1

- **Absolute Convergence** Within its interval of convergence a power series converges absolutely. In other words, for x in the interval of convergence the series of absolute values $\sum_{n=0}^{\infty} |c_n||(x-a)^n|$ converges.
- **Finding the Interval of Convergence** Convergence of a power series can often be determined by the ratio test:

$$\lim_{n \to \infty} \left| \frac{c_{n+1}}{c_n} \right| |x - a| = L.$$

The series will converge absolutely for those values of x for which $L < 1$. From this test we see that the radius of convergence is given by

$$R = \lim_{n \to \infty} \left| \frac{c_n}{c_{n+1}} \right| \tag{1}$$

provided the limit exists.
- **A Power Series Defines a Function** A power series defines a function

$$f(x) = \sum_{n=0}^{\infty} c_n(x-a)^n = c_0 + c_1(x-a) + c_2(x-a)^2 + c_3(x-a)^3 + \cdots$$

whose domain is the interval of convergence of the series. If the series has a radius of convergence $R > 0$, then f is continuous, differentiable, and integrable on the interval $(a - R, a + R)$. Moreover, $f'(x)$ and $\int f(x)\,dx$ can be found from term-by-term differentiation and integration:

$$f'(x) = c_1 + 2c_2(x-a) + 3c_3(x-a)^2 + \cdots = \sum_{n=1}^{\infty} nc_n(x-a)^{n-1}$$

$$\int f(x)\,dx = C + c_0(x-a) + c_1 \frac{(x-a)^2}{2} + c_2 \frac{(x-a)^3}{3} + \cdots = C + \sum_{n=0}^{\infty} c_n \frac{(x-a)^{n+1}}{n+1}.$$

Although the radius of convergence for both these series is R, the interval of convergence may differ from the original series in that convergence at an endpoint may be either lost by differentiation or gained through integration.
- **Series That Are Identically Zero** If $\sum_{n=0}^{\infty} c_n(x-a)^n = 0$, $R > 0$, for all real numbers x in the interval of convergence, then $c_n = 0$ for all n.
- **Analytic at a Point** In calculus it is seen that functions such as e^x, $\cos x$, and $\ln(x-1)$ can be represented by power series by expansions in either Maclaurin or Taylor series. We say that a function f is analytic at point a if it can be represented by a power series in $x - a$ with a positive radius of convergence. The notion of analyticity at a point will be important in Sections 6.2 and 6.3.
- **Arithmetic of Power Series** Power series can be combined through the operations of addition, multiplication, and division. The procedures for power series are similar to the way in which

two polynomials are added, multiplied, and divided—that is, we add coefficients of like powers of x, use the distributive law and collect like terms, and perform long division. For example, if the power series $f(x) = \sum_{n=0}^{\infty} c_n x^n$ and $g(x) = \sum_{n=0}^{\infty} b_n x^n$ both converge for $|x| < R$, then

$$f(x) + g(x) = (c_0 + b_0) + (c_1 + b_1)x + (c_2 + b_2)x^2 + \cdots$$

$$f(x)g(x) = c_0 b_0 + (c_0 b_1 + c_1 b_0)x + (c_0 b_2 + c_1 b_1 + c_2 b_0)x^2 + \cdots.$$

EXAMPLE 1   **Interval of Convergence**

Find the interval of convergence of the power series $\displaystyle\sum_{n=1}^{\infty} \frac{(x-3)^n}{2^n n}$.

SOLUTION   The power series is centered at 3. From (1) the radius of convergence is

$$R = \lim_{n \to \infty} \frac{2^{n+1}(n+1)}{2^n n} = 2.$$

The series converges absolutely for $|x - 3| < 2$ or $1 < x < 5$. At the left endpoint $x = 1$ we find that the series of constants $\sum_{n=1}^{\infty}((-1)^n/n)$ is convergent by the alternating series test. At the right endpoint $x = 5$ we find that the series is the divergent harmonic series $\sum_{n=1}^{\infty}(1/n)$. Thus the interval of convergence is $[1, 5)$. ∎

EXAMPLE 2   **Multiplication of Two Power Series**

Find the first four terms of a power series in x for $e^x \cos x$.

SOLUTION   From calculus the Maclaurin series for e^x and $\cos x$ are, respectively,

$$e^x = 1 + x + \frac{x^2}{2} + \frac{x^3}{6} + \frac{x^4}{24} + \cdots \quad \text{and} \quad \cos x = 1 - \frac{x^2}{2} + \frac{x^4}{24} - \cdots.$$

Multiplying out and collecting like terms yields

$$e^x \cos x = \left(1 + x + \frac{x^2}{2} + \frac{x^3}{6} + \frac{x^4}{24} + \cdots\right)\left(1 - \frac{x^2}{2} + \frac{x^4}{24} - \cdots\right)$$

$$= 1 + (1)x + \left(-\frac{1}{2} + \frac{1}{2}\right)x^2 + \left(-\frac{1}{2} + \frac{1}{6}\right)x^3 + \left(\frac{1}{24} - \frac{1}{4} + \frac{1}{24}\right)x^4 + \cdots$$

$$= 1 + x - \frac{x^3}{3} - \frac{x^4}{6} + \cdots. \quad \blacksquare$$

In Example 2 the interval of convergence for the Maclaurin series for both e^x and $\cos x$ is $(-\infty, \infty)$. Consequently the interval of convergence for the power series for $e^x \cos x$ is also $(-\infty, \infty)$.

EXAMPLE 3   **Division by a Power Series**

Find the first four terms of a power series in x for $\sec x$.

SOLUTION   One way of proceeding is to use the Maclaurin series for $\cos x$ given in Example 2 and then to use long division. Since $\sec x = 1/\cos x$, we have

$$\cos x = 1 - \frac{x^2}{2} + \frac{x^4}{24} - \frac{x^6}{720} + \cdots \overline{\smash{\big)}\ 1}$$

$$1 + \frac{x^2}{2} + \frac{5x^4}{24} + \frac{61x^6}{720} + \cdots$$

$$1 - \frac{x^2}{2} + \frac{x^4}{24} - \frac{x^6}{720} + \cdots$$

$$\frac{x^2}{2} - \frac{x^4}{24} + \frac{x^6}{720} - \cdots$$

$$\frac{x^2}{2} - \frac{x^4}{4} + \frac{x^6}{48} - \cdots$$

$$\frac{5x^4}{24} - \frac{7x^6}{360} + \cdots$$

$$\frac{5x^4}{24} - \frac{5x^6}{48} + \cdots$$

$$\frac{61x^6}{720} - \cdots$$

Thus

$$\sec x = 1 + \frac{x^2}{2} + \frac{5x^4}{24} + \frac{61x^6}{720} + \cdots. \tag{2}$$

The interval of convergence of this series is $(-\pi/2, \pi/2)$. (Why?)   ∎

The procedures illustrated in Examples 2 and 3 are obviously tedious to do by hand. Problems of this sort can be done with minimal fuss using a computer algebra system such as Mathematica or Maple. In Mathematica the division in Example 3 is avoided by using the command **Series[Sec[x], {x, 0, 8}].** See Problems 11–14 in Exercises 6.1.

For the remainder of this section, as well as this chapter, it is important that you become adept at simplifying the sum of two or more power series, each series expressed in summation (sigma) notation, to an expression with a single Σ. This often requires a shift of the summation indices.

EXAMPLE 4   **Adding Two Power Series**

Write $\sum_{n=1}^{\infty} 2nc_n x^{n-1} + \sum_{n=0}^{\infty} 6c_n x^{n+1}$ as one series.

SOLUTION   In order to add the series, we require that both summation indices start with the same number and that the powers of x in each series be "in phase"; that is, if one series starts with a multiple of, say, x to the first power, then we want the other series to start with the same power. By writing

series starts with x for $n = 2$ ↓

series starts with x for $n = 0$ ↓

$$\sum_{n=1}^{\infty} 2nc_n x^{n-1} + \sum_{n=0}^{\infty} 6c_n x^{n+1} = 2 \cdot 1 \cdot c_1 x^0 + \sum_{n=2}^{\infty} 2nc_n x^{n-1} + \sum_{n=0}^{\infty} 6c_n x^{n+1}, \quad \textbf{(3)}$$

we have both series on the right side start with x^1. To get the same summation index we are inspired by the exponents of x; we let $k = n - 1$ in the first series and at the same time let $k = n + 1$ in the second series. Thus the right side of (3) becomes

$$2c_1 + \sum_{k=1}^{\infty} 2(k+1)c_{k+1} x^k + \sum_{k=1}^{\infty} 6c_{k-1} x^k. \quad \textbf{(4)}$$

Recall that the summation index is a "dummy" variable. The fact that $k = n - 1$ in one case and $k = n + 1$ in the other should cause no confusion if you keep in mind that it is the *value* of the summation index that is important. In both cases k takes on the same successive values $1, 2, 3, \ldots$ for $n = 2, 3, 4, \ldots$ (for $k = n - 1$) and $n = 0, 1, 2, \ldots$ (for $k = n + 1$).

We are now in a position to add the series in (4) term by term:

$$\sum_{n=1}^{\infty} 2nc_n x^{n-1} + \sum_{n=0}^{\infty} 6c_n x^{n+1} = 2c_1 + \sum_{k=1}^{\infty} [2(k+1)c_{k+1} + 6c_{k-1}]x^k. \quad \textbf{(5)}$$

If you are not convinced, then write out a few terms on both sides of (5). ∎

Powers Series Solution of a DE We saw in Section 1.1 that the function $y = e^{x^2}$ is an explicit solution of the linear first-order differential equation

$$\frac{dy}{dx} - 2xy = 0. \quad \textbf{(6)}$$

By replacing x by x^2 in the Maclaurin series for e^x we can write the solution of (6) as $y = \sum_{n=0}^{\infty} (x^{2n}/n!)$. This last series converges for all real values of x. In other words, when we know the solution in advance, we can find an infinite series solution of the differential equation.

We now propose to obtain a **power series solution** of equation (6) directly; the method of attack is similar to the technique of undetermined coefficients.

EXAMPLE 5 **Using a Power Series to Solve a DE**

Find a solution of $\dfrac{dy}{dx} - 2xy = 0$ in the form of a power series in x.

SOLUTION If we assume that a solution of the given equation exists in the form

$$y = \sum_{n=0}^{\infty} c_n x^n, \quad \textbf{(7)}$$

we pose the question: Can we determine coefficients c_n for which the power series converges to a function satisfying (6)? Formal* term-by-term differentiation of (7) gives

$$\frac{dy}{dx} = \sum_{n=0}^{\infty} nc_n x^{n-1} = \sum_{n=1}^{\infty} nc_n x^{n-1}.$$

Note that since the first term in the first series (corresponding to $n = 0$) is zero, we begin the summation with $n = 1$. Using the last result and assumption (7), we find

$$\frac{dy}{dx} - 2xy = \sum_{n=1}^{\infty} nc_n x^{n-1} - \sum_{n=0}^{\infty} 2c_n x^{n+1}. \qquad (8)$$

We would like to add the two series in (8). To this end we write

$$\frac{dy}{dx} - 2xy = 1 \cdot c_1 x^0 + \sum_{n=2}^{\infty} nc_n x^{n-1} - \sum_{n=0}^{\infty} 2c_n x^{n+1} \qquad (9)$$

and then proceed as in Example 4 by letting $k = n - 1$ in the first series and $k = n + 1$ in the second. The right side of (9) becomes

$$c_1 + \sum_{k=1}^{\infty} (k + 1)c_{k+1} x^k - \sum_{k=1}^{\infty} 2c_{k-1} x^k.$$

After we add the series termwise, it follows that

$$\frac{dy}{dx} - 2xy = c_1 + \sum_{k=1}^{\infty} [(k + 1)c_{k+1} - 2c_{k-1}]x^k = 0. \qquad (10)$$

Hence in order to have (10) identically zero it is necessary that the coefficients of like powers of x be zero—that is,

$$c_1 = 0 \qquad \text{and} \qquad (k + 1)c_{k+1} - 2c_{k-1} = 0, \quad k = 1, 2, 3, \ldots . \qquad (11)$$

Equation (11) provides a **recurrence relation** that determines the c_k. Since $k + 1 \neq 0$ for all the indicated values of k, we can write (11) as

$$c_{k+1} = \frac{2c_{k-1}}{k + 1}. \qquad (12)$$

Iteration of this formula then gives

$$k = 1, \qquad c_2 = \frac{2}{2}c_0 = c_0$$

$$k = 2, \qquad c_3 = \frac{2}{3}c_1 = 0$$

$$k = 3, \qquad c_4 = \frac{2}{4}c_2 = \frac{1}{2}c_0 = \frac{1}{2!}c_0$$

$$k = 4, \qquad c_5 = \frac{2}{5}c_3 = 0$$

$$k = 5, \qquad c_6 = \frac{2}{6}c_4 = \frac{1}{3 \cdot 2!}c_0 = \frac{1}{3!}c_0$$

---

*At this point we do not know the interval of convergence.

$$k = 6, \qquad c_7 = \frac{2}{7}c_5 = 0$$

$$k = 7, \qquad c_8 = \frac{2}{8}c_6 = \frac{1}{4 \cdot 3!}c_0 = \frac{1}{4!}c_0$$

and so on. Thus from the original assumption (7) we find

$$y = \sum_{n=0}^{\infty} c_n x^n = c_0 + c_1 x + c_2 x^2 + c_3 x^3 + c_4 x^4 + c_5 x^5 + c_6 x^6 + \cdots$$

$$= c_0 + 0 + c_0 x^2 + 0 + \frac{1}{2!}c_0 x^4 + 0 + \frac{1}{3!}c_0 x^6 + 0 + \cdots$$

$$= c_0 \left[1 + x^2 + \frac{1}{2!}x^4 + \frac{1}{3!}x^6 + \cdots \right] = c_0 \sum_{n=0}^{\infty} \frac{x^{2n}}{n!}. \qquad \textbf{(13)}$$

Since the iteration of (12) leaves c_0 completely undetermined, we have in fact found the general solution of (6). ∎

The differential equation in Example 5, like the differential equation in the following example, can be easily solved by prior methods. The point of these two examples is to prepare you for the techniques considered in Sections 6.2 and 6.3.

EXAMPLE 6 **Using a Power Series to Solve a DE**

Find solutions of $4y'' + y = 0$ in the form of a power series in x.

SOLUTION If $y = \sum_{n=0}^{\infty} c_n x^n$, we have seen that $y' = \sum_{n=1}^{\infty} nc_n x^{n-1}$ and so

$$y'' = \sum_{n=1}^{\infty} n(n-1)c_n x^{n-2} = \sum_{n=2}^{\infty} n(n-1)c_n x^{n-2}.$$

Substituting the expressions for y'' and y back into the differential equation gives

$$4y'' + y = \underbrace{\sum_{n=2}^{\infty} 4n(n-1)c_n x^{n-2} + \sum_{n=0}^{\infty} c_n x^n}_{\text{both series start with } x^0}.$$

If we substitute $k = n - 2$ in the first series and $k = n$ in the second, we get (after using, in turn, $n = k + 2$ and $n = k$)

$$4y'' + y = \sum_{k=0}^{\infty} 4(k+2)(k+1)c_{k+2}x^k + \sum_{k=0}^{\infty} c_k x^k$$

$$= \sum_{k=0}^{\infty} [4(k+2)(k+1)c_{k+2} + c_k]x^k = 0.$$

From this last identity it is apparent that for $k = 0, 1, 2, \ldots$

$$4(k+2)(k+1)c_{k+2} + c_k = 0 \qquad \text{or} \qquad c_{k+2} = \frac{-c_k}{4(k+2)(k+1)}.$$

By iterating the last formula we get

$$c_2 = \frac{-c_0}{4 \cdot 2 \cdot 1} = -\frac{c_0}{2^2 \cdot 2!}$$

$$c_3 = \frac{-c_1}{4 \cdot 3 \cdot 2} = -\frac{c_1}{2^2 \cdot 3!}$$

$$c_4 = \frac{-c_2}{4 \cdot 4 \cdot 3} = \frac{c_0}{2^4 \cdot 4!}$$

$$c_5 = \frac{-c_3}{4 \cdot 5 \cdot 4} = \frac{c_1}{2^4 \cdot 5!}$$

$$c_6 = \frac{-c_4}{4 \cdot 6 \cdot 5} = -\frac{c_0}{2^6 \cdot 6!}$$

$$c_7 = \frac{-c_5}{4 \cdot 7 \cdot 6} = -\frac{c_1}{2^6 \cdot 7!}$$

and so forth. This iteration leaves both c_0 and c_1 arbitrary. From the original assumption we have

$$y = c_0 + c_1 x + c_2 x^2 + c_3 x^3 + c_4 x^4 + c_5 x^5 + c_6 x^6 + c_7 x^7 + \cdots$$

$$= c_0 + c_1 x - \frac{c_0}{2^2 \cdot 2!} x^2 - \frac{c_1}{2^2 \cdot 3!} x^3 + \frac{c_0}{2^4 \cdot 4!} x^4 + \frac{c_1}{2^4 \cdot 5!} x^5 - \frac{c_0}{2^6 \cdot 6!} x^6 - \frac{c_1}{2^6 \cdot 7!} x^7 + \cdots$$

or

$$y = c_0 \left[1 - \frac{1}{2^2 \cdot 2!} x^2 + \frac{1}{2^4 \cdot 4!} x^4 - \frac{1}{2^6 \cdot 6!} x^6 + \cdots \right]$$

$$+ c_1 \left[x - \frac{1}{2^2 \cdot 3!} x^3 + \frac{1}{2^4 \cdot 5!} x^5 - \frac{1}{2^6 \cdot 7!} x^7 + \cdots \right]$$

as a general solution. When the series are written in summation notation,

$$y_1(x) = c_0 \sum_{k=0}^{\infty} \frac{(-1)^k}{(2k)!} \left(\frac{x}{2} \right)^{2k} \quad \text{and} \quad y_2(x) = 2c_1 \sum_{k=0}^{\infty} \frac{(-1)^k}{(2k+1)!} \left(\frac{x}{2} \right)^{2k+1},$$

the ratio test can be applied to show that both series converge for all x. You might also recognize the Maclaurin series as $y_1(x) = c_0 \cos(x/2)$ and $y_2(x) = 2c_1 \sin(x/2)$. ∎

SECTION 6.1 EXERCISES

Answers to odd-numbered problems begin on page A-8.

In Problems 1–10 find the interval of convergence of the given power series.

1. $\displaystyle\sum_{n=1}^{\infty} \frac{(-1)^n}{n} x^n$

2. $\displaystyle\sum_{n=1}^{\infty} \frac{x^n}{n^2}$

3. $\displaystyle\sum_{k=1}^{\infty} \frac{2^k}{k} x^k$

4. $\displaystyle\sum_{k=0}^{\infty} \frac{5^k}{k!} x^k$

5. $\displaystyle\sum_{n=1}^{\infty} \frac{(x-3)^n}{n^3}$

6. $\displaystyle\sum_{n=1}^{\infty} \frac{(x+7)^n}{\sqrt{n}}$

7. $\displaystyle\sum_{k=1}^{\infty} \frac{(-1)^k}{10^k}(x-5)^k$ **8.** $\displaystyle\sum_{k=1}^{\infty} \frac{k}{(k+2)^2}(x-4)^k$

9. $\displaystyle\sum_{k=0}^{\infty} k!2^k x^k$ **10.** $\displaystyle\sum_{k=2}^{\infty} \frac{k-1}{k^{2k}}x^k$

In Problems 11–14 find the first four terms of a power series in x for the given function. Calculate the series by hand or use a CAS, as instructed.

11. $e^x \sin x$ **12.** $e^{-x} \cos x$

13. $\sin x \cos x$ **14.** $e^x \ln(1-x)$

In Problems 15–24 solve each differential equation in the manner of the previous chapters and then compare the results with the solutions obtained by assuming a power series solution $y = \sum_{n=0}^{\infty} c_n x^n$.

15. $y' + y = 0$ **16.** $y' = 2y$

17. $y' - x^2 y = 0$ **18.** $y' + x^3 y = 0$

19. $(1-x)y' - y = 0$ **20.** $(1+x)y' - 2y = 0$

21. $y'' + y = 0$ **22.** $y'' - y = 0$

23. $y'' = y'$ **24.** $2y'' + y' = 0$

25. Consider the function $y = J_0(x)$ defined by the power series

$$J_0(x) = \sum_{n=0}^{\infty} \frac{(-1)^n}{2^{2n}(n!)^2}x^{2n}$$

that converges for all x. Show that $J_0(x)$ is a particular solution of the differential equation $xy'' + y' + xy = 0$.

Discussion Problem

26. Suppose the power series $\sum_{k=0}^{\infty} c_k(x-4)^k$ is known to converge at -2 and diverge at 13. Discuss whether the series converges at 10, 7, -7, and 11. Possible answers are *does, does not, might*.

6.2   SOLUTIONS ABOUT ORDINARY POINTS

- *Ordinary points of a DE* ■ *Singular points of a DE*
- *Existence of a power series solution about an ordinary point*
- *Finding a power series solution*

Suppose the linear second-order differential equation

$$a_2(x)y'' + a_1(x)y' + a_0(x)y = 0 \tag{1}$$

is put into the standard form

$$y'' + P(x)y' + Q(x)y = 0 \tag{2}$$

by dividing by the leading coefficient $a_2(x)$. We make the following definition.

> **DEFINITION 6.1** **Ordinary and Singular Points**
>
> A point x_0 is said to be an **ordinary point** of the differential equation (1) if both $P(x)$ and $Q(x)$ are analytic at x_0. A point that is not an ordinary point is said to be a **singular point** of the equation.

EXAMPLE 1 **Ordinary Points**

Every finite value of x is an ordinary point of $y'' + (e^x)y' + (\sin x)y = 0$. In particular we see that $x = 0$ is an ordinary point since both e^x and $\sin x$ are analytic at this point; that is, both functions can be represented by a power series centered at 0. Recall from calculus that

$$e^x = 1 + \frac{x}{1!} + \frac{x^2}{2!} + \cdots \qquad \text{and} \qquad \sin x = x - \frac{x^3}{3!} + \frac{x^5}{5!} - \cdots$$

converge for all finite values of x. ∎

EXAMPLE 2 **Ordinary/Singular Points**

(a) The differential equation $xy'' + (\sin x)y = 0$ has an ordinary point at $x = 0$ since $Q(x) = (\sin x)/x$ possesses the power series expansion

$$Q(x) = 1 - \frac{x^2}{3!} + \frac{x^4}{5!} - \frac{x^6}{7!} + \cdots$$

that converges for all finite values of x.

(b) The differential equation $y'' + (\ln x)y = 0$ has a singular point at $x = 0$ because $Q(x) = \ln x$ possesses no power series in x. ∎

Polynomial Coefficients Primarily we shall be concerned with the case when (1) has *polynomial* coefficients. As a consequence of Definition 6.1 we note that when $a_2(x)$, $a_1(x)$, and $a_0(x)$ are polynomials with *no common factors*, a point $x = x_0$ is

(*i*) an ordinary point if $a_2(x_0) \neq 0$ or

(*ii*) a singular point if $a_2(x_0) = 0$.

EXAMPLE 3 **Singular/Ordinary Points**

(a) The singular points of the equation $(x^2 - 1)y'' + 2xy' + 6y = 0$ are the solutions of $x^2 - 1 = 0$ or $x = \pm 1$. All other finite values of x are ordinary points.

(b) Singular points need not be real numbers. The equation $(x^2 + 1)y'' + xy' - y = 0$ has singular points at the solutions of $x^2 + 1 = 0$, namely, $x = \pm i$. All other finite values of x, real or complex, are ordinary points.

(c) The Cauchy-Euler equation $ax^2y'' + bxy' + cy = 0$, where a, b, and c are constants, has a singular point at $x = 0$. All other finite values of x, real or complex, are ordinary points. ∎

For our purpose ordinary points and singular points will always be finite points. It is possible for a differential equation to have, say, a singular point at infinity. (See Remarks on pages 241–242.)

We state the following theorem about the existence of power series solutions without proof.

THEOREM 6.1   **Existence of Power Series Solutions**

If $x = x_0$ is an ordinary point of the differential equation (2), we can always find two linearly independent solutions in the form of a power series centered at x_0:

$$y = \sum_{n=0}^{\infty} c_n(x - x_0)^n.$$ **(3)**

A series solution converges at least for $|x - x_0| < R$, where R is the distance from x_0 to the closest singular point (real or complex).

A solution of a differential equation of the form given in (3) is said to be a solution *about* the ordinary point x_0. The distance R given in Theorem 6.1 is the minimum value for the radius of convergence. A differential equation could have a finite singular point and yet a solution could be valid for all x; for example, the differential equation may possess a polynomial solution.

To solve a linear second-order equation such as (1) we find two sets of coefficients c_n so that we have two distinct power series $y_1(x)$ and $y_2(x)$, both expanded about the same ordinary point x_0. The procedure used to solve a second-order equation is the same as that used in Example 6 of Section 6.1; that is, we assume a solution $y = \sum_{n=0}^{\infty} c_n(x - x_0)^n$ and then determine the c_n. The general solution of the differential equation is $y = C_1y_1(x) + C_2y_2(x)$; in fact, it can be shown that $C_1 = c_0$ and $C_2 = c_1$, where c_0 and c_1 are arbitrary.

Note

For the sake of simplicity, we assume an ordinary point is always located at $x = 0$, since, if not, the substitution $t = x - x_0$ translates the value $x = x_0$ to $t = 0$.

EXAMPLE 4   **Power Series Solution About an Ordinary Point**

Solve $y'' + xy = 0$.

SOLUTION   We see that $x = 0$ is an ordinary point of the equation. Since there are no finite singular points, Theorem 6.1 guarantees two power series solutions, centered at 0, convergent for $|x| < \infty$. Substituting

$$y = \sum_{n=0}^{\infty} c_n x^n \qquad \text{and} \qquad y'' = \sum_{n=2}^{\infty} n(n-1)c_n x^{n-2}$$

into the differential equation gives

$$y'' + xy = \sum_{n=2}^{\infty} n(n-1)c_n x^{n-2} + \sum_{n=0}^{\infty} c_n x^{n+1}$$

$$= 2 \cdot 1 c_2 x^0 + \underbrace{\sum_{n=3}^{\infty} n(n-1)c_n x^{n-2} + \sum_{n=0}^{\infty} c_n x^{n+1}}_{\text{both series start with } x}.$$

Letting $k = n - 2$ in the first series and $k = n + 1$ in the second, we have

$$y'' + xy = 2c_2 + \sum_{k=1}^{\infty} (k+2)(k+1)c_{k+2} x^k + \sum_{k=1}^{\infty} c_{k-1} x^k$$

$$= 2c_2 + \sum_{k=1}^{\infty} [(k+2)(k+1)c_{k+2} + c_{k-1}]x^k = 0.$$

We must then have $2c_2 = 0$, which implies $c_2 = 0$, and

$$(k+2)(k+1)c_{k+2} + c_{k-1} = 0.$$

The last expression is the same as

$$c_{k+2} = -\frac{c_{k-1}}{(k+2)(k+1)}, \quad k = 1, 2, 3, \ldots.$$

Iteration gives $\quad c_3 = -\dfrac{c_0}{3 \cdot 2}$

$$c_4 = -\frac{c_1}{4 \cdot 3}$$

$$c_5 = -\frac{c_2}{5 \cdot 4} = 0$$

$$c_6 = -\frac{c_3}{6 \cdot 5} = \frac{1}{6 \cdot 5 \cdot 3 \cdot 2}c_0$$

$$c_7 = -\frac{c_4}{7 \cdot 6} = \frac{1}{7 \cdot 6 \cdot 4 \cdot 3}c_1$$

$$c_8 = -\frac{c_5}{8 \cdot 7} = 0$$

$$c_9 = -\frac{c_6}{9 \cdot 8} = -\frac{1}{9 \cdot 8 \cdot 6 \cdot 5 \cdot 3 \cdot 2}c_0$$

$$c_{10} = -\frac{c_7}{10 \cdot 9} = -\frac{1}{10 \cdot 9 \cdot 7 \cdot 6 \cdot 4 \cdot 3}c_1$$

$$c_{11} = -\frac{c_8}{11 \cdot 10} = 0$$

and so on. It should be apparent that both c_0 and c_1 are arbitrary. Now

$$y = c_0 + c_1 x + c_2 x^2 + c_3 x^3 + c_4 x^4 + c_5 x^5 + c_6 x^6 + c_7 x^7 + c_8 x^8$$
$$+ c_9 x^9 + c_{10} x^{10} + c_{11} x^{11} + \cdots$$

$$= c_0 + c_1 x + 0 - \frac{1}{3 \cdot 2} c_0 x^3 - \frac{1}{4 \cdot 3} c_1 x^4 + 0 + \frac{1}{6 \cdot 5 \cdot 3 \cdot 2} c_0 x^6$$

$$+ \frac{1}{7 \cdot 6 \cdot 4 \cdot 3} c_1 x^7 + 0 - \frac{1}{9 \cdot 8 \cdot 6 \cdot 5 \cdot 3 \cdot 2} c_0 x^9 - \frac{1}{10 \cdot 9 \cdot 7 \cdot 6 \cdot 4 \cdot 3} c_1 x^{10} + 0 + \cdots$$

$$= c_0 \left[1 - \frac{1}{3 \cdot 2} x^3 + \frac{1}{6 \cdot 5 \cdot 3 \cdot 2} x^6 - \frac{1}{9 \cdot 8 \cdot 6 \cdot 5 \cdot 3 \cdot 2} x^9 + \cdots \right]$$

$$+ c_1 \left[x - \frac{1}{4 \cdot 3} x^4 + \frac{1}{7 \cdot 6 \cdot 4 \cdot 3} x^7 - \frac{1}{10 \cdot 9 \cdot 7 \cdot 6 \cdot 4 \cdot 3} x^{10} + \cdots \right]. \qquad \blacksquare$$

Although the pattern of the coefficients in Example 4 should be clear, it is sometimes useful to write the solutions in terms of summation notation. By using the properties of the factorial we can write

$$y_1(x) = c_0 \left[1 + \sum_{k=1}^{\infty} \frac{(-1)^k [1 \cdot 4 \cdot 7 \cdots (3k-2)]}{(3k)!} x^{3k} \right]$$

and

$$y_2(x) = c_1 \left[x + \sum_{k=1}^{\infty} \frac{(-1)^k [2 \cdot 5 \cdot 8 \cdots (3k-1)]}{(3k+1)!} x^{3k+1} \right].$$

In this form the ratio test can be used to show that each series converges for $|x| < \infty$.

The differential equation in Example 4 is called **Airy's equation** and is encountered in the study of diffraction of light, diffraction of radio waves around the surface of the earth, aerodynamics, and the deflection of a uniform thin vertical column that bends under its own weight. Other common forms of Airy's equation are $y'' - xy = 0$ and $y'' + \alpha^2 xy = 0$. (See Problem 43 in Exercises 6.4 for an application of the last equation.)

EXAMPLE 5   **Power Series Solution About an Ordinary Point**

Solve $(x^2 + 1)y'' + xy' - y = 0$.

SOLUTION   Since the singular points are $x = \pm i$, a power series solution will converge at least for $|x| < 1$.* The assumption $y = \sum_{n=0}^{\infty} c_n x^n$ leads to

$$(x^2 + 1) \sum_{n=2}^{\infty} n(n-1)c_n x^{n-2} + x \sum_{n=1}^{\infty} nc_n x^{n-1} - \sum_{n=0}^{\infty} c_n x^n$$

$$= \sum_{n=2}^{\infty} n(n-1)c_n x^n + \sum_{n=2}^{\infty} n(n-1)c_n x^{n-2} + \sum_{n=1}^{\infty} nc_n x^n - \sum_{n=0}^{\infty} c_n x^n$$

$$= 2c_2 x^0 - c_0 x^0 + 6c_3 x + c_1 x - c_1 x + \underbrace{\sum_{n=2}^{\infty} n(n-1)c_n x^n}_{k = n}$$

$$+ \underbrace{\sum_{n=4}^{\infty} n(n-1)c_n x^{n-2}}_{k = n-2} + \underbrace{\sum_{n=2}^{\infty} nc_n x^n}_{k = n} - \underbrace{\sum_{n=2}^{\infty} c_n x^n}_{k = n}$$

---

*The **modulus,** or magnitude, of the complex number $x = i$ is $|x| = 1$. If $x = a + bi$ is a singular point, then $|x| = \sqrt{a^2 + b^2}$.

$$= 2c_2 - c_0 + 6c_3x + \sum_{k=2}^{\infty} [k(k-1)c_k + (k+2)(k+1)c_{k+2} + kc_k - c_k]x^k$$

$$= 2c_2 - c_0 + 6c_3x + \sum_{k=2}^{\infty} [(k+1)(k-1)c_k + (k+2)(k+1)c_{k+2}]x^k = 0.$$

Thus
$$2c_2 - c_0 = 0, \qquad c_3 = 0$$

$$(k+1)(k-1)c_k + (k+2)(k+1)c_{k+2} = 0$$

or
$$c_2 = \frac{1}{2}c_0, \qquad c_3 = 0$$

$$c_{k+2} = \frac{1-k}{k+2}c_k, \quad k = 2, 3, 4, \ldots .$$

Iteration of the last formula gives

$$c_4 = -\frac{1}{4}c_2 = -\frac{1}{2\cdot 4}c_0 = -\frac{1}{2^2 2!}c_0$$

$$c_5 = -\frac{2}{5}c_3 = 0$$

$$c_6 = -\frac{3}{6}c_4 = \frac{3}{2\cdot 4\cdot 6}c_0 = \frac{1\cdot 3}{2^3 3!}c_0$$

$$c_7 = -\frac{4}{7}c_5 = 0$$

$$c_8 = -\frac{5}{8}c_6 = -\frac{3\cdot 5}{2\cdot 4\cdot 6\cdot 8}c_0 = -\frac{1\cdot 3\cdot 5}{2^4 4!}c_0$$

$$c_9 = -\frac{6}{9}c_7 = 0$$

$$c_{10} = -\frac{7}{10}c_8 = \frac{3\cdot 5\cdot 7}{2\cdot 4\cdot 6\cdot 8\cdot 10}c_0 = \frac{1\cdot 3\cdot 5\cdot 7}{2^5 5!}c_0$$

and so on. Therefore

$$y = c_0 + c_1x + c_2x^2 + c_3x^3 + c_4x^4 + c_5x^5 + c_6x^6 + c_7x^7 + c_8x^8 + \cdots$$

$$= c_1x + c_0\left[1 + \frac{1}{2}x^2 - \frac{1}{2^2 2!}x^4 + \frac{1\cdot 3}{2^3 3!}x^6 - \frac{1\cdot 3\cdot 5}{2^4 4!}x^8 + \frac{1\cdot 3\cdot 5\cdot 7}{2^5 5!}x^{10} - \cdots\right].$$

The solutions are the polynomial $y_2(x) = c_1x$ and the series

$$y_1(x) = c_0\left[1 + \frac{1}{2}x^2 + \sum_{n=2}^{\infty}(-1)^{n-1}\frac{1\cdot 3\cdot 5 \cdots (2n-3)}{2^n n!}x^{2n}\right], \quad |x| < 1. \quad \blacksquare$$

EXAMPLE 6   **Three-Term Recurrence Relation**

If we seek a solution $y = \sum_{n=0}^{\infty} c_nx^n$ for the equation

$$y'' - (1+x)y = 0,$$

we obtain $c_2 = c_0/2$ and the three-term recurrence relation

$$c_{k+2} = \frac{c_k + c_{k-1}}{(k+1)(k+2)}, \quad k = 1, 2, 3, \ldots .$$

To simplify the iteration we can first choose $c_0 \neq 0$, $c_1 = 0$; this yields one solution. The other solution follows from next choosing $c_0 = 0$, $c_1 \neq 0$. With the first assumption we find

$$c_2 = \frac{1}{2} c_0$$

$$c_3 = \frac{c_1 + c_0}{2 \cdot 3} = \frac{c_0}{2 \cdot 3} = \frac{1}{6} c_0$$

$$c_4 = \frac{c_2 + c_1}{3 \cdot 4} = \frac{c_0}{2 \cdot 3 \cdot 4} = \frac{1}{24} c_0$$

$$c_5 = \frac{c_3 + c_2}{4 \cdot 5} = \frac{c_0}{4 \cdot 5} \left[\frac{1}{2 \cdot 3} + \frac{1}{2} \right] = \frac{1}{30} c_0$$

and so on. Thus one solution is

$$y_1(x) = c_0 \left[1 + \frac{1}{2} x^2 + \frac{1}{6} x^3 + \frac{1}{24} x^4 + \frac{1}{30} x^5 + \cdots \right].$$

Similarly, if we choose $c_0 = 0$, then

$$c_2 = 0$$

$$c_3 = \frac{c_1 + c_0}{2 \cdot 3} = \frac{c_1}{2 \cdot 3} = \frac{1}{6} c_1$$

$$c_4 = \frac{c_2 + c_1}{3 \cdot 4} = \frac{c_1}{3 \cdot 4} = \frac{1}{12} c_1$$

$$c_5 = \frac{c_3 + c_2}{4 \cdot 5} = \frac{c_1}{2 \cdot 3 \cdot 4 \cdot 5} = \frac{1}{120} c_1$$

and so on. Hence another solution is

$$y_2(x) = c_1 \left[x + \frac{1}{6} x^3 + \frac{1}{12} x^4 + \frac{1}{120} x^5 + \cdots \right].$$

Each series converges for all finite values of x.   ∎

Nonpolynomial Coefficients

The next example illustrates how to find a power series solution about an ordinary point of a differential equation when its coefficients are not polynomials. In this example we see an application of multiplication of two power series that was discussed in Section 6.1.

EXAMPLE 7   **DE with Nonpolynomial Coefficients**

Solve $y'' + (\cos x) y = 0$.

SOLUTION   Since $\cos x = 1 - \dfrac{x^2}{2!} + \dfrac{x^4}{4!} - \dfrac{x^6}{6!} + \cdots$, it is seen that $x = 0$

is an ordinary point. Thus the assumption $y = \sum_{n=0}^{\infty} c_n x^n$ leads to

$$y'' + (\cos x)y = \sum_{n=2}^{\infty} n(n-1)c_n x^{n-2} + \left(1 - \frac{x^2}{2!} + \frac{x^4}{4!} - \cdots\right)\sum_{n=0}^{\infty} c_n x^n$$

$$= (2c_2 + 6c_3 x + 12c_4 x^2 + 20c_5 x^3 + \cdots)$$

$$+ \left(1 - \frac{x^2}{2} + \frac{x^4}{24} - \cdots\right)(c_0 + c_1 x + c_2 x^2 + c_3 x^3 + \cdots)$$

$$= 2c_2 + c_0 + (6c_3 + c_1)x + \left(12c_4 + c_2 - \frac{1}{2}c_0\right)x^2 + \left(20c_5 + c_3 - \frac{1}{2}c_1\right)x^3 + \cdots.$$

The last line is to be identically zero, so we must have

$$2c_2 + c_0 = 0, \qquad 6c_3 + c_1 = 0, \qquad 12c_4 + c_2 - \frac{1}{2}c_0 = 0, \qquad 20c_5 + c_3 - \frac{1}{2}c_1 = 0,$$

and so on. Because c_0 and c_1 are arbitrary, we find

$$y_1(x) = c_0\left[1 - \frac{1}{2}x^2 + \frac{1}{12}x^4 - \cdots\right] \qquad \text{and} \qquad y_2(x) = c_1\left[x - \frac{1}{6}x^3 + \frac{1}{30}x^5 - \cdots\right].$$

Since the differential equation has no singular points, both series converge for all finite values of x. ∎

SECTION 6.2 EXERCISES

Answers to odd-numbered problems begin on page A-9.

In Problems 1–14 find two linearly independent power series solutions for each differential equation about the ordinary point $x = 0$.

1. $y'' - xy = 0$
2. $y'' + x^2 y = 0$
3. $y'' - 2xy' + y = 0$
4. $y'' - xy' + 2y = 0$
5. $y'' + x^2 y' + xy = 0$
6. $y'' + 2xy' + 2y = 0$
7. $(x - 1)y'' + y' = 0$
8. $(x + 2)y'' + xy' - y = 0$
9. $(x^2 - 1)y'' + 4xy' + 2y = 0$
10. $(x^2 + 1)y'' - 6y = 0$
11. $(x^2 + 2)y'' + 3xy' - y = 0$
12. $(x^2 - 1)y'' + xy' - y = 0$
13. $y'' - (x + 1)y' - y = 0$
14. $y'' - xy' - (x + 2)y = 0$

In Problems 15–18 use the power series method to solve the given differential equation subject to the indicated initial conditions.

15. $(x - 1)y'' - xy' + y = 0, \quad y(0) = -2, y'(0) = 6$
16. $(x + 1)y'' - (2 - x)y' + y = 0, \quad y(0) = 2, y'(0) = -1$
17. $y'' - 2xy' + 8y = 0, \quad y(0) = 3, y'(0) = 0$
18. $(x^2 + 1)y'' + 2xy' = 0, \quad y(0) = 0, y'(0) = 1$

In Problems 19–22 use the procedure illustrated in Example 7 to find two power series solutions of the given differential equation about the ordinary point $x = 0$.

19. $y'' + (\sin x)y = 0$

20. $xy'' + (\sin x)y = 0$ [*Hint:* See Example 2.]

21. $y'' + e^{-x}y = 0$ **22.** $y'' + e^x y' - y = 0$

In Problems 23 and 24 use the power series method to solve the nonhomogeneous equation.

23. $y'' - xy = 1$ **24.** $y'' - 4xy' - 4y = e^x$

6.3   SOLUTIONS ABOUT SINGULAR POINTS

■ *Regular singular points of a DE* ■ *Irregular singular points of a DE*
■ *Existence of a series solution about a regular singular point* ■ *Method of Frobenius*
■ *Indicial equation* ■ *Indicial roots*

In the preceding section we saw that there is no basic problem in finding two linearly independent power series solutions of

$$a_2(x)y'' + a_1(x)y' + a_0(x)y = 0 \tag{1}$$

about an ordinary point $x = x_0$. However, when $x = x_0$ is a singular point, it is not always possible to find a solution in the form of $y = \sum_{n=0}^{\infty} c_n(x - x_0)^n$; it turns out that we *may* be able to find a series solution of the form $y = \sum_{n=0}^{\infty} c_n(x - x_0)^{n+r}$, where r is a constant to be determined. If r is found to be a number that is not a nonnegative integer, then the last series is not a power series.

Regular and Irregular Singular Points   Singular points are further classified as either regular or irregular. To define these concepts we again put (1) into the standard form

$$y'' + P(x)y' + Q(x)y = 0. \tag{2}$$

DEFINITION 6.2   **Regular and Irregular Singular Points**

A singular point $x = x_0$ of equation (1) is said to be a **regular singular point** if both $(x - x_0)P(x)$ and $(x - x_0)^2 Q(x)$ are analytic at x_0. A singular point that is not regular is said to be an **irregular singular point** of the equation.

Polynomial Coefficients   In the case in which the coefficients in (1) are polynomials with no common factors, Definition 6.2 is equivalent to the following.

Let $a_2(x_0) = 0$. Form $P(x)$ and $Q(x)$ by reducing $a_1(x)/a_2(x)$ and $a_0(x)/a_2(x)$ to lowest terms, respectively. If the factor $(x - x_0)$ appears at most to the first power in the denominator of $P(x)$ and at most to the second power in the denominator of $Q(x)$, then $x = x_0$ is a regular singular point.

EXAMPLE 1   **Classification of Singular Points**

It should be clear that $x = -2$ and $x = 2$ are singular points of the equation

$$(x^2 - 4)^2 y'' + (x - 2)y' + y = 0.$$

Dividing the equation by $(x^2 - 4)^2 = (x - 2)^2 (x + 2)^2$, we find that

$$P(x) = \frac{1}{(x - 2)(x + 2)^2} \quad \text{and} \quad Q(x) = \frac{1}{(x - 2)^2(x + 2)^2}.$$

We now test $P(x)$ and $Q(x)$ at each singular point.

In order for $x = -2$ to be a regular singular point, the factor $x + 2$ can appear at most to the first power in the denominator of $P(x)$ and can appear at most to the second power in the denominator of $Q(x)$. Inspection of $P(x)$ and $Q(x)$ shows that the first condition does not hold, and so we conclude that $x = -2$ is an irregular singular point.

In order for $x = 2$ to be a regular singular point, the factor $x - 2$ can appear at most to the first power in the denominator of $P(x)$ and can appear at most to the second power in the denominator of $Q(x)$. Further inspection of $P(x)$ and $Q(x)$ shows that both these conditions are satisfied, so $x = 2$ is a regular singular point. ∎

EXAMPLE 2   **Classification of Singular Points**

Both $x = 0$ and $x = -1$ are singular points of the differential equation

$$x^2(x + 1)^2 y'' + (x^2 - 1)y' + 2y = 0.$$

Inspection of

$$P(x) = \frac{x - 1}{x^2(x + 1)} \quad \text{and} \quad Q(x) = \frac{2}{x^2(x + 1)^2}$$

shows that $x = 0$ is an irregular singular point since $(x - 0)$ appears to the second power in the denominator of $P(x)$. Note, however, that $x = -1$ is a regular singular point. ∎

EXAMPLE 3   **Classification of Singular Points**

(a) $x = 1$ and $x = -1$ are regular singular points of

$$(1 - x^2)y'' - 2xy' + 30y = 0.$$

(b) $x = 0$ is an irregular singular point of $x^3 y'' - 2xy' + 5y = 0$ since

$$P(x) = -\frac{2}{x^2} \quad \text{and} \quad Q(x) = \frac{5}{x^3}.$$

(c) $x = 0$ is a regular singular point of $xy'' - 2xy' + 5y = 0$ since

$$P(x) = -2 \quad \text{and} \quad Q(x) = \frac{5}{x}.$$ ∎

In part (c) of Example 3 notice that $(x - 0)$ and $(x - 0)^2$ do not even appear in the denominators of $P(x)$ and $Q(x)$, respectively. Remember, these factors can appear at most in this fashion. For a singular point $x = x_0$, any nonnegative power of $(x - x_0)$ less than one (namely, zero) and nonnegative power less than two (namely, zero and one) in the denominators of $P(x)$ and $Q(x)$, respectively, imply that x_0 is a regular singular point.

Also, recall that singular points can be complex numbers. It should be apparent that both $x = 3i$ and $x = -3i$ are regular singular points of the equation $(x^2 + 9)y'' - 3xy' + (1 - x)y = 0$ since

$$P(x) = \frac{-3x}{(x - 3i)(x + 3i)} \quad \text{and} \quad Q(x) = \frac{1 - x}{(x - 3i)(x + 3i)}.$$

EXAMPLE 4   **Cauchy-Euler Equation**

From our discussion of the Cauchy-Euler equation in Section 4.7 we can show that $y_1 = x^2$ and $y_2 = x^2 \ln x$ are solutions of the equation $x^2 y'' - 3xy' + 4y = 0$ on the interval $(0, \infty)$. If the procedure of Theorem 6.1 were attempted at the regular singular point $x = 0$ (that is, an assumed solution of the form $y = \sum_{n=0}^{\infty} c_n x^n$), we would succeed in obtaining only the solution $y_1 = x^2$. The fact that we would not obtain the second solution is not really surprising since $\ln x$ does not possess a Taylor series expansion about $x = 0$. It follows that $y_2 = x^2 \ln x$ does not have a power series in x.   ∎

EXAMPLE 5   **DE with No Power Series Solution**

The differential equation $6x^2 y'' + 5xy' + (x^2 - 1)y = 0$ has a regular singular point at $x = 0$ but does not possess any solution of the form $y = \sum_{n=0}^{\infty} c_n x^n$. By the procedure that we shall now consider it can be shown, however, that there exist two series solutions of the form

$$y = \sum_{n=0}^{\infty} c_n x^{n+1/2} \quad \text{and} \quad y = \sum_{n=0}^{\infty} c_n x^{n-1/3}.$$ ∎

Method of Frobenius   To solve a differential equation such as (1) about a regular singular point we employ the following theorem due to Frobenius.

THEOREM 6.2   **Frobenius' Theorem**

If $x = x_0$ is a regular singular point of equation (1), then there exists at least one series solution of the form

$$y = (x - x_0)^r \sum_{n=0}^{\infty} c_n(x - x_0)^n = \sum_{n=0}^{\infty} c_n(x - x_0)^{n+r}, \tag{3}$$

where the number r is a constant that must be determined. The series will converge at least on some interval $0 < x - x_0 < R$.

Note the words *at least* in the first sentence of Theorem 6.2. This means that, in contrast to Theorem 6.1, Theorem 6.2 does *not* guarantee two solutions of the indicated form. The **method of Frobenius** consists of identifying a regular singular point x_0, substituting $y = \sum_{n=0}^{\infty} c_n (x - x_0)^{n+r}$ into the differential equation, and determining the unknown exponent r and the coefficients c_n.

As in the preceding section, for the sake of simplicity we shall always assume $x_0 = 0$.

EXAMPLE 6 **Series Solution About a Regular Singular Point**

Since $x = 0$ is a regular singular point of the differential equation

$$3xy'' + y' - y = 0, \tag{4}$$

we try a solution of the form $y = \sum_{n=0}^{\infty} c_n x^{n+r}$. Now

$$y' = \sum_{n=0}^{\infty} (n + r)c_n x^{n+r-1} \qquad \text{and} \qquad y'' = \sum_{n=0}^{\infty} (n + r)(n + r - 1)c_n x^{n+r-2}$$

so that

$$3xy'' + y' - y = 3 \sum_{n=0}^{\infty} (n + r)(n + r - 1)c_n x^{n+r-1} + \sum_{n=0}^{\infty} (n + r)c_n x^{n+r-1} - \sum_{n=0}^{\infty} c_n x^{n+r}$$

$$= \sum_{n=0}^{\infty} (n + r)(3n + 3r - 2)c_n x^{n+r-1} - \sum_{n=0}^{\infty} c_n x^{n+r}$$

$$= x^r \left[r(3r - 2)c_0 x^{-1} + \underbrace{\sum_{n=1}^{\infty} (n + r)(3n + 3r - 2)c_n x^{n-1}}_{k = n - 1} - \underbrace{\sum_{n=0}^{\infty} c_n x^n}_{k = n} \right]$$

$$= x^r \left[r(3r - 2)c_0 x^{-1} + \sum_{k=0}^{\infty} [(k + r + 1)(3k + 3r + 1)c_{k+1} - c_k]x^k \right] = 0,$$

which implies $\qquad r(3r - 2)c_0 = 0$

$$(k + r + 1)(3k + 3r + 1)c_{k+1} - c_k = 0, \quad k = 0, 1, 2, \ldots . \tag{5}$$

Since nothing is gained by taking $c_0 = 0$, we must then have

$$r(3r - 2) = 0 \tag{6}$$

and $\qquad c_{k+1} = \dfrac{c_k}{(k + r + 1)(3k + 3r + 1)}, \quad k = 0, 1, 2, \ldots . \tag{7}$

The two values of r that satisfy (6), $r_1 = \frac{2}{3}$ and $r_2 = 0$, when substituted into (7) give two different recurrence relations:

$$r_1 = \tfrac{2}{3}: \quad c_{k+1} = \dfrac{c_k}{(3k + 5)(k + 1)}, \quad k = 0, 1, 2, \ldots; \tag{8}$$

$$r_2 = 0: \quad c_{k+1} = \dfrac{c_k}{(k + 1)(3k + 1)}, \quad k = 0, 1, 2, \ldots . \tag{9}$$

Iteration of (8) gives

$$c_1 = \frac{c_0}{5 \cdot 1}$$

$$c_2 = \frac{c_1}{8 \cdot 2} = \frac{c_0}{2!5 \cdot 8}$$

$$c_3 = \frac{c_2}{11 \cdot 3} = \frac{c_0}{3!5 \cdot 8 \cdot 11}$$

$$c_4 = \frac{c_3}{14 \cdot 4} = \frac{c_0}{4!5 \cdot 8 \cdot 11 \cdot 14}$$

$$\vdots$$

$$c_n = \frac{c_0}{n!5 \cdot 8 \cdot 11 \cdots (3n + 2)}, \quad n = 1, 2, 3, \ldots,$$

whereas iteration of (9) yields

$$c_1 = \frac{c_0}{1 \cdot 1}$$

$$c_2 = \frac{c_1}{2 \cdot 4} = \frac{c_0}{2!1 \cdot 4}$$

$$c_3 = \frac{c_2}{3 \cdot 7} = \frac{c_0}{3!1 \cdot 4 \cdot 7}$$

$$c_4 = \frac{c_3}{4 \cdot 10} = \frac{c_0}{4!1 \cdot 4 \cdot 7 \cdot 10}$$

$$\vdots$$

$$c_n = \frac{c_0}{n!1 \cdot 4 \cdot 7 \cdots (3n - 2)}, \quad n = 1, 2, 3, \ldots.$$

Thus we obtain two series solutions

$$y_1 = c_0 x^{2/3} \left[1 + \sum_{n=1}^{\infty} \frac{1}{n!5 \cdot 8 \cdot 11 \cdots (3n + 2)} x^n \right] \tag{10}$$

and

$$y_2 = c_0 x^0 \left[1 + \sum_{n=1}^{\infty} \frac{1}{n!1 \cdot 4 \cdot 7 \cdots (3n - 2)} x^n \right]. \tag{11}$$

By the ratio test it can be demonstrated that both (10) and (11) converge for all finite values of x. Also it should be clear from the form of (10) and (11) that neither series is a constant multiple of the other and, therefore, $y_1(x)$ and $y_2(x)$ are linearly independent solutions on the x-axis. Hence by the superposition principle

$$y = C_1 y_1(x) + C_2 y_2(x) = C_1 \left[x^{2/3} + \sum_{n=1}^{\infty} \frac{1}{n!5 \cdot 8 \cdot 11 \cdots (3n + 2)} x^{n+2/3} \right]$$

$$+ C_2 \left[1 + \sum_{n=1}^{\infty} \frac{1}{n!1 \cdot 4 \cdot 7 \cdots (3n - 2)} x^n \right]$$

is another solution of (4). On any interval not containing the origin, this combination represents the general solution of the differential equation.

Although Example 6 illustrates the general procedure for using the method of Frobenius, we hasten to point out that we may not always be able to find two solutions so readily or for that matter find two solutions that are infinite series consisting entirely of powers of x.

Indicial Equation   Equation (6) is called the **indicial equation** of the problem, and the values $r_1 = \frac{2}{3}$ and $r_2 = 0$ are called the **indicial roots** or **exponents** of the singularity. In general, if $x = 0$ is a regular singular point of (1), then the functions $xP(x)$ and $x^2Q(x)$ obtained from (2) are analytic at zero; that is, the expansions

$$xP(x) = p_0 + p_1x + p_2x^2 + \cdots \quad \text{and} \quad x^2Q(x) = q_0 + q_1x + q_2x^2 + \cdots \quad \textbf{(12)}$$

are valid on intervals that have a positive radius of convergence. After substituting $y = \sum_{n=0}^{\infty} c_nx^{n+r}$ into (1) or (2) and simplifying, the indicial equation is a quadratic equation in r that results from equating the *total coefficient of the lowest power of x to zero*. It is readily shown that the general indicial equation is

$$r(r - 1) + p_0r + q_0 = 0. \quad \textbf{(13)}$$

We then solve the latter equation for the two values of the exponents and substitute these values into a recurrence relation such as (7). Theorem 6.2 guarantees that at least one solution of the assumed series form can be found.

Cases of Indicial Roots   When using the method of Frobenius we usually distinguish three cases corresponding to the nature of the indicial roots. For the sake of discussion let us suppose that r_1 and r_2 are the *real* solutions of the indicial equation and that, when appropriate, r_1 *denotes the largest root*.

CASE I: Roots Not Differing by an Integer   If r_1 and r_2 are distinct and do not differ by an integer, then there exist two linearly independent solutions of equation (1) of the form

$$y_1 = \sum_{n=0}^{\infty} c_nx^{n+r_1}, \quad c_0 \neq 0 \quad \textbf{(14a)}$$

$$y_2 = \sum_{n=0}^{\infty} b_nx^{n+r_2}, \quad b_0 \neq 0. \quad \textbf{(14b)}$$

EXAMPLE 7   **Case I: Two Solutions of Form (3)**

Solve $2xy'' + (1 + x)y' + y = 0.$ $\qquad$ **(15)**

SOLUTION   If $y = \sum_{n=0}^{\infty} c_nx^{n+r}$, then

$$2xy'' + (1 + x)y' + y = 2\sum_{n=0}^{\infty}(n + r)(n + r - 1)c_nx^{n+r-1} + \sum_{n=0}^{\infty}(n + r)c_nx^{n+r-1}$$

$$+ \sum_{n=0}^{\infty}(n + r)c_nx^{n+r} + \sum_{n=0}^{\infty}c_nx^{n+r}$$

$$= \sum_{n=0}^{\infty} (n+r)(2n+2r-1)c_n x^{n+r-1} + \sum_{n=0}^{\infty} (n+r+1)c_n x^{n+r}$$

$$= x^r \left[r(2r-1)c_0 x^{-1} + \underbrace{\sum_{n=1}^{\infty} (n+r)(2n+2r-1)c_n x^{n-1}}_{k=n-1} + \underbrace{\sum_{n=0}^{\infty} (n+r+1)c_n x^n}_{k=n} \right]$$

$$= x^r \left[r(2r-1)c_0 x^{-1} + \sum_{k=0}^{\infty} [(k+r+1)(2k+2r+1)c_{k+1} + (k+r+1)c_k]x^k \right] = 0,$$

which implies $\qquad\qquad r(2r-1) = 0 \qquad\qquad$ **(16)**

$$(k+r+1)(2k+2r+1)c_{k+1} + (k+r+1)c_k = 0, \quad k = 0, 1, 2, \ldots. \quad \textbf{(17)}$$

From (16) we see that the indicial roots are $r_1 = \frac{1}{2}$ and $r_2 = 0$. Because the difference of the indicial roots $r_1 - r_2$ is not an integer, we are guaranteed, as indicated in (14a) and (14b), two linearly independent solutions of the form $y_1 = \sum_{n=0}^{\infty} c_n x^{n+1/2}$ and $y_2 = \sum_{n=0}^{\infty} c_n x^n$.

For $r_1 = \frac{1}{2}$, we can divide by $k + \frac{3}{2}$ in (17) to obtain

$$c_{k+1} = \frac{-c_k}{2(k+1)}$$

$$c_1 = \frac{-c_0}{2 \cdot 1}$$

$$c_2 = \frac{-c_1}{2 \cdot 2} = \frac{c_0}{2^2 \cdot 2!}$$

$$c_3 = \frac{-c_2}{2 \cdot 3} = \frac{-c_0}{2^3 \cdot 3!}$$

$$\vdots$$

$$c_n = \frac{(-1)^n c_0}{2^n n!}, \quad n = 1, 2, 3, \ldots.$$

Thus $\qquad y_1 = c_0 x^{1/2} \left[1 + \sum_{n=1}^{\infty} \frac{(-1)^n}{2^n n!} x^n \right] = c_0 \sum_{n=0}^{\infty} \frac{(-1)^n}{2^n n!} x^{n+1/2}, \qquad$ **(18)**

which converges for $x \geq 0$. As given, the series is not meaningful for $x < 0$ because of the presence of $x^{1/2}$.

Now for $r_2 = 0$, (17) becomes

$$c_{k+1} = \frac{-c_k}{2k+1}$$

$$c_1 = \frac{-c_0}{1}$$

$$c_2 = \frac{-c_1}{3} = \frac{c_0}{1 \cdot 3}$$

$$c_3 = \frac{-c_2}{5} = \frac{-c_0}{1 \cdot 3 \cdot 5}$$

$$c_4 = \frac{-c_3}{7} = \frac{c_0}{1 \cdot 3 \cdot 5 \cdot 7}$$

$$\vdots$$

$$c_n = \frac{(-1)^n c_0}{1 \cdot 3 \cdot 5 \cdot 7 \cdots (2n-1)}, \quad n = 1, 2, 3, \ldots.$$

We conclude that a second solution of (15) is

$$y_2 = c_0 \left[1 + \sum_{n=1}^{\infty} \frac{(-1)^n}{1 \cdot 3 \cdot 5 \cdot 7 \cdots (2n-1)} x^n \right], \quad |x| < \infty. \quad \textbf{(19)}$$

On the interval $(0, \infty)$, the general solution is $y = C_1 y_1(x) + C_2 y_2(x)$. ∎

When the roots of the indicial equation differ by a positive integer, we may or may not be able to find two solutions of (1) having form (3). If not, then one solution corresponding to the smaller root contains a logarithmic term. When the indicial roots are equal, a second solution *always* contains a logarithm. This latter situation is analogous to the solutions of the Cauchy-Euler differential equation when the roots of the auxiliary equation are equal. We have the next two cases.

CASE II: Roots Differing by a Positive Integer   If $r_1 - r_2 = N$, where N is a positive integer, then there exist two linearly independent solutions of equation (1) of the form

$$y_1 = \sum_{n=0}^{\infty} c_n x^{n+r_1}, \quad c_0 \neq 0 \quad \textbf{(20a)}$$

$$y_2 = C y_1(x) \ln x + \sum_{n=0}^{\infty} b_n x^{n+r_2}, \quad b_0 \neq 0, \quad \textbf{(20b)}$$

where C is a constant that could be zero.

CASE III: Equal Indicial Roots   If $r_1 = r_2$, there always exist two linearly independent solutions of equation (1) of the form

$$y_1 = \sum_{n=0}^{\infty} c_n x^{n+r_1}, \quad c_0 \neq 0 \quad \textbf{(21a)}$$

$$y_2 = y_1(x) \ln x + \sum_{n=1}^{\infty} b_n x^{n+r_1}. \quad \textbf{(21b)}$$

EXAMPLE 8   **Case II: Two Solutions of Form (3)**

Solve $xy'' + (x - 6)y' - 3y = 0$. $\qquad\qquad$ **(22)**

SOLUTION   The assumption $y = \sum_{n=0}^{\infty} c_n x^{n+r}$ leads to

$$xy'' + (x - 6)y' - 3y$$

$$= \sum_{n=0}^{\infty} (n+r)(n+r-1)c_n x^{n+r-1} - 6 \sum_{n=0}^{\infty} (n+r)c_n x^{n+r-1} + \sum_{n=0}^{\infty} (n+r)c_n x^{n+r} - 3 \sum_{n=0}^{\infty} c_n x^{n+r}$$

$$= x^r \left[r(r-7)c_0 x^{-1} + \underbrace{\sum_{n=1}^{\infty} (n+r)(n+r-7)c_n x^{n-1}}_{k = n-1} + \underbrace{\sum_{n=0}^{\infty} (n+r-3)c_n x^n}_{k = n} \right]$$

$$= x^r \left[r(r-7)c_0 x^{-1} + \sum_{k=0}^{\infty} [(k+r+1)(k+r-6)c_{k+1} + (k+r-3)c_k]x^k \right] = 0.$$

Thus $r(r-7) = 0$ so that $r_1 = 7$, $r_2 = 0$, $r_1 - r_2 = 7$, and

$$(k+r+1)(k+r-6)c_{k+1} + (k+r-3)c_k = 0, \quad k = 0, 1, 2, \ldots. \quad \textbf{(23)}$$

For the smaller root $r_2 = 0$, (23) becomes

$$(k+1)(k-6)c_{k+1} + (k-3)c_k = 0. \quad \textbf{(24)}$$

Since $k - 6 = 0$ when $k = 6$, we do not divide by this term until $k > 6$. We find

$$1 \cdot (-6)c_1 + (-3)c_0 = 0$$
$$2 \cdot (-5)c_2 + (-2)c_1 = 0$$
$$3 \cdot (-4)c_3 + (-1)c_2 = 0$$
$$\left. \begin{array}{l} 4 \cdot (-3)c_4 + 0 \cdot c_3 = 0 \\ 5 \cdot (-2)c_5 + 1 \cdot c_4 = 0 \\ 6 \cdot (-1)c_6 + 2 \cdot c_5 = 0 \\ 7 \cdot 0c_7 + 3 \cdot c_6 = 0 \end{array} \right]$$

implies $c_4 = c_5 = c_6 = 0$
← but c_0 and c_7 can be chosen arbitrarily

Hence

$$c_1 = -\frac{1}{2}c_0$$

$$c_2 = -\frac{1}{5}c_1 = \frac{1}{10}c_0 \qquad \textbf{(25)}$$

$$c_3 = -\frac{1}{12}c_2 = -\frac{1}{120}c_0.$$

Now for $k \geq 7$,
$$c_{k+1} = \frac{-(k-3)}{(k+1)(k-6)}c_k.$$

Iterating this last formula gives

$$c_8 = \frac{-4}{8 \cdot 1}c_7$$

$$c_9 = \frac{-5}{9 \cdot 2}c_8 = \frac{4 \cdot 5}{2! 8 \cdot 9}c_7$$

$$c_{10} = \frac{-6}{10 \cdot 3}c_9 = \frac{-4 \cdot 5 \cdot 6}{3! 8 \cdot 9 \cdot 10}c_7$$

$$\vdots$$

$$c_n = \frac{(-1)^{n+1} 4 \cdot 5 \cdot 6 \cdots (n-4)}{(n-7)! 8 \cdot 9 \cdot 10 \cdots n}c_7, \quad n = 8, 9, 10, \ldots. \quad \textbf{(26)}$$

If we choose $c_7 = 0$ and $c_0 \neq 0$, we obtain the polynomial solution

$$y_1 = c_0 \left[1 - \frac{1}{2}x + \frac{1}{10}x^2 - \frac{1}{120}x^3 \right], \quad \textbf{(27)}$$

but when $c_7 \neq 0$ and $c_0 = 0$, it follows that a second, though infinite series, solution is

$$y_2 = c_7 \left[x^7 + \sum_{n=8}^{\infty} \frac{(-1)^{n+1} 4 \cdot 5 \cdot 6 \cdots (n-4)}{(n-7)! \, 8 \cdot 9 \cdot 10 \cdots n} x^n \right]$$

$$= c_7 \left[x^7 + \sum_{k=1}^{\infty} \frac{(-1)^k 4 \cdot 5 \cdot 6 \cdots (k+3)}{k! \, 8 \cdot 9 \cdot 10 \cdots (k+7)} x^{k+7} \right], \quad |x| < \infty. \quad \textbf{(28)}$$

Finally, the general solution of (22) on the interval $(0, \infty)$ is

$$y = C_1 y_1(x) + C_2 y_2(x)$$

$$= C_1 \left[1 - \frac{1}{2} x + \frac{1}{10} x^2 - \frac{1}{120} x^3 \right] + C_2 \left[x^7 + \sum_{k=1}^{\infty} \frac{(-1)^k 4 \cdot 5 \cdot 6 \cdots (k+3)}{k! \, 8 \cdot 9 \cdot 10 \cdots (k+7)} x^{k+7} \right]. \quad \blacksquare$$

It is interesting to observe that in Example 8 we did not use the larger root $r_1 = 7$. Had we done so, we would have obtained a series solution of the form* $y = \sum_{n=0}^{\infty} c_n x^{n+7}$, where the c_n are defined by (23) with $r_1 = 7$:

$$c_{k+1} = \frac{-(k+4)}{(k+8)(k+1)} c_k, \quad k = 0, 1, 2, \ldots.$$

Iteration of this latter recurrence relation then would yield only *one* solution—namely, the solution given by (28) (with c_0 playing the part of c_7).

When the roots of the indicial equation differ by a positive integer, the second solution *may* contain a logarithm. In practice this is something we do not know in advance but is determined after we have found the indicial roots and have carefully examined the recurrence relation that defines the coefficients c_n. As the foregoing example shows, we just may be lucky enough to find two solutions that involve only powers of x. On the other hand, if we fail to find a second series-type solution, we can always use the fact that

$$y_2 = y_1(x) \int \frac{e^{-\int P(x) \, dx}}{y_1^2(x)} \, dx \quad \textbf{(29)}$$

is also a solution of the equation $y'' + P(x) y' + Q(x) y = 0$ whenever y_1 is a known solution (see Section 4.2).

EXAMPLE 9   Case II: One Solution of Form (3)

Find the general solution of $xy'' + 3y' - y = 0$.

SOLUTION   You should verify that the indicial roots are $r_1 = 0$, $r_2 = -2$, $r_1 - r_2 = 2$ and that the method of Frobenius yields only one solution:

$$y_1 = \sum_{n=0}^{\infty} \frac{2}{n! \, (n+2)!} x^n = 1 + \frac{1}{3} x + \frac{1}{24} x^2 + \frac{1}{360} x^3 + \cdots. \quad \textbf{(30)}$$

---

*Observe that both (28) and this series start with the power x^7. In Case II it is always a good idea to work with the smaller root first.

From (29) we obtain a second solution:

$$y_2 = y_1(x) \int \frac{e^{-\int (3/x)dx}}{y_1^2(x)} dx = y_1(x) \int \frac{dx}{x^3 \left[1 + \dfrac{1}{3}x + \dfrac{1}{24}x^2 + \dfrac{1}{360}x^3 + \cdots \right]^2}$$

$$= y_1(x) \int \frac{dx}{x^3 \left[1 + \dfrac{2}{3}x + \dfrac{7}{36}x^2 + \dfrac{1}{30}x^3 + \cdots \right]} \qquad \leftarrow \text{squaring}$$

$$= y_1(x) \int \frac{1}{x^3} \left[1 - \dfrac{2}{3}x + \dfrac{1}{4}x^2 - \dfrac{19}{270}x^3 + \cdots \right] dx \qquad \leftarrow \text{long division}$$

$$= y_1(x) \int \left[\dfrac{1}{x^3} - \dfrac{2}{3x^2} + \dfrac{1}{4x} - \dfrac{19}{270} + \cdots \right] dx$$

$$= y_1(x) \left[-\dfrac{1}{2x^2} + \dfrac{2}{3x} + \dfrac{1}{4} \ln x - \dfrac{19}{270}x + \cdots \right]$$

or $\qquad y_2 = \dfrac{1}{4} y_1(x) \ln x + y_1(x) \left[-\dfrac{1}{2x^2} + \dfrac{2}{3x} - \dfrac{19}{270}x + \cdots \right].$ $\qquad$ **(31)**

Hence on the interval $(0, \infty)$ the general solution is

$$y = C_1 y_1(x) + C_2 \left[\dfrac{1}{4} y_1(x) \ln x + y_1(x) \left(-\dfrac{1}{2x^2} + \dfrac{2}{3x} - \dfrac{19}{270}x + \cdots \right) \right], \qquad \textbf{(32)}$$

where $y_1(x)$ is defined by (30). $\qquad$ ∎

EXAMPLE 10 **Case III: Finding the Second Solution**

Find the general solution of $xy'' + y' - 4y = 0$.

SOLUTION $\qquad$ The assumption $y = \sum_{n=0}^{\infty} c_n x^{n+r}$ leads to

$$xy'' + y' - 4y = \sum_{n=0}^{\infty} (n + r)(n + r - 1)c_n x^{n+r-1} + \sum_{n=0}^{\infty} (n + r)c_n x^{n+r-1} - 4 \sum_{n=0}^{\infty} c_n x^{n+r}$$

$$= \sum_{n=0}^{\infty} (n + r)^2 c_n x^{n+r-1} - 4 \sum_{n=0}^{\infty} c_n x^{n+r}$$

$$= x^r \left[r^2 c_0 x^{-1} + \underbrace{\sum_{n=1}^{\infty} (n + r)^2 c_n x^{n-1}}_{k = n - 1} - 4 \underbrace{\sum_{n=0}^{\infty} c_n x^n}_{k = n} \right]$$

$$= x^r \left[r^2 c_0 x^{-1} + \sum_{k=0}^{\infty} [(k + r + 1)^2 c_{k+1} - 4c_k]x^k \right] = 0.$$

Therefore $r^2 = 0$, and so the indicial roots are equal: $r_1 = r_2 = 0$.
Moreover, we have

$$(k + r + 1)^2 c_{k+1} - 4c_k = 0, \quad k = 0, 1, 2, \ldots . \qquad \textbf{(33)}$$

Clearly the root $r_1 = 0$ will yield only one solution corresponding to the coefficients defined by the iteration of

$$c_{k+1} = \frac{4c_k}{(k+1)^2}, \quad k = 0, 1, 2, \ldots.$$

The result is
$$y_1 = c_0 \sum_{n=0}^{\infty} \frac{4^n}{(n!)^2} x^n, \quad |x| < \infty. \tag{34}$$

To obtain the second linearly independent solution we set $c_0 = 1$ in (34) and then use (29):

$$y_2 = y_1(x) \int \frac{e^{-\int (1/x)\,dx}}{y_1^2(x)}\,dx = y_1(x) \int \frac{dx}{x\left[1 + 4x + 4x^2 + \frac{16}{9}x^3 + \cdots\right]^2}$$

$$= y_1(x) \int \frac{dx}{x\left[1 + 8x + 24x^2 + \frac{16}{9}x^3 + \cdots\right]}$$

$$= y_1(x) \int \frac{1}{x}\left[1 - 8x + 40x^2 - \frac{1472}{9}x^3 + \cdots\right]dx$$

$$= y_1(x) \int \left[\frac{1}{x} - 8 + 40x - \frac{1472}{9}x^2 + \cdots\right]dx$$

$$= y_1(x)\left[\ln x - 8x + 20x^2 - \frac{1472}{27}x^3 + \cdots\right].$$

Thus on the interval $(0, \infty)$ the general solution is

$$y = C_1 y_1(x) + C_2\left[y_1(x)\ln x + y_1(x)\left(-8x + 20x^2 - \frac{1472}{27}x^3 + \cdots\right)\right],$$

where $y_1(x)$ is defined by (34). ∎

Use of Computers The operations performed in Examples 9 and 10—squaring a series, long division by a series, and integration of the quotient—can, of course, be carried out by hand. But life can be simplified since all these operations, including the indicated multiplication in (31), can be done with relative ease with the help of a CAS such as Mathematica, Maple, or Derive.

Remarks

(i) We deliberately have not addressed several additional complications of solving a differential equation such as (1) about a singular point x_0. The indicial roots r_1 and r_2 could be complex numbers. In this case the inequality $r_1 > r_2$ is meaningless and must be replaced with $\text{Re}(r_1) > \text{Re}(r_2)$ (if $r = \alpha + i\beta$, then $\text{Re}(r) = \alpha$). In particular, when the indicial equation has real coefficients, the complex roots will be a conjugate pair $r_1 = \alpha + i\beta$, $r_2 = \alpha - i\beta$, and $r_1 - r_2 = 2i\beta \neq$ integer. Thus for $x_0 = 0$ there always exist two solutions, $y_1 = \sum_{n=0}^{\infty} c_n x^{n+r_1}$ and $y_2 = \sum_{n=0}^{\infty} c_n x^{n+r_2}$. Both solutions give complex values of y for each real choice of x. We can overcome this difficulty by using the superposition principle and forming appropriate linear combinations of $y_1(x)$ and $y_2(x)$ to yield real solutions (see Case III in Section 4.7).

(*ii*) If $x_0 = 0$ is an irregular singular point, we may not be able to find *any* solution of the form $y = \sum_{n=0}^{\infty} c_n x^{n+r}$.

(*iii*) In the advanced study of differential equations it is sometimes important to examine the nature of a singular point at ∞. A differential equation is said to have a singular point at ∞ if, after the substitution $z = 1/x$, the resulting equation has a singular point at $z = 0$. For example, the differential equation $y'' + xy = 0$ has no finite singular points. However, by the Chain Rule the substitution $z = 1/x$ transforms the equation into

$$z^5 \frac{d^2y}{dz^2} + 2z^4 \frac{dy}{dz} + y = 0.$$

(Verify this.) Inspection of $P(z) = 2/z$ and $Q(z) = 1/z^5$ shows that $z = 0$ is an irregular singular point of the equation. Hence ∞ is an irregular singular point.

SECTION 6.3 EXERCISES

Answers to odd-numbered problems begin on page A-9.

In Problems 1–10 determine the singular points of each differential equation. Classify each singular point as regular or irregular.

1. $x^3y'' + 4x^2y' + 3y = 0$ **2.** $xy'' - (x + 3)^{-2}y = 0$

3. $(x^2 - 9)^2y'' + (x + 3)y' + 2y = 0$

4. $y'' - \frac{1}{x}y' + \frac{1}{(x-1)^3}y = 0$

5. $(x^3 + 4x)y'' - 2xy' + 6y = 0$

6. $x^2(x - 5)^2y'' + 4xy' + (x^2 - 25)y = 0$

7. $(x^2 + x - 6)y'' + (x + 3)y' + (x - 2)y = 0$

8. $x(x^2 + 1)^2y'' + y = 0$

9. $x^3(x^2 - 25)(x - 2)^2y'' + 3x(x - 2)y' + 7(x + 5)y = 0$

10. $(x^3 - 2x^2 - 3x)^2y'' + x(x - 3)^2y' - (x + 1)y = 0$

In Problems 11–22 show that the indicial roots do not differ by an integer. Use the method of Frobenius to obtain two linearly independent series solutions about the regular singular point $x_0 = 0$. Form the general solution on $(0, \infty)$.

11. $2xy'' - y' + 2y = 0$ **12.** $2xy'' + 5y' + xy = 0$

13. $4xy'' + \frac{1}{2}y' + y = 0$ **14.** $2x^2y'' - xy' + (x^2 + 1)y = 0$

15. $3xy'' + (2 - x)y' - y = 0$ **16.** $x^2y'' - \left(x - \frac{2}{9}\right)y = 0$

17. $2xy'' - (3 + 2x)y' + y = 0$ **18.** $x^2y'' + xy' + \left(x^2 - \frac{4}{9}\right)y = 0$

19. $9x^2y'' + 9x^2y' + 2y = 0$ **20.** $2x^2y'' + 3xy' + (2x - 1)y = 0$

21. $2x^2y'' - x(x - 1)y' - y = 0$ **22.** $x(x - 2)y'' + y' - 2y = 0$

In Problems 23–30 show that the indicial roots differ by an integer. Use the method of Frobenius to obtain two linearly independent series solutions about the regular singular point $x_0 = 0$. Form the general solution on $(0, \infty)$.

23. $xy'' + 2y' - xy = 0$ **24.** $x^2y'' + xy' + \left(x^2 - \dfrac{1}{4}\right)y = 0$

25. $x(x - 1)y'' + 3y' - 2y = 0$ **26.** $y'' + \dfrac{3}{x}y' - 2y = 0$

27. $xy'' + (1 - x)y' - y = 0$ **28.** $xy'' + y = 0$

29. $xy'' + y' + y = 0$ **30.** $xy'' - xy' + y = 0$

6.4   TWO SPECIAL EQUATIONS

■ *Bessel's equation* ■ *Legendre's equation* ■ *Solution of Bessel's equation*
■ *Bessel functions of the first kind* ■ *Bessel functions of the second kind*
■ *Parametric Bessel equation* ■ *Recurrence relations* ■ *Spherical Bessel functions*
■ *Solution of Legendre's equation* ■ *Legendre polynomials*

The two equations

$$x^2y'' + xy' + (x^2 - \nu^2)y = 0 \tag{1}$$

$$(1 - x^2)y'' - 2xy' + n(n + 1)y = 0 \tag{2}$$

occur frequently in advanced studies in applied mathematics, physics, and engineering. They are called **Bessel's equation** and **Legendre's equation,** respectively. In solving (1) we shall assume $\nu \geq 0$, whereas in (2) we shall consider only the case when n is a nonnegative integer. Since we seek series solutions of each equation about $x = 0$, we observe that the origin is a regular singular point of Bessel's equation, but it is an ordinary point of Legendre's equation.

Solution of Bessel's Equation   If we assume $y = \sum_{n=0}^{\infty} c_n x^{n+r}$, then

$$x^2y'' + xy' + (x^2 - \nu^2)y = \sum_{n=0}^{\infty} c_n(n + r)(n + r - 1)x^{n+r} + \sum_{n=0}^{\infty} c_n(n + r)x^{n+r} + \sum_{n=0}^{\infty} c_n x^{n+r+2}$$

$$- \nu^2 \sum_{n=0}^{\infty} c_n x^{n+r}$$

$$= c_0(r^2 - r + r - \nu^2)x^r$$

$$+ x^r \sum_{n=1}^{\infty} c_n[(n + r)(n + r - 1) + (n + r) - \nu^2]x^n + x^r \sum_{n=0}^{\infty} c_n x^{n+2}$$

$$= c_0(r^2 - \nu^2)x^r + x^r \sum_{n=1}^{\infty} c_n[(n + r)^2 - \nu^2]x^n + x^r \sum_{n=0}^{\infty} c_n x^{n+2}. \tag{3}$$

From (3) we see that the indicial equation is $r^2 - \nu^2 = 0$ so that the indicial roots are $r_1 = \nu$ and $r_2 = -\nu$. When $r_1 = \nu$, (3) becomes

$$x^\nu \sum_{n=1}^{\infty} c_n n(n + 2\nu)x^n + x^\nu \sum_{n=0}^{\infty} c_n x^{n+2}$$

$$= x^\nu \left[(1 + 2\nu)c_1 x + \underbrace{\sum_{n=2}^{\infty} c_n n(n + 2\nu)x^n}_{k = n - 2} + \underbrace{\sum_{n=0}^{\infty} c_n x^{n+2}}_{k = n} \right]$$

$$= x^\nu \left[(1 + 2\nu)c_1 x + \sum_{k=0}^{\infty} [(k + 2)(k + 2 + 2\nu)c_{k+2} + c_k]x^{k+2} \right] = 0.$$

Therefore by the usual argument we can write

$$(1 + 2\nu)c_1 = 0$$

$$(k + 2)(k + 2 + 2\nu)c_{k+2} + c_k = 0$$

or
$$c_{k+2} = \frac{-c_k}{(k + 2)(k + 2 + 2\nu)}, \quad k = 0, 1, 2, \ldots. \tag{4}$$

The choice $c_1 = 0$ in (4) implies $c_3 = c_5 = c_7 = \cdots = 0$, so for $k = 0, 2, 4, \ldots$ we find, after letting $k + 2 = 2n$, $n = 1, 2, 3, \ldots$, that

$$c_{2n} = -\frac{c_{2n-2}}{2^2 n(n + \nu)}. \tag{5}$$

Thus
$$c_2 = -\frac{c_0}{2^2 \cdot 1 \cdot (1 + \nu)}$$

$$c_4 = -\frac{c_2}{2^2 \cdot 2(2 + \nu)} = \frac{c_0}{2^4 \cdot 1 \cdot 2(1 + \nu)(2 + \nu)}$$

$$c_6 = -\frac{c_4}{2^2 \cdot 3(3 + \nu)} = -\frac{c_0}{2^6 \cdot 1 \cdot 2 \cdot 3(1 + \nu)(2 + \nu)(3 + \nu)}$$

$$\vdots$$

$$c_{2n} = \frac{(-1)^n c_0}{2^{2n} n!(1 + \nu)(2 + \nu) \cdots (n + \nu)}, \quad n = 1, 2, 3, \ldots. \tag{6}$$

It is standard practice to choose c_0 to be a specific value—namely,

$$c_0 = \frac{1}{2^\nu \Gamma(1 + \nu)},$$

where $\Gamma(1 + \nu)$ is the gamma function. See Appendix I. Since this latter function possesses the convenient property $\Gamma(1 + \alpha) = \alpha\Gamma(\alpha)$, we can reduce the indicated product in the denominator of (6) to one term. For example,

$$\Gamma(1 + \nu + 1) = (1 + \nu)\Gamma(1 + \nu)$$

$$\Gamma(1 + \nu + 2) = (2 + \nu)\Gamma(2 + \nu) = (2 + \nu)(1 + \nu)\Gamma(1 + \nu).$$

Hence we can write (6) as

$$c_{2n} = \frac{(-1)^n}{2^{2n+\nu} n!(1 + \nu)(2 + \nu) \cdots (n + \nu)\Gamma(1 + \nu)} = \frac{(-1)^n}{2^{2n+\nu} n!\Gamma(1 + \nu + n)}$$

for $n = 0, 1, 2, \ldots$.

Bessel Functions of the First Kind The series solution $y = \sum_{n=0}^{\infty} c_{2n}x^{2n+\nu}$ is usually denoted by $J_\nu(x)$:

$$J_\nu(x) = \sum_{n=0}^{\infty} \frac{(-1)^n}{n!\,\Gamma(1+\nu+n)} \left(\frac{x}{2}\right)^{2n+\nu}. \tag{7}$$

If $\nu \geq 0$, the series converges at least on the interval $[0, \infty)$. Also, for the second exponent $r_2 = -\nu$ we obtain, in exactly the same manner,

$$J_{-\nu}(x) = \sum_{n=0}^{\infty} \frac{(-1)^n}{n!\,\Gamma(1-\nu+n)} \left(\frac{x}{2}\right)^{2n-\nu}. \tag{8}$$

The functions $J_\nu(x)$ and $J_{-\nu}(x)$ are called **Bessel functions of the first kind** of order ν and $-\nu$, respectively. Depending on the value of ν, (8) may contain negative powers of x and hence converge on $(0, \infty)$.*

Now some care must be taken in writing the general solution of (1). When $\nu = 0$, it is apparent that (7) and (8) are the same. If $\nu > 0$ and $r_1 - r_2 = \nu - (-\nu) = 2\nu$ is not a positive integer, it follows from Case I of Section 6.3 that $J_\nu(x)$ and $J_{-\nu}(x)$ are linearly independent solutions of (1) on $(0, \infty)$, and so the general solution of the interval is $y = c_1 J_\nu(x) + c_2 J_{-\nu}(x)$. But we also know from Case II of Section 6.3 that when $r_1 - r_2 = 2\nu$ is a positive integer, a second series solution of (1) *may* exist. In this second case we distinguish two possibilities. When $\nu = m = $ positive integer, $J_{-m}(x)$ defined by (8) and $J_m(x)$ are not linearly independent solutions. It can be shown that J_{-m} is a constant multiple of J_m (see Property (*i*) on page 247). In addition, $r_1 - r_2 = 2\nu$ can be a positive integer when ν is half an odd positive integer. It can be shown in this latter event that $J_\nu(x)$ and $J_{-\nu}(x)$ are linearly independent. In other words, the general solution of (1) on $(0, \infty)$ is

$$y = c_1 J_\nu(x) + c_2 J_{-\nu}(x), \quad \nu \neq \text{integer}. \tag{9}$$

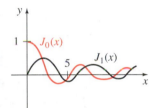

FIGURE 6.2

The graphs of $y = J_0(x)$ and $y = J_1(x)$ are given in Figure 6.2.

EXAMPLE 1 **General Solution: ν Not an Integer**

By identifying $\nu^2 = \frac{1}{4}$ and $\nu = \frac{1}{2}$ we can see from (9) that the general solution of the equation $x^2 y'' + xy' + (x^2 - \frac{1}{4})y = 0$ on $(0, \infty)$ is $y = c_1 J_{1/2}(x) + c_2 J_{-1/2}(x)$. ∎

Bessel Functions of the Second Kind If $\nu \neq$ integer, the function defined by the linear combination

$$Y_\nu(x) = \frac{\cos \nu\pi\, J_\nu(x) - J_{-\nu}(x)}{\sin \nu\pi} \tag{10}$$

and the function $J_\nu(x)$ are linearly independent solutions of (1). Thus another form of the general solution of (1) is $y = c_1 J_\nu(x) + c_2 Y_\nu(x)$, provided $\nu \neq$ integer. As $\nu \to m$, m an integer, (10) has the indeterminate form 0/0. However, it can be shown by L'Hôpital's rule that $\lim_{\nu \to m} Y_\nu(x)$

---

*When we replace x by $|x|$, the series given in (7) and (8) converge for $0 < |x| < \infty$.

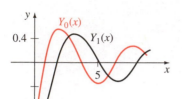

FIGURE 6.3

exists. Moreover, the function

$$Y_m(x) = \lim_{\nu \to m} Y_\nu(x)$$

and $J_m(x)$ are linearly independent solutions of $x^2 y'' + xy' + (x^2 - m^2)y = 0$. Hence for *any* value of ν the general solution of (1) on $(0, \infty)$ can be written as

$$y = c_1 J_\nu(x) + c_2 Y_\nu(x). \tag{11}$$

$Y_\nu(x)$ is called the **Bessel function of the second kind** of order ν. Figure 6.3 shows the graphs of $Y_0(x)$ and $Y_1(x)$.

EXAMPLE 2   **General Solution: ν an Integer**

By identifying $\nu^2 = 9$ and $\nu = 3$ we see from (11) that the general solution of the equation $x^2 y'' + xy' + (x^2 - 9)y = 0$ on $(0, \infty)$ is $y = c_1 J_3(x) + c_2 Y_3(x)$.   ∎

Sometimes it is possible to transform a given differential equation into equation (1) by means of a change of variable. We can then express the general solution of the original equation in terms of Bessel functions. Example 3 illustrates this technique.

EXAMPLE 3   **The Aging Spring Revisited**

Recall that in Section 5.1 we saw that one mathematical model for the free undamped motion of a mass on an aging spring is given by $mx'' + ke^{-\alpha t}x = 0$, $\alpha > 0$. We are now in a position to find the general solution of the equation. It is left as a problem to show that the change of variables $s = \dfrac{2}{\alpha}\sqrt{\dfrac{k}{m}}\, e^{-\alpha t/2}$ transforms the differential equation of the aging spring into

$$s^2 \frac{d^2 x}{ds^2} + s \frac{dx}{ds} + s^2 x = 0.$$

The last equation is recognized as (1) with $\nu = 0$ and with the symbols x and s playing the roles of y and x, respectively. The general solution of the new equation is $x = c_1 J_0(s) + c_2 Y_0(s)$. If we resubstitute s, the general solution of $mx'' + ke^{-\alpha t}x = 0$ is then seen to be

$$x(t) = c_1 J_0\left(\frac{2}{\alpha}\sqrt{\frac{k}{m}}\, e^{-\alpha t/2}\right) + c_2 Y_0\left(\frac{2}{\alpha}\sqrt{\frac{k}{m}}\, e^{-\alpha t/2}\right).$$

See Problems 39 and 40 in Exercises 6.4.   ∎

The other model discussed in Section 5.1 of a spring whose characteristics change with time was $mx'' + ktx = 0$. By dividing through by m we see that this equation is Airy's equation, $y'' + \alpha^2 xy = 0$. See Example 4 in Section 6.2. The general solution of Airy's differential equation can also be written in terms of Bessel functions. See Problems 41–43 in Exercises 6.4.

Parametric Bessel Equation   By replacing x by λx in (1) and using the Chain Rule we obtain an alternative form of Bessel's equation known as the **parametric Bessel equation:**

$$x^2 y'' + xy' + (\lambda^2 x^2 - \nu^2)y = 0. \tag{12}$$

The general solution of (12) is

$$y = c_1 J_\nu(\lambda x) + c_2 Y_\nu(\lambda x). \tag{13}$$

Properties   We list below a few of the more useful properties of Bessel functions of order m, $m = 0, 1, 2, \ldots$:

$$(i)\ J_{-m}(x) = (-1)^m J_m(x) \qquad (ii)\ J_m(-x) = (-1)^m J_m(x)$$

$$(iii)\ J_m(0) = \begin{cases} 0, & m > 0 \\ 1, & m = 0 \end{cases} \qquad (iv)\ \lim_{x \to 0^+} Y_m(x) = -\infty$$

Note that Property (*ii*) indicates that $J_m(x)$ is an even function if m is an even integer and an odd function if m is an odd integer. The graphs of $Y_0(x)$ and $Y_1(x)$ in Figure 6.3 illustrate Property (*iv*): $Y_m(x)$ is unbounded at the origin. This last fact is not obvious from (10). It can be shown either from (10) or by the methods of Section 6.3 that for $x > 0$,

$$Y_0(x) = \frac{2}{\pi} J_0(x) \left[\gamma + \ln \frac{x}{2} \right] - \frac{2}{\pi} \sum_{k=1}^{\infty} \frac{(-1)^k}{(k!)^2} \left(1 + \frac{1}{2} + \cdots + \frac{1}{k} \right) \left(\frac{x}{2} \right)^{2k},$$

where $\gamma = 0.57721566\ldots$ is **Euler's constant.** Because of the presence of the logarithmic term, $Y_0(x)$ is discontinuous at $x = 0$.

Numerical Values   Some functional values of $J_0(x)$, $J_1(x)$, $Y_0(x)$, and $Y_1(x)$ for selected values of x are given in Table 6.1. The first five nonnegative zeros of $J_0(x)$, $J_1(x)$, $Y_0(x)$, and $Y_1(x)$ are given in Table 6.2.

TABLE 6.1   Numerical Values of J_0, J_1, Y_0, and Y_1

x	$J_0(x)$	$J_1(x)$	$Y_0(x)$	$Y_1(x)$
0	1.0000	0.0000	—	—
1	0.7652	0.4401	0.0883	−0.7812
2	0.2239	0.5767	0.5104	−0.1070
3	−0.2601	0.3391	0.3769	0.3247
4	−0.3971	−0.0660	−0.0169	0.3979
5	−0.1776	−0.3276	−0.3085	0.1479
6	0.1506	−0.2767	−0.2882	−0.1750
7	0.3001	−0.0047	−0.0259	−0.3027
8	0.1717	0.2346	0.2235	−0.1581
9	−0.0903	0.2453	0.2499	0.1043
10	−0.2459	0.0435	0.0557	0.2490
11	−0.1712	−0.1768	−0.1688	0.1637
12	0.0477	−0.2234	−0.2252	−0.0571
13	0.2069	−0.0703	−0.0782	−0.2101
14	0.1711	0.1334	0.1272	−0.1666
15	−0.0142	0.2051	0.2055	0.0211

TABLE 6.2 Zeros of J_0, J_1, Y_0, and Y_1

$J_0(x)$	$J_1(x)$	$Y_0(x)$	$Y_1(x)$
2.4048	0.0000	0.8936	2.1971
5.5201	3.8317	3.9577	5.4297
8.6537	7.0156	7.0861	8.5960
11.7915	10.1735	10.2223	11.7492
14.9309	13.3237	13.3611	14.8974

Differential Recurrence Relation Recurrence formulas that relate Bessel functions of different orders are important in theory and in applications. In the next example we derive a **differential recurrence relation.**

EXAMPLE 4 **Derivation Using the Series Definition**

Derive the formula $xJ'_\nu(x) = \nu J_\nu(x) - xJ_{\nu+1}(x)$.

SOLUTION   It follows from (7) that

$$xJ'_\nu(x) = \sum_{n=0}^{\infty} \frac{(-1)^n(2n+\nu)}{n!\,\Gamma(1+\nu+n)}\left(\frac{x}{2}\right)^{2n+\nu}$$

$$= \nu\sum_{n=0}^{\infty} \frac{(-1)^n}{n!\,\Gamma(1+\nu+n)}\left(\frac{x}{2}\right)^{2n+\nu} + 2\sum_{n=0}^{\infty} \frac{(-1)^n n}{n!\,\Gamma(1+\nu+n)}\left(\frac{x}{2}\right)^{2n+\nu}$$

$$= \nu J_\nu(x) + x\underbrace{\sum_{n=1}^{\infty} \frac{(-1)^n}{(n-1)!\,\Gamma(1+\nu+n)}\left(\frac{x}{2}\right)^{2n+\nu-1}}_{k = n - 1}$$

$$= \nu J_\nu(x) - x\sum_{k=0}^{\infty} \frac{(-1)^k}{k!\,\Gamma(2+\nu+k)}\left(\frac{x}{2}\right)^{2k+\nu+1} = \nu J_\nu(x) - xJ_{\nu+1}(x). \qquad \blacksquare$$

The result in Example 4 can be written in an alternative form. Dividing $xJ'_\nu(x) - \nu J_\nu(x) = -xJ_{\nu+1}(x)$ by x gives

$$J'_\nu(x) - \frac{\nu}{x}J_\nu(x) = -J_{\nu+1}(x).$$

This last expression is recognized as a linear first-order differential equation in $J_\nu(x)$. Multiplying both sides of the equality by the integrating factor $x^{-\nu}$ then yields

$$\frac{d}{dx}[x^{-\nu}J_\nu(x)] = -x^{-\nu}J_{\nu+1}(x). \tag{14}$$

It can be shown in a similar manner that

$$\frac{d}{dx}[x^{\nu}J_\nu(x)] = x^{\nu}J_{\nu-1}(x). \tag{15}$$

See Problem 20 in Exercises 6.4. The differential recurrence relations (14) and (15) are also valid for the Bessel function of the second kind $Y_\nu(x)$. Observe that when $\nu = 0$ it follows from (14) that

$$J_0'(x) = -J_1(x) \qquad \text{and} \qquad Y_0'(x) = -Y_1(x). \tag{16}$$

An application of these results is given in Problem 40 of Exercises 6.4.

When $\nu = $ half an odd integer, $J_\nu(x)$ can be expressed in terms of $\sin x$, $\cos x$, and powers of x. Such Bessel functions are called **spherical Bessel functions.**

EXAMPLE 5 **Spherical Bessel Function with $\nu = \frac{1}{2}$**

Find an alternative expression for $J_{1/2}(x)$. Use the fact that $\Gamma(\frac{1}{2}) = \sqrt{\pi}$.

SOLUTION   With $\nu = \frac{1}{2}$ we have from (7)

$$J_{1/2}(x) = \sum_{n=0}^{\infty} \frac{(-1)^n}{n!\,\Gamma(1 + \frac{1}{2} + n)} \left(\frac{x}{2}\right)^{2n+1/2}.$$

Now in view of the property $\Gamma(1 + \alpha) = \alpha\Gamma(\alpha)$ we obtain

$$n = 0: \quad \Gamma\left(1 + \frac{1}{2}\right) = \frac{1}{2}\Gamma\left(\frac{1}{2}\right) = \frac{1}{2}\sqrt{\pi}$$

$$n = 1: \quad \Gamma\left(1 + \frac{3}{2}\right) = \frac{3}{2}\Gamma\left(\frac{3}{2}\right) = \frac{3}{2^2}\sqrt{\pi}$$

$$n = 2: \quad \Gamma\left(1 + \frac{5}{2}\right) = \frac{5}{2}\Gamma\left(\frac{5}{2}\right) = \frac{5 \cdot 3}{2^3}\sqrt{\pi} = \frac{5 \cdot 4 \cdot 3 \cdot 2 \cdot 1}{2^3 4 \cdot 2}\sqrt{\pi} = \frac{5!}{2^5 2!}\sqrt{\pi}$$

$$n = 3: \quad \Gamma\left(1 + \frac{7}{2}\right) = \frac{7}{2}\Gamma\left(\frac{7}{2}\right) = \frac{7 \cdot 5!}{2^6 2!}\sqrt{\pi} = \frac{7 \cdot 6 \cdot 5!}{2^6 \cdot 6 \cdot 2!}\sqrt{\pi} = \frac{7!}{2^7 3!}\sqrt{\pi}.$$

In general, $\Gamma\left(1 + \dfrac{1}{2} + n\right) = \dfrac{(2n+1)!}{2^{2n+1}n!}\sqrt{\pi}.$

Hence

$$J_{1/2}(x) = \sum_{n=0}^{\infty} \frac{(-1)^n}{n!\,\dfrac{(2n+1)!\sqrt{\pi}}{2^{2n+1}n!}} \left(\frac{x}{2}\right)^{2n+1/2} = \sqrt{\frac{2}{\pi x}} \sum_{n=0}^{\infty} \frac{(-1)^n}{(2n+1)!} x^{2n+1}.$$

Since the series in the last line is the Maclaurin series for $\sin x$, we have shown that

$$J_{1/2}(x) = \sqrt{\frac{2}{\pi x}} \sin x. \qquad\blacksquare$$

Solution of Legendre's Equation Since $x = 0$ is an ordinary point of equation (2), we assume a solution of the form $y = \sum_{k=0}^{\infty} c_k x^k$. Therefore

$$(1 - x^2)y'' - 2xy' + n(n+1)y = (1 - x^2) \sum_{k=0}^{\infty} c_k k(k-1)x^{k-2} - 2\sum_{k=0}^{\infty} c_k k x^k + n(n+1)\sum_{k=0}^{\infty} c_k x^k$$

$$= \sum_{k=2}^{\infty} c_k k(k-1)x^{k-2} - \sum_{k=2}^{\infty} c_k k(k-1)x^k - 2\sum_{k=1}^{\infty} c_k k x^k + n(n+1)\sum_{k=0}^{\infty} c_k x^k$$

$$= [n(n+1)c_0 + 2c_2]x^0 + [n(n+1)c_1 - 2c_1 + 6c_3]x$$

$$+ \underbrace{\sum_{k=4}^{\infty} c_k k(k-1)x^{k-2}}_{j\,=\,k\,-\,2} - \underbrace{\sum_{k=2}^{\infty} c_k k(k-1)x^k}_{j\,=\,k} - 2\underbrace{\sum_{k=2}^{\infty} c_k k x^k}_{j\,=\,k} + n(n+1)\underbrace{\sum_{k=2}^{\infty} c_k x^k}_{j\,=\,k}$$

$$= [n(n+1)c_0 + 2c_2] + [(n-1)(n+2)c_1 + 6c_3]x$$

$$+ \sum_{j=2}^{\infty} [(j+2)(j+1)c_{j+2} + (n-j)(n+j+1)c_j]x^j = 0$$

implies that

$$n(n+1)c_0 + 2c_2 = 0$$

$$(n-1)(n+2)c_1 + 6c_3 = 0$$

$$(j+2)(j+1)c_{j+2} + (n-j)(n+j+1)c_j = 0$$

or

$$c_2 = -\frac{n(n+1)}{2!}c_0$$

$$c_3 = -\frac{(n-1)(n+2)}{3!}c_1$$

$$c_{j+2} = -\frac{(n-j)(n+j+1)}{(j+2)(j+1)}c_j, \quad j = 2, 3, 4, \ldots . \tag{17}$$

Iterating (17) gives

$$c_4 = -\frac{(n-2)(n+3)}{4 \cdot 3}c_2 = \frac{(n-2)n(n+1)(n+3)}{4!}c_0$$

$$c_5 = -\frac{(n-3)(n+4)}{5 \cdot 4}c_3 = \frac{(n-3)(n-1)(n+2)(n+4)}{5!}c_1$$

$$c_6 = -\frac{(n-4)(n+5)}{6 \cdot 5}c_4 = -\frac{(n-4)(n-2)n(n+1)(n+3)(n+5)}{6!}c_0$$

$$c_7 = -\frac{(n-5)(n+6)}{7 \cdot 6}c_5$$

$$= -\frac{(n-5)(n-3)(n-1)(n+2)(n+4)(n+6)}{7!}c_1$$

and so on. Thus for at least $|x| < 1$ we obtain two linearly independent power series solutions:

$$y_1(x) = c_0 \left[1 - \frac{n(n+1)}{2!}x^2 + \frac{(n-2)n(n+1)(n+3)}{4!}x^4 \right.$$

$$\left. - \frac{(n-4)(n-2)n(n+1)(n+3)(n+5)}{6!}x^6 + \cdots \right] \tag{18}$$

$$y_2(x) = c_1 \left[x - \frac{(n-1)(n+2)}{3!}x^3 + \frac{(n-3)(n-1)(n+2)(n+4)}{5!}x^5 \right.$$

$$\left. - \frac{(n-5)(n-3)(n-1)(n+2)(n+4)(n+6)}{7!}x^7 + \cdots \right].$$

Notice that if n is an even integer, the first series terminates, whereas $y_2(x)$ is an infinite series. For example, if $n = 4$, then

$$y_1(x) = c_0 \left[1 - \frac{4 \cdot 5}{2!} x^2 + \frac{2 \cdot 4 \cdot 5 \cdot 7}{4!} x^4 \right] = c_0 \left[1 - 10x^2 + \frac{35}{3} x^4 \right].$$

Similarly, when n is an odd integer, the series for $y_2(x)$ terminates with x^n; that is, *when n is a nonnegative integer, we obtain an nth-degree polynomial solution* of Legendre's equation.

Since we know that a constant multiple of a solution of Legendre's equation is also a solution, it is traditional to choose specific values for c_0 or c_1, depending on whether n is an even or odd positive integer, respectively. For $n = 0$ we choose $c_0 = 1$, and for $n = 2, 4, 6, \ldots,$

$$c_0 = (-1)^{n/2} \frac{1 \cdot 3 \cdots (n-1)}{2 \cdot 4 \cdots n};$$

whereas for $n = 1$ we choose $c_1 = 1$, and for $n = 3, 5, 7, \ldots,$

$$c_1 = (-1)^{(n-1)/2} \frac{1 \cdot 3 \cdots n}{2 \cdot 4 \cdots (n-1)}.$$

For example, when $n = 4$ we have

$$y_1(x) = (-1)^{4/2} \frac{1 \cdot 3}{2 \cdot 4} \left[1 - 10x^2 + \frac{35}{3} x^4 \right] = \frac{1}{8} (35x^4 - 30x^2 + 3).$$

Legendre Polynomials   These specific nth-degree polynomial solutions are called **Legendre polynomials** and are denoted by $P_n(x)$. From the series for $y_1(x)$ and $y_2(x)$ and from the above choices of c_0 and c_1 we find that the first several Legendre polynomials are

$$P_0(x) = 1, \qquad\qquad P_1(x) = x,$$

$$P_2(x) = \frac{1}{2}(3x^2 - 1), \qquad P_3(x) = \frac{1}{2}(5x^3 - 3x), \qquad \textbf{(19)}$$

$$P_4(x) = \frac{1}{8}(35x^4 - 30x^2 + 3), \qquad P_5(x) = \frac{1}{8}(63x^5 - 70x^3 + 15x).$$

Remember, $P_0(x), P_1(x), P_2(x), P_3(x), \ldots$ are, in turn, particular solutions of the differential equations

$$
\begin{array}{lll}
n = 0: & (1 - x^2)y'' - 2xy' = 0 & \\
n = 1: & (1 - x^2)y'' - 2xy' + 2y = 0 & \\
n = 2: & (1 - x^2)y'' - 2xy' + 6y = 0 & \textbf{(20)} \\
n = 3: & (1 - x^2)y'' - 2xy' + 12y = 0 & \\
& \vdots \qquad\qquad \vdots &
\end{array}
$$

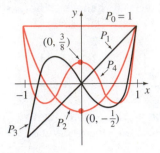

FIGURE 6.4

The graphs of the first four Legendre polynomials on the interval $-1 \leq x \leq 1$ are given in Figure 6.4.

Properties   The following properties of the Legendre polynomials are apparent in (19) and Figure 6.4.

$$(i) \ \ P_n(-x) = (-1)^n P_n(x)$$

$$(ii) \ \ P_n(1) = 1 \qquad\qquad (iii) \ \ P_n(-1) = (-1)^n$$

$$(iv) \ \ P_n(0) = 0, \ \ n \text{ odd} \qquad (v) \ \ P'_n(0) = 0, \ \ n \text{ even}$$

Property (i) indicates that $P_n(x)$ is an even or odd function according to whether n is even or odd.

Recurrence Relation   Recurrence relations that relate Legendre polynomials of different degrees are also very important in some aspects of their application. We shall derive one such relation using the formula

$$(1 - 2xt + t^2)^{-1/2} = \sum_{n=0}^{\infty} P_n(x)t^n. \tag{21}$$

The function on the left is called a **generating function** for the Legendre polynomials. Its derivation follows from binomial series and is left as an exercise. See Problem 49 in Exercises 6.4.

Differentiating both sides of (21) with respect to t gives

$$(1 - 2xt + t^2)^{-3/2}(x - t) = \sum_{n=0}^{\infty} nP_n(x)t^{n-1} = \sum_{n=1}^{\infty} nP_n(x)t^{n-1}$$

so that after multiplying by $1 - 2xt + t^2$ we have

$$(x - t)(1 - 2xt + t^2)^{-1/2} = (1 - 2xt + t^2)\sum_{n=1}^{\infty} nP_n(x)t^{n-1}$$

or $\qquad\qquad (x - t)\sum_{n=0}^{\infty} P_n(x)t^n = (1 - 2xt + t^2)\sum_{n=1}^{\infty} nP_n(x)t^{n-1}. \tag{22}$

We multiply out and rewrite (22) as

$$\sum_{n=0}^{\infty} xP_n(x)t^n - \sum_{n=0}^{\infty} P_n(x)t^{n+1} - \sum_{n=1}^{\infty} nP_n(x)t^{n-1} + 2x\sum_{n=1}^{\infty} nP_n(x)t^n - \sum_{n=1}^{\infty} nP_n(x)t^{n+1} = 0$$

or $\quad x + x^2t + \sum_{n=2}^{\infty} xP_n(x)t^n - t - \sum_{n=1}^{\infty} P_n(x)t^{n+1} - x - 2\left(\dfrac{3x^2 - 1}{2}\right)t$

$$- \sum_{n=3}^{\infty} nP_n(x)t^{n-1} + 2x^2t + 2x\sum_{n=2}^{\infty} nP_n(x)t^n - \sum_{n=1}^{\infty} nP_n(x)t^{n+1} = 0.$$

Observing the appropriate cancellations, simplifying, and changing the summation indices gives

$$\sum_{k=2}^{\infty} [-(k + 1)P_{k+1}(x) + (2k + 1)xP_k(x) - kP_{k-1}(x)]t^k = 0.$$

Equating the total coefficient of t^k to zero gives the three-term recurrence relation

$$(k + 1)P_{k+1}(x) - (2k + 1)xP_k(x) + kP_{k-1}(x) = 0, \quad k = 2, 3, 4, \ldots. \quad \text{(23)}$$

This formula is also valid when $k = 1$.

In (19) we listed the first six Legendre polynomials. If, say, we wished to find $P_6(x)$, we could use (23) with $k = 5$. This relation then expresses $P_6(x)$ in terms of the known quantities $P_4(x)$ and $P_5(x)$. See Problem 51 in Exercises 6.4.

SECTION 6.4 EXERCISES

Answers to odd-numbered problems begin on page A-10.

In Problems 1–8 find the general solution of the given differential equation on $(0, \infty)$.

1. $x^2y'' + xy' + \left(x^2 - \dfrac{1}{9}\right)y = 0$ **2.** $x^2y'' + xy' + (x^2 - 1)y = 0$

3. $4x^2y'' + 4xy' + (4x^2 - 25)y = 0$

4. $16x^2y'' + 16xy' + (16x^2 - 1)y = 0$

5. $xy'' + y' + xy = 0$ **6.** $\dfrac{d}{dx}[xy'] + \left(x - \dfrac{4}{x}\right)y = 0$

7. $x^2y'' + xy' + (9x^2 - 4)y = 0$ **8.** $x^2y'' + xy' + \left(36x^2 - \dfrac{1}{4}\right)y = 0$

9. Use the change of variables $y = x^{-1/2}v(x)$ to find the general solution of the equation

$$x^2y'' + 2xy' + \lambda^2x^2y = 0, \quad x > 0.$$

10. Verify that the differential equation

$$xy'' + (1 - 2n)y' + xy = 0, \quad x > 0$$

possesses the particular solution $y = x^nJ_n(x)$.

11. Verify that the differential equation

$$xy'' + (1 + 2n)y' + xy = 0, \quad x > 0$$

possesses the particular solution $y = x^{-n}J_n(x)$.

12. Verify that the differential equation

$$x^2y'' + \left(\lambda^2x^2 - \nu^2 + \dfrac{1}{4}\right)y = 0, \quad x > 0$$

possesses the particular solution $y = \sqrt{x}J_\nu(\lambda x)$, where $\lambda > 0$.

In Problems 13–18 use the results of Problems 10, 11, and 12 to find a particular solution of the given differential equation on $(0, \infty)$.

13. $y'' + y = 0$ **14.** $xy'' - y' + xy = 0$

15. $xy'' + 3y' + xy = 0$ **16.** $4x^2y'' + (16x^2 + 1)y = 0$

17. $x^2y'' + (x^2 - 2)y = 0$ **18.** $xy'' - 5y' + xy = 0$

In Problems 19–22 derive the given recurrence relation.

19. $xJ_\nu'(x) = -\nu J_\nu(x) + xJ_{\nu-1}(x)$ [*Hint:* $2n + \nu = 2(n + \nu) - \nu.$]

20. $\dfrac{d}{dx}[x^\nu J_\nu(x)] = x^\nu J_{\nu-1}(x)$

21. $2\nu J_\nu(x) = xJ_{\nu+1}(x) + xJ_{\nu-1}(x)$ **22.** $2J_\nu'(x) = J_{\nu-1}(x) - J_{\nu+1}(x)$

In Problems 23–26 use (14) or (15) to obtain the given result.

23. $\displaystyle\int_0^x rJ_0(r)\,dr = xJ_1(x)$ **24.** $J_0'(x) = J_{-1}(x) = -J_1(x)$

25. $\displaystyle\int x^n J_0(x)\,dx = x^n J_1(x) + (n-1)x^{n-1}J_0(x) - (n-1)^2\int x^{n-2}J_0(x)\,dx$

26. $\displaystyle\int x^3 J_0(x)\,dx = x^3 J_1(x) + 2x^2 J_0(x) - 4xJ_1(x) + c$

27. Proceed as in Example 5 and express $J_{-1/2}(x)$ in terms of cos x and a power of x.

In Problems 28–33 use the recurrence relation given in Problem 21 and the results obtained in Problem 27 and Example 5 to express the given Bessel function in terms of sin x, cos x, and powers of x.

28. $J_{3/2}(x)$ **29.** $J_{-3/2}(x)$

30. $J_{5/2}(x)$ **31.** $J_{-5/2}(x)$

32. $J_{7/2}(x)$ **33.** $J_{-7/2}(x)$

34. Show that $i^{-\nu}J_\nu(ix)$, $i^2 = -1$ is a real function. The function defined by $I_\nu(x) = i^{-\nu}J_\nu(ix)$ is called a **modified Bessel function of the first kind** of order ν.

35. Find the general solution of the differential equation
$$x^2 y'' + xy' - (x^2 + \nu^2)y = 0, \quad x > 0, \nu \neq \text{integer}.$$
[*Hint:* $i^2 x^2 = -x^2.$]

36. If $y_1 = J_0(x)$ is one solution of the zero-order Bessel equation, verify that another solution is
$$y_2 = J_0(x)\ln x + \frac{x^2}{4} - \frac{3x^4}{128} + \frac{11x^6}{13{,}824} - \cdots.$$

37. Use (8) with $\nu = m$, where m is a positive integer, and the fact that $1/\Gamma(N) = 0$, where N is a negative integer, to show that
$$J_{-m}(x) = (-1)^m J_m(x).$$

38. Use (7) with $\nu = m$, where m is a nonnegative integer, to show that
$$J_m(-x) = (-1)^m J_m(x).$$

39. Use the change of variables $s = \dfrac{2}{\alpha}\sqrt{\dfrac{k}{m}}\,e^{-\alpha t/2}$ to show that the differential equation of an aging spring $mx'' + ke^{-\alpha t}x = 0$, $\alpha > 0$, becomes
$$s^2\frac{d^2 x}{ds^2} + s\frac{dx}{ds} + s^2 x = 0.$$

40. (a) Use the general solution given in Example 3 to solve the initial-value problem

$$4x'' + e^{-0.1t}x = 0, \qquad x(0) = 1, \quad x'(0) = -\frac{1}{2}.$$

Use Table 6.1 and (16) or a CAS to evaluate coefficients.

(b) Use a CAS to plot the graph of the solution obtained in part (a) over the interval $0 \le t \le 200$. Does the graph corroborate your conjecture in Problem 15 in Exercises 5.1?

41. Show that $y = x^{1/2}w(\frac{2}{3}\alpha x^{3/2})$ is a solution of **Airy's differential equation** $y'' + \alpha^2 xy = 0$, $x > 0$ whenever w is a solution of Bessel's equation $t^2w'' + tw' + (t^2 - \frac{1}{9})w = 0, t > 0$. [*Hint:* After differentiating, substituting, and simplifying, let $t = \frac{2}{3}\alpha x^{3/2}$.]

42. Use the result of Problem 41 to express the general solution of Airy's equation for $x > 0$ in terms of Bessel functions.

43. (a) Use the general solution obtained in Problem 42 to solve the initial-value problem

$$4x'' + tx = 0, \qquad x(0.1) = 1, \quad x'(0.1) = -\frac{1}{2}.$$

Use a CAS to evaluate coefficients.

(b) Use a CAS to plot the graph of the solution obtained in part (a) over the interval $0 \le t \le 200$. Does the graph corroborate your conjecture in Problem 16 in Exercises 5.1?

44. A uniform thin column, positioned vertically with one end embedded in the ground, will deflect, or bend away, from the vertical under the influence of its own weight when its length is greater than a certain critical height. It can be shown that the angular deflection $\theta(x)$ of the column from the vertical at a point $P(x)$ is a solution of the boundary-value problem

$$EI\frac{d^2\theta}{dx^2} + \delta g(L - x)\theta = 0, \qquad \theta(0) = 0, \quad \theta'(L) = 0,$$

where E is Young's modulus, I is the cross-sectional moment of inertia, δ is the constant linear density, and x is distance along the column measured from its base. See Figure 6.5. The column will bend only if this boundary-value problem has a nontrivial solution.

(a) First make the change of variables $t = L - x$, and state the resulting boundary-value problem. Then use the result of Problem 42 to express the general solution of the differential equation in terms of Bessel functions.

(b) With the aid of a CAS, find the critical length L of a solid steel rod of radius $r = 0.05$ in., $\delta g = 0.28A$ lb/in., $E = 2.6 \times 10^7$ lb/in.^{2}, $A = \pi r^2$, and $I = \frac{1}{4}\pi r^4$.

45. (a) Use the explicit solutions $y_1(x)$ and $y_2(x)$ of Legendre's equation and the appropriate choices of c_0 and c_1 to find the Legendre polynomials $P_6(x)$ and $P_7(x)$.

(b) Write the differential equations for which $P_6(x)$ and $P_7(x)$ are particular solutions.

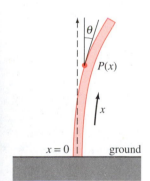

FIGURE 6.5

46. Show that Legendre's equation has the alternative form

$$\frac{d}{dx}\left[(1 - x^2)\frac{dy}{dx}\right] + n(n + 1)y = 0.$$

47. Show that the equation

$$\sin \theta \frac{d^2y}{d\theta^2} + \cos \theta \frac{dy}{d\theta} + n(n + 1)(\sin \theta)y = 0$$

can be transformed in Legendre's equation by means of the substitution $x = \cos \theta$.

48. The general Legendre polynomial can be written as

$$P_n(x) = \sum_{k=0}^{[n/2]} \frac{(-1)^k(2n - 2k)!}{2^n k!(n - k)!(n - 2k)!} x^{n-2k},$$

where $[n/2]$ is the greatest integer not greater than $n/2$. Verify the results for $n = 0, 1, 2, 3, 4, 5$.

49. Use binomial series to formally show that

$$(1 - 2xt + t^2)^{-1/2} = \sum_{n=0}^{\infty} P_n(x)t^n.$$

50. Use Problem 49 to show that $P_n(1) = 1$ and $P_n(-1) = (-1)^n$.

51. Use the recurrence relation (23) and $P_0(x) = 1$, $P_1(x) = x$ to generate the next five Legendre polynomials.

52. The Legendre polynomials are also generated by **Rodrigues' formula**

$$P_n(x) = \frac{1}{2^n n!} \frac{d^n}{dx^n}(x^2 - 1)^n.$$

Verify the results for $n = 0, 1, 2, 3$.

Discussion Problems

53. For the purposes of this problem ignore the graphs given in Figure 6.2. Use the substitution $y = u/\sqrt{x}$ to show that Bessel's equation (1) has the alternative form

$$\frac{d^2u}{dx^2} + \left(1 - \frac{\nu^2 - \frac{1}{4}}{x^2}\right)u = 0.$$

This is a form of the differential equation in Problem 12. For a fixed value of ν, discuss how this last equation enables us to discern the qualitative behavior of a solution of (1) as $x \to \infty$.

54. As a consequence of Problem 46 we observe that

$$\frac{d}{dx}[(1 - x^2)P_n'(x)] = -n(n + 1)P_n(x) \quad \text{and} \quad \frac{d}{dx}[(1 - x^2)P_m'(x)] = -m(m + 1)P_m(x).$$

Discuss how these two identities can be used to show

$$\int_{-1}^{1} P_m(x)P_n(x)\, dx = 0, \quad m \neq n.$$

CHAPTER 6 REVIEW EXERCISES

Answers to odd-numbered problems begin on page A-10.

1. Specify the ordinary points of $(x^3 - 8)y'' - 2xy' + y = 0$.

2. Specify the singular points of $(x^4 - 16)y'' + 2y = 0$.

In Problems 3–6 specify the regular and irregular singular points of the given differential equation.

3. $(x^3 - 10x^2 + 25x)y'' + y' = 0$ **4.** $(x^3 - 10x^2 + 25x)y'' + y = 0$

5. $x^2(x^2 - 9)^2 y'' - (x^2 - 9)y' + xy = 0$

6. $x(x^2 + 1)^3 y'' + y' - 8xy = 0$

In Problems 7 and 8 specify an interval around $x = 0$ for which a power series solution of the given differential equation will converge.

7. $y'' - xy' + 6y = 0$ **8.** $(x^2 - 4)y'' - 2xy' + 9y = 0$

In Problems 9–12 find two power series solutions for each differential equation about the ordinary point $x = 0$.

9. $y'' - xy' - y = 0$ **10.** $y'' - x^2 y' + xy = 0$

11. $(x - 1)y'' + 3y = 0$ **12.** $(\cos x)y'' + y = 0$

In Problems 13 and 14 solve the given initial-value problem.

13. $y'' + xy' + 2y = 0$, $y(0) = 3, y'(0) = -2$

14. $(x + 2)y'' + 3y = 0$, $y(0) = 0, y'(0) = 1$

In Problems 15–20 find two linearly independent solutions of the given differential equation.

15. $2x^2 y'' + xy' - (x + 1)y = 0$ **16.** $2xy'' + y' + y = 0$

17. $x(1 - x)y'' - 2y' + y = 0$ **18.** $x^2 y'' - xy' + (x^2 + 1)y = 0$

19. $xy'' - (2x - 1)y' + (x - 1)y = 0$

20. $x^2 y'' - x^2 y' + (x^2 - 2)y = 0$

THE LAPLACE TRANSFORM

INTRODUCTION

In the linear mathematical model for a physical system such as a spring/mass system or a series electrical circuit, the right-hand member of the differential equation

$$m\frac{d^2x}{dt^2} + \beta\frac{dx}{dt} + kx = f(t) \qquad \text{or} \qquad L\frac{d^2q}{dt^2} + R\frac{dq}{dt} + \frac{1}{C}q = E(t)$$

is a driving function and represents either an external force $f(t)$ or an impressed voltage $E(t)$. In Section 5.1 we considered problems in which the functions f and E were continuous. Piecewise continuous driving functions, however, are not uncommon. For example, the impressed voltage on a circuit could be as shown in Figure 7.1. Solving the differential equation of the circuit in this case is difficult but not impossible. The Laplace transform studied in this chapter is an invaluable tool in solving problems such as these.

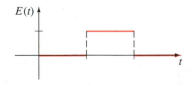

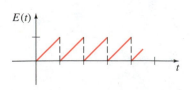

FIGURE 7.1

7.1 DEFINITION OF THE LAPLACE TRANSFORM

■ *Linearity property* ■ *Integral transform* ■ *Definition of the Laplace transform*
■ *Piecewise continuous functions* ■ *Functions of exponential order*
■ *Existence of the Laplace transform* ■ *Transforms of some basic functions*

Linearity Property   In elementary calculus you learned that differentiation and integration transform a function into another function. For example, the function $f(x) = x^2$ is transformed, in turn, into a linear function, a family of cubic polynomial functions, and a constant by the operations of differentiation, indefinite integration, and definite integration:

$$\frac{d}{dx}x^2 = 2x, \qquad \int x^2\,dx = \frac{x^3}{3} + c, \qquad \int_0^3 x^2\,dx = 9.$$

Moreover, these three operations possess the **linearity property.** This means that for any constants α and β

$$\frac{d}{dx}[\alpha f(x) + \beta g(x)] = \alpha\frac{d}{dx}f(x) + \beta\frac{d}{dx}g(x)$$

$$\int [\alpha f(x) + \beta g(x)]\,dx = \alpha\int f(x)\,dx + \beta\int g(x)\,dx \qquad \textbf{(1)}$$

$$\int_a^b [\alpha f(x) + \beta g(x)]\,dx = \alpha\int_a^b f(x)\,dx + \beta\int_a^b g(x)\,dx$$

provided each derivative and integral exists.

If $f(x, y)$ is a function of two variables, then a definite integral of f with respect to one of the variables leads to a function of the other variable. For example, by holding y constant we see that $\int_1^2 2xy^2\,dx = 3y^2$. Similarly, a definite integral such as $\int_a^b K(s, t)f(t)\,dt$ transforms a function $f(t)$ into a function of the variable s. We are particularly interested in **integral transforms** of this last kind, where the interval of integration is the unbounded interval $[0, \infty)$.

Basic Definition   If $f(t)$ is defined for $t \geq 0$, then the improper integral $\int_0^\infty K(s, t)f(t)\,dt$ is defined as a limit:

$$\int_0^\infty K(s, t)f(t)\,dt = \lim_{b\to\infty}\int_0^b K(s, t)f(t)\,dt.$$

If the limit exists, the integral is said to exist or to be convergent; if the limit does not exist, the integral does not exist and is said to be divergent. The foregoing limit will, in general, exist for only certain values of the variable s. The choice $K(s, t) = e^{-st}$ gives us an especially important integral transform.

DEFINITION 7.1   **Laplace Transform**

Let f be a function defined for $t \geq 0$. Then the integral

$$\mathcal{L}\{f(t)\} = \int_0^{\infty} e^{-st}f(t)\,dt \qquad (2)$$

is said to be the **Laplace transform** of f, provided the integral converges.

When the defining integral (2) converges, the result is a function of s. In general discussion we shall use a lowercase letter to denote the function being transformed and the corresponding capital letter to denote its Laplace transform; for example,

$$\mathcal{L}\{f(t)\} = F(s), \qquad \mathcal{L}\{g(t)\} = G(s), \qquad \mathcal{L}\{y(t)\} = Y(s).$$

EXAMPLE 1   **Applying Definition 7.1**

Evaluate $\mathcal{L}\{1\}$.

SOLUTION $\displaystyle \mathcal{L}\{1\} = \int_0^{\infty} e^{-st}(1)\,dt = \lim_{b \to \infty} \int_0^b e^{-st}\,dt$

$$= \lim_{b \to \infty} \frac{-e^{-st}}{s}\Big|_0^b = \lim_{b \to \infty} \frac{-e^{-sb} + 1}{s} = \frac{1}{s}$$

provided $s > 0$. In other words, when $s > 0$ the exponent $-sb$ is negative and $e^{-sb} \to 0$ as $b \to \infty$. When $s < 0$ the integral is divergent. ■

The use of the limit sign becomes somewhat tedious, so we shall adopt the notation $\big|_0^{\infty}$ as a shorthand to writing $\lim_{b \to \infty}(\;\;)\big|_0^b$. For example,

$$\mathcal{L}\{1\} = \int_0^{\infty} e^{-st}\,dt = \frac{-e^{-st}}{s}\Big|_0^{\infty} = \frac{1}{s}, \quad s > 0.$$

At the upper limit it is understood we mean $e^{-st} \to 0$ as $t \to \infty$ for $s > 0$.

$\mathcal{L}$ Is a Linear Transform   For a sum of functions we can write

$$\int_0^{\infty} e^{-st}[\alpha f(t) + \beta g(t)]\,dt = \alpha \int_0^{\infty} e^{-st}f(t)\,dt + \beta \int_0^{\infty} e^{-st}g(t)\,dt$$

whenever both integrals converge. Hence it follows that

$$\mathcal{L}\{\alpha f(t) + \beta g(t)\} = \alpha \mathcal{L}\{f(t)\} + \beta \mathcal{L}\{g(t)\} = \alpha F(s) + \beta G(s). \qquad (3)$$

Because of the property given in (3), $\mathcal{L}$ is said to be a **linear transform.**

Sufficient Conditions for Existence of $\mathcal{L}\{f(t)\}$ The integral that defines the Laplace transform does not have to converge. For example, neither $\mathcal{L}\{1/t\}$ nor $\mathcal{L}\{e^{t^2}\}$ exists. Sufficient conditions guaranteeing the existence of $\mathcal{L}\{f(t)\}$ are that f be piecewise continuous on $[0, \infty)$ and that

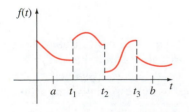

$f(t)$

FIGURE 7.2

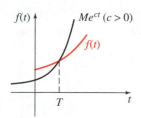

FIGURE 7.3

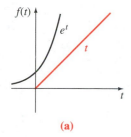

(a)

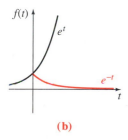

(b)

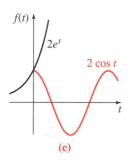

(c)

FIGURE 7.4

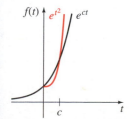

FIGURE 7.5

f be of exponential order for $t > T$. Recall that a function f is **piecewise continuous** on $[0, \infty)$ if, in any interval $0 \le a \le t \le b$, there are at most a finite number of points t_k, $k = 1, 2, \ldots, n$ $(t_{k-1} < t_k)$ at which f has finite discontinuities and is continuous on each open interval $t_{k-1} < t < t_k$. See Figure 7.2. The concept of **exponential order** is defined in the following manner.

DEFINITION 7.2   Exponential Order

A function f is said to be of **exponential order** c if there exist constants c, $M > 0$, and $T > 0$ such that $|f(t)| \le Me^{ct}$ for all $t > T$.

If f is an *increasing* function, then the condition $|f(t)| \le Me^{ct}$, $t > T$ simply states that the graph of f on the interval (T, ∞) does not grow faster than the graph of the exponential function Me^{ct}, where c is a positive constant. See Figure 7.3. The functions $f(t) = t$, $f(t) = e^{-t}$, and $f(t) = 2\cos t$ are all of exponential order $c = 1$ for $t > 0$ since we have, respectively,

$$|t| \le e^t, \qquad |e^{-t}| \le e^t, \qquad |2\cos t| \le 2e^t.$$

A comparison of the graphs on the interval $[0, \infty)$ is given in Figure 7.4.

A function such as $f(t) = e^{t^2}$ is not of exponential order since, as shown in Figure 7.5, its graph grows faster than any positive linear power of e for $t > c > 0$.

A positive integral power of t is always of exponential order since, for $c > 0$,

$$|t^n| \le Me^{ct} \qquad \text{or} \qquad \left|\frac{t^n}{e^{ct}}\right| \le M \quad \text{for } t > T$$

is equivalent to showing that $\lim_{t \to \infty} t^n/e^{ct}$ is finite for $n = 1, 2, 3, \ldots$. The result follows by n applications of L'Hôpital's rule.

THEOREM 7.1   Sufficient Conditions for Existence

If $f(t)$ is piecewise continuous on the interval $[0, \infty)$ and of exponential order c for $t > T$, then $\mathscr{L}\{f(t)\}$ exists for $s > c$.

PROOF $\displaystyle \mathscr{L}\{f(t)\} = \int_0^T e^{-st} f(t)\, dt + \int_T^\infty e^{-st} f(t)\, dt = I_1 + I_2.$

The integral I_1 exists because it can be written as a sum of integrals over intervals on which $e^{-st}f(t)$ is continuous. Now

$$|I_2| \le \int_T^\infty |e^{-st} f(t)|\, dt \le M \int_T^\infty e^{-st} e^{ct}\, dt$$

$$= M \int_T^\infty e^{-(s-c)t}\, dt = -M \left. \frac{e^{-(s-c)t}}{s-c} \right|_T^\infty = M \frac{e^{-(s-c)T}}{s-c}$$

for $s > c$. Since $\int_T^\infty Me^{-(s-c)t}\, dt$ converges, the integral $\int_T^\infty |e^{-st}f(t)|\, dt$ converges by the comparison test for improper integrals. This, in turn, implies that I_2 exists for $s > c$. The existence of I_1 and I_2 implies that $\mathscr{L}\{f(t)\} = \int_0^\infty e^{-st}f(t)\, dt$ exists for $s > c$. ∎

Throughout this entire chapter we shall be concerned only with functions that are both piecewise continuous and of exponential order. We note, however, that these conditions are sufficient but not necessary for the existence of a Laplace transform. The function $f(t) = t^{-1/2}$ is not piecewise continuous on the interval $[0, \infty)$, but its Laplace transform exists. See Problem 40 in Exercises 7.1.

EXAMPLE 2   **Applying Definition 7.1**

Evaluate $\mathscr{L}\{t\}$.

SOLUTION   From Definition 7.1 we have $\mathscr{L}\{t\} = \int_0^\infty e^{-st}t\, dt$. Integrating by parts and using $\lim_{t \to \infty} te^{-st} = 0$, $s > 0$, along with the result given in Example 1, we obtain

$$\mathscr{L}\{t\} = \frac{-te^{-st}}{s}\Big|_0^\infty + \frac{1}{s}\int_0^\infty e^{-st}\, dt$$

$$= \frac{1}{s}\mathscr{L}\{1\}$$

$$= \frac{1}{s}\left(\frac{1}{s}\right)$$

$$= \frac{1}{s^2}.$$ ∎

EXAMPLE 3   **Applying Definition 7.1**

Evaluate $\mathscr{L}\{e^{-3t}\}$.

SOLUTION   From Definition 7.1 we have

$$\mathscr{L}\{e^{-3t}\} = \int_0^\infty e^{-st}e^{-3t}\, dt$$

$$= \int_0^\infty e^{-(s+3)t}\, dt$$

$$= \frac{-e^{-(s+3)t}}{s+3}\Big|_0^\infty$$

$$= \frac{1}{s+3}, \quad s > -3.$$

The result follows from the fact that $\lim_{t \to \infty} e^{-(s+3)t} = 0$ for $s + 3 > 0$ or $s > -3$. ∎

EXAMPLE 4   **Applying Definition 7.1**

Evaluate $\mathscr{L}\{\sin 2t\}$.

SOLUTION   From Definition 7.1 and integration by parts we have

$$\mathscr{L}\{\sin 2t\} = \int_0^\infty e^{-st} \sin 2t \, dt = \left.\frac{-e^{-st} \sin 2t}{s}\right|_0^\infty + \frac{2}{s}\int_0^\infty e^{-st}\cos 2t \, dt$$

$$= \frac{2}{s}\int_0^\infty e^{-st}\cos 2t \, dt, \quad s > 0$$

$\underset{t\to\infty}{\lim}\, e^{-st}\cos 2t = 0,\, s > 0$ $\qquad\qquad\qquad\qquad$ Laplace transform of $\sin 2t$
$\qquad\qquad\qquad\qquad\downarrow \qquad\qquad\qquad\qquad\qquad\qquad\qquad\qquad\downarrow$

$$= \frac{2}{s}\left[\left.\frac{-e^{-st}\cos 2t}{s}\right|_0^\infty - \frac{2}{s}\int_0^\infty e^{-st}\sin 2t \, dt\right]$$

$$= \frac{2}{s^2} - \frac{4}{s^2}\,\mathscr{L}\{\sin 2t\}.$$

At this point we have an equation with $\mathscr{L}\{\sin 2t\}$ on both sides of the equality. Solving for that quantity yields the result

$$\mathscr{L}\{\sin 2t\} = \frac{2}{s^2 + 4}, \quad s > 0. \qquad\blacksquare$$

EXAMPLE 5   **Using Linearity**

Evaluate $\mathscr{L}\{3t - 5\sin 2t\}$.

SOLUTION   From Examples 2 and 4 and the linearity property of the Laplace transform we can write

$$\mathscr{L}\{3t - 5\sin 2t\} = 3\mathscr{L}\{t\} - 5\mathscr{L}\{\sin 2t\}$$

$$= 3\cdot\frac{1}{s^2} - 5\cdot\frac{2}{s^2 + 4}$$

$$= \frac{-7s^2 + 12}{s^2(s^2 + 4)}, \quad s > 0. \qquad\blacksquare$$

EXAMPLE 6   **Applying Definition 7.1**

Evaluate   (a) $\mathscr{L}\{te^{-2t}\}$ (b) $\mathscr{L}\{t^2 e^{-2t}\}$

SOLUTION   (a) From Definition 7.1 and integration by parts we have

$$\mathscr{L}\{te^{-2t}\} = \int_0^\infty e^{-st}(te^{-2t}) \, dt = \int_0^\infty te^{-(s+2)t} \, dt$$

$$= \left.\frac{-te^{-(s+2)t}}{s+2}\right|_0^\infty + \frac{1}{s+2}\int_0^\infty e^{-(s+2)t} \, dt$$

$$= \left.\frac{-e^{-(s+2)t}}{(s+2)^2}\right|_0^\infty, \quad s > -2$$

$$= \frac{1}{(s+2)^2}, \quad s > -2.$$

(b) Again, integration by parts gives

$$\mathscr{L}\{t^2 e^{-2t}\} = \frac{-t^2 e^{-(s+2)t}}{s+2}\Bigg|_0^\infty + \frac{2}{s+2}\int_0^\infty t e^{-(s+2)t}\, dt$$

$$= \frac{2}{s+2}\int_0^\infty e^{-st}(te^{-2t})\, dt, \quad s > -2$$

$$= \frac{2}{s+2}\mathscr{L}\{te^{-2t}\} = \frac{2}{s+2}\left[\frac{1}{(s+2)^2}\right] \qquad \leftarrow \text{from part (a)}$$

$$= \frac{2}{(s+2)^3}, \quad s > -2. \qquad \blacksquare$$

EXAMPLE 7   **Transform of a Piecewise Defined Function**

Evaluate $\mathscr{L}\{f(t)\}$ for $f(t) = \begin{cases} 0, & 0 \le t < 3 \\ 2, & t \ge 3. \end{cases}$

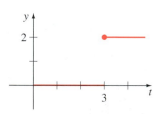

FIGURE 7.6

SOLUTION   This piecewise continuous function appears in Figure 7.6. Since f is defined in two pieces, $\mathscr{L}\{f(t)\}$ is expressed as the sum of two integrals:

$$\mathscr{L}\{f(t)\} = \int_0^\infty e^{-st} f(t)\, dt = \int_0^3 e^{-st}(0)\, dt + \int_3^\infty e^{-st}(2)\, dt$$

$$= -\frac{2e^{-st}}{s}\Bigg|_3^\infty$$

$$= \frac{2e^{-3s}}{s}, \quad s > 0. \qquad \blacksquare$$

We state the generalization of some of the preceding examples by means of the next theorem. From this point on we shall also refrain from stating any restrictions on s; it is understood that s is sufficiently restricted to guarantee the convergence of the appropriate Laplace transform.

THEOREM 7.2   **Transforms of Some Basic Functions**

$$\textbf{(a)} \ \ \mathscr{L}\{1\} = \frac{1}{s}$$

$$\textbf{(b)} \ \ \mathscr{L}\{t^n\} = \frac{n!}{s^{n+1}}, \quad n = 1, 2, 3, \ldots \qquad \textbf{(c)} \ \ \mathscr{L}\{e^{at}\} = \frac{1}{s-a}$$

$$\textbf{(d)} \ \ \mathscr{L}\{\sin kt\} = \frac{k}{s^2 + k^2} \qquad\qquad \textbf{(e)} \ \ \mathscr{L}\{\cos kt\} = \frac{s}{s^2 + k^2}$$

$$\textbf{(f)} \ \ \mathscr{L}\{\sinh kt\} = \frac{k}{s^2 - k^2} \qquad\qquad \textbf{(g)} \ \ \mathscr{L}\{\cosh kt\} = \frac{s}{s^2 - k^2}$$

Part (b) of Theorem 7.2 can be justified in the following manner. Integration by parts yields

$$\mathscr{L}\{t^n\} = \int_0^\infty e^{-st}t^n \, dt = -\frac{1}{s}e^{-st}t^n \Big|_0^\infty + \frac{n}{s}\int_0^\infty e^{-st}t^{n-1} \, dt = \frac{n}{s}\int_0^\infty e^{-st}t^{n-1} \, dt$$

or

$$\mathscr{L}\{t^n\} = \frac{n}{s}\mathscr{L}\{t^{n-1}\}, \quad n = 1, 2, 3, \dots.$$

Now $\mathscr{L}\{1\} = 1/s$, so it follows by iteration that

$$\mathscr{L}\{t\} = \frac{1}{s}\mathscr{L}\{1\} = \frac{1}{s^2}, \qquad \mathscr{L}\{t^2\} = \frac{2}{s}\mathscr{L}\{t\} = \frac{2!}{s^3}, \qquad \mathscr{L}\{t^3\} = \frac{3}{s}\mathscr{L}\{t^2\} = \frac{3 \cdot 2}{s^4} = \frac{3!}{s^4}.$$

Although a rigorous proof requires mathematical induction, it seems reasonable to conclude from the foregoing results that in general

$$\mathscr{L}\{t^n\} = \frac{n}{s}\mathscr{L}\{t^{n-1}\} = \frac{n}{s}\left[\frac{(n-1)!}{s^n}\right] = \frac{n!}{s^{n+1}}.$$

The justifications of parts (f) and (g) of Theorem 7.2 are left to you. See Problems 33 and 34 in Exercises 7.1.

EXAMPLE 8 **Trigonometric Identity and Linearity**

Evaluate $\mathscr{L}\{\sin^2 t\}$.

SOLUTION With the aid of a trigonometric identity, linearity, and parts (a) and (e) of Theorem 7.2, we obtain

$$\mathscr{L}\{\sin^2 t\} = \mathscr{L}\left\{\frac{1-\cos 2t}{2}\right\} = \frac{1}{2}\mathscr{L}\{1\} - \frac{1}{2}\mathscr{L}\{\cos 2t\}$$

$$= \frac{1}{2}\cdot\frac{1}{s} - \frac{1}{2}\cdot\frac{s}{s^2+4}$$

$$= \frac{2}{s(s^2+4)}.$$

■

SECTION 7.1 EXERCISES

Answers to odd-numbered problems begin on page A-10.

In Problems 1–18 use Definition 7.1 to find $\mathscr{L}\{f(t)\}$.

1. $f(t) = \begin{cases} -1, & 0 \le t < 1 \\ 1, & t \ge 1 \end{cases}$

2. $f(t) = \begin{cases} 4, & 0 \le t < 2 \\ 0, & t \ge 2 \end{cases}$

3. $f(t) = \begin{cases} t, & 0 \le t < 1 \\ 1, & t \ge 1 \end{cases}$

4. $f(t) = \begin{cases} 2t+1, & 0 \le t < 1 \\ 0, & t \ge 1 \end{cases}$

5. $f(t) = \begin{cases} \sin t, & 0 \le t < \pi \\ 0, & t \ge \pi \end{cases}$

6. $f(t) = \begin{cases} 0, & 0 \le t < \pi/2 \\ \cos t, & t \ge \pi/2 \end{cases}$

7. $f(t)$

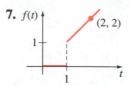

FIGURE 7.7

8. $f(t)$

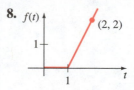

FIGURE 7.8

9. $f(t)$

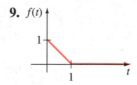

FIGURE 7.9

10. $f(t)$

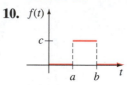

FIGURE 7.10

11. $f(t) = e^{t+7}$

12. $f(t) = e^{-2t-5}$

13. $f(t) = te^{4t}$

14. $f(t) = t^2 e^{3t}$

15. $f(t) = e^{-t} \sin t$

16. $f(t) = e^t \cos t$

17. $f(t) = t \cos t$

18. $f(t) = t \sin t$

In Problems 19–38 use Theorem 7.2 to find $\mathcal{L}\{f(t)\}$.

19. $f(t) = 2t^4$

20. $f(t) = t^5$

21. $f(t) = 4t - 10$

22. $f(t) = 7t + 3$

23. $f(t) = t^2 + 6t - 3$

24. $f(t) = -4t^2 + 16t + 9$

25. $f(t) = (t + 1)^3$

26. $f(t) = (2t - 1)^3$

27. $f(t) = 1 + e^{4t}$

28. $f(t) = t^2 - e^{-9t} + 5$

29. $f(t) = (1 + e^{2t})^2$

30. $f(t) = (e^t - e^{-t})^2$

31. $f(t) = 4t^2 - 5 \sin 3t$

32. $f(t) = \cos 5t + \sin 2t$

33. $f(t) = \sinh kt$

34. $f(t) = \cosh kt$

35. $f(t) = e^t \sinh t$

36. $f(t) = e^{-t} \cosh t$

37. $f(t) = \sin 2t \cos 2t$

38. $f(t) = \cos^2 t$

39. The **gamma function** is defined by the integral

$$\Gamma(\alpha) = \int_0^\infty t^{\alpha-1} e^{-t}\, dt, \quad \alpha > 0.$$

See Appendix I. Show that $\mathcal{L}\{t^\alpha\} = \dfrac{\Gamma(\alpha + 1)}{s^{\alpha+1}}, \alpha > -1$.

In Problems 40–42 use the result of Problem 39 to find $\mathcal{L}\{f(t)\}$.

40. $f(t) = t^{-1/2}$ **41.** $f(t) = t^{1/2}$ **42.** $f(t) = t^{3/2}$

43. Show that the function $f(t) = 1/t^2$ does not possess a Laplace transform. [*Hint:* $\mathcal{L}\{f(t)\} = \int_0^1 e^{-st}f(t)\, dt + \int_1^\infty e^{-st}f(t)\, dt$. Use the definition of an improper integral to show that $\int_0^1 e^{-st}f(t)\, dt$ does not exist.]

Discussion Problem

44. Make up a function $F(t)$ that is of exponential order but $f(t) = F'(t)$ is not of exponential order. Make up a function f that is not of exponential order but whose Laplace transform exists.

7.2   INVERSE TRANSFORM

■ *Inverse Laplace transform* ■ *Linearity* ■ *Some inverse transforms* ■ *Use of partial fractions*

In the preceding section we were concerned with the problem of transforming a function $f(t)$ into another function $F(s)$ by means of the integral $\int_0^\infty e^{-st} f(t)\, dt$. We denoted this symbolically by $\mathscr{L}\{f(t)\} = F(s)$. We now turn the problem around; namely, given $F(s)$, find the function $f(t)$ corresponding to this transform. We say $f(t)$ is the **inverse Laplace transform** of $F(s)$ and write

$$f(t) = \mathscr{L}^{-1}\{F(s)\}.$$

The analogue of Theorem 7.2 for the inverse transform is Theorem 7.3.

THEOREM 7.3   **Some Inverse Transforms**

$$\textbf{(a)}\ \ 1 = \mathscr{L}^{-1}\left\{\frac{1}{s}\right\}$$

$$\textbf{(b)}\ \ t^n = \mathscr{L}^{-1}\left\{\frac{n!}{s^{n+1}}\right\}, \quad n = 1, 2, 3, \ldots \qquad \textbf{(c)}\ \ e^{at} = \mathscr{L}^{-1}\left\{\frac{1}{s-a}\right\}$$

$$\textbf{(d)}\ \ \sin kt = \mathscr{L}^{-1}\left\{\frac{k}{s^2 + k^2}\right\} \qquad \textbf{(e)}\ \ \cos kt = \mathscr{L}^{-1}\left\{\frac{s}{s^2 + k^2}\right\}$$

$$\textbf{(f)}\ \ \sinh kt = \mathscr{L}^{-1}\left\{\frac{k}{s^2 - k^2}\right\} \qquad \textbf{(g)}\ \ \cosh kt = \mathscr{L}^{-1}\left\{\frac{s}{s^2 - k^2}\right\}$$

$\mathscr{L}^{-1}$ Is a Linear Transform   We assume that the inverse Laplace transform is itself a linear transform; that is, for constants α and β

$$\mathscr{L}^{-1}\{\alpha F(s) + \beta G(s)\} = \alpha \mathscr{L}^{-1}\{F(s)\} + \beta \mathscr{L}^{-1}\{G(s)\},$$

where F and G are the transforms of some functions f and g.

The inverse Laplace transform of a function $F(s)$ may not be unique. It is possible that $\mathscr{L}\{f_1(t)\} = \mathscr{L}\{f_2(t)\}$ and yet $f_1 \neq f_2$, but for our purposes this is not anything to be concerned about. If f_1 and f_2 are piecewise continuous on $[0, \infty)$ and of exponential order for $t > 0$ and if $\mathscr{L}\{f_1(t)\} = \mathscr{L}\{f_2(t)\}$, then the functions f_1 and f_2 are *essentially* the same. See Problem 35 in Exercises 7.2. However, if f_1 and f_2 are continuous on $[0, \infty)$ and $\mathscr{L}\{f_1(t)\} = \mathscr{L}\{f_2(t)\}$, then $f_1 = f_2$ on the interval.

EXAMPLE 1   **Applying Theorem 7.3**

Evaluate $\mathscr{L}^{-1}\left\{\dfrac{1}{s^5}\right\}$.

SOLUTION   To match the form given in part (b) of Theorem 7.3, we identify $n = 4$ and then multiply and divide by 4!. It follows that

$$\mathscr{L}^{-1}\left\{\frac{1}{s^5}\right\} = \frac{1}{4!}\,\mathscr{L}^{-1}\left\{\frac{4!}{s^5}\right\} = \frac{1}{24}\,t^4.$$ ∎

EXAMPLE 2   **Applying Theorem 7.3**

Evaluate $\mathscr{L}^{-1}\left\{\dfrac{1}{s^2 + 64}\right\}$.

SOLUTION   Since $k^2 = 64$, we fix up the expression by multiplying and dividing by 8. From part (d) of Theorem 7.3,

$$\mathscr{L}^{-1}\left\{\frac{1}{s^2 + 64}\right\} = \frac{1}{8}\,\mathscr{L}^{-1}\left\{\frac{8}{s^2 + 64}\right\} = \frac{1}{8}\sin 8t.$$ ∎

EXAMPLE 3   **Termwise Division and Linearity**

Evaluate $\mathscr{L}^{-1}\left\{\dfrac{3s + 5}{s^2 + 7}\right\}$.

SOLUTION   The given function of s can be written as two expressions by means of termwise division:

$$\frac{3s + 5}{s^2 + 7} = \frac{3s}{s^2 + 7} + \frac{5}{s^2 + 7}.$$

From the linearity property of the inverse transform and parts (e) and (d) of Theorem 7.3 we then have

$$\mathscr{L}^{-1}\left\{\frac{3s + 5}{s^2 + 7}\right\} = 3\,\mathscr{L}^{-1}\left\{\frac{s}{s^2 + 7}\right\} + \frac{5}{\sqrt{7}}\,\mathscr{L}^{-1}\left\{\frac{\sqrt{7}}{s^2 + 7}\right\}$$

$$= 3\cos\sqrt{7}t + \frac{5}{\sqrt{7}}\sin\sqrt{7}t.$$ ∎

Partial Fractions   Partial fractions play an important role in finding inverse Laplace transforms. As mentioned in Section 2.1, this fraction decomposition can be done quickly by means of a single command on some computer algebra systems. Indeed, some CASs have packages that implement Laplace transform and inverse Laplace transform commands. But for those of you without access to such software, in the next three examples we review the basic algebra in three cases of partial fraction decomposition. For example, the denominators of

$(i)\ F(s) = \dfrac{1}{(s - 1)(s + 2)(s + 4)}$ $(ii)\ F(s) = \dfrac{s + 1}{s^2(s + 2)^3}$ $(iii)\ F(s) = \dfrac{3s - 2}{s^3(s^2 + 4)}$

contain, respectively, distinct linear factors, repeated linear factors, and a quadratic expression with no real factors. You should consult a calculus text for a more complete review of this theory.

EXAMPLE 4   **Partial Fractions and Linearity**

Evaluate $\mathscr{L}^{-1}\left\{\dfrac{1}{(s-1)(s+2)(s+4)}\right\}$.

SOLUTION   There exist unique constants A, B, and C so that

$$\frac{1}{(s-1)(s+2)(s+4)} = \frac{A}{s-1} + \frac{B}{s+2} + \frac{C}{s+4}$$

$$= \frac{A(s+2)(s+4) + B(s-1)(s+4) + C(s-1)(s+2)}{(s-1)(s+2)(s+4)}.$$

Since the denominators are identical, the numerators are identical:

$$1 = A(s+2)(s+4) + B(s-1)(s+4) + C(s-1)(s+2).$$

By comparing coefficients of powers of s on both sides of the equality we know that the last equation is equivalent to a system of three equations in the three unknowns A, B, and C. However, you might recall the following shortcut for determining these unknowns. If we set $s = 1$, $s = -2$, and $s = -4$, the zeros of the common denominator $(s-1)(s+2)(s+4)$, we obtain, in turn,

$$1 = A(3)(5), \qquad 1 = B(-3)(2), \qquad 1 = C(-5)(-2)$$

or $A = \frac{1}{15}$, $B = -\frac{1}{6}$, and $C = \frac{1}{10}$. Hence we can write

$$\frac{1}{(s-1)(s+2)(s+4)} = \frac{1/15}{s-1} - \frac{1/6}{s+2} + \frac{1/10}{s+4}$$

and thus, from part (c) of Theorem 7.3,

$$\mathscr{L}^{-1}\left\{\frac{1}{(s-1)(s+2)(s+4)}\right\} = \frac{1}{15}\mathscr{L}^{-1}\left\{\frac{1}{s-1}\right\} - \frac{1}{6}\mathscr{L}^{-1}\left\{\frac{1}{s+2}\right\} + \frac{1}{10}\mathscr{L}^{-1}\left\{\frac{1}{s+4}\right\}$$

$$= \frac{1}{15}e^{t} - \frac{1}{6}e^{-2t} + \frac{1}{10}e^{-4t}. \qquad \blacksquare$$

EXAMPLE 5   **Partial Fractions and Linearity**

Evaluate $\mathscr{L}^{-1}\left\{\dfrac{s+1}{s^2(s+2)^3}\right\}$.

SOLUTION   Assume

$$\frac{s+1}{s^2(s+2)^3} = \frac{A}{s} + \frac{B}{s^2} + \frac{C}{s+2} + \frac{D}{(s+2)^2} + \frac{E}{(s+2)^3}$$

so that

$$s + 1 = As(s + 2)^3 + B(s + 2)^3 + Cs^2(s + 2)^2 + Ds^2(s + 2) + Es^2.$$

Setting $s = 0$ and $s = -2$ gives $B = \frac{1}{8}$ and $E = -\frac{1}{4}$, respectively. By equating the coefficients of s^4, s^3, and s, we obtain

$$0 = A + C, \qquad 0 = 6A + B + 4C + D, \qquad 1 = 8A + 12B,$$

from which it follows that $A = -\frac{1}{16}$, $C = \frac{1}{16}$, and $D = 0$. Hence from parts (a), (b), and (c) of Theorem 7.3

$$\mathscr{L}^{-1}\left\{\frac{s+1}{s^2(s+2)^3}\right\} = \mathscr{L}^{-1}\left\{-\frac{1/16}{s} + \frac{1/8}{s^2} + \frac{1/16}{s+2} - \frac{1/4}{(s+2)^3}\right\}$$

$$= -\frac{1}{16}\mathscr{L}^{-1}\left\{\frac{1}{s}\right\} + \frac{1}{8}\mathscr{L}^{-1}\left\{\frac{1}{s^2}\right\} + \frac{1}{16}\mathscr{L}^{-1}\left\{\frac{1}{s+2}\right\} - \frac{1}{8}\mathscr{L}^{-1}\left\{\frac{2}{(s+2)^3}\right\}$$

$$= -\frac{1}{16} + \frac{1}{8}t + \frac{1}{16}e^{-2t} - \frac{1}{8}t^2e^{-2t}.$$

Here we have also used $\mathscr{L}^{-1}\{2/(s + 2)^3\} = t^2e^{-2t}$ from Example 6 of Section 7.1. ∎

EXAMPLE 6   **Partial Fractions and Linearity**

Evaluate $\mathscr{L}^{-1}\left\{\dfrac{3s - 2}{s^3(s^2 + 4)}\right\}$.

SOLUTION   Assume

$$\frac{3s - 2}{s^3(s^2 + 4)} = \frac{A}{s} + \frac{B}{s^2} + \frac{C}{s^3} + \frac{Ds + E}{s^2 + 4}$$

so that

$$3s - 2 = As^2(s^2 + 4) + Bs(s^2 + 4) + C(s^2 + 4) + (Ds + E)s^3.$$

Setting $s = 0$ gives immediately $C = -\frac{1}{2}$. Now the coefficients of s^4, s^3, s^2, and s are, respectively,

$$0 = A + D, \qquad 0 = B + E, \qquad 0 = 4A + C, \qquad 3 = 4B,$$

from which we obtain $B = \frac{3}{4}$, $E = -\frac{3}{4}$, $A = \frac{1}{8}$, and $D = -\frac{1}{8}$. Therefore from parts (a), (b), (e), and (d) of Theorem 7.3 we have

$$\mathscr{L}^{-1}\left\{\frac{3s - 2}{s^3(s^2 + 4)}\right\} = \mathscr{L}^{-1}\left\{\frac{1/8}{s} + \frac{3/4}{s^2} - \frac{1/2}{s^3} + \frac{-s/8 - 3/4}{s^2 + 4}\right\}$$

$$= \frac{1}{8}\mathscr{L}^{-1}\left\{\frac{1}{s}\right\} + \frac{3}{4}\mathscr{L}^{-1}\left\{\frac{1}{s^2}\right\} - \frac{1}{4}\mathscr{L}^{-1}\left\{\frac{2}{s^3}\right\}$$

$$- \frac{1}{8}\mathscr{L}^{-1}\left\{\frac{s}{s^2 + 4}\right\} - \frac{3}{8}\mathscr{L}^{-1}\left\{\frac{2}{s^2 + 4}\right\}$$

$$= \frac{1}{8} + \frac{3}{4}t - \frac{1}{4}t^2 - \frac{1}{8}\cos 2t - \frac{3}{8}\sin 2t.$$
∎

As the next theorem indicates, not every arbitrary function of s is a Laplace transform of a piecewise continuous function of exponential order.

THEOREM 7.4   **Behavior of $F(s)$ as $s \to \infty$**

If $f(t)$ is piecewise continuous on $[0, \infty)$ and of exponential order for $t > T$, then $\lim_{s \to \infty} \mathcal{L}\{f(t)\} = 0$.

PROOF   Since $f(t)$ is piecewise continuous on $0 \le t \le T$, it is necessarily bounded on the interval. That is, $|f(t)| \le M_1 = M_1 e^{0t}$. Also, $|f(t)| \le M_2 e^{\gamma t}$ for $t > T$. If M denotes the maximum of $\{M_1, M_2\}$ and c denotes the maximum of $\{0, \gamma\}$, then

$$|\mathcal{L}\{f(t)\}| \le \int_0^\infty e^{-st}|f(t)|\, dt \le M \int_0^\infty e^{-st} \cdot e^{ct}\, dt = -M \frac{e^{-(s-c)t}}{s-c}\bigg|_0^\infty = \frac{M}{s-c}$$

for $s > c$. As $s \to \infty$, we have $|\mathcal{L}\{f(t)\}| \to 0$ and so $\mathcal{L}\{f(t)\} \to 0$.   ∎

In view of Theorem 7.4 we can say that $F_1(s) = 1$ and $F_2(s) = s/(s+1)$ are not the Laplace transforms of piecewise continuous functions of exponential order since $F_1(s) \not\to 0$ and $F_2(s) \not\to 0$ as $s \to \infty$. You should not conclude from this, for example, that $\mathcal{L}^{-1}\{F_1(s)\}$ does not exist. There are other kinds of functions.

Remark

This remark is for those of you who will be required to do partial fraction decompositions by hand. There is another way of determining the coefficients in a partial fraction decomposition in the special case when $\mathcal{L}\{f(t)\} = P(s)/Q(s)$, P and Q are polynomials, and Q is a product of *distinct* factors:

$$F(s) = \frac{P(s)}{(s-r_1)(s-r_2)\cdots(s-r_n)}.$$

Let us illustrate by means of a specific example. From the theory of partial fractions we know there exist unique constants A, B, and C such that

$$\frac{s^2 + 4s - 1}{(s-1)(s-2)(s+3)} = \frac{A}{s-1} + \frac{B}{s-2} + \frac{C}{s+3}. \tag{1}$$

Suppose we multiply both sides of this last expression by, say, $s - 1$, simplify, and then set $s = 1$. Since the coefficients of B and C are zero, we get

$$\frac{s^2 + 4s - 1}{(s-2)(s+3)}\bigg|_{s=1} = A \quad \text{or} \quad A = -1.$$

Written another way,

$$\left.\frac{s^2 + 4s - 1}{\boxed{(s-1)}(s-2)(s+3)}\right|_{s=1} = A,$$

where we have shaded or *covered up* the factor that canceled when the left side of (1) was multiplied by $s - 1$. We *do not evaluate this covered-up factor* at $s = 1$. Now to obtain B and C we simply evaluate the left member of (1) while covering, in turn, $s - 2$ and $s + 3$:

$$\left.\frac{s^2 + 4s - 1}{(s-1)\boxed{(s-2)}(s+3)}\right|_{s=2} = B \quad \text{or} \quad B = \frac{11}{5}$$

$$\left.\frac{s^2 + 4s - 1}{(s-1)(s-2)\boxed{(s+3)}}\right|_{s=-3} = C \quad \text{or} \quad C = -\frac{1}{5}.$$

Note carefully that in the calculation of C we evaluated at $s = -3$. By filling in the details in arriving at this last expression you will see why this is so. You should also verify by other means that

$$\frac{s^2 + 4s - 1}{(s-1)(s-2)(s+3)} = \frac{-1}{s-1} + \frac{11/5}{s-2} + \frac{-1/5}{s+3}.$$

This **cover-up method** is a simplified version of a result known as **Heaviside's expansion theorem.**

SECTION 7.2 EXERCISES

Answers to odd-numbered problems begin on page A-10.

In Problems 1–34 use Theorem 7.3 to find the given inverse transform.

1. $\mathscr{L}^{-1}\left\{\dfrac{1}{s^3}\right\}$

2. $\mathscr{L}^{-1}\left\{\dfrac{1}{s^4}\right\}$

3. $\mathscr{L}^{-1}\left\{\dfrac{1}{s^2} - \dfrac{48}{s^5}\right\}$

4. $\mathscr{L}^{-1}\left\{\left(\dfrac{2}{s} - \dfrac{1}{s^3}\right)^2\right\}$

5. $\mathscr{L}^{-1}\left\{\dfrac{(s+1)^3}{s^4}\right\}$

6. $\mathscr{L}^{-1}\left\{\dfrac{(s+2)^2}{s^3}\right\}$

7. $\mathscr{L}^{-1}\left\{\dfrac{1}{s^2} - \dfrac{1}{s} + \dfrac{1}{s-2}\right\}$

8. $\mathscr{L}^{-1}\left\{\dfrac{4}{s} + \dfrac{6}{s^5} - \dfrac{1}{s+8}\right\}$

9. $\mathscr{L}^{-1}\left\{\dfrac{1}{4s+1}\right\}$

10. $\mathscr{L}^{-1}\left\{\dfrac{1}{5s-2}\right\}$

11. $\mathscr{L}^{-1}\left\{\dfrac{5}{s^2+49}\right\}$

12. $\mathscr{L}^{-1}\left\{\dfrac{10s}{s^2+16}\right\}$

13. $\mathscr{L}^{-1}\left\{\dfrac{4s}{4s^2+1}\right\}$

14. $\mathscr{L}^{-1}\left\{\dfrac{1}{4s^2+1}\right\}$

15. $\mathscr{L}^{-1}\left\{\dfrac{1}{s^2-16}\right\}$

16. $\mathscr{L}^{-1}\left\{\dfrac{10s}{s^2-25}\right\}$

17. $\mathscr{L}^{-1}\left\{\dfrac{2s-6}{s^2+9}\right\}$

18. $\mathscr{L}^{-1}\left\{\dfrac{s-1}{s^2+2}\right\}$

19. $\mathcal{L}^{-1}\left\{\dfrac{1}{s^2 + 3s}\right\}$

20. $\mathcal{L}^{-1}\left\{\dfrac{s + 1}{s^2 - 4s}\right\}$

21. $\mathcal{L}^{-1}\left\{\dfrac{s}{s^2 + 2s - 3}\right\}$

22. $\mathcal{L}^{-1}\left\{\dfrac{1}{s^2 + s - 20}\right\}$

23. $\mathcal{L}^{-1}\left\{\dfrac{0.9s}{(s - 0.1)(s + 0.2)}\right\}$

24. $\mathcal{L}^{-1}\left\{\dfrac{s - 3}{(s - \sqrt{3})(s + \sqrt{3})}\right\}$

25. $\mathcal{L}^{-1}\left\{\dfrac{s}{(s - 2)(s - 3)(s - 6)}\right\}$

26. $\mathcal{L}^{-1}\left\{\dfrac{s^2 + 1}{s(s - 1)(s + 1)(s - 2)}\right\}$

27. $\mathcal{L}^{-1}\left\{\dfrac{2s + 4}{(s - 2)(s^2 + 4s + 3)}\right\}$

28. $\mathcal{L}^{-1}\left\{\dfrac{s + 1}{(s^2 - 4s)(s + 5)}\right\}$

29. $\mathcal{L}^{-1}\left\{\dfrac{1}{s^2(s^2 + 4)}\right\}$

30. $\mathcal{L}^{-1}\left\{\dfrac{s - 1}{s^2(s^2 + 1)}\right\}$

31. $\mathcal{L}^{-1}\left\{\dfrac{s}{(s^2 + 4)(s + 2)}\right\}$

32. $\mathcal{L}^{-1}\left\{\dfrac{1}{s^4 - 9}\right\}$

33. $\mathcal{L}^{-1}\left\{\dfrac{1}{(s^2 + 1)(s^2 + 4)}\right\}$

34. $\mathcal{L}^{-1}\left\{\dfrac{6s + 3}{(s^2 + 1)(s^2 + 4)}\right\}$

Discussion Problem

35. Make up two functions f and g that have the same Laplace transform. Do not think profound thoughts.

7.3 TRANSLATION THEOREMS AND DERIVATIVES OF A TRANSFORM

■ *First translation theorem* ■ *Inverse form of first translation theorem*
■ *Unit step function* ■ *Functions expressed in terms of unit step functions*
■ *Second translation theorem* ■ *Transform of a unit step function*
■ *Inverse form of second translation theorem* ■ *Derivatives of a transform*

It is not convenient to use Definition 7.1 each time we wish to find the Laplace transform of a function $f(t)$. For example, the integration by parts involved in evaluating, say, $\mathcal{L}\{e^t t^2 \sin 3t\}$ is formidable to say the least. In the discussion that follows we present several labor-saving theorems; these, in turn, enable us to build up a more extensive list of transforms without the necessity of using the definition of the Laplace transform. Indeed, we shall see that evaluating transforms such as $\mathcal{L}\{e^{4t} \cos 6t\}$, $\mathcal{L}\{t^3 \sin 2t\}$, and $\mathcal{L}\{t^{10}e^{-t}\}$ is fairly straightforward, provided we know $\mathcal{L}\{\cos 6t\}$, $\mathcal{L}\{\sin 2t\}$, and $\mathcal{L}\{t^{10}\}$, respectively. Though extensive tables can be constructed and a table is included in Appendix III, it is nonetheless a good idea to know the Laplace transforms of basic functions such as t^n, e^{at}, $\sin kt$, $\cos kt$, $\sinh kt$, and $\cosh kt$.

If we know $\mathcal{L}\{f(t)\} = F(s)$, we can compute the Laplace transform $\mathcal{L}\{e^{at}f(t)\}$ with no additional effort other than *translating*, or *shifting*, $F(s)$ to $F(s - a)$. This result is known as the **first translation theorem** or **first shifting theorem**.

> **THEOREM 7.5**   **First Translation Theorem**
>
> If $F(s) = \mathcal{L}\{f(t)\}$ and a is any real number, then
>
> $$\mathcal{L}\{e^{at}f(t)\} = F(s - a).$$

PROOF   The proof is immediate, since by Definition 7.1

$$\mathcal{L}\{e^{at}f(t)\} = \int_0^\infty e^{-st}e^{at}f(t)\,dt = \int_0^\infty e^{-(s-a)t}f(t)\,dt = F(s - a). \quad \blacksquare$$

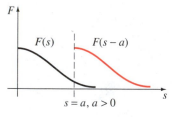

FIGURE 7.11

If we consider s a real variable, then the graph of $F(s - a)$ is the graph of $F(s)$ shifted on the s-axis by the amount $|a|$ units. If $a > 0$, the graph of $F(s)$ is shifted a units to the right, whereas if $a < 0$, the graph is shifted $|a|$ units to the left. See Figure 7.11.

For emphasis it is sometimes useful to use the symbolism

$$\mathcal{L}\{e^{at}f(t)\} = \mathcal{L}\{f(t)\}_{s\to s-a},$$

where $s \to s - a$ means that we replace s in $F(s)$ by $s - a$.

EXAMPLE 1   **First Translation Theorem**

Evaluate   (a) $\mathcal{L}\{e^{5t}t^3\}$ (b) $\mathcal{L}\{e^{-2t}\cos 4t\}$

SOLUTION   The results follow from Theorem 7.5.

(a) $\mathcal{L}\{e^{5t}t^3\} = \mathcal{L}\{t^3\}_{s\to s-5} = \dfrac{3!}{s^4}\bigg|_{s\to s-5} = \dfrac{6}{(s - 5)^4}.$

(b) $\mathcal{L}\{e^{-2t}\cos 4t\} = \mathcal{L}\{\cos 4t\}_{s\to s+2}$ $\leftarrow a = -2 \text{ so } s - a = s - (-2) = s + 2$

$$= \frac{s}{s^2 + 16}\bigg|_{s\to s+2} = \frac{s + 2}{(s + 2)^2 + 16}. \quad \blacksquare$$

Inverse Form of the First Translation Theorem   If $f(t) = \mathcal{L}^{-1}\{F(s)\}$, the inverse form of Theorem 7.5 is

$$\mathcal{L}^{-1}\{F(s - a)\} = \mathcal{L}^{-1}\{F(s)|_{s\to s-a}\} = e^{at}f(t). \tag{1}$$

EXAMPLE 2   **Completing the Square to Find $\mathcal{L}^{-1}$**

Evaluate $\mathcal{L}^{-1}\left\{\dfrac{s}{s^2 + 6s + 11}\right\}.$

SOLUTION   If $s^2 + 6s + 11$ had real factors, we would use partial fractions. Since this quadratic term does not factor, we complete the square.

$$\mathcal{L}^{-1}\left\{\frac{s}{s^2 + 6s + 11}\right\} = \mathcal{L}^{-1}\left\{\frac{s}{(s+3)^2 + 2}\right\} \qquad \leftarrow \text{completion of square}$$

$$= \mathcal{L}^{-1}\left\{\frac{s + 3 - 3}{(s+3)^2 + 2}\right\} \qquad \leftarrow \text{adding zero in the numerator}$$

$$= \mathcal{L}^{-1}\left\{\frac{s + 3}{(s+3)^2 + 2} - \frac{3}{(s+3)^2 + 2}\right\} \qquad \leftarrow \text{termwise division}$$

$$= \mathcal{L}^{-1}\left\{\frac{s + 3}{(s+3)^2 + 2}\right\} - 3\mathcal{L}^{-1}\left\{\frac{1}{(s+3)^2 + 2}\right\} \qquad \leftarrow \text{linearity of } \mathcal{L}^{-1}$$

$$= \mathcal{L}^{-1}\left\{\frac{s}{s^2 + 2}\bigg|_{s \to s+3}\right\} - \frac{3}{\sqrt{2}}\,\mathcal{L}^{-1}\left\{\frac{\sqrt{2}}{s^2 + 2}\bigg|_{s \to s+3}\right\}$$

$$= e^{-3t}\cos\sqrt{2}t - \frac{3}{\sqrt{2}}e^{-3t}\sin\sqrt{2}t. \qquad \leftarrow \text{from (1) and Theorem 7.3} \quad \blacksquare$$

EXAMPLE 3   **Completing the Square and Linearity**

Evaluate $\mathcal{L}^{-1}\left\{\dfrac{1}{(s-1)^3} + \dfrac{1}{s^2 + 2s - 8}\right\}$.

SOLUTION   Completing the square in the second denominator and using linearity yields

$$\mathcal{L}^{-1}\left\{\frac{1}{(s-1)^3} + \frac{1}{s^2 + 2s - 8}\right\} = \mathcal{L}^{-1}\left\{\frac{1}{(s-1)^3} + \frac{1}{(s+1)^2 - 9}\right\}$$

$$= \frac{1}{2!}\,\mathcal{L}^{-1}\left\{\frac{2!}{(s-1)^3}\right\} + \frac{1}{3}\,\mathcal{L}^{-1}\left\{\frac{3}{(s+1)^2 - 9}\right\}$$

$$= \frac{1}{2!}\,\mathcal{L}^{-1}\left\{\frac{2!}{s^3}\bigg|_{s \to s-1}\right\} + \frac{1}{3}\,\mathcal{L}^{-1}\left\{\frac{3}{s^2 - 9}\bigg|_{s \to s+1}\right\}$$

$$= \frac{1}{2}e^t t^2 + \frac{1}{3}e^{-t}\sinh 3t. \qquad \blacksquare$$

Unit Step Function   In engineering one frequently encounters functions that can be either "on" or "off." For example, an external force acting on a mechanical system or a voltage impressed on a circuit can be turned off after a period of time. It is thus convenient to define a special function called the **unit step function.**

DEFINITION 7.3   **Unit Step Function**

The function $\mathcal{U}(t - a)$ is defined to be

$$\mathcal{U}(t - a) = \begin{cases} 0, & 0 \leq t < a \\ 1, & t \geq a. \end{cases}$$

Notice that we define $\mathcal{U}(t - a)$ only on the nonnegative t-axis since this is all that we are concerned with in the study of the Laplace transform. In a broader sense $\mathcal{U}(t - a) = 0$ for $t < a$.

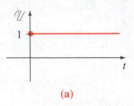

(a)

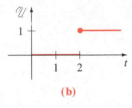

(b)

FIGURE 7.12

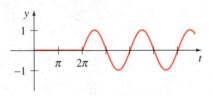

FIGURE 7.13

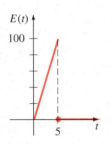

FIGURE 7.14

EXAMPLE 4　　**Graphs of Unit Step Functions**

Graph　　(a) $\mathcal{U}(t)$　(b) $\mathcal{U}(t-2)$

SOLUTION　　(a) $\mathcal{U}(t) = 1, \quad t \geq 0$　(b) $\mathcal{U}(t-2) = \begin{cases} 0, & 0 \leq t < 2 \\ 1, & t \geq 2. \end{cases}$

The respective graphs are given in Figure 7.12.　　∎

When multiplied by another function defined for $t \geq 0$, the unit step function "turns off" a portion of the graph of the function. For example, Figure 7.13 illustrates the graph of $\sin t, t \geq 0$ when multiplied by $\mathcal{U}(t - 2\pi)$:

$$f(t) = \sin t\, \mathcal{U}(t - 2\pi) = \begin{cases} 0, & 0 \leq t < 2\pi \\ \sin t, & t \geq 2\pi. \end{cases}$$

The unit step function can also be used to write piecewise-defined functions in a compact form. For instance, the piecewise-defined function

$$f(t) = \begin{cases} g(t), & 0 \leq t < a \\ h(t), & t \geq a \end{cases} = g(t) + \begin{cases} 0, & 0 \leq t < a \\ -g(t) + h(t), & t \geq a \end{cases} \quad \text{(2)}$$

is the same as

$$f(t) = g(t) - g(t)\, \mathcal{U}(t - a) + h(t)\, \mathcal{U}(t - a). \quad \text{(3)}$$

Similarly, a function of the type

$$f(t) = \begin{cases} 0, & 0 \leq t < a \\ g(t), & a \leq t < b \\ 0, & t \geq b \end{cases} \quad \text{(4)}$$

can be written　　$f(t) = g(t)[\mathcal{U}(t - a) - \mathcal{U}(t - b)]. \quad \text{(5)}$

EXAMPLE 5　　**Function Expressed in Terms of a Unit Step Function**

The voltage in a circuit is given by $E(t) = \begin{cases} 20t, & 0 \leq t < 5 \\ 0, & t \geq 5 \end{cases}$. Graph $E(t)$. Express $E(t)$ in terms of unit step functions.

SOLUTION　　The graph of this piecewise-defined function is given in Figure 7.14. Now from (2) and (3), with $g(t) = 20t$ and $h(t) = 0$, we get

$$E(t) = 20t - 20t\, \mathcal{U}(t - 5).$$

∎

EXAMPLE 6　　**Comparison of Functions**

Consider the function $y = f(t)$ defined by $f(t) = t^3$. Compare the graphs of

(a) $f(t), \quad -\infty < t < \infty$　　(b) $f(t), \quad t \geq 0$

(c) $f(t - 2), \quad t \geq 0$　　　　(d) $f(t - 2)\, \mathcal{U}(t - 2), \quad t \geq 0$

SOLUTION　　The respective graphs are given in Figure 7.15.

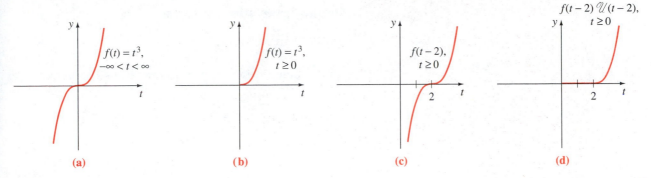

FIGURE 7.15

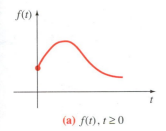

(a) $f(t), t \geq 0$

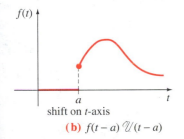

shift on t-axis

(b) $f(t - a)\, \mathcal{U}(t - a)$

FIGURE 7.16

In general, if $a > 0$, then the graph of $y = f(t - a)$ is the graph of $y = f(t)$, $t \geq 0$, shifted a units to the right on the t-axis. However, when $y = f(t - a)$ is multiplied by the unit step function $\mathcal{U}(t - a)$ in the manner illustrated in part (d) of Example 6, then the graph of the function

$$y = f(t - a)\mathcal{U}(t - a) \tag{6}$$

coincides with the graph of $y = f(t - a)$ for $t \geq a$ but is identically zero for $0 \leq t < a$. See Figure 7.16.

We saw in Theorem 7.5 that an exponential multiple of $f(t)$ results in a translation, or shift, of the transform $F(s)$ on the s-axis. In the next theorem we see that whenever $F(s)$ is multiplied by an appropriate exponential function, the inverse transform of this product is the shifted function given in (6). This result is called the **second translation theorem** or **second shifting theorem**.

> **THEOREM 7.6**   **Second Translation Theorem**
>
> If $F(s) = \mathcal{L}\{f(t)\}$ and $a > 0$, then
>
> $$\mathcal{L}\{f(t - a)\mathcal{U}(t - a)\} = e^{-as}F(s).$$

PROOF   We express $\int_0^\infty e^{-st} f(t - a)\mathcal{U}(t - a)\, dt$ as the sum of two integrals:

$$\mathcal{L}\{f(t - a)\mathcal{U}(t - a)\} = \int_0^a e^{-st} f(t - a)\underbrace{\mathcal{U}(t - a)}_{\substack{\text{zero for} \\ 0 \leq t < a}}\, dt + \int_a^\infty e^{-st} f(t - a)\underbrace{\mathcal{U}(t - a)}_{\substack{\text{one for} \\ t \geq a}}\, dt$$

$$= \int_a^\infty e^{-st} f(t - a)\, dt.$$

Now let $v = t - a$, $dv = dt$; then

$$\mathcal{L}\{f(t - a)\mathcal{U}(t - a)\} = \int_0^\infty e^{-s(v + a)} f(v)\, dv$$

$$= e^{-as} \int_0^\infty e^{-sv} f(v)\, dv = e^{-as}\mathcal{L}\{f(t)\}. \quad \blacksquare$$

EXAMPLE 7   **Second Translation Theorem**

Evaluate $\mathscr{L}\{(t-2)^3\mathscr{U}(t-2)\}$.

SOLUTION   With the identification $a = 2$ it follows from Theorem 7.6 that

$$\mathscr{L}\{(t-2)^3\mathscr{U}(t-2)\} = e^{-2s}\mathscr{L}\{t^3\} = e^{-2s}\frac{3!}{s^4} = \frac{6}{s^4}e^{-2s}. \quad \blacksquare$$

We often wish to find the Laplace transform of just the unit step function. This can be found from either Definition 7.1 or Theorem 7.6. If we identify $f(t) = 1$ in Theorem 7.6, then $f(t-a) = 1$, $F(s) = \mathscr{L}\{1\} = 1/s$, and so

$$\mathscr{L}\{\mathscr{U}(t-a)\} = \frac{e^{-as}}{s}. \tag{7}$$

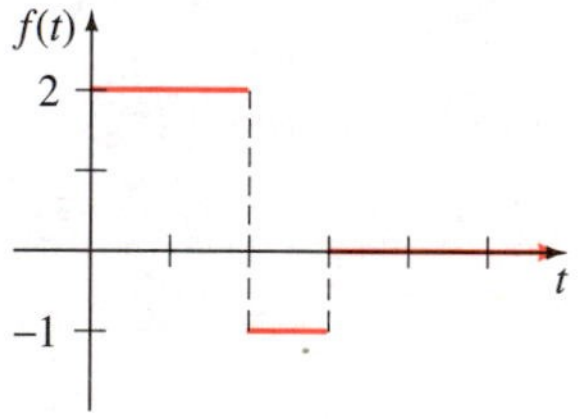

EXAMPLE 8   **Function Expressed in Terms of Unit Step Functions**

Find the Laplace transform of the function shown in Figure 7.17.

SOLUTION   With the aid of the unit step function we can write

$$f(t) = 2 - 3\mathscr{U}(t-2) + \mathscr{U}(t-3).$$

Using linearity and the result in (7) it follows that

$$\mathscr{L}\{f(t)\} = \mathscr{L}\{2\} - 3\mathscr{L}\{\mathscr{U}(t-2)\} + \mathscr{L}\{\mathscr{U}(t-3)\}$$
$$= \frac{2}{s} - 3\frac{e^{-2s}}{s} + \frac{e^{-3s}}{s}. \quad \blacksquare$$

FIGURE 7.17

Alternative Form of the Second Translation Theorem   We are frequently confronted with the problem of finding the Laplace transform of a product of a function g and a unit step function $\mathscr{U}(t-a)$, where the function g lacks the precise shifted form $f(t-a)$ required in Theorem 7.6. To find the Laplace transform of $g(t)\mathscr{U}(t-a)$ it is possible to "fix up" $g(t)$ by algebraic manipulations to force it into the desired form $f(t-a)$. But since these manipulations are time consuming and often not obvious, it is simpler to devise an alternative version of Theorem 7.6. Using Definition 7.1, the definition of $\mathscr{U}(t-a)$, and the substitution $u = t - a$, we obtain

$$\mathscr{L}\{g(t)\mathscr{U}(t-a)\} = \int_a^\infty e^{-st}g(t)\,dt = \int_0^\infty e^{-s(u+a)}g(u+a)\,du.$$

That is, $$\mathscr{L}\{g(t)\mathscr{U}(t-a)\} = e^{-as}\mathscr{L}\{g(t+a)\}. \tag{8}$$

EXAMPLE 9   **Second Translation Theorem—Alternative Form**

Evaluate $\mathscr{L}\{\sin t\,\mathscr{U}(t-2\pi)\}$.

SOLUTION With $g(t) = \sin t$, $a = 2\pi$, note that $g(t + 2\pi) = \sin(t + 2\pi) = \sin t$ because the sine function has period 2π. Now, by (8),

$$\mathscr{L}\{\sin t\, \mathscr{U}(t - 2\pi)\} = e^{-2\pi s}\mathscr{L}\{\sin t\} = \frac{e^{-2\pi s}}{s^2 + 1}. \qquad \blacksquare$$

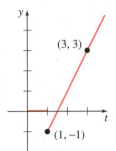

FIGURE 7.18

EXAMPLE 10 **Second Translation Theorem—Alternative Form**

Find the Laplace transform of the function shown in Figure 7.18.

SOLUTION An equation of the straight line through the two points is found to be $y = 2t - 3$. To "turn off" the graph $y = 2t - 3$ on the interval $0 \le t < 1$ we employ the product $(2t - 3)\,\mathscr{U}(t - 1)$. In this case, with $g(t) = 2t - 3$, $a = 1$, and $g(t + 1) = 2(t + 1) - 3 = 2t - 1$, it follows from (8) that

$$\mathscr{L}\{(2t - 3)\,\mathscr{U}(t - 1)\} = e^{-s}\mathscr{L}\{2t - 1\} = e^{-s}\left(\frac{2}{s^2} - \frac{1}{s}\right). \qquad \blacksquare$$

Inverse Form of the Second Translation Theorem If $f(t) = \mathscr{L}^{-1}\{F(s)\}$, the inverse form of Theorem 7.6, $a > 0$, is

$$\mathscr{L}^{-1}\{e^{-as}F(s)\} = f(t - a)\,\mathscr{U}(t - a). \qquad (9)$$

EXAMPLE 11 **Inverse by Formula (9)**

Evaluate $\mathscr{L}^{-1}\left\{\dfrac{e^{-\pi s/2}}{s^2 + 9}\right\}$.

SOLUTION We identify $a = \dfrac{\pi}{2}$ and $f(t) = \mathscr{L}^{-1}\left\{\dfrac{1}{s^2 + 9}\right\} = \dfrac{1}{3}\sin 3t$. Thus from (9)

$$\mathscr{L}^{-1}\left\{\frac{e^{-\pi s/2}}{s^2 + 9}\right\} = \frac{1}{3}\,\mathscr{L}^{-1}\left\{\frac{3}{s^2 + 9}\right\}_{t \to t - \pi/2}\mathscr{U}\left(t - \frac{\pi}{2}\right)$$

$$= \frac{1}{3}\sin 3\left(t - \frac{\pi}{2}\right)\mathscr{U}\left(t - \frac{\pi}{2}\right)$$

trigonometric identity $\rightarrow$
$$= \frac{1}{3}\cos 3t\,\mathscr{U}\left(t - \frac{\pi}{2}\right). \qquad \blacksquare$$

If $F(s) = \mathscr{L}\{f(t)\}$ and if we assume that interchanging of differentiation and integration is possible, then

$$\frac{d}{ds}F(s) = \frac{d}{ds}\int_0^\infty e^{-st}f(t)\,dt = \int_0^\infty \frac{\partial}{\partial s}[e^{-st}f(t)]\,dt = -\int_0^\infty e^{-st}tf(t)\,dt = -\mathscr{L}\{tf(t)\};$$

that is,
$$\mathscr{L}\{tf(t)\} = -\frac{d}{ds}\mathscr{L}\{f(t)\}.$$

Similarly, $\mathscr{L}\{t^2 f(t)\} = \mathscr{L}\{t \cdot tf(t)\} = -\dfrac{d}{ds}\mathscr{L}\{tf(t)\}$

$$= -\frac{d}{ds}\left(-\frac{d}{ds}\mathscr{L}\{f(t)\}\right) = \frac{d^2}{ds^2}\mathscr{L}\{f(t)\}.$$

The preceding two cases suggest the general result for $\mathscr{L}\{t^n f(t)\}$.

THEOREM 7.7   **Derivatives of Transforms**

If $F(s) = \mathscr{L}\{f(t)\}$ and $n = 1, 2, 3, \ldots$, then

$$\mathscr{L}\{t^n f(t)\} = (-1)^n \frac{d^n}{ds^n} F(s).$$

EXAMPLE 12   **Applying Theorem 7.7**

Evaluate

(a) $\mathscr{L}\{te^{3t}\}$ (b) $\mathscr{L}\{t \sin kt\}$ (c) $\mathscr{L}\{t^2 \sin kt\}$ (d) $\mathscr{L}\{te^{-t} \cos t\}$

SOLUTION   We make use of results (c), (d), and (e) of Theorem 7.2.

(a) Note in this first example that we could also use the first translation theorem. To apply Theorem 7.7 we identify $n = 1$ and $f(t) = e^{3t}$:

$$\mathscr{L}\{te^{3t}\} = -\frac{d}{ds}\mathscr{L}\{e^{3t}\} = -\frac{d}{ds}\left(\frac{1}{s-3}\right) = \frac{1}{(s-3)^2}.$$

(b) $\mathscr{L}\{t \sin kt\} = -\dfrac{d}{ds}\mathscr{L}\{\sin kt\} = -\dfrac{d}{ds}\left(\dfrac{k}{s^2 + k^2}\right) = \dfrac{2ks}{(s^2 + k^2)^2}$

(c) With $n = 2$ in Theorem 7.7 this transform can be written

$$\mathscr{L}\{t^2 \sin kt\} = \frac{d^2}{ds^2}\mathscr{L}\{\sin kt\},$$

and so by carrying out the two derivatives we obtain the result. Alternatively, we can make use of the result already obtained in part (b). Since $t^2 \sin kt = t(t \sin kt)$, we have

$$\mathscr{L}\{t^2 \sin kt\} = -\frac{d}{ds}\mathscr{L}\{t \sin kt\} = -\frac{d}{ds}\left(\frac{2ks}{(s^2 + k^2)^2}\right). \qquad \leftarrow \text{from part (b)}$$

Differentiating and simplifying then gives

$$\mathscr{L}\{t^2 \sin kt\} = \frac{6ks^2 - 2k^3}{(s^2 + k^2)^3}.$$

(d) $\mathscr{L}\{te^{-t} \cos t\} = -\dfrac{d}{ds}\mathscr{L}\{e^{-t} \cos t\}$

$$= -\frac{d}{ds}\mathscr{L}\{\cos t\}_{s \to s+1} \qquad \leftarrow \text{first translation theorem}$$

$$= -\frac{d}{ds}\left(\frac{s+1}{(s+1)^2 + 1}\right)$$

$$= \frac{(s+1)^2 - 1}{[(s+1)^2 + 1]^2}$$

Answers to odd-numbered problems begin on page A-11.

In Problems 1–44 find either $F(s)$ or $f(t)$, as indicated.

1. $\mathscr{L}\{te^{10t}\}$ **2.** $\mathscr{L}\{te^{-6t}\}$

3. $\mathscr{L}\{t^3 e^{-2t}\}$ **4.** $\mathscr{L}\{t^{10}e^{-7t}\}$

5. $\mathscr{L}\{e^t \sin 3t\}$ **6.** $\mathscr{L}\{e^{-2t} \cos 4t\}$

7. $\mathscr{L}\{e^{5t} \sinh 3t\}$ **8.** $\mathscr{L}\left\{\dfrac{\cosh t}{e^t}\right\}$

9. $\mathscr{L}\{t(e^t + e^{2t})^2\}$ **10.** $\mathscr{L}\{e^{2t}(t-1)^2\}$

11. $\mathscr{L}\{e^{-t} \sin^2 t\}$ **12.** $\mathscr{L}\{e^t \cos^2 3t\}$

13. $\mathscr{L}^{-1}\left\{\dfrac{1}{(s+2)^3}\right\}$ **14.** $\mathscr{L}^{-1}\left\{\dfrac{1}{(s-1)^4}\right\}$

15. $\mathscr{L}^{-1}\left\{\dfrac{1}{s^2 - 6s + 10}\right\}$ **16.** $\mathscr{L}^{-1}\left\{\dfrac{1}{s^2 + 2s + 5}\right\}$

17. $\mathscr{L}^{-1}\left\{\dfrac{s}{s^2 + 4s + 5}\right\}$ **18.** $\mathscr{L}^{-1}\left\{\dfrac{2s + 5}{s^2 + 6s + 34}\right\}$

19. $\mathscr{L}^{-1}\left\{\dfrac{s}{(s+1)^2}\right\}$ **20.** $\mathscr{L}^{-1}\left\{\dfrac{5s}{(s-2)^2}\right\}$

21. $\mathscr{L}^{-1}\left\{\dfrac{2s - 1}{s^2(s+1)^3}\right\}$ **22.** $\mathscr{L}^{-1}\left\{\dfrac{(s+1)^2}{(s+2)^4}\right\}$

23. $\mathscr{L}\{(t-1)\,\mathscr{U}(t-1)\}$ **24.** $\mathscr{L}\{e^{2-t}\,\mathscr{U}(t-2)\}$

25. $\mathscr{L}\{t\,\mathscr{U}(t-2)\}$ **26.** $\mathscr{L}\{(3t+1)\,\mathscr{U}(t-3)\}$

27. $\mathscr{L}\{\cos 2t\,\mathscr{U}(t-\pi)\}$ **28.** $\mathscr{L}\left\{\sin t\,\mathscr{U}\left(t-\dfrac{\pi}{2}\right)\right\}$

29. $\mathscr{L}\{(t-1)^3 e^{t-1}\,\mathscr{U}(t-1)\}$ **30.** $\mathscr{L}\{te^{t-5}\,\mathscr{U}(t-5)\}$

31. $\mathscr{L}^{-1}\left\{\dfrac{e^{-2s}}{s^3}\right\}$ **32.** $\mathscr{L}^{-1}\left\{\dfrac{(1+e^{-2s})^2}{s+2}\right\}$

33. $\mathscr{L}^{-1}\left\{\dfrac{e^{-\pi s}}{s^2 + 1}\right\}$ **34.** $\mathscr{L}^{-1}\left\{\dfrac{se^{-\pi s/2}}{s^2 + 4}\right\}$

35. $\mathscr{L}^{-1}\left\{\dfrac{e^{-s}}{s(s+1)}\right\}$ **36.** $\mathscr{L}^{-1}\left\{\dfrac{e^{-2s}}{s^2(s-1)}\right\}$

37. $\mathscr{L}\{t \cos 2t\}$ **38.** $\mathscr{L}\{t \sinh 3t\}$

39. $\mathscr{L}\{t^2 \sinh t\}$ **40.** $\mathscr{L}\{t^2 \cos t\}$

41. $\mathscr{L}\{te^{2t} \sin 6t\}$ **42.** $\mathscr{L}\{te^{-3t} \cos 3t\}$

43. $\mathscr{L}^{-1}\left\{\dfrac{s}{(s^2 + 1)^2}\right\}$ **44.** $\mathscr{L}^{-1}\left\{\dfrac{s+1}{(s^2 + 2s + 2)^2}\right\}$

In Problems 45–50 match the given graph with one of the functions in (a)–(f). The graph of $f(t)$ is given in Figure 7.19.

(a) $f(t) - f(t)\,\mathcal{U}(t - a)$

(b) $f(t - b)\,\mathcal{U}(t - b)$

(c) $f(t)\,\mathcal{U}(t - a)$

(d) $f(t) - f(t)\,\mathcal{U}(t - b)$

(e) $f(t)\,\mathcal{U}(t - a) - f(t)\,\mathcal{U}(t - b)$

(f) $f(t - a)\,\mathcal{U}(t - a) - f(t - a)\,\mathcal{U}(t - b)$

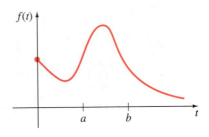

FIGURE 7.19

45.

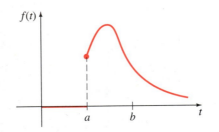

FIGURE 7.20

46.

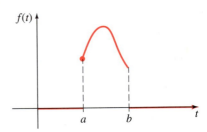

FIGURE 7.21

47.

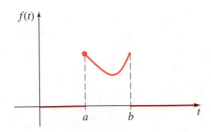

FIGURE 7.22

48.

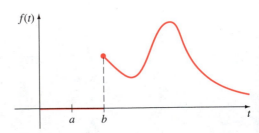

FIGURE 7.23

49.

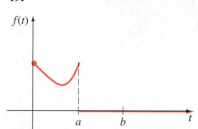

FIGURE 7.24

50.

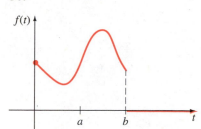

FIGURE 7.25

In Problems 51–58 write each function in terms of unit step functions. Find the Laplace transform of the given function.

51. $f(t) = \begin{cases} 2, & 0 \leq t < 3 \\ -2, & t \geq 3 \end{cases}$

52. $f(t) = \begin{cases} 1, & 0 \leq t < 4 \\ 0, & 4 \leq t < 5 \\ 1, & t \geq 5 \end{cases}$

53. $f(t) = \begin{cases} 0, & 0 \leq t < 1 \\ t^2, & t \geq 1 \end{cases}$

54. $f(t) = \begin{cases} 0, & 0 \leq t < 3\pi/2 \\ \sin t, & t \geq 3\pi/2 \end{cases}$

55. $f(t) = \begin{cases} t, & 0 \leq t < 2 \\ 0, & t \geq 2 \end{cases}$

56. $f(t) = \begin{cases} \sin t, & 0 \leq t < 2\pi \\ 0, & t \geq 2\pi \end{cases}$

57.

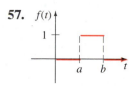

rectangular pulse

FIGURE 7.26

58.

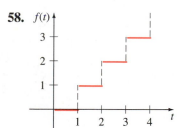

staircase function

FIGURE 7.27

In Problems 59 and 60 sketch the graph of the given function.

59. $f(t) = \mathscr{L}^{-1} \left\{ \dfrac{1}{s^2} - \dfrac{e^{-s}}{s^2} \right\}$

60. $f(t) = \mathscr{L}^{-1} \left\{ \dfrac{2}{s} - \dfrac{3e^{-s}}{s^2} + \dfrac{5e^{-2s}}{s^2} \right\}$

In Problems 61 and 62 use Theorem 7.7 in the form ($n = 1$)

$$f(t) = -\frac{1}{t} \mathscr{L}^{-1} \left\{ \frac{d}{ds} F(s) \right\}$$

to evaluate the given inverse Laplace transform.

61. $\mathscr{L}^{-1} \left\{ \ln \dfrac{s-3}{s+1} \right\}$

62. $\mathscr{L}^{-1} \left\{ \ln \dfrac{s^2+1}{s^2+4} \right\}$

63. Use Theorem 7.6 to find $\mathcal{L}\{(t^2 - 3t)\mathcal{U}(t - 2)\}$. You will first need to "fix up" $g(t) = t^2 - 3t$ by rewriting it in terms of powers of $t - 2$. Check your answer using (8) of this section.

Discussion Problem

64. How would you "fix up" each function so that Theorem 7.6 can be used directly to find the given Laplace transform?

(a) $\mathcal{L}\{(2t + 1)\mathcal{U}(t - 1)\}$ (b) $\mathcal{L}\{e^t\mathcal{U}(t - 5)\}$

(c) $\mathcal{L}\{\cos t\,\mathcal{U}(t - \pi)\}$ (d) $\mathcal{L}\{t \sin t\,\mathcal{U}(t - 2\pi)\}$

7.4   TRANSFORMS OF DERIVATIVES, INTEGRALS, AND PERIODIC FUNCTIONS

■ *Transform of a derivative* ■ *Convolution of two functions* ■ *Convolution theorem*
■ *Inverse form of convolution theorem* ■ *Transform of an integral*
■ *Transform of a periodic function*

Our goal is to use the Laplace transform to solve certain kinds of differential equations. To that end we need to evaluate quantities such as $\mathcal{L}\{dy/dt\}$ and $\mathcal{L}\{d^2y/dt^2\}$. For example, if f' is continuous for $t \geq 0$, then integration by parts gives

$$\mathcal{L}\{f'(t)\} = \int_0^\infty e^{-st}f'(t)\,dt = e^{-st}f(t)\Big|_0^\infty + s\int_0^\infty e^{-st}f(t)\,dt$$

$$= -f(0) + s\mathcal{L}\{f(t)\}$$

or
$$\mathcal{L}\{f'(t)\} = sF(s) - f(0). \tag{1}$$

Here we have assumed that $e^{-st}f(t) \to 0$ as $t \to \infty$. Similarly, the transform of the second derivative is

$$\mathcal{L}\{f''(t)\} = \int_0^\infty e^{-st}f''(t)\,dt = e^{-st}f'(t)\Big|_0^\infty + s\int_0^\infty e^{-st}f'(t)\,dt$$

$$= -f'(0) + s\mathcal{L}\{f'(t)\}$$
$$= s[sF(s) - f(0)] - f'(0)$$

or
$$\mathcal{L}\{f''(t)\} = s^2F(s) - sf(0) - f'(0). \tag{2}$$

In like manner it can be shown that

$$\mathcal{L}\{f'''(t)\} = s^3F(s) - s^2f(0) - sf'(0) - f''(0). \tag{3}$$

The recursive nature of the Laplace transform of the derivatives of a function f should be apparent from the results in (1), (2), and (3). The next theorem gives the Laplace transform of the nth derivative of f. The proof is omitted.

> **THEOREM 7.8**   **Transform of a Derivative**
>
> If $f(t), f'(t), \ldots, f^{(n-1)}(t)$ are continuous on $[0, \infty)$ and are of exponential order and if $f^{(n)}(t)$ is piecewise continuous on $[0, \infty)$, then
>
> $$\mathcal{L}\{f^{(n)}(t)\} = s^n F(s) - s^{n-1} f(0) - s^{n-2} f'(0) - \cdots - f^{(n-1)}(0),$$
>
> where $F(s) = \mathcal{L}\{f(t)\}$.

> **EXAMPLE 1**   **Applying Theorem 7.8**

Observe that the sum $kt \cos kt + \sin kt$ is the derivative of $t \sin kt$. Hence

$$\mathcal{L}\{kt \cos kt + \sin kt\} = \mathcal{L}\left\{\frac{d}{dt}(t \sin kt)\right\}$$

$$= s\mathcal{L}\{t \sin kt\} \qquad \leftarrow \text{ by (1)}$$

$$= s\left(-\frac{d}{ds}\mathcal{L}\{\sin kt\}\right) \qquad \leftarrow \text{ from Theorem 7.7}$$

$$= s\left(\frac{2ks}{(s^2 + k^2)^2}\right) = \frac{2ks^2}{(s^2 + k^2)^2}. \qquad \blacksquare$$

Convolution   If functions f and g are piecewise continuous on $[0, \infty)$, then the **convolution** of f and g, denoted by $f * g$, is given by the integral

$$f * g = \int_0^t f(\tau)g(t - \tau)\, d\tau.$$

For example, the convolution of $f(t) = e^t$ and $g(t) = \sin t$ is

$$e^t * \sin t = \int_0^t e^\tau \sin(t - \tau)\, d\tau = \frac{1}{2}(-\sin t - \cos t + e^t). \qquad \textbf{(4)}$$

It is left as an exercise to show that

$$\int_0^t f(\tau)g(t - \tau)\, d\tau = \int_0^t f(t - \tau)g(\tau)\, d\tau;$$

that is, $$f * g = g * f.$$

See Problem 29 in Exercises 7.4. This means that the convolution of two functions is commutative.

It is possible to find the Laplace transform of the convolution of two functions without actually evaluating the integral as we did in (4). The result that follows is known as the **convolution theorem.**

> **THEOREM 7.9**   **Convolution Theorem**
>
> If $f(t)$ and $g(t)$ are piecewise continuous on $[0, \infty)$ and of exponential order, then
>
> $$\mathcal{L}\{f * g\} = \mathcal{L}\{f(t)\}\mathcal{L}\{g(t)\} = F(s)G(s).$$

PROOF   Let $F(s) = \mathcal{L}\{f(t)\} = \int_0^\infty e^{-s\tau} f(\tau)\, d\tau$

and $G(s) = \mathcal{L}\{g(t)\} = \int_0^\infty e^{-s\beta} g(\beta)\, d\beta.$

Proceeding formally we have

$$F(s)G(s) = \left(\int_0^\infty e^{-s\tau} f(\tau)\, d\tau\right)\left(\int_0^\infty e^{-s\beta} g(\beta)\, d\beta\right)$$

$$= \int_0^\infty \int_0^\infty e^{-s(\tau+\beta)} f(\tau) g(\beta)\, d\tau\, d\beta$$

$$= \int_0^\infty f(\tau)\, d\tau \int_0^\infty e^{-s(\tau+\beta)} g(\beta)\, d\beta.$$

Holding τ fixed, we let $t = \tau + \beta$, $dt = d\beta$ so that

$$F(s)G(s) = \int_0^\infty f(\tau)\, d\tau \int_\tau^\infty e^{-st} g(t - \tau)\, dt.$$

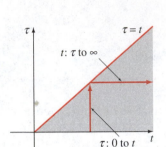

FIGURE 7.28

In the $t\tau$-plane we are integrating over the shaded region in Figure 7.28. Since f and g are piecewise continuous on $[0, \infty)$ and of exponential order, it is possible to interchange the order of integration:

$$F(s)G(s) = \int_0^\infty e^{-st}\, dt \int_0^t f(\tau) g(t - \tau)\, d\tau = \int_0^\infty e^{-st} \left\{ \int_0^t f(\tau) g(t - \tau)\, d\tau \right\} dt = \mathcal{L}\{f * g\}. \quad \blacksquare$$

EXAMPLE 2   **Transform of a Convolution**

Evaluate $\mathcal{L}\left\{ \int_0^t e^\tau \sin(t - \tau)\, d\tau \right\}.$

SOLUTION   With $f(t) = e^t$ and $g(t) = \sin t$ the convolution theorem states that the Laplace transform of the convolution of f and g is the product of their Laplace transforms:

$$\mathcal{L}\left\{ \int_0^t e^\tau \sin(t - \tau)\, d\tau \right\} = \mathcal{L}\{e^t\} \cdot \mathcal{L}\{\sin t\} = \frac{1}{s - 1} \cdot \frac{1}{s^2 + 1} = \frac{1}{(s - 1)(s^2 + 1)}. \quad \blacksquare$$

Inverse Form of the Convolution Theorem   The convolution theorem is sometimes useful in finding the inverse Laplace transform of a product of two Laplace transforms. From Theorem 7.9 we have

$$\mathcal{L}^{-1}\{F(s)G(s)\} = f * g. \tag{5}$$

EXAMPLE 3   **Inverse Transform as a Convolution**

Evaluate $\mathcal{L}^{-1}\left\{ \dfrac{1}{(s - 1)(s + 4)} \right\}.$

SOLUTION   Partial fractions could be used, but if we identify

$$F(s) = \frac{1}{s - 1} \quad \text{and} \quad G(s) = \frac{1}{s + 4},$$

then

$$\mathscr{L}^{-1}\{F(s)\} = f(t) = e^t \quad \text{and} \quad \mathscr{L}^{-1}\{G(s)\} = g(t) = e^{-4t}.$$

Hence from (4) we obtain

$$\mathscr{L}^{-1}\left\{\frac{1}{(s-1)(s+4)}\right\} = \int_0^t f(\tau)g(t-\tau)\,d\tau = \int_0^t e^{\tau}e^{-4(t-\tau)}\,d\tau$$

$$= e^{-4t}\int_0^t e^{5\tau}\,d\tau$$

$$= e^{-4t}\frac{1}{5}e^{5\tau}\bigg|_0^t$$

$$= \frac{1}{5}e^t - \frac{1}{5}e^{-4t}. \qquad \blacksquare$$

EXAMPLE 4 **Inverse Transform as a Convolution**

Evaluate $\mathscr{L}^{-1}\left\{\dfrac{1}{(s^2+k^2)^2}\right\}$.

SOLUTION Let $F(s) = G(s) = \dfrac{1}{s^2+k^2}$

so that $f(t) = g(t) = \dfrac{1}{k}\mathscr{L}^{-1}\left\{\dfrac{k}{s^2+k^2}\right\} = \dfrac{1}{k}\sin kt.$

In this case (5) gives

$$\mathscr{L}^{-1}\left\{\frac{1}{(s^2+k^2)^2}\right\} = \frac{1}{k^2}\int_0^t \sin k\tau \sin k(t-\tau)\,d\tau. \qquad (6)$$

Now recall from trigonometry that

$$\cos(A+B) = \cos A \cos B - \sin A \sin B$$

and

$$\cos(A-B) = \cos A \cos B + \sin A \sin B.$$

Subtracting the first from the second gives the identity

$$\sin A \sin B = \frac{1}{2}[\cos(A-B) - \cos(A+B)].$$

If we set $A = k\tau$ and $B = k(t-\tau)$, we can carry out the integration in (6):

$$\mathscr{L}^{-1}\left\{\frac{1}{(s^2+k^2)^2}\right\} = \frac{1}{2k^2}\int_0^t [\cos k(2\tau - t) - \cos kt]\,d\tau$$

$$= \frac{1}{2k^2}\left[\frac{1}{2k}\sin k(2\tau - t) - \tau\cos kt\right]_0^t$$

$$= \frac{\sin kt - kt\cos kt}{2k^3}. \qquad \blacksquare$$

Transform of an Integral When $g(t) = 1$ and $\mathscr{L}\{g(t)\} = G(s) = 1/s$, the convolution theorem implies that the Laplace transform of the integral of f is

$$\mathscr{L}\left\{\int_0^t f(\tau)\,d\tau\right\} = \frac{F(s)}{s}. \tag{7}$$

The inverse form of (7),

$$\int_0^t f(\tau)\,d\tau = \mathscr{L}^{-1}\left\{\frac{F(s)}{s}\right\}, \tag{8}$$

can be used at times in lieu of partial fractions when s^n is a factor of the denominator and $f(t) = \mathscr{L}^{-1}\{F(s)\}$ is easy to integrate. For example, we know for $f(t) = \sin t$ that $F(s) = 1/(s^2 + 1)$, and so by (8)

$$\mathscr{L}^{-1}\left\{\frac{1}{s(s^2 + 1)}\right\} = \int_0^t \sin \tau\,d\tau = 1 - \cos t$$

$$\mathscr{L}^{-1}\left\{\frac{1}{s^2(s^2 + 1)}\right\} = \int_0^t (1 - \cos \tau)\,d\tau = t - \sin t$$

$$\mathscr{L}^{-1}\left\{\frac{1}{s^3(s^2 + 1)}\right\} = \int_0^t (\tau - \sin \tau)\,d\tau = \frac{1}{2}t^2 - 1 + \cos t$$

and so on. We shall also use (7) in the next section on applications.

Transform of a Periodic Function   If a periodic function has period T, $T > 0$, then $f(t + T) = f(t)$. The Laplace transform of a **periodic function** can be obtained by an integration over one period.

> **THEOREM 7.10**   **Transform of a Periodic Function**
>
> If $f(t)$ is piecewise continuous on $[0, \infty)$, of exponential order, and periodic with period T, then
>
> $$\mathscr{L}\{f(t)\} = \frac{1}{1 - e^{-sT}}\int_0^T e^{-st}f(t)\,dt. \tag{9}$$

PROOF   Write the Laplace transform as two integrals:

$$\mathscr{L}\{f(t)\} = \int_0^T e^{-st}f(t)\,dt + \int_T^\infty e^{-st}f(t)\,dt. \tag{10}$$

When we let $t = u + T$, the last integral in (9) becomes

$$\int_T^\infty e^{-st}f(t)\,dt = \int_0^\infty e^{-s(u+T)}f(u + T)\,du = e^{-sT}\int_0^\infty e^{-su}f(u)\,du = e^{-sT}\mathscr{L}\{f(t)\}.$$

Hence (10) is $\mathscr{L}\{f(t)\} = \int_0^T e^{-st}f(t)\,dt + e^{-sT}\mathscr{L}\{f(t)\}.$

Solving for $\mathscr{L}\{f(t)\}$ yields the result given in (9).   ∎

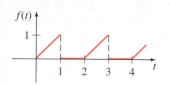

FIGURE 7.29

EXAMPLE 5 **Transform of a Periodic Function**

Find the Laplace transform of the periodic function shown in Figure 7.29.

SOLUTION The function can be defined on the interval $0 \leq t < 2$ by

$$f(t) = \begin{cases} t, & 0 \leq t < 1 \\ 0, & 1 \leq t < 2 \end{cases}$$

and outside the interval by $f(t + 2) = f(t)$. Identifying $T = 2$, we use (9) and integration by parts to obtain

$$\mathcal{L}\{f(t)\} = \frac{1}{1 - e^{-2s}} \int_0^2 e^{-st} f(t) \, dt = \frac{1}{1 - e^{-2s}} \left[\int_0^1 e^{-st} t \, dt + \int_1^2 e^{-st} 0 \, dt \right]$$

$$= \frac{1}{1 - e^{-2s}} \left[-\frac{e^{-s}}{s} + \frac{1 - e^{-s}}{s^2} \right] \qquad \textbf{(11)}$$

$$= \frac{1 - (s + 1)e^{-s}}{s^2(1 - e^{-2s})}. \qquad ■$$

The result in (11) of Example 5 can be obtained without actually integrating by making use of the second translation theorem. If we define

$$g(t) = \begin{cases} t, & 0 \leq t < 1 \\ 0, & t \geq 1, \end{cases}$$

then $f(t) = g(t)$ on the interval $[0, T]$, where $T = 2$. But we can express g in terms of a unit step function as $g(t) = t - t \mathcal{U}(t - 1)$. Thus

$$\mathcal{L}\{f(t)\} = \frac{1}{1 - e^{-2s}} \mathcal{L}\{g(t)\}$$

$$= \frac{1}{1 - e^{-2s}} \mathcal{L}\{t - t \mathcal{U}(t - 1)\}$$

$$= \frac{1}{1 - e^{-2s}} \left[\frac{1}{s^2} - \frac{1}{s^2} e^{-s} - \frac{1}{s} e^{-s} \right]. \qquad \leftarrow \text{by (8) of Section 7.3}$$

Inspection of the expression inside the brackets reveals that it is identical to (11).

SECTION 7.4 EXERCISES

Answers to odd-numbered problems begin on page A-11.

1. Use the result $(d/dt)e^t = e^t$ and (1) of this section to evaluate $\mathcal{L}\{e^t\}$.

2. Use the result $(d/dt)\cos^2 t = -\sin 2t$ and (1) of this section to evaluate $\mathcal{L}\{\cos^2 t\}$.

In Problems 3 and 4 suppose a function $y(t)$ has the properties that $y(0) = 1$ and $y'(0) = -1$. Find the Laplace transform of the given expression.

3. $y'' + 3y'$

4. $y'' - 4y' + 5y$

In Problems 5 and 6 suppose a function $y(t)$ has the properties that $y(0) = 2$ and $y'(0) = 3$. Solve for the Laplace transform $\mathscr{L}\{y(t)\} = Y(s)$.

5. $y'' - 2y' + y = 0$ **6.** $y'' + y = 1$

In Problems 7–20 evaluate the given Laplace transform without evaluating the integral.

7. $\mathscr{L}\left\{\int_0^t e^\tau \, d\tau\right\}$ **8.** $\mathscr{L}\left\{\int_0^t \cos \tau \, d\tau\right\}$

9. $\mathscr{L}\left\{\int_0^t e^{-\tau} \cos \tau \, d\tau\right\}$ **10.** $\mathscr{L}\left\{\int_0^t \tau \sin \tau \, d\tau\right\}$

11. $\mathscr{L}\left\{\int_0^t \tau e^{t-\tau} \, d\tau\right\}$ **12.** $\mathscr{L}\left\{\int_0^t \sin \tau \cos(t - \tau) \, d\tau\right\}$

13. $\mathscr{L}\left\{t \int_0^t \sin \tau \, d\tau\right\}$ **14.** $\mathscr{L}\left\{t \int_0^t \tau e^{-\tau} \, d\tau\right\}$

15. $\mathscr{L}\{1 * t^3\}$ **16.** $\mathscr{L}\{1 * e^{-2t}\}$

17. $\mathscr{L}\{t^2 * t^4\}$ **18.** $\mathscr{L}\{t^2 * te^t\}$

19. $\mathscr{L}\{e^{-t} * e^t \cos t\}$ **20.** $\mathscr{L}\{e^{2t} * \sin t\}$

In Problems 21 and 22 suppose $\mathscr{L}^{-1}\{F(s)\} = f(t)$. Find the inverse Laplace transform of the given function.

21. $\dfrac{1}{s + 5} F(s)$ **22.** $\dfrac{s}{s^2 + 4} F(s)$

In Problems 23–28 use (4) or (7) to find $f(t)$.

23. $\mathscr{L}^{-1}\left\{\dfrac{1}{s(s + 1)}\right\}$ **24.** $\mathscr{L}^{-1}\left\{\dfrac{1}{s^3(s - 1)}\right\}$

25. $\mathscr{L}^{-1}\left\{\dfrac{1}{(s + 1)(s - 2)}\right\}$ **26.** $\mathscr{L}^{-1}\left\{\dfrac{1}{(s + 1)^2}\right\}$

27. $\mathscr{L}^{-1}\left\{\dfrac{s}{(s^2 + 4)^2}\right\}$ **28.** $\mathscr{L}^{-1}\left\{\dfrac{1}{(s^2 + 4s + 5)^2}\right\}$

29. Prove the commutative property of the convolution integral

$$f * g = g * f.$$

30. Prove the distributive property of the convolution integral

$$f * (g + h) = f * g + f * h.$$

In Problems 31–38 use Theorem 7.10 to find the Laplace transform of the given periodic function.

31.

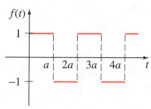

meander function

FIGURE 7.30

32.

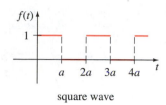

square wave

FIGURE 7.31

33.

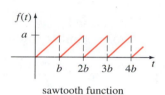

sawtooth function

FIGURE 7.32

34.

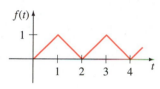

triangular wave

FIGURE 7.33

35.

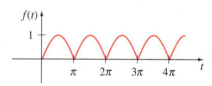

full-wave rectification of sin t

FIGURE 7.34

36.

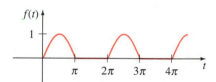

half-wave rectification of sin t

FIGURE 7.35

37. $f(t) = \sin t$
   $f(t + 2\pi) = f(t)$

38. $f(t) = \cos t$
   $f(t + 2\pi) = f(t)$

Discussion Problems

39. Explain $t * \mathcal{U}(t - a) = \frac{1}{2}(t - a)^2 \mathcal{U}(t - a)$.

40. In (7) we saw that the result $\mathcal{L}\{\int_0^t f(\tau)\, d\tau\} = F(s)/s$, where $F(s) = \mathcal{L}\{f(t)\}$, follows from the convolution theorem when $g(t) = 1$. Using the definitions and theorems of this chapter, find two more ways of obtaining the same result.

7.5   APPLICATIONS

■ *Using the Laplace transform to solve an initial-value problem* ■ *Volterra integral equation*
■ *Integrodifferential equation* ■ *Using the Laplace transform to solve a boundary-value problem*

Since $\mathscr{L}\{y^{(n)}(t)\}, n > 1$ depends on $y(t)$ and its $n - 1$ derivatives evaluated at $t = 0$, the Laplace transform is ideally suited to initial-value problems for linear differential equations with constant coefficients. This kind of differential equation can be reduced to an *algebraic equation* in the transformed function $Y(s)$. To see this, consider the initial-value problem

$$a_n \frac{d^n y}{dt^n} + a_{n-1} \frac{d^{n-1}y}{dt^{n-1}} + \cdots + a_1 \frac{dy}{dt} + a_0 y = g(t)$$

$$y(0) = y_0, \quad y'(0) = y_1, \quad \ldots, \quad y^{(n-1)}(0) = y_{n-1},$$

where $a_i, i = 0, 1, \ldots, n$ and $y_0, y_1, \ldots, y_{n-1}$ are constants. By the linearity property of the Laplace transform we can write

$$a_n \mathscr{L}\left\{\frac{d^n y}{dt^n}\right\} + a_{n-1}\mathscr{L}\left\{\frac{d^{n-1}y}{dt^{n-1}}\right\} + \cdots + a_0 \mathscr{L}\{y\} = \mathscr{L}\{g(t)\}. \qquad \textbf{(1)}$$

From Theorem 7.8, (1) becomes

$$a_n[s^n Y(s) - s^{n-1}y(0) - \cdots - y^{(n-1)}(0)]$$
$$+ a_{n-1}[s^{n-1}Y(s) - s^{n-2}y(0) - \cdots - y^{(n-2)}(0)] + \cdots + a_0 Y(s) = G(s)$$

or

$$[a_n s^n + a_{n-1}s^{n-1} + \cdots + a_0]Y(s) = a_n[s^{n-1}y_0 + \cdots + y_{n-1}]$$
$$+ a_{n-1}[s^{n-2}y_0 + \cdots + y_{n-2}] + \cdots + G(s), \qquad \textbf{(2)}$$

where $Y(s) = \mathscr{L}\{y(t)\}$ and $G(s) = \mathscr{L}\{g(t)\}$. By solving (2) for $Y(s)$ we find $y(t)$ by determining the inverse transform

$$y(t) = \mathscr{L}^{-1}\{Y(s)\}.$$

The procedure is outlined in Figure 7.36. Note that this method incorporates the prescribed initial conditions directly into the solution. Hence there is no need for the separate operations of determining constants in the general solution of the differential equation.

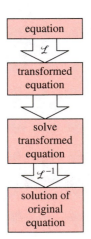

FIGURE 7.36

EXAMPLE 1   **DE Transformed into an Algebraic Equation**

Solve $\dfrac{dy}{dt} - 3y = e^{2t}, y(0) = 1$.

SOLUTION   We first take the transform of each member of the given differential equation:

$$\mathscr{L}\left\{\frac{dy}{dt}\right\} - 3\mathscr{L}\{y\} = \mathscr{L}\{e^{2t}\}.$$

We then use $\mathcal{L}\{dy/dt\} = sY(s) - y(0) = sY(s) - 1$ and $\mathcal{L}\{e^{2t}\} = 1/(s - 2)$. Solving

$$sY(s) - 1 - 3Y(s) = \frac{1}{s - 2}$$

for $Y(s)$ and carrying out partial fraction decomposition gives

$$Y(s) = \frac{s - 1}{(s - 2)(s - 3)} = \frac{-1}{s - 2} + \frac{2}{s - 3},$$

and so

$$y(t) = -\mathcal{L}^{-1}\left\{\frac{1}{s - 2}\right\} + 2\mathcal{L}^{-1}\left\{\frac{1}{s - 3}\right\}.$$

From part (c) of Theorem 7.3 it follows that

$$y(t) = -e^{2t} + 2e^{3t}.$$

■

EXAMPLE 2 **An Initial-Value Problem**

Solve $y'' - 6y' + 9y = t^2 e^{3t}$, $y(0) = 2$, $y'(0) = 6$.

SOLUTION $\mathcal{L}\{y''\} - 6\mathcal{L}\{y'\} + 9\mathcal{L}\{y\} = \mathcal{L}\{t^2 e^{3t}\}$

$$\underbrace{s^2 Y(s) - sy(0) - y'(0)}_{\mathcal{L}\{y''\}} - \underbrace{6[sY(s) - y(0)]}_{\mathcal{L}\{y'\}} + \underbrace{9Y(s)}_{\mathcal{L}\{y\}} = \underbrace{\frac{2}{(s - 3)^3}}_{\mathcal{L}\{t^2 e^{3t}\}}.$$

Using the initial conditions and simplifying gives

$$(s^2 - 6s + 9)Y(s) = 2s - 6 + \frac{2}{(s - 3)^3}$$

$$(s - 3)^2 Y(s) = 2(s - 3) + \frac{2}{(s - 3)^3}$$

$$Y(s) = \frac{2}{s - 3} + \frac{2}{(s - 3)^5}.$$

Thus

$$y(t) = 2\mathcal{L}^{-1}\left\{\frac{1}{s - 3}\right\} + \frac{2}{4!}\mathcal{L}^{-1}\left\{\frac{4!}{(s - 3)^5}\right\}.$$

Recall from the first translation theorem that

$$\mathcal{L}^{-1}\left\{\frac{4!}{s^5}\bigg|_{s \to s-3}\right\} = t^4 e^{3t}.$$

Hence we have

$$y(t) = 2e^{3t} + \frac{1}{12}t^4 e^{3t}.$$

■

EXAMPLE 3 **Using the First Translation Theorem**

Solve $y'' + 4y' + 6y = 1 + e^{-t}$, $y(0) = 0$, $y'(0) = 0$.

SOLUTION $$\mathcal{L}\{y''\} + 4\mathcal{L}\{y'\} + 6\mathcal{L}\{y\} = \mathcal{L}\{1\} + \mathcal{L}\{e^{-t}\}$$

$$s^2 Y(s) - sy(0) - y'(0) + 4[sY(s) - y(0)] + 6Y(s) = \frac{1}{s} + \frac{1}{s+1}$$

$$(s^2 + 4s + 6)Y(s) = \frac{2s+1}{s(s+1)}$$

$$Y(s) = \frac{2s+1}{s(s+1)(s^2+4s+6)}.$$

The partial fraction decomposition for $Y(s)$ is

$$Y(s) = \frac{1/6}{s} + \frac{1/3}{s+1} + \frac{-s/2 - 5/3}{s^2 + 4s + 6}.$$

In preparation for taking the inverse transform we fix up $Y(s)$ in the following manner:

$$Y(s) = \frac{1/6}{s} + \frac{1/3}{s+1} + \frac{(-1/2)(s+2) - 2/3}{(s+2)^2 + 2}$$

$$= \frac{1/6}{s} + \frac{1/3}{s+1} - \frac{1}{2}\frac{s+2}{(s+2)^2 + 2} - \frac{2}{3}\frac{1}{(s+2)^2 + 2}.$$

Finally, from parts (a) and (c) of Theorem 7.3 and the first translation theorem we obtain

$$y(t) = \frac{1}{6}\mathcal{L}^{-1}\left\{\frac{1}{s}\right\} + \frac{1}{3}\mathcal{L}^{-1}\left\{\frac{1}{s+1}\right\} - \frac{1}{2}\mathcal{L}^{-1}\left\{\frac{s+2}{(s+2)^2 + 2}\right\} - \frac{2}{3\sqrt{2}}\mathcal{L}^{-1}\left\{\frac{\sqrt{2}}{(s+2)^2 + 2}\right\}$$

$$= \frac{1}{6} + \frac{1}{3}e^{-t} - \frac{1}{2}e^{-2t}\cos\sqrt{2}t - \frac{\sqrt{2}}{3}e^{-2t}\sin\sqrt{2}t. \qquad\blacksquare$$

EXAMPLE 4   **Using Theorems 7.3/7.7**

Solve $x'' + 16x = \cos 4t$, $x(0) = 0$, $x'(0) = 1$.

SOLUTION   Recall that this initial-value problem could describe the forced, undamped, and resonant motion of a mass on a spring. The mass starts with an initial velocity of 1 ft/s in the downward direction from the equilibrium position.

Transforming the equation gives

$$(s^2 + 16)X(s) = 1 + \frac{s}{s^2 + 16}$$

$$X(s) = \frac{1}{s^2 + 16} + \frac{s}{(s^2 + 16)^2}.$$

With the aid of part (d) of Theorem 7.3 and Theorem 7.7 we find

$$x(t) = \frac{1}{4}\mathcal{L}^{-1}\left\{\frac{4}{s^2 + 16}\right\} + \frac{1}{8}\mathcal{L}^{-1}\left\{\frac{8s}{(s^2 + 16)^2}\right\}$$

$$= \frac{1}{4}\sin 4t + \frac{1}{8}t\sin 4t. \qquad\blacksquare$$

EXAMPLE 5 **Using a Unit Step Function**

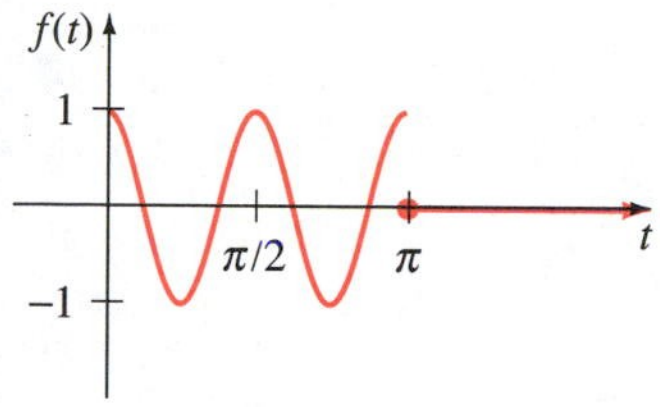

Solve $x'' + 16x = f(t)$, $x(0) = 0$, $x'(0) = 1$,

where $f(t) = \begin{cases} \cos 4t, & 0 \le t < \pi \\ 0, & t \ge \pi. \end{cases}$

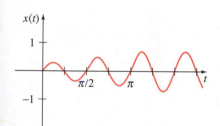

FIGURE 7.37

SOLUTION The function $f(t)$ can be interpreted as an external force that acts on a mechanical system for only a short period of time and then is removed. See Figure 7.37. Although this problem could be solved by conventional means, the procedure is not at all convenient when $f(t)$ is defined in a piecewise manner. With the aid of (2) and (3) of Section 7.3 and the periodicity of the cosine we can rewrite f in terms of the unit step function as

$$f(t) = \cos 4t - \cos 4t \, \mathcal{U}(t - \pi).$$

Using (8) of Section 7.3 to transform f, we obtain

$$\mathcal{L}\{x''\} + 16\mathcal{L}\{x\} = \mathcal{L}\{f(t)\}$$

$$s^2 X(s) - sx(0) - x'(0) + 16X(s) = \frac{s}{s^2 + 16} - \frac{s}{s^2 + 16} e^{-\pi s}$$

$$(s^2 + 16)X(s) = 1 + \frac{s}{s^2 + 16} - \frac{s}{s^2 + 16} e^{-\pi s}$$

$$X(s) = \frac{1}{s^2 + 16} + \frac{s}{(s^2 + 16)^2} - \frac{s}{(s^2 + 16)^2} e^{-\pi s}.$$

From part (b) of Example 12 in Section 7.3 (with $k = 4$), along with (8) of that section, we find

$$x(t) = \frac{1}{4} \mathcal{L}^{-1}\left\{\frac{4}{s^2 + 16}\right\} + \frac{1}{8} \mathcal{L}^{-1}\left\{\frac{8s}{(s^2 + 16)^2}\right\} - \frac{1}{8} \mathcal{L}^{-1}\left\{\frac{8s}{(s^2 + 16)^2} e^{-\pi s}\right\}$$

$$= \frac{1}{4} \sin 4t + \frac{1}{8} t \sin 4t - \frac{1}{8}(t - \pi) \sin 4(t - \pi) \, \mathcal{U}(t - \pi).$$

The foregoing solution is the same as

$$x(t) = \begin{cases} \dfrac{1}{4} \sin 4t + \dfrac{1}{8} t \sin 4t, & 0 \le t < \pi \\ \dfrac{2 + \pi}{8} \sin 4t, & t \ge \pi. \end{cases}$$

FIGURE 7.38

Observe from the graph of $x(t)$ in Figure 7.38 that the amplitudes of vibration become steady as soon as the external force is turned off.

■

Volterra Integral Equation The convolution theorem is useful in solving other types of equations in which an unknown function appears under an integral sign. In the next example we solve a **Volterra integral equation**,

$$f(t) = g(t) + \int_0^t f(\tau)h(t - \tau) \, d\tau,$$

for $f(t)$. The functions $g(t)$ and $h(t)$ are known.

EXAMPLE 6   **An Integral Equation**

Solve $f(t) = 3t^2 - e^{-t} - \int_0^t f(\tau)e^{t-\tau}\,d\tau$ for $f(t)$.

SOLUTION   It follows from Theorem 7.9 that

$$\mathscr{L}\{f(t)\} = 3\mathscr{L}\{t^2\} - \mathscr{L}\{e^{-t}\} - \mathscr{L}\{f(t)\}\mathscr{L}\{e^t\}$$

$$F(s) = 3 \cdot \frac{2}{s^3} - \frac{1}{s+1} - F(s) \cdot \frac{1}{s-1}.$$

Solving the last equation for $F(s)$ gives

$$F(s) = \frac{6(s-1)}{s^4} - \frac{s-1}{s(s+1)}$$

$$= \frac{6}{s^3} - \frac{6}{s^4} + \frac{1}{s} - \frac{2}{s+1}. \quad \leftarrow \text{termwise division and partial fractions}$$

The inverse transform is

$$f(t) = 3\mathscr{L}^{-1}\left\{\frac{2!}{s^3}\right\} - \mathscr{L}^{-1}\left\{\frac{3!}{s^4}\right\} + \mathscr{L}^{-1}\left\{\frac{1}{s}\right\} - 2\mathscr{L}^{-1}\left\{\frac{1}{s+1}\right\}$$

$$= 3t^2 - t^3 + 1 - 2e^{-t}. \qquad\blacksquare$$

Series Circuits

In a single-loop or series circuit Kirchhoff's second law states that the sum of the voltage drops across an inductor, resistor, and capacitor is equal to the impressed voltage $E(t)$. Now it is known that the voltage drops across an inductor, resistor, and capacitor are, respectively,

$$L\frac{di}{dt}, \qquad Ri(t), \qquad \text{and} \qquad \frac{1}{C}\int_0^t i(\tau)\,d\tau,$$

where $i(t)$ is the current and L, R, and C are constants. It follows that the current in a circuit, such as that shown in Figure 7.39, is governed by the **integrodifferential equation**

$$L\frac{di}{dt} + Ri + \frac{1}{C}\int_0^t i(\tau)\,d\tau = E(t). \tag{3}$$

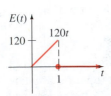

FIGURE 7.39

EXAMPLE 7   **An Integrodifferential Equation**

Determine the current $i(t)$ in a single-loop LRC circuit when $L = 0.1$ h, $R = 20\ \Omega$, $C = 10^{-3}$ f, $i(0) = 0$, and the impressed voltage $E(t)$ is as given in Figure 7.40.

SOLUTION   Since the voltage is off for $t \geq 1$, we can write

$$E(t) = 120t - 120t\,\mathscr{U}(t-1). \tag{4}$$

Equation (3) then becomes

$$0.1\frac{di}{dt} + 20i + 10^3 \int_0^t i(\tau)\,d\tau = 120t - 120t\,\mathscr{U}(t-1). \tag{5}$$

FIGURE 7.40

Now recall from (7) of Section 7.4 that $\mathscr{L}\{\int_0^t i(\tau)\,d\tau\} = I(s)/s$, where $I(s) = \mathscr{L}\{i(t)\}$. Thus the transform of equation (5) is

$$0.1sI(s) + 20I(s) + 10^3\frac{I(s)}{s} = 120\left[\frac{1}{s^2} - \frac{1}{s^2}e^{-s} - \frac{1}{s}e^{-s}\right]. \qquad \leftarrow \text{by (8) of Section 7.3}$$

Multiplying this equation by $10s$ and then solving for $I(s)$ gives

$$I(s) = 1200\left[\frac{1}{s(s+100)^2} - \frac{1}{s(s+100)^2}e^{-s} - \frac{1}{(s+100)^2}e^{-s}\right].$$

By partial fractions we can write

$$I(s) = 1200\left[\frac{1/10,000}{s} - \frac{1/10,000}{s+100} - \frac{1/100}{(s+100)^2} - \frac{1/10,000}{s}e^{-s}\right.$$
$$\left. + \frac{1/10,000}{s+100}e^{-s} + \frac{1/100}{(s+100)^2}e^{-s} - \frac{1}{(s+100)^2}e^{-s}\right].$$

Employing the inverse form of the second translation theorem, we obtain

$$i(t) = \frac{3}{25}[1 - \mathscr{U}(t-1)] - \frac{3}{25}[e^{-100t} - e^{-100(t-1)}\mathscr{U}(t-1)]$$
$$- 12te^{-100t} - 1188(t-1)e^{-100(t-1)}\mathscr{U}(t-1). \qquad \blacksquare$$

EXAMPLE 8   **A Periodic Impressed Voltage**

The differential equation for the current $i(t)$ in a single-loop LR series circuit is

$$L\frac{di}{dt} + Ri = E(t). \qquad (6)$$

FIGURE 7.41

Determine the current $i(t)$ when $i(0) = 0$ and $E(t)$ is the square wave function shown in Figure 7.41.

SOLUTION   The Laplace transform of the equation is

$$LsI(s) + RI(s) = \mathscr{L}\{E(t)\}. \qquad (7)$$

Since $E(t)$ is periodic with period $T = 2$, we use (9) of Section 7.4:

$$\mathscr{L}\{E(t)\} = \frac{1}{1 - e^{-2s}}\left(\int_0^1 1\cdot e^{-st}\,dt + \int_1^2 0\cdot e^{-st}\,dt\right)$$

$$= \frac{1}{1 - e^{-2s}}\frac{1 - e^{-s}}{s} \qquad \leftarrow 1 - e^{-2s} = (1 + e^{-s})(1 - e^{-s})$$

$$= \frac{1}{s(1 + e^{-s})}.$$

Hence from (7) we find

$$I(s) = \frac{1/L}{s(s + R/L)(1 + e^{-s})}. \qquad (8)$$

To find the inverse Laplace transform of this function we first make use of a geometric series. Recall that, for $|x| < 1$,

$$\frac{1}{1+x} = 1 - x + x^2 - x^3 + \cdots.$$

With the identification $x = e^{-s}$ we then have, for $s > 0$,

$$\frac{1}{1+e^{-s}} = 1 - e^{-s} + e^{-2s} - e^{-3s} + \cdots.$$

If we write

$$\frac{1}{s(s+R/L)} = \frac{L/R}{s} - \frac{L/R}{s+R/L},$$

(8) becomes

$$I(s) = \frac{1}{R}\left(\frac{1}{s} - \frac{1}{s+R/L}\right)(1 - e^{-s} + e^{-2s} - e^{-3s} + \cdots)$$

$$= \frac{1}{R}\left(\frac{1}{s} - \frac{e^{-s}}{s} + \frac{e^{-2s}}{s} - \frac{e^{-3s}}{s} + \cdots\right) - \frac{1}{R}\left(\frac{1}{s+R/L} - \frac{e^{-s}}{s+R/L} + \frac{e^{-2s}}{s+R/L} - \frac{e^{-3s}}{s+R/L} + \cdots\right).$$

By applying the inverse form of the second translation theorem to each term of both series we obtain

$$i(t) = \frac{1}{R}\left(1 - \mathscr{U}(t-1) + \mathscr{U}(t-2) - \mathscr{U}(t-3) + \cdots\right)$$

$$- \frac{1}{R}\left(e^{-Rt/L} - e^{-R(t-1)/L}\,\mathscr{U}(t-1) + e^{-R(t-2)/L}\,\mathscr{U}(t-2) - e^{-R(t-3)/L}\,\mathscr{U}(t-3) + \cdots\right)$$

or, equivalently,

$$i(t) = \frac{1}{R}\left(1 - e^{-Rt/L}\right) + \frac{1}{R}\sum_{n=1}^{\infty}(-1)^n\left(1 - e^{-R(t-n)/L}\right)\mathscr{U}(t-n). \qquad \blacksquare$$

To interpret the solution in Example 8 let us assume for the sake of illustration that $R = 1$, $L = 1$, and $0 \le t < 4$. In this case

$$i(t) = 1 - e^{-t} - (1 - e^{t-1})\,\mathscr{U}(t-1) + (1 - e^{-(t-2)})\,\mathscr{U}(t-2) - (1 - e^{-(t-3)})\,\mathscr{U}(t-3);$$

in other words,

$$i(t) = \begin{cases} 1 - e^{-t}, & 0 \le t < 1 \\ -e^{-t} + e^{-(t-1)}, & 1 \le t < 2 \\ 1 - e^{-t} + e^{-(t-1)} - e^{-(t-2)}, & 2 \le t < 3 \\ -e^{-t} + e^{-(t-1)} - e^{-(t-2)} + e^{-(t-3)}, & 3 \le t < 4. \end{cases}$$

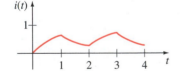

FIGURE 7.42

The graph of $i(t)$ on the interval $0 \le t < 4$ is given in Figure 7.42.

Beams   In Section 5.2 we saw that the static deflection $y(x)$ of a uniform beam of length L carrying a load $w(x)$ per unit length is found from the fourth-order differential equation

$$EI\frac{d^4y}{dx^4} = w(x), \qquad (9)$$

where E is Young's modulus of elasticity and I is the moment of inertia of a cross-section of the beam. The Laplace transform method is especially useful in solving (9) when $w(x)$ is piecewise defined. To apply the Laplace transform we tacitly assume that $w(x)$ and $y(x)$ are defined on $(0, \infty)$ rather than on $(0, L)$. Note, too, that the next example is a boundary-value problem rather than an initial-value problem.

EXAMPLE 9   **A Boundary-Value Problem**

A beam of length L is embedded at both ends as shown in Figure 7.43. Find the deflection of the beam when the load is given by

$$w(x) = \begin{cases} w_0\left(1 - \dfrac{2}{L}x\right), & 0 < x < L/2 \\ 0, & L/2 < x < L. \end{cases}$$

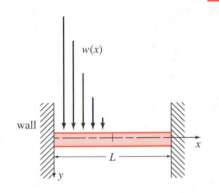

wall

FIGURE 7.43

SOLUTION   Recall that since the beam is embedded at both ends the boundary conditions are $y(0) = 0$, $y'(0) = 0$, $y(L) = 0$, $y'(L) = 0$. Also, you should verify

$$w(x) = w_0\left(1 - \frac{2}{L}x\right) - w_0\left(1 - \frac{2}{L}x\right)\mathcal{U}\left(x - \frac{L}{2}\right)$$

$$= \frac{2w_0}{L}\left[\frac{L}{2} - x + \left(x - \frac{L}{2}\right)\mathcal{U}\left(x - \frac{L}{2}\right)\right].$$

Transforming (9) with respect to the variable x gives

$$EI\left(s^4 Y(s) - s^3 y(0) - s^2 y'(0) - s y''(0) - y'''(0)\right) = \frac{2w_0}{EIL}\left[\frac{L/2}{s} - \frac{1}{s^2} + \frac{1}{s^2}e^{-Ls/2}\right]$$

or

$$s^4 Y(s) - s y''(0) - y'''(0) = \frac{2w_0}{EIL}\left[\frac{L/2}{s} - \frac{1}{s^2} + \frac{1}{s^2}e^{-Ls/2}\right].$$

If we let $c_1 = y''(0)$ and $c_2 = y'''(0)$, then

$$Y(s) = \frac{c_1}{s^3} + \frac{c_2}{s^4} + \frac{2w_0}{EIL}\left[\frac{L/2}{s^5} - \frac{1}{s^6} + \frac{1}{s^6}e^{-Ls/2}\right],$$

and consequently

$$y(x) = \frac{c_1}{2!}\mathcal{L}^{-1}\left\{\frac{2!}{s^3}\right\} + \frac{c_2}{3!}\mathcal{L}^{-1}\left\{\frac{3!}{s^4}\right\}$$

$$+ \frac{2w_0}{EIL}\left[\frac{L/2}{4!}\mathcal{L}^{-1}\left\{\frac{4!}{s^5}\right\} - \frac{1}{5!}\mathcal{L}^{-1}\left\{\frac{5!}{s^6}\right\} + \frac{1}{5!}\mathcal{L}^{-1}\left\{\frac{5!}{s^6}e^{-Ls/2}\right\}\right]$$

$$= \frac{c_1}{2}x^2 + \frac{c_2}{6}x^3 + \frac{w_0}{60EIL}\left[\frac{5L}{2}x^4 - x^5 + \left(x - \frac{L}{2}\right)^5\mathcal{U}\left(x - \frac{L}{2}\right)\right].$$

Applying the conditions $y(L) = 0$ and $y'(L) = 0$ to the last result yields a system of equations for c_1 and c_2:

$$c_1\frac{L^2}{2} + c_2\frac{L^3}{6} + \frac{49w_0 L^4}{1920EI} = 0$$

$$c_1 L + c_2\frac{L^2}{2} + \frac{85w_0 L^3}{960EI} = 0.$$

Solving, we find $c_1 = 23w_0L^2/960EI$ and $c_2 = -9w_0L/40EI$. Thus the deflection is given by

$$y(x) = \frac{23w_0L^2}{1920EI}x^2 - \frac{9w_0L}{240EI}x^3 + \frac{w_0}{60EIL}\left[\frac{5L}{2}x^4 - x^5 + \left(x - \frac{L}{2}\right)^5 \mathcal{U}\left(x - \frac{L}{2}\right)\right].$$

■

Remark

This remark continues the introduction to the terminology of dynamical systems.

In light of (1) and (2) of this section the Laplace transform is well adapted to the analysis of *linear* dynamical systems. If we solve the general transformed equation (2) for the symbol $Y(s)$, we obtain the expression

$$Y(s) = \frac{G(s)}{P(s)} + \frac{Q(s)}{P(s)}. \tag{10}$$

Here $P(s) = a_ns^n + a_{n-1}s_{n-1} + \cdots + a_0$ and is the same as the nth-degree auxiliary polynomial with the usual symbol m replaced by s, $G(s)$ is the Laplace transform of $g(t)$, and $Q(s)$ is a polynomial in s of degree $n - 1$ consisting of the various products of the coefficients a_i, $i = 1, 2, \ldots, n$ and the prescribed initial-conditions $y_0, y_1, \ldots, y_{n-1}$. For example, for $n = 2$ you should verify that $Q(s)/P(s) = (a_2y_0s + a_2y_1 + a_1y_0)/(a_2s^2 + a_1s + s_0)$. It is usual practice to call the reciprocal of $P(s)$, namely, $W(s) = 1/P(s)$, the **transfer function** of the system and write (10) as

$$Y(s) = W(s)G(s) + W(s)Q(s). \tag{11}$$

In this manner we have separated, in an additive sense, the effects on the response that are due to the input function g (that is, $W(s)G(s)$) and to the initial conditions (that is, $W(s)Q(s)$). Hence the response $y(t)$ of the system is a superposition of two responses:

$$y(t) = \mathcal{L}^{-1}\{W(s)G(s)\} + \mathcal{L}^{-1}\{W(s)Q(s)\} = y_0(t) + y_1(t).$$

The function $y_0(t) = \mathcal{L}^{-1}\{W(s)G(s)\}$ is the output due to the input $g(t)$. Now if the initial state of the system is the zero state (all the initial conditions are zero: $y_0 = 0$, $y_1 = 0$, $\ldots$, $y_{n-1} = 0$), then $Q(s) = 0$ and so the only solution of the initial-value problem is $y_0(t)$. This solution is called the **zero-state response** of the system. If you think in terms of solving the differential equation by, say, undetermined coefficients, the particular solution obtained would be $y_0(t)$. Note also that by the convolution theorem the zero-state response can be written as a weighted integral of the input: $y_0(t) = \int_0^t w(\tau)g(t - \tau)\,d\tau = w(t) * g(t)$. Consequently the inverse of the transfer function, $w(t) = \mathcal{L}^{-1}\{W(s)\}$, is called the **weight function** of the system. Finally, if the input is $g(t) = 0$, then the solution of the problem is $y_1(t) = \mathcal{L}^{-1}\{Q(s)W(s)\}$, called the **zero-input response** of the system. In Example 2 the zero-state response is $y_0(t) = t^4e^{3t}/12$, the zero-input response is $y_1(t) = 2e^{3t}$, the transfer function is $W(s) = 1/(s^2 - 6s + 9)$, and the weight function of the system is $w(t) = \mathcal{L}^{-1}\{W(s)\} = te^{3t}$.

Answers to odd-numbered problems begin on page A-11.

A table of the transforms of some basic functions is given in Appendix III. In Problems 1–26 use the Laplace transform to solve the given differential equation subject to the indicated initial conditions. Where appropriate, write f in terms of unit step functions.

1. $\dfrac{dy}{dt} - y = 1, \quad y(0) = 0$

2. $\dfrac{dy}{dt} + 2y = t, \quad y(0) = -1$

3. $y' + 4y = e^{-4t}, \quad y(0) = 2$

4. $y' - y = \sin t, \quad y(0) = 0$

5. $y'' + 5y' + 4y = 0, \quad y(0) = 1, y'(0) = 0$

6. $y'' - 6y' + 13y = 0, \quad y(0) = 0, y'(0) = -3$

7. $y'' - 6y' + 9y = t, \quad y(0) = 0, y'(0) = 1$

8. $y'' - 4y' + 4y = t^3, \quad y(0) = 1, y'(0) = 0$

9. $y'' - 4y' + 4y = t^3 e^{2t}, \quad y(0) = 0, y'(0) = 0$

10. $y'' - 2y' + 5y = 1 + t, \quad y(0) = 0, y'(0) = 4$

11. $y'' + y = \sin t, \quad y(0) = 1, y'(0) = -1$

12. $y'' + 16y = 1, \quad y(0) = 1, y'(0) = 2$

13. $y'' - y' = e^t \cos t, \quad y(0) = 0, y'(0) = 0$

14. $y'' - 2y' = e^t \sinh t, \quad y(0) = 0, y'(0) = 0$

15. $2y''' + 3y'' - 3y' - 2y = e^{-t}, \quad y(0) = 0, y'(0) = 0, y''(0) = 1$

16. $y''' + 2y'' - y' - 2y = \sin 3t, \quad y(0) = 0, y'(0) = 0, y''(0) = 1$

17. $y^{(4)} - y = 0, \quad y(0) = 1, y'(0) = 0, y''(0) = -1, y'''(0) = 0$

18. $y^{(4)} - y = t, \quad y(0) = 0, y'(0) = 0, y''(0) = 0, y'''(0) = 0$

19. $y' + y = f(t), \quad y(0) = 0$, where $f(t) = \begin{cases} 0, & 0 \le t < 1 \\ 5, & t \ge 1 \end{cases}$

20. $y' + y = f(t), \quad y(0) = 0$, where $f(t) = \begin{cases} 1, & 0 \le t < 1 \\ -1, & t \ge 1 \end{cases}$

21. $y' + 2y = f(t), \quad y(0) = 0$, where $f(t) = \begin{cases} t, & 0 \le t < 1 \\ 0, & t \ge 1 \end{cases}$

22. $y'' + 4y = f(t), \quad y(0) = 0, y'(0) = -1$, where $f(t) = \begin{cases} 1, & 0 \le t < 1 \\ 0, & t \ge 1 \end{cases}$

23. $y'' + 4y = \sin t \, \mathcal{U}(t - 2\pi), \quad y(0) = 1, y'(0) = 0$

24. $y'' - 5y' + 6y = \mathcal{U}(t - 1), \quad y(0) = 0, y'(0) = 1$

25. $y'' + y = f(t), \quad y(0) = 0, y'(0) = 1$, where $f(t) = \begin{cases} 0, & 0 \le t < \pi \\ 1, & \pi \le t < 2\pi \\ 0, & t \ge 2\pi \end{cases}$

26. $y'' + 4y' + 3y = 1 - \mathcal{U}(t - 2) - \mathcal{U}(t - 4) + \mathcal{U}(t - 6), \quad y(0) = 0, \; y'(0) = 0$

In Problems 27 and 28 use the Laplace transform to solve the given differential equation subject to the indicated boundary conditions.

27. $y'' + 2y' + y = 0, \quad y'(0) = 2, y(1) = 2$

28. $y'' - 9y' + 20y = 1, \quad y(0) = 0, y'(1) = 0$

In Problems 29–38 use the Laplace transform to solve the given integral equation or integrodifferential equation.

29. $f(t) + \displaystyle\int_0^t (t - \tau) f(\tau)\, d\tau = t$

30. $f(t) = 2t - 4 \displaystyle\int_0^t \sin \tau f(t - \tau)\, d\tau$

31. $f(t) = t e^t + \displaystyle\int_0^t \tau f(t - \tau)\, d\tau$

32. $f(t) + 2 \displaystyle\int_0^t f(\tau) \cos (t - \tau)\, d\tau = 4e^{-t} + \sin t$

33. $f(t) + \displaystyle\int_0^t f(\tau)\, d\tau = 1$

34. $f(t) = \cos t + \displaystyle\int_0^t e^{-\tau} f(t - \tau)\, d\tau$

35. $f(t) = 1 + t - \dfrac{8}{3} \displaystyle\int_0^t (\tau - t)^3 f(\tau)\, d\tau$

36. $t - 2f(t) = \displaystyle\int_0^t (e^\tau - e^{-\tau}) f(t - \tau)\, d\tau$

37. $y'(t) = 1 - \sin t - \displaystyle\int_0^t y(\tau)\, d\tau, \quad y(0) = 0$

38. $\dfrac{dy}{dt} + 6y(t) + 9 \displaystyle\int_0^t y(\tau)\, d\tau = 1, \quad y(0) = 0$

39. Use equation (3) to determine the current $i(t)$ in a single-loop LRC circuit when $L = 0.005$ h, $R = 1\ \Omega$, $C = 0.02$ f, $E(t) = 100[1 - \mathcal{U}(t - 1)]$ V, and $i(0) = 0$.

40. Solve Problem 39 when $E(t) = 100[t - (t - 1)\mathcal{U}(t - 1)]$.

41. Recall that the differential equation for the charge $q(t)$ on the capacitor in an RC series circuit is

$$R\frac{dq}{dt} + \frac{1}{C}q = E(t),$$

where $E(t)$ is the impressed voltage. See Section 3.1. Use the Laplace transform to determine the charge $q(t)$ when $q(0) = 0$ and $E(t) = E_0 e^{-kt}$, $k > 0$. Consider two cases: $k \neq 1/RC$ and $k = 1/RC$.

42. Use the Laplace transform to determine the charge on the capacitor in an RC series circuit if $q(0) = q_0$, $R = 10\ \Omega$, $C = 0.1$ f, and $E(t)$ is as given in Figure 7.44.

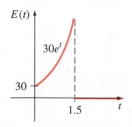

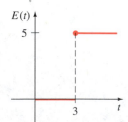

FIGURE 7.44   **FIGURE 7.45**

43. Use the Laplace transform to determine the charge on the capacitor in an RC series circuit if $q(0) = 0$, $R = 2.5\ \Omega$, $C = 0.08$ f, and $E(t)$ is as given in Figure 7.45.

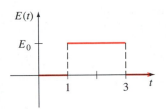

FIGURE 7.46

44. (a) Use the Laplace transform to determine the charge $q(t)$ on the capacitor in an RC series circuit when $q(0) = 0$, $R = 50\ \Omega$, $C = 0.01$ f, and $E(t)$ is as given in Figure 7.46.

(b) Assume $E_0 = 100$ V. Use a computer graphing program to graph $q(t)$ on the interval $0 \le t \le 6$. Use the graph to estimate q_{max}, the maximum value of the charge.

45. (a) Use the Laplace transform to determine the current $i(t)$ in a single-loop LR series circuit when $i(0) = 0$, $L = 1$ h, $R = 10\ \Omega$, and $E(t)$ is as given in Figure 7.47.

(b) Use a computer graphing program to graph $i(t)$ on the interval $0 \le t \le 6$. Use the graph to estimate i_{max} and i_{min}, the maximum and minimum values of the current.

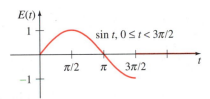

FIGURE 7.47

46. Solve equation (6) subject to $i(0) = 0$, where $E(t)$ is the meander function given in Figure 7.48. [*Hint:* See Problem 31 in Exercises 7.4.]

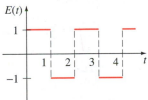

FIGURE 7.48

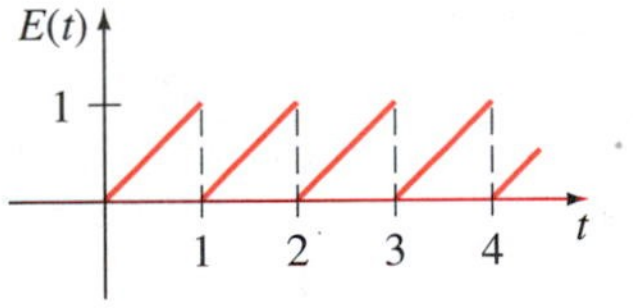

FIGURE 7.49

47. Solve equation (6) subject to $i(0) = 0$, where $E(t)$ is the sawtooth function given in Figure 7.49. Specify the solution for $0 \le t < 2$. [*Hint:* See Problem 33 in Exercises 7.4.]

48. Recall that the differential equation for the instantaneous charge $q(t)$ on the capacitor in an LRC series circuit is given by

$$L\frac{d^2q}{dt^2} + R\frac{dq}{dt} + \frac{1}{C}q = E(t). \qquad (12)$$

See Section 5.1. Use the Laplace transform to determine $q(t)$ when $L = 1$ h, $R = 20\ \Omega$, $C = 0.005$ f, $E(t) = 150$ V, $t > 0$, $q(0) = 0$, and $i(0) = 0$. What is the current $i(t)$? What is the charge $q(t)$ if the same constant voltage is turned off for $t \ge 2$?

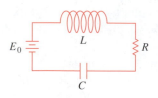

FIGURE 7.50

49. Determine the charge $q(t)$ and current $i(t)$ for a series circuit in which $L = 1$ h, $R = 20$ Ω, $C = 0.01$ f, $E(t) = 120 \sin 10t$ V, $q(0) = 0$, and $i(0) = 0$. What is the steady-state current?

50. Consider the battery of constant voltage E_0 that charges the capacitor shown in Figure 7.50. If we divide by L and define $\lambda = R/2L$ and $\omega^2 = 1/LC$, then (12) becomes

$$\frac{d^2q}{dt^2} + 2\lambda \frac{dq}{dt} + \omega^2 q = \frac{E_0}{L}.$$

Use the Laplace transform to show that the solution of this equation, subject to $q(0) = 0$ and $i(0) = 0$, is

$$q(t) = \begin{cases} E_0 C\left[1 - e^{-\lambda t}\left(\cosh\sqrt{\lambda^2 - \omega^2}\,t + \dfrac{\lambda}{\sqrt{\lambda^2 - \omega^2}}\sinh\sqrt{\lambda^2 - \omega^2}\,t\right)\right], & \lambda > \omega \\ E_0 C[1 - e^{-\lambda t}(1 + \lambda t)], & \lambda = \omega \\ E_0 C\left[1 - e^{-\lambda t}\left(\cos\sqrt{\omega^2 - \lambda^2}\,t + \dfrac{\lambda}{\sqrt{\omega^2 - \lambda^2}}\sin\sqrt{\omega^2 - \lambda^2}\,t\right)\right], & \lambda < \omega. \end{cases}$$

51. Use the Laplace transform to determine the charge $q(t)$ on the capacitor in an LC series circuit when $q(0) = 0$, $i(0) = 0$, and $E(t) = E_0 e^{-kt}$, $k > 0$.

52. Suppose a 32-pound weight stretches a spring 2 feet. If the weight is released from rest at the equilibrium position, determine the equation of motion if an impressed force $f(t) = \sin t$ acts on the system for $0 \le t < 2\pi$ and is then removed. Ignore any damping forces. [*Hint:* Write the impressed force in terms of the unit step function.]

53. A 4-pound weight stretches a spring 2 feet. The weight is released from rest 18 inches above the equilibrium position, and the resulting motion takes place in a medium offering a damping force numerically equal to $\frac{7}{8}$ times the instantaneous velocity. Use the Laplace transform to determine the equation of motion.

54. A 16-pound weight is attached to a spring whose constant is $k = 4.5$ lb/ft. Beginning at $t = 0$, a force equal to $f(t) = 4 \sin 3t + 2 \cos 3t$ acts on the system. Assuming that no damping forces are present, use the Laplace transform to find the equation of motion if the weight is released from rest from the equilibrium position.

55. A cantilever beam is embedded at its left end and free at its right end. Find the deflection $y(x)$ when the load is given by

$$w(x) = \begin{cases} w_0, & 0 < x < L/2 \\ 0, & L/2 < x < L. \end{cases}$$

56. Solve Problem 55 when the load is given by

$$w(x) = \begin{cases} 0, & 0 < x < L/3 \\ w_0, & L/3 < x < 2L/3 \\ 0, & 2L/3 < x < L. \end{cases}$$

57. Find the deflection $y(x)$ of a cantilever beam when the load is as given in Example 9.

58. A beam is embedded at its left end and simply supported at its right end. Find the deflection $y(x)$ when the load is as given in Problem 55.

Discussion Problem

59. The Laplace transform is not well suited to solving linear differential equations with variable coefficients; nevertheless it can be used in some circumstances. Which theorems of Sections 7.3 and 7.4 are appropriate for transforming $ty'' + 2ty' + 2y = t$? Find a solution of the differential equation that satisfies $y(0) = 0$.

CAS Programming Problem

60. In part (a) of this problem you are led through the commands in Mathematica that enable you to obtain the symbolic Laplace transform of a differential equation and the solution of the initial-value problem by finding the inverse transform. In Mathematica the Laplace transform of a function $y(t)$ is obtained using

LaplaceTransform [y[t], t, s].

In line two of the syntax we replace

LaplaceTransform [y[t], t, s]

by the symbol **Y**. (*If you do not have Mathematica, adapt the given procedure by finding the corresponding syntax for the CAS you have on hand.*)

(a) Consider the initial-value problem

$$y'' + 6y' + 9y = t \sin t, \qquad y(0) = 2, \quad y'(0) = -1.$$

Load the Laplace transform package. Precisely reproduce and then, in turn, execute each line in the given sequence of commands. Either copy the output by hand or print out the results.

diffequat = y″[t] + 6y′[t] + 9y[t] == t Sin[t]
transformdeq = LaplaceTransform [diffequat, t, s] /. {y[0] − > 2,
          y′[0] − > −1, LaplaceTransform [y[t], t, s] − > Y}
soln = Solve[transformdeq, Y] // Flatten
Y = Y/. soln
InverseLaplaceTransform[Y, s, t]

  (b) Appropriately modify the procedure of part (a) to find a solution of

$$y''' + 3y' - 4y = 0, \qquad y(0) = 0, \quad y'(0) = 0, \quad y''(0) = 1.$$

  (c) The charge $q(t)$ on a capacitor in an LC series circuit is given by

$$\frac{d^2q}{dt^2} + q = 1 - 4\,\mathcal{U}(t - \pi) + 6\,\mathcal{U}(t - 3\pi), \qquad q(0) = 0, \quad q'(0) = 0.$$

Appropriately modify the procedure in part (a) to find $q(t)$. (In Mathematica the unit step function $\mathcal{U}(t - a)$ is written in the form **UnitStep[t − a]**.) Graph your solution.

7.6   DIRAC DELTA FUNCTION

■ *Unit impulse* ■ *Dirac delta function* ■ *Transform of the Dirac delta function*
■ *Sifting property*

Unit Impulse   Mechanical systems are often acted upon by an external force (or emf in an electrical circuit) of large magnitude that acts only for a very short period of time. For example, a vibrating airplane wing could be struck by lightning, a mass on a spring could be given a sharp blow by a ball peen hammer, a ball (baseball, golf ball, tennis ball) could be sent soaring when struck violently by some kind of club (baseball bat, golf club, tennis racket). The function

$$\delta_a(t - t_0) = \begin{cases} 0, & 0 \le t < t_0 - a \\ \dfrac{1}{2a}, & t_0 - a \le t < t_0 + a \\ 0, & t \ge t_0 + a, \end{cases} \tag{1}$$

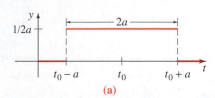

(a)

$a > 0$, $t_0 > 0$, shown in Figure 7.51(a), could serve as a mathematical model for such a force. For a small value of a, $\delta_a(t - t_0)$ is essentially a constant function of large magnitude that is "on" for just a very short period of time, around t_0. The behavior of $\delta_a(t - t_0)$ as $a \to 0$ is illustrated in Figure 7.51(b). The function $\delta_a(t - t_0)$ is called a **unit impulse** since it possesses the integration property $\int_0^\infty \delta_a(t - t_0) \, dt = 1$.

Dirac Delta Function   In practice it is convenient to work with another type of unit impulse, a "function" that approximates $\delta_a(t - t_0)$ and is defined by the limit

$$\delta(t - t_0) = \lim_{a \to 0} \delta_a(t - t_0). \tag{2}$$

The latter expression, which is not a function at all, can be characterized by the two properties

$$(i) \ \ \delta(t - t_0) = \begin{cases} \infty, & t = t_0 \\ 0, & t \ne t_0 \end{cases} \quad \text{and} \quad (ii) \ \int_0^\infty \delta(t - t_0) \, dt = 1.$$

(b) Behavior of δ_a as $a \to 0$

FIGURE 7.51

The unit impulse $\delta(t - t_0)$ is called the **Dirac delta function.**

It is possible to obtain the Laplace transform of the Dirac delta function by the formal assumption that $\mathscr{L}\{\delta(t - t_0)\} = \lim_{a \to 0} \mathscr{L}\{\delta_a(t - t_0)\}$.

THEOREM 7.11   **Transform of the Dirac Delta Function**

For $t_0 > 0$, $$\mathscr{L}\{\delta(t - t_0)\} = e^{-st_0}. \tag{3}$$

PROOF   To begin we can write $\delta_a(t - t_0)$ in terms of the unit step function by virtue of (4) and (5) of Section 7.3:

$$\delta_a(t - t_0) = \frac{1}{2a} [\mathscr{U}(t - (t_0 - a)) - \mathscr{U}(t - (t_0 + a))].$$

By linearity and (7) of Section 7.3 the Laplace transform of this last expression is

$$\mathcal{L}\{\delta_a(t - t_0)\} = \frac{1}{2a}\left[\frac{e^{-s(t_0-a)}}{s} - \frac{e^{-s(t_0+a)}}{s}\right] = e^{-st_0}\left(\frac{e^{sa} - e^{-sa}}{2sa}\right). \qquad \textbf{(4)}$$

Since (4) has the indeterminate form $0/0$ as $a \to 0$, we apply L'Hôpital's rule:

$$\mathcal{L}\{\delta(t - t_0)\} = \lim_{a \to 0}\mathcal{L}\{\delta_a(t - t_0)\} = e^{-st_0}\lim_{a \to 0}\left(\frac{e^{sa} - e^{-sa}}{2sa}\right) = e^{-st_0}. \qquad \blacksquare$$

Now when $t_0 = 0$, it seems plausible to conclude from (3) that

$$\mathcal{L}\{\delta(t)\} = 1.$$

The last result emphasizes the fact that $\delta(t)$ is not the usual type of function that we have been considering since we expect from Theorem 7.4 that $\mathcal{L}\{f(t)\} \to 0$ as $s \to \infty$.

EXAMPLE 1 Two Initial-Value Problems

Solve $y'' + y = 4\delta(t - 2\pi)$ subject to

(a) $y(0) = 1$, $y'(0) = 0$ (b) $y(0) = 0$, $y'(0) = 0$

The two initial-value problems could serve as models for describing the motion of a mass on a spring moving in a medium in which damping is negligible. At $t = 2\pi$ the mass is given a sharp blow. In (a) the mass is released from rest 1 unit below the equilibrium position. In (b) the mass is at rest in the equilibrium position.

SOLUTION   (a) From (3) the Laplace transform of the differential equation is

$$s^2Y(s) - s + Y(s) = 4e^{-2\pi s} \qquad \text{or} \qquad Y(s) = \frac{s}{s^2 + 1} + \frac{4e^{-2\pi s}}{s^2 + 1}.$$

Using the inverse form of the second translation theorem, we find

$$y(t) = \cos t + 4\sin(t - 2\pi)\,\mathcal{U}(t - 2\pi).$$

Since $\sin(t - 2\pi) = \sin t$, the foregoing solution can be written as

$$y(t) = \begin{cases} \cos t, & 0 \le t < 2\pi \\ \cos t + 4\sin t, & t \ge 2\pi. \end{cases} \qquad \textbf{(5)}$$

In Figure 7.52 we see from the graph of (5) that the mass is exhibiting simple harmonic motion until it is struck at $t = 2\pi$. The influence of the unit impulse is to increase the amplitude of vibration to $\sqrt{17}$ for $t > 2\pi$.

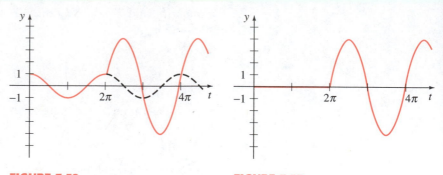

FIGURE 7.52   **FIGURE 7.53**

(b) In this case the transform of the equation is simply

$$Y(s) = \frac{4e^{-2\pi s}}{s^2 + 1},$$

and so

$$y(t) = 4 \sin(t - 2\pi)\,\mathcal{U}(t - 2\pi)$$

$$= \begin{cases} 0, & 0 \le t < 2\pi \\ 4 \sin t, & t \ge 2\pi. \end{cases} \tag{6}$$

The graph of (6) in Figure 7.53 shows, as we would expect from the initial conditions, that the mass exhibits no motion until it is struck at $t = 2\pi$. ∎

Remarks

(*i*) If $\delta(t - t_0)$ were a function in the usual sense, then property (*i*) on page 306 would imply $\int_0^\infty \delta(t - t_0)\, dt = 0$ rather than $\int_0^\infty \delta(t - t_0)\, dt = 1$. Since the Dirac delta function did not "behave" like an ordinary function, even though its users produced correct results, it was met initially with great scorn by mathematicians. However, in the 1940s Dirac's controversial function was put on a rigorous footing by the French mathematician Laurent Schwartz in his book *La Théorie de distribution,* and this, in turn, led to an entirely new branch of mathematics known as the **theory of distributions** or **generalized functions**. In this theory, (2) is not an accepted definition of $\delta(t - t_0)$, nor does one speak of a function whose values are either ∞ or 0. Although we shall not pursue this topic any further, suffice it to say that the Dirac delta function is best characterized by its effect on other functions. If f is a continuous function, then

$$\int_0^\infty f(t)\, \delta(t - t_0)\, dt = f(t_0) \tag{7}$$

can be taken as the *definition* of $\delta(t - t_0)$. This result is known as the **sifting property** since $\delta(t - t_0)$ has the effect of sifting the value $f(t_0)$ out of the set of values of f on $[0, \infty)$. Note that property (*ii*) (with $f(t) = 1$) and (3) (with $f(t) = e^{-st}$) are consistent with (7).

(*ii*) In the remark in Section 7.5 we indicated that the transfer function of a general linear nth-order differential equation with constant coefficients is $W(s) = 1/P(s)$, where $P(s) = a_n s^n + a_{n-1}s^{n-1} + \cdots + a_0$. The transfer function is the Laplace transform of function $w(t)$, called the **weight function** of a linear system. But $w(t)$ can also be characterized in terms of the

discussion at hand. For simplicity let us consider a second-order linear system in which the input is a unit impulse at $t = 0$:

$$a_2 y'' + a_1 y' + a_0 y = \delta(t), \qquad y(0) = 0, \quad y'(0) = 0.$$

Applying the Laplace transform and using $\mathcal{L}\{\delta(t)\} = 1$ shows that the transform of the response y in this case is the transfer function

$$Y(s) = \frac{1}{a_2 s^2 + a_1 s + a_0} = \frac{1}{P(s)} = W(s) \qquad \text{and so} \qquad y = \mathcal{L}^{-1}\left\{\frac{1}{P(s)}\right\} = w(t).$$

From this we can see, in general, that the weight function $y = w(t)$ of an nth-order linear system is the zero-state response of the system to a unit impulse. For this reason $w(t)$ is called as well the **impulse response** of the system.

SECTION 7.6 EXERCISES

Answers to odd-numbered problems begin on page A-12.

In Problems 1–12 use the Laplace transform to solve the given differential equation subject to the indicated initial conditions.

1. $y' - 3y = \delta(t - 2), \quad y(0) = 0$

2. $y' + y = \delta(t - 1), \quad y(0) = 2$

3. $y'' + y = \delta(t - 2\pi), \quad y(0) = 0, y'(0) = 1$

4. $y'' + 16y = \delta(t - 2\pi), \quad y(0) = 0, y'(0) = 0$

5. $y'' + y = \delta\left(t - \frac{\pi}{2}\right) + \delta\left(t - \frac{3\pi}{2}\right), \quad y(0) = 0, y'(0) = 0$

6. $y'' + y = \delta(t - 2\pi) + \delta(t - 4\pi), \quad y(0) = 1, y'(0) = 0$

7. $y'' + 2y' = \delta(t - 1), \quad y(0) = 0, y'(0) = 1$

8. $y'' - 2y' = 1 + \delta(t - 2), \quad y(0) = 0, y'(0) = 1$

9. $y'' + 4y' + 5y = \delta(t - 2\pi), \quad y(0) = 0, y'(0) = 0$

10. $y'' + 2y' + y = \delta(t - 1), \quad y(0) = 0, y'(0) = 0$

11. $y'' + 4y' + 13y = \delta(t - \pi) + \delta(t - 3\pi), \quad y(0) = 1, y'(0) = 0$

12. $y'' - 7y' + 6y = e^t + \delta(t - 2) + \delta(t - 4), \quad y(0) = 0, y'(0) = 0$

13. A uniform beam of length L carries a concentrated load w_0 at $x = L/2$. The beam is embedded at its left end and is free at its right end. Use the Laplace transform to determine the deflection $y(x)$ from

$$EI\frac{d^4 y}{dx^4} = w_0\delta\left(x - \frac{L}{2}\right),$$

where $y(0) = 0$, $y'(0) = 0$, $y''(L) = 0$, and $y'''(L) = 0$.

14. Solve the differential equation in Problem 13 subject to $y(0) = 0$, $y'(0) = 0$, $y(L) = 0$, $y'(L) = 0$. In this case the beam is embedded at both ends. See Figure 7.54.

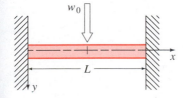

FIGURE 7.54

Discussion Problem

15. Use the Laplace transform to solve the initial-value problem $y'' + \omega^2 y = \delta(t), y(0) = 0, y'(0) = 0$. Do you see anything unusual about the solution?

7.7 SYSTEMS OF LINEAR EQUATIONS

■ *Using the Laplace transform to solve a system of DEs* ■ *Coupled springs* ■ *Networks*

When initial conditions are specified, the Laplace transform reduces a system of linear differential equations with constant coefficients to a set of simultaneous algebraic equations in the transformed functions.

EXAMPLE 1 **System of DEs Transformed into an Algebraic System**

Solve
$$2x' + y' - y = t$$
$$x' + y' \quad = t^2 \tag{1}$$

subject to $x(0) = 1$, $y(0) = 0$.

SOLUTION If $X(s) = \mathcal{L}\{x(t)\}$ and $Y(s) = \mathcal{L}\{y(t)\}$, then after transforming each equation we obtain

$$2[sX(s) - x(0)] + sY(s) - y(0) - Y(s) = \frac{1}{s^2}$$

$$sX(s) - x(0) + sY(s) - y(0) \qquad = \frac{2}{s^3}$$

or
$$2sX(s) + (s - 1)Y(s) = 2 + \frac{1}{s^2}$$
$$sX(s) + \qquad sY(s) = 1 + \frac{2}{s^3}. \tag{2}$$

Multiplying the second equation of (2) by 2 and subtracting yields

$$(-s - 1)Y(s) = \frac{1}{s^2} - \frac{4}{s^3} \quad \text{or} \quad Y(s) = \frac{4 - s}{s^3(s + 1)}. \tag{3}$$

Now by partial fractions

$$Y(s) = \frac{5}{s} - \frac{5}{s^2} + \frac{4}{s^3} - \frac{5}{s + 1},$$

and so

$$y(t) = 5\mathcal{L}^{-1}\left\{\frac{1}{s}\right\} - 5\mathcal{L}^{-1}\left\{\frac{1}{s^2}\right\} + 2\mathcal{L}^{-1}\left\{\frac{2!}{s^3}\right\} - 5\mathcal{L}^{-1}\left\{\frac{1}{s + 1}\right\}$$

$$= 5 - 5t + 2t^2 - 5e^{-t}.$$

By the second equation of (2)

$$X(s) = -Y(s) + \frac{1}{s} + \frac{2}{s^4},$$

from which it follows that

$$x(t) = -\mathcal{L}^{-1}\{Y(s)\} + \mathcal{L}^{-1}\left\{\frac{1}{s}\right\} + \frac{2}{3!}\mathcal{L}^{-1}\left\{\frac{3!}{s^4}\right\}$$

$$= -4 + 5t - 2t^2 + \frac{1}{3}t^3 + 5e^{-t}.$$

Hence we conclude that the solution of the given system (1) is

$$x(t) = -4 + 5t - 2t^2 + \frac{1}{3}t^3 + 5e^{-t}$$

$$y(t) = 5 - 5t + 2t^2 - 5e^{-t}. \tag{4}$$

Applications   Let us turn now to some elementary applications involving systems of linear differential equations. The solutions of the problems that we shall consider can be obtained either by the method of Section 4.8 or through the use of the Laplace transform.

Coupled Springs   Two masses m_1 and m_2 are connected to two springs A and B of negligible mass having spring constants k_1 and k_2, respectively. In turn, the two springs are attached as shown in Figure 7.55. Let $x_1(t)$ and $x_2(t)$ denote the vertical displacements of the masses from their equilibrium positions. When the system is in motion, spring B is subject to both an elongation and a compression; hence its net elongation is $x_2 - x_1$. Therefore it follows from Hooke's law that springs A and B exert forces $-k_1x_1$ and $k_2(x_2 - x_1)$, respectively, on m_1. If no external force is impressed on the system and if no damping force is present, then the net force on m_1 is $-k_1x_1 + k_2(x_2 - x_1)$. By Newton's second law we can write

$$m_1 \frac{d^2x_1}{dt^2} = -k_1x_1 + k_2(x_2 - x_1).$$

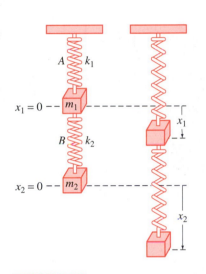

FIGURE 7.55

Similarly, the net force exerted on mass m_2 is due solely to the net elongation of B; that is, $-k_2(x_2 - x_1)$. Hence we have

$$m_2 \frac{d^2x_2}{dt^2} = -k_2(x_2 - x_1).$$

In other words, the motion of the coupled system is represented by the system of simultaneous second-order differential equations

$$m_1x_1'' = -k_1x_1 + k_2(x_2 - x_1)$$

$$m_2x_2'' = -k_2(x_2 - x_1). \tag{5}$$

In the next example we solve (5) under the assumption that $k_1 = 6$, $k_2 = 4$, $m_1 = 1$, $m_2 = 1$ and that the masses start from their equilibrium positions with opposite unit velocities.

EXAMPLE 2   **Coupled Springs**

Solve

$$x_1'' + 10x_1 \qquad - 4x_2 = 0$$

$$-4x_1 + x_2'' + 4x_2 = 0 \tag{6}$$

subject to $x_1(0) = 0$, $x_1'(0) = 1$, $x_2(0) = 0$, $x_2'(0) = -1$.

SOLUTION   The Laplace transform of each equation is

$$s^2X_1(s) - sx_1(0) - x_1'(0) + 10X_1(s) - 4X_2(s) = 0$$

$$-4X_1(s) + s^2X_2(s) - sx_2(0) - x_2'(0) + 4X_2(s) = 0,$$

where $X_1(s) = \mathcal{L}\{x_1(t)\}$ and $X_2(s) = \mathcal{L}\{x_2(t)\}$. The preceding system is the same as

$$(s^2 + 10)X_1(s) - \qquad 4X_2(s) = 1$$
$$-4X_1(s) + (s^2 + 4)X_2(s) = -1. \qquad \text{(7)}$$

Solving (7) for $X_1(s)$ and using partial fractions on the result yields

$$X_1(s) = \frac{s^2}{(s^2 + 2)(s^2 + 12)} = -\frac{1/5}{s^2 + 2} + \frac{6/5}{s^2 + 12},$$

and therefore

$$x_1(t) = -\frac{1}{5\sqrt{2}}\mathcal{L}^{-1}\left\{\frac{\sqrt{2}}{s^2 + 2}\right\} + \frac{6}{5\sqrt{12}}\mathcal{L}^{-1}\left\{\frac{\sqrt{12}}{s^2 + 12}\right\}$$

$$= -\frac{\sqrt{2}}{10}\sin\sqrt{2}t + \frac{\sqrt{3}}{5}\sin 2\sqrt{3}t.$$

Substituting the expression for $X_1(s)$ into the first equation of (7) gives us

$$X_2(s) = -\frac{s^2 + 6}{(s^2 + 2)(s^2 + 12)} = -\frac{2/5}{s^2 + 2} - \frac{3/5}{s^2 + 12}$$

and
$$x_2(t) = -\frac{2}{5\sqrt{2}}\mathcal{L}^{-1}\left\{\frac{\sqrt{2}}{s^2 + 2}\right\} - \frac{3}{5\sqrt{12}}\mathcal{L}^{-1}\left\{\frac{\sqrt{12}}{s^2 + 12}\right\}$$

$$= -\frac{\sqrt{2}}{5}\sin\sqrt{2}t - \frac{\sqrt{3}}{10}\sin 2\sqrt{3}t.$$

Finally the solution to the given system (6) is

$$x_1(t) = -\frac{\sqrt{2}}{10}\sin\sqrt{2}t + \frac{\sqrt{3}}{5}\sin 2\sqrt{3}t$$

$$x_2(t) = -\frac{\sqrt{2}}{5}\sin\sqrt{2}t - \frac{\sqrt{3}}{10}\sin 2\sqrt{3}t. \qquad \text{(8)} \qquad \blacksquare$$

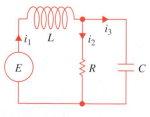

FIGURE 7.56

Networks   In (18) of Section 3.3 we saw that currents $i_1(t)$ and $i_2(t)$ in the network containing an inductor, a resistor, and a capacitor shown in Figure 7.56 were governed by the system of first-order differential equations

$$L\frac{di_1}{dt} + Ri_2 = E(t)$$

$$RC\frac{di_2}{dt} + i_2 - i_1 = 0. \qquad \text{(9)}$$

We solve this system by the Laplace transform in the next example.

EXAMPLE 3   **An Electrical Network**

Solve the system in (9) under the conditions $E(t) = 60$ V, $L = 1$ h, $R = 50\ \Omega$, $C = 10^{-4}$ f, and the currents i_1 and i_2 are initially zero.

SOLUTION   We must solve

$$\frac{di_1}{dt} + 50i_2 = 60$$

$$50(10^{-4})\frac{di_2}{dt} + i_2 - i_1 = 0$$

subject to $i_1(0) = 0$, $i_2(0) = 0$.

Applying the Laplace transform to each equation of the system and simplifying gives

$$sI_1(s) + \qquad 50I_2(s) = \frac{60}{s}$$

$$-200I_1(s) + (s + 200)I_2(s) = 0,$$

where $I_1(s) = \mathcal{L}\{i_1(t)\}$ and $I_2(s) = \mathcal{L}\{i_2(t)\}$. Solving the system for I_1 and I_2 and decomposing the results into partial fractions gives

$$I_1(s) = \frac{60s + 12{,}000}{s(s + 100)^2} = \frac{6/5}{s} - \frac{6/5}{s + 100} - \frac{60}{(s + 100)^2}$$

$$I_2(s) = \frac{12{,}000}{s(s + 100)^2} = \frac{6/5}{s} - \frac{6/5}{s + 100} - \frac{120}{(s + 100)^2}.$$

Taking the inverse Laplace transform, we find the currents to be

$$i_1(t) = \frac{6}{5} - \frac{6}{5}e^{-100t} - 60te^{-100t}$$

$$i_2(t) = \frac{6}{5} - \frac{6}{5}e^{-100t} - 120te^{-100t}.$$

∎

Note that both $i_1(t)$ and $i_2(t)$ in Example 3 tend toward the value $E/R = \frac{6}{5}$ as $t \to \infty$. Furthermore, since the current through the capacitor is $i_3(t) = i_1(t) - i_2(t) = 60te^{-100t}$, we observe that $i_3(t) \to 0$ as $t \to \infty$.

SECTION 7.7 EXERCISES

Answers to odd-numbered problems begin on page A-12.

In Problems 1–12 use the Laplace transform to solve the given system of differential equations.

1. $\dfrac{dx}{dt} = -x + y$

$\dfrac{dy}{dt} = 2x$

$x(0) = 0, y(0) = 1$

2. $\dfrac{dx}{dt} = 2y + e^t$

$\dfrac{dy}{dt} = 8x - t$

$x(0) = 1, y(0) = 1$

3. $\dfrac{dx}{dt} = x - 2y$

$\dfrac{dy}{dt} = 5x - y$

$x(0) = -1, y(0) = 2$

4. $\dfrac{dx}{dt} + 3x + \dfrac{dy}{dt} = 1$

$\dfrac{dx}{dt} - x + \dfrac{dy}{dt} - y = e^t$

$x(0) = 0, y(0) = 0$

5. $2\dfrac{dx}{dt} + \dfrac{dy}{dt} - 2x = 1$

$\dfrac{dx}{dt} + \dfrac{dy}{dt} - 3x - 3y = 2$

$x(0) = 0,\ y(0) = 0$

6. $\dfrac{dx}{dt} + x - \dfrac{dy}{dt} + y = 0$

$\dfrac{dx}{dt} + \dfrac{dy}{dt} + 2y = 0$

$x(0) = 0,\ y(0) = 1$

7. $\dfrac{d^2x}{dt^2} + x - y = 0$

$\dfrac{d^2y}{dt^2} + y - x = 0$

$x(0) = 0,\ x'(0) = -2$

$y(0) = 0,\ y'(0) = 1$

8. $\dfrac{d^2x}{dt^2} + \dfrac{dx}{dt} + \dfrac{dy}{dt} = 0$

$\dfrac{d^2y}{dt^2} + \dfrac{dy}{dt} - 4\dfrac{dx}{dt} = 0$

$x(0) = 1,\ x'(0) = 0,$

$y(0) = -1,\ y'(0) = 5$

9. $\dfrac{d^2x}{dt^2} + \dfrac{d^2y}{dt^2} = t^2$

$\dfrac{d^2x}{dt^2} - \dfrac{d^2y}{dt^2} = 4t$

$x(0) = 8,\ x'(0) = 0,$

$y(0) = 0,\ y'(0) = 0$

10. $\dfrac{dx}{dt} - 4x + \dfrac{d^3y}{dt^3} = 6\sin t$

$\dfrac{dx}{dt} + 2x - 2\dfrac{d^3y}{dt^3} = 0$

$x(0) = 0,\ y(0) = 0,$

$y'(0) = 0,\ y''(0) = 0$

11. $\dfrac{d^2x}{dt^2} + 3\dfrac{dy}{dt} + 3y = 0$

$\dfrac{d^2x}{dt^2} + 3y = te^{-t}$

$x(0) = 0,\ x'(0) = 2,\ y(0) = 0$

12. $\dfrac{dx}{dt} = 4x - 2y + 2\mathcal{U}(t - 1)$

$\dfrac{dy}{dt} = 3x - y + \mathcal{U}(t - 1)$

$x(0) = 0,\ y(0) = 1/2$

13. Solve system (5) when $k_1 = 3$, $k_2 = 2$, $m_1 = 1$, $m_2 = 1$ and $x_1(0) = 0$, $x_1'(0) = 1$, $x_2(0) = 1$, $x_2'(0) = 0$.

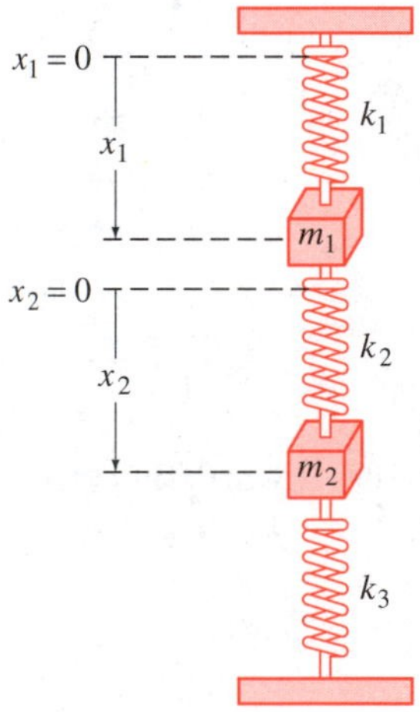

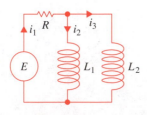

FIGURE 7.57

14. Derive the system of differential equations describing the straight-line vertical motion of the coupled springs shown in Figure 7.57. Use the Laplace transform to solve the system when $k_1 = 1$, $k_2 = 1$, $k_3 = 1$, $m_1 = 1$, $m_2 = 1$ and $x_1(0) = 0$, $x_1'(0) = -1$, $x_2(0) = 0$, $x_2'(0) = 1$.

15. (a) Show that the system of differential equations for the currents $i_2(t)$ and $i_3(t)$ in the electrical network shown in Figure 7.58 is

$$L_1\frac{di_2}{dt} + Ri_2 + Ri_3 = E(t)$$

$$L_2\frac{di_3}{dt} + Ri_2 + Ri_3 = E(t).$$

FIGURE 7.58

(b) Solve the system in part (a) if $R = 5\ \Omega$, $L_1 = 0.01$ h, $L_2 = 0.0125$ h, $E = 100$ V, $i_2(0) = 0$, and $i_3(0) = 0$.

(c) Determine the current $i_1(t)$.

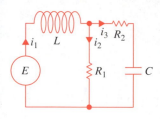

FIGURE 7.59

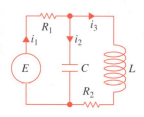

FIGURE 7.60

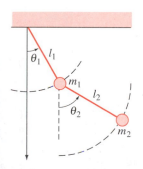

FIGURE 7.61

16. **(a)** In Problem 12 in Exercises 3.3 you were asked to show that the currents $i_2(t)$ and $i_3(t)$ in the electrical network shown in Figure 7.59 satisfy

$$L \frac{di_2}{dt} + L \frac{di_3}{dt} + R_1 i_2 = E(t)$$

$$-R_1 \frac{di_2}{dt} + R_2 \frac{di_3}{dt} + \frac{1}{C} i_3 = 0.$$

Solve the system if $R_1 = 10 \ \Omega$, $R_2 = 5 \ \Omega$, $L = 1$ h, $C = 0.2$ f,

$$E(t) = \begin{cases} 120, & 0 \le t < 2 \\ 0, & t \ge 2, \end{cases}$$

$i_2(0) = 0$, and $i_3(0) = 0$.

(b) Determine the current $i_1(t)$.

17. Solve the system given in (17) of Section 3.3 when $R_1 = 6 \ \Omega$, $R_2 = 5 \ \Omega$, $L_1 = 1$ h, $L_2 = 1$ h, $E(t) = 50 \sin t$ V, $i_2(0) = 0$, and $i_3(0) = 0$.

18. Solve (9) when $E = 60$ V, $L = \frac{1}{2}$ h, $R = 50 \ \Omega$, $C = 10^{-4}$ f, $i_1(0) = 0$, and $i_2(0) = 0$.

19. Solve (9) when $E = 60$ V, $L = 2$ h, $R = 50 \ \Omega$, $C = 10^{-4}$ f, $i_1(0) = 0$, and $i_2(0) = 0$.

20. **(a)** Show that the system of differential equations for the charge on the capacitor $q(t)$ and the current $i_3(t)$ in the electrical network shown in Figure 7.60 is

$$R_1 \frac{dq}{dt} + \frac{1}{C} q + R_1 i_3 = E(t)$$

$$L \frac{di_3}{dt} + R_2 i_3 - \frac{1}{C} q = 0.$$

(b) Find the charge on the capacitor when $L = 1$ h, $R_1 = 1 \ \Omega$, $R_2 = 1 \ \Omega$, $C = 1$ f,

$$E(t) = \begin{cases} 0, & 0 < t < 1 \\ 50e^{-t}, & t \ge 1, \end{cases}$$

$i_3(0) = 0$, and $q(0) = 0$.

21. A double pendulum oscillates in a vertical plane under the influence of gravity. See Figure 7.61. For small displacements $\theta_1(t)$ and $\theta_2(t)$ it can be shown that the differential equations of motion are

$$(m_1 + m_2) l_1^2 \theta_1'' + m_2 l_1 l_2 \theta_2'' + (m_1 + m_2) l_1 g \theta_1 = 0$$

$$m_2 l_2^2 \theta_2'' + m_2 l_1 l_2 \theta_1'' + m_2 l_2 g \theta_2 = 0.$$

Use the Laplace transform to solve the system when $m_1 = 3$, $m_2 = 1$, $l_1 = l_2 = 16$, $\theta_1(0) = 1$, $\theta_2(0) = -1$, $\theta_1'(0) = 0$, and $\theta_2'(0) = 0$.

CHAPTER 7 REVIEW EXERCISES

Answers to odd-numbered problems begin on page A-12.

In Problems 1 and 2 use the definition of the Laplace transform to find
$\mathcal{L}\{f(t)\}$.

1. $f(t) = \begin{cases} t, & 0 \le t < 1 \\ 2 - t, & t \ge 1 \end{cases}$ **2.** $f(t) = \begin{cases} 0, & 0 \le t < 2 \\ 1, & 2 \le t < 4 \\ 0, & t \ge 4 \end{cases}$

In Problems 3–24 fill in the blanks or answer true/false.

3. If f is not piecewise continuous on $[0, \infty)$, then $\mathcal{L}\{f(t)\}$ will not exist. _____

4. The function $f(t) = (e^t)^{10}$ is not of exponential order. _____

5. $F(s) = s^2/(s^2 + 4)$ is not the Laplace transform of a function that is piecewise continuous and of exponential order. _____

6. If $\mathcal{L}\{f(t)\} = F(s)$ and $\mathcal{L}\{g(t)\} = G(s)$, then $\mathcal{L}^{-1}\{F(s)G(s)\} = f(t)g(t)$. _____

7. $\mathcal{L}\{e^{-7t}\} = $ _____   **8.** $\mathcal{L}\{te^{-7t}\} = $ _____

9. $\mathcal{L}\{\sin 2t\} = $ _____   **10.** $\mathcal{L}\{e^{-3t} \sin 2t\} = $ _____

11. $\mathcal{L}\{t \sin 2t\} = $ _____   **12.** $\mathcal{L}\{\sin 2t\,\mathcal{U}(t - \pi)\} = $ _____

13. $\mathcal{L}^{-1}\left\{\dfrac{20}{s^6}\right\} = $ _____   **14.** $\mathcal{L}^{-1}\left\{\dfrac{1}{3s - 1}\right\} = $ _____

15. $\mathcal{L}^{-1}\left\{\dfrac{1}{(s - 5)^3}\right\} = $ _____   **16.** $\mathcal{L}^{-1}\left\{\dfrac{1}{s^2 - 5}\right\} = $ _____

17. $\mathcal{L}^{-1}\left\{\dfrac{s}{s^2 - 10s + 29}\right\} = $ _____   **18.** $\mathcal{L}^{-1}\left\{\dfrac{e^{-5s}}{s^2}\right\} = $ _____

19. $\mathcal{L}^{-1}\left\{\dfrac{s + \pi}{s^2 + \pi^2}e^{-s}\right\} = $ _____   **20.** $\mathcal{L}^{-1}\left\{\dfrac{1}{L^2 s^2 + n^2 \pi^2}\right\} = $ _____

21. $\mathcal{L}\{e^{-5t}\}$ exists for $s > $ _____.

22. If $\mathcal{L}\{f(t)\} = F(s)$, then $\mathcal{L}\{te^{8t}f(t)\} = $ _____.

23. If $\mathcal{L}\{f(t)\} = F(s)$ and $k > 0$, then $\mathcal{L}\{e^{at}f(t - k)\mathcal{U}(t - k)\} = $ _____.

24. $1 * 1 = $ _____.

In Problems 25–28
(a) express f in terms of unit step functions,
(b) find $\mathcal{L}\{f(t)\}$, and
(c) find $\mathcal{L}\{e^t f(t)\}$.

25.

FIGURE 7.62

26.

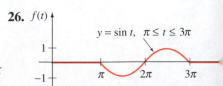

FIGURE 7.63

27.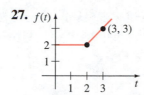

FIGURE 7.64

28.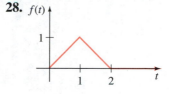

FIGURE 7.65

In Problems 29–36 use the Laplace transform to solve the given equation.

29. $y'' - 2y' + y = e^t, \quad y(0) = 0, y'(0) = 5$

30. $y'' - 8y' + 20y = te^t, \quad y(0) = 0, y'(0) = 0$

31. $y'' - 4y' + 6y = 30\,\mathcal{U}(t - \pi), \quad y(0) = 0, y'(0) = 0$

32. $y'' + 6y' + 5y = t - t\,\mathcal{U}(t - 2), \quad y(0) = 1, y'(0) = 0$

33. $y' - 5y = f(t),$ where $f(t) = \begin{cases} t^2, & 0 \le t < 1 \\ 0, & t \ge 1 \end{cases}, \quad y(0) = 1$

34. $f(t) = 1 - 2\displaystyle\int_0^t e^{-3\tau} f(t - \tau)\, d\tau$

35. $y'(t) = \cos t + \displaystyle\int_0^t y(\tau) \cos(t - \tau)\, d\tau, \quad y(0) = 1$

36. $\displaystyle\int_0^t f(\tau) f(t - \tau)\, d\tau = 6t^3$

In Problems 37 and 38 use the Laplace transform to solve each system.

37. $x' + y = t$
$4x + y' = 0$
$x(0) = 1, y(0) = 2$

38. $x'' + y'' = e^{2t}$
$2x' + y'' = -e^{2t}$
$x(0) = 0, y(0) = 0$
$x'(0) = 0, y'(0) = 0$

39. The current $i(t)$ in an RC series circuit can be determined from the integral equation

$$Ri + \frac{1}{C}\int_0^t i(\tau)\, d\tau = E(t),$$

where $E(t)$ is the impressed voltage. Determine $i(t)$ when $R = 10\ \Omega$, $C = 0.5$ f, and $E(t) = 2(t^2 + t)$.

40. A series circuit contains an inductor, a resistor, and a capacitor for which $L = \frac{1}{2}$ h, $R = 10\ \Omega$, and $C = 0.01$ f, respectively. The voltage

$$E(t) = \begin{cases} 10, & 0 \le t < 5 \\ 0, & t \ge 5 \end{cases}$$

is applied to the circuit. Determine the instantaneous charge $q(t)$ on the capacitor for $t > 0$ if $q(0) = 0$ and $q'(0) = 0$.

41. A uniform cantilever beam of length L is embedded at its left end ($x = 0$) and free at its right end. Find the deflection $y(x)$ if the load per unit length is given by

$$w(x) = \frac{2w_0}{L}\left[\frac{L}{2} - x + \left(x - \frac{L}{2}\right)\mathcal{U}\left(x - \frac{L}{2}\right)\right].$$

42. When a uniform beam is supported by an elastic foundation, the differential equation for its deflection $y(x)$ is

$$\frac{d^4y}{dx^4} + 4a^4y = \frac{w(x)}{EI},$$

where a is a constant. In the case when $a = 1$, find the deflection $y(x)$ of an elastically supported beam of length π that is embedded in concrete at both ends when a concentrated load w_0 is applied at $x = \pi/2$. [*Hint:* Use the table of Laplace transforms in Appendix III.]

SYSTEMS OF LINEAR FIRST-ORDER DIFFERENTIAL EQUATIONS

INTRODUCTION

We have seen systems of differential equations in Sections 3.3, 4.8, and 7.7, and we were able to solve some of these systems by means of either systematic elimination or the Laplace transform. In this chapter we are going to concentrate on only systems of linear first-order equations. Although most of the systems considered could be solved using elimination or the Laplace transform, we are going to develop a general theory for these kinds of systems and, in the case of systems with constant coefficients, a method of solution that utilizes some basic concepts from the algebra of matrices. We shall see that this general theory and solution procedure is similar to that of linear higher-order differential equations considered in Sections 4.1, 4.3, and 4.6. This material is fundamental to the analysis of systems of nonlinear first-order equations.

8.1 PRELIMINARY THEORY

- *Linear systems* ■ *Homogeneous and nonhomogeneous systems* ■ *Solution vector*
- *Initial-value problem* ■ *Superposition principle* ■ *Linear dependence* ■ *Linear independence*
- *Wronskian* ■ *Fundamental set of solutions* ■ *General solution* ■ *Complementary solution*
- *Particular solution*

Note

Matrix notation and properties are used extensively throughout this chapter. You should review Appendix II if you are unfamiliar with these concepts.

In Section 4.8 of Chapter 4 we dealt with systems of differential equations of the form

$$
\begin{aligned}
P_{11}(D)x_1 + P_{12}(D)x_2 + \cdots + P_{1n}(D)x_n &= b_1(t) \\
P_{21}(D)x_1 + P_{22}(D)x_2 + \cdots + P_{2n}(D)x_n &= b_2(t) \\
&\ \vdots \\
P_{n1}(D)x_1 + P_{n2}(D)x_2 + \cdots + P_{nn}(D)x_n &= b_n(t),
\end{aligned}
\tag{1}
$$

where the P_{ij} were polynomials of various degrees in the differential operator D. In this chapter we confine our study to systems of first-order differential equations

$$
\begin{aligned}
\frac{dx_1}{dt} &= g_1(t, x_1, x_2, \ldots, x_n) \\
\frac{dx_2}{dt} &= g_2(t, x_1, x_2, \ldots, x_n) \\
&\ \vdots \\
\frac{dx_n}{dt} &= g_n(t, x_1, x_2, \ldots, x_n).
\end{aligned}
\tag{2}
$$

System (2) of n first-order equations is called an **nth-order system.**

Linear Systems   If each of the functions $g_1, g_2, \ldots, g_n$ is linear in the dependent variables $x_1, x_2, \ldots, x_n$, then (2) is a **system of linear first-order equations.** Such a system has the standard form

$$
\begin{aligned}
\frac{dx_1}{dt} &= a_{11}(t)x_1 + a_{12}(t)x_2 + \cdots + a_{1n}(t)x_n + f_1(t) \\
\frac{dx_2}{dt} &= a_{21}(t)x_1 + a_{22}(t)x_2 + \cdots + a_{2n}(t)x_n + f_2(t) \\
&\ \vdots \\
\frac{dx_n}{dt} &= a_{n1}(t)x_1 + a_{n2}(t)x_2 + \cdots + a_{nn}(t)x_n + f_n(t).
\end{aligned}
\tag{3}
$$

We also refer to a system of the form (3) as an **nth-order linear system** or simply as a **linear system.** We assume that the coefficients a_{ij} and the functions f_i are continuous on a common interval I. When $f_i(t) = 0$,

$i = 1, 2, \ldots, n$, the linear system is said to be **homogeneous;** otherwise it is **nonhomogeneous.**

Matrix Form of a Linear System

If $\mathbf{X}$, $\mathbf{A}(t)$, and $\mathbf{F}(t)$ denote the respective matrices

$$
\mathbf{X} = \begin{pmatrix} x_1(t) \\ x_2(t) \\ \vdots \\ x_n(t) \end{pmatrix}, \qquad
\mathbf{A}(t) = \begin{pmatrix} a_{11}(t) & a_{12}(t) & \cdots & a_{1n}(t) \\ a_{21}(t) & a_{22}(t) & \cdots & a_{2n}(t) \\ \vdots & & & \vdots \\ a_{n1}(t) & a_{n2}(t) & \cdots & a_{nn}(t) \end{pmatrix}, \qquad
\mathbf{F}(t) = \begin{pmatrix} f_1(t) \\ f_2(t) \\ \vdots \\ f_n(t) \end{pmatrix},
$$

then the system of linear first-order differential equations (3) can be written as

$$
\frac{d}{dt} \begin{pmatrix} x_1 \\ x_2 \\ \vdots \\ x_n \end{pmatrix} = \begin{pmatrix} a_{11}(t) & a_{12}(t) & \cdots & a_{1n}(t) \\ a_{21}(t) & a_{22}(t) & \cdots & a_{2n}(t) \\ \vdots & & & \vdots \\ a_{n1}(t) & a_{n2}(t) & \cdots & a_{nn}(t) \end{pmatrix} \begin{pmatrix} x_1 \\ x_2 \\ \vdots \\ x_n \end{pmatrix} + \begin{pmatrix} f_1(t) \\ f_2(t) \\ \vdots \\ f_n(t) \end{pmatrix}
$$

or simply
$$
\mathbf{X}' = \mathbf{AX} + \mathbf{F}. \tag{4}
$$

If the system is homogeneous, its matrix form is then

$$
\mathbf{X}' = \mathbf{AX}. \tag{5}
$$

EXAMPLE 1   Systems Written in Matrix Notation

(a) If $\mathbf{X} = \begin{pmatrix} x \\ y \end{pmatrix}$, then the matrix form of the homogeneous system

$$
\begin{aligned} \frac{dx}{dt} &= 3x + 4y \\ \frac{dy}{dt} &= 5x - 7y \end{aligned} \qquad \text{is} \qquad \mathbf{X}' = \begin{pmatrix} 3 & 4 \\ 5 & -7 \end{pmatrix} \mathbf{X}.
$$

(b) If $\mathbf{X} = \begin{pmatrix} x \\ y \\ z \end{pmatrix}$, then the matrix form of the nonhomogeneous system

$$
\begin{aligned} \frac{dx}{dt} &= 6x + y + z + t \\ \frac{dy}{dt} &= 8x + 7y - z + 10t \\ \frac{dz}{dt} &= 2x + 9y - z + 6t \end{aligned} \qquad \text{is} \qquad \mathbf{X}' = \begin{pmatrix} 6 & 1 & 1 \\ 8 & 7 & -1 \\ 2 & 9 & -1 \end{pmatrix} \mathbf{X} + \begin{pmatrix} t \\ 10t \\ 6t \end{pmatrix}.
$$

∎

> **DEFINITION 8.1**   **Solution Vector**
>
> A **solution vector** on an interval I is any column matrix
>
> $$\mathbf{X} = \begin{pmatrix} x_1(t) \\ x_2(t) \\ \vdots \\ x_n(t) \end{pmatrix}$$
>
> whose entries are differentiable functions satisfying the system (4) on the interval.

EXAMPLE 2   **Verification of Solutions**

Verify that on the interval $(-\infty, \infty)$

$$\mathbf{X}_1 = \begin{pmatrix} 1 \\ -1 \end{pmatrix} e^{-2t} = \begin{pmatrix} e^{-2t} \\ -e^{-2t} \end{pmatrix} \qquad \text{and} \qquad \mathbf{X}_2 = \begin{pmatrix} 3 \\ 5 \end{pmatrix} e^{6t} = \begin{pmatrix} 3e^{6t} \\ 5e^{6t} \end{pmatrix}$$

are solutions of $\qquad \mathbf{X}' = \begin{pmatrix} 1 & 3 \\ 5 & 3 \end{pmatrix} \mathbf{X}.$ **(6)**

SOLUTION   From $\mathbf{X}_1' = \begin{pmatrix} -2e^{-2t} \\ 2e^{-2t} \end{pmatrix}$ and $\mathbf{X}_2' = \begin{pmatrix} 18e^{6t} \\ 30e^{6t} \end{pmatrix}$ we see that

$$\mathbf{AX}_1 = \begin{pmatrix} 1 & 3 \\ 5 & 3 \end{pmatrix} \begin{pmatrix} e^{-2t} \\ -e^{-2t} \end{pmatrix} = \begin{pmatrix} e^{-2t} - 3e^{-2t} \\ 5e^{-2t} - 3e^{-2t} \end{pmatrix} = \begin{pmatrix} -2e^{-2t} \\ 2e^{-2t} \end{pmatrix} = \mathbf{X}_1'$$

and $\qquad \mathbf{AX}_2 = \begin{pmatrix} 1 & 3 \\ 5 & 3 \end{pmatrix} \begin{pmatrix} 3e^{6t} \\ 5e^{6t} \end{pmatrix} = \begin{pmatrix} 3e^{6t} + 15e^{6t} \\ 15e^{6t} + 15e^{6t} \end{pmatrix} = \begin{pmatrix} 18e^{6t} \\ 30e^{6t} \end{pmatrix} = \mathbf{X}_2'.$ ∎

Much of the theory of systems of n linear first-order differential equations is similar to that of linear nth-order differential equations.

Initial-Value Problem   Let t_0 denote a point on an interval I and

$$\mathbf{X}(t_0) = \begin{pmatrix} x_1(t_0) \\ x_2(t_0) \\ \vdots \\ x_n(t_0) \end{pmatrix} \qquad \text{and} \qquad \mathbf{X}_0 = \begin{pmatrix} \gamma_1 \\ \gamma_2 \\ \vdots \\ \gamma_n \end{pmatrix},$$

where the γ_i, $i = 1, 2, \ldots, n$ are given constants. Then the problem

$$\begin{array}{ll} Solve: & \mathbf{X}' = \mathbf{A}(t)\mathbf{X} + \mathbf{F}(t) \\ Subject\ to: & \mathbf{X}(t_0) = \mathbf{X}_0 \end{array}$$ **(7)**

is an **initial-value problem** on the interval.

THEOREM 8.1 **Existence of a Unique Solution**

Let the entries of the matrices $\mathbf{A}(t)$ and $\mathbf{F}(t)$ be functions continuous on a common interval I that contains the point t_0. Then there exists a unique solution of the initial-value problem (7) on the interval.

Homogeneous Systems In the next several definitions and theorems we are concerned only with homogeneous systems. Without stating it, we shall always assume that the a_{ij} and the f_i are continuous functions of t on some common interval I.

Superposition Principle The following result is a **superposition principle** for solutions of linear systems.

THEOREM 8.2 **Superposition Principle**

Let $\mathbf{X}_1, \mathbf{X}_2, \dots, \mathbf{X}_k$ be a set of solution vectors of the homogeneous system (5) on an interval I. Then the linear combination

$$\mathbf{X} = c_1 \mathbf{X}_1 + c_2 \mathbf{X}_2 + \cdots + c_k \mathbf{X}_k,$$

where the c_i, $i = 1, 2, \dots, k$ are arbitrary constants, is also a solution on the interval.

It follows from Theorem 8.2 that a constant multiple of any solution vector of a homogeneous system of linear first-order differential equations is also a solution.

EXAMPLE 3 **Using the Superposition Principle**

You should practice by verifying that the two vectors

$$\mathbf{X}_1 = \begin{pmatrix} \cos t \\ -\frac{1}{2}\cos t + \frac{1}{2}\sin t \\ -\cos t - \sin t \end{pmatrix} \quad \text{and} \quad \mathbf{X}_2 = \begin{pmatrix} 0 \\ e^t \\ 0 \end{pmatrix}$$

are solutions of the system

$$\mathbf{X}' = \begin{pmatrix} 1 & 0 & 1 \\ 1 & 1 & 0 \\ -2 & 0 & -1 \end{pmatrix} \mathbf{X}. \tag{8}$$

By the superposition principle the linear combination

$$\mathbf{X} = c_1 \mathbf{X}_1 + c_2 \mathbf{X}_2 = c_1 \begin{pmatrix} \cos t \\ -\frac{1}{2}\cos t + \frac{1}{2}\sin t \\ -\cos t - \sin t \end{pmatrix} + c_2 \begin{pmatrix} 0 \\ e^t \\ 0 \end{pmatrix}$$

is yet another solution of the system. ∎

Linear Dependence and Linear Independence We are primarily interested in linearly independent solutions of the homogeneous system (5).

DEFINITION 8.2 **Linear Dependence/Independence**

Let $\mathbf{X}_1, \mathbf{X}_2, \ldots, \mathbf{X}_k$ be a set of solution vectors of the homogeneous system (5) on an interval I. We say that the set is **linearly dependent** on the interval if there exist constants $c_1, c_2, \ldots, c_k$, not all zero, such that

$$c_1\mathbf{X}_1 + c_2\mathbf{X}_2 + \cdots + c_k\mathbf{X}_k = \mathbf{0}$$

for every t in the interval. If the set of vectors is not linearly dependent on the interval, it is said to be **linearly independent.**

The case when $k = 2$ should be clear; two solution vectors $\mathbf{X}_1$ and $\mathbf{X}_2$ are linearly dependent if one is a constant multiple of the other, and conversely. For $k > 2$ a set of solution vectors is linearly dependent if we can express at least one solution vector as a linear combination of the remaining vectors.

Wronskian As in our earlier consideration of the theory of a single ordinary differential equation, we can introduce the concept of the **Wronskian** determinant as a test for linear independence. We state the following theorem without proof.

THEOREM 8.3 **Criterion for Linearly Independent Solutions**

Let
$$\mathbf{X}_1 = \begin{pmatrix} x_{11} \\ x_{21} \\ \vdots \\ x_{n1} \end{pmatrix}, \quad \mathbf{X}_2 = \begin{pmatrix} x_{12} \\ x_{22} \\ \vdots \\ x_{n2} \end{pmatrix}, \quad \ldots, \quad \mathbf{X}_n = \begin{pmatrix} x_{1n} \\ x_{2n} \\ \vdots \\ x_{nn} \end{pmatrix}$$

be n solution vectors of the homogeneous system (5) on an interval I. Then the set of solution vectors is linearly independent on I if and only if the **Wronskian**

$$W(\mathbf{X}_1, \mathbf{X}_2, \ldots, \mathbf{X}_n) = \begin{vmatrix} x_{11} & x_{12} & \cdots & x_{1n} \\ x_{21} & x_{22} & \cdots & x_{2n} \\ \vdots & & & \vdots \\ x_{n1} & x_{n2} & \cdots & x_{nn} \end{vmatrix} \neq 0 \qquad (9)$$

for every t in the interval.

It can be shown that if $\mathbf{X}_1, \mathbf{X}_2, \ldots, \mathbf{X}_n$ are solution vectors of (5), then for every t in I either $W(\mathbf{X}_1, \mathbf{X}_2, \ldots, \mathbf{X}_n) \neq 0$ or $W(\mathbf{X}_1, \mathbf{X}_2, \ldots, \mathbf{X}_n) = 0$.

Thus if we can show that $W \neq 0$ for some t_0 in I, then $W \neq 0$ for every t, and hence the solutions are linearly independent on the interval.

Notice that, unlike our definition of the Wronskian in Section 4.1, here the definition of the determinant (9) does not involve differentiation.

EXAMPLE 4 Linearly Independent Solutions

In Example 2 we saw that $\mathbf{X}_1 = \begin{pmatrix} 1 \\ -1 \end{pmatrix} e^{-2t}$ and $\mathbf{X}_2 = \begin{pmatrix} 3 \\ 5 \end{pmatrix} e^{6t}$ are solutions of system (6). Clearly $\mathbf{X}_1$ and $\mathbf{X}_2$ are linearly independent on the interval $(-\infty, \infty)$ since neither vector is a constant multiple of the other. In addition, we have

$$W(\mathbf{X}_1, \mathbf{X}_2) = \begin{vmatrix} e^{-2t} & 3e^{6t} \\ -e^{-2t} & 5e^{6t} \end{vmatrix} = 8e^{4t} \neq 0$$

for all real values of t. ∎

DEFINITION 8.3 Fundamental Set of Solutions

Any set $\mathbf{X}_1, \mathbf{X}_2, \ldots, \mathbf{X}_n$ of n linearly independent solution vectors of the homogeneous system (5) on an interval I is said to be a **fundamental set of solutions** on the interval.

THEOREM 8.4 Existence of a Fundamental Set

There exists a fundamental set of solutions for the homogeneous system (5) on an interval I.

The next two theorems are the linear system equivalents of Theorems 4.5 and 4.6.

THEOREM 8.5 General Solution–Homogeneous Systems

Let $\mathbf{X}_1, \mathbf{X}_2, \ldots, \mathbf{X}_n$ be a fundamental set of solutions of the homogeneous system (5) on an interval I. Then the **general solution** of the system on the interval is

$$\mathbf{X} = c_1\mathbf{X}_1 + c_2\mathbf{X}_2 + \cdots + c_n\mathbf{X}_n,$$

where the c_i, $i = 1, 2, \ldots, n$ are arbitrary constants.

EXAMPLE 5 General Solution of System (6)

From Example 2 we know that $\mathbf{X}_1 = \begin{pmatrix} 1 \\ -1 \end{pmatrix} e^{-2t}$ and $\mathbf{X}_2 = \begin{pmatrix} 3 \\ 5 \end{pmatrix} e^{6t}$ are linearly independent solutions of (6) on $(-\infty, \infty)$. Hence $\mathbf{X}_1$ and $\mathbf{X}_2$

form a fundamental set of solutions on the interval. The general solution of the system on the interval is then

$$\mathbf{X} = c_1\mathbf{X}_1 + c_2\mathbf{X}_2 = c_1 \begin{pmatrix} 1 \\ -1 \end{pmatrix} e^{-2t} + c_2 \begin{pmatrix} 3 \\ 5 \end{pmatrix} e^{6t}. \qquad \textbf{(10)}$$

EXAMPLE 6   **General Solution of System (8)**

The vectors

$$\mathbf{X}_1 = \begin{pmatrix} \cos t \\ -\frac{1}{2}\cos t + \frac{1}{2}\sin t \\ -\cos t - \sin t \end{pmatrix}, \qquad \mathbf{X}_2 = \begin{pmatrix} 0 \\ 1 \\ 0 \end{pmatrix} e^t, \qquad \mathbf{X}_3 = \begin{pmatrix} \sin t \\ -\frac{1}{2}\sin t - \frac{1}{2}\cos t \\ -\sin t + \cos t \end{pmatrix}$$

are solutions of the system (8) in Example 3 (see Problem 16 in Exercises 8.1). Now

$$W(\mathbf{X}_1, \mathbf{X}_2, \mathbf{X}_3) = \begin{vmatrix} \cos t & 0 & \sin t \\ -\frac{1}{2}\cos t + \frac{1}{2}\sin t & e^t & -\frac{1}{2}\sin t - \frac{1}{2}\cos t \\ -\cos t - \sin t & 0 & -\sin t + \cos t \end{vmatrix} = e^t \neq 0$$

for all real values of t. We conclude that $\mathbf{X}_1$, $\mathbf{X}_2$, and $\mathbf{X}_3$ form a fundamental set of solutions on $(-\infty, \infty)$. Thus the general solution of the system on the interval is the linear combination $\mathbf{X} = c\mathbf{X}_1 + c_2\mathbf{X}_2 + c_3\mathbf{X}_3$, that is,

$$\mathbf{X} = c_1 \begin{pmatrix} \cos t \\ -\frac{1}{2}\cos t + \frac{1}{2}\sin t \\ -\cos t - \sin t \end{pmatrix} + c_2 \begin{pmatrix} 0 \\ 1 \\ 0 \end{pmatrix} e^t + c_3 \begin{pmatrix} \sin t \\ -\frac{1}{2}\sin t - \frac{1}{2}\cos t \\ -\sin t + \cos t \end{pmatrix}.$$

Nonhomogeneous Systems   For nonhomogeneous systems a **particular solution** $\mathbf{X}_p$ on an interval I is any vector, free of arbitrary parameters, whose entries are functions that satisfy the system (4).

THEOREM 8.6   **General Solution—Nonhomogeneous Systems**

Let $\mathbf{X}_p$ be a given solution of the nonhomogeneous system (4) on an interval I, and let

$$\mathbf{X}_c = c_1\mathbf{X}_1 + c_2\mathbf{X}_2 + \cdots + c_n\mathbf{X}_n$$

denote the general solution on the same interval of the corresponding homogeneous system (5). Then the **general solution** of the nonhomogeneous system on the interval is

$$\mathbf{X} = \mathbf{X}_c + \mathbf{X}_p.$$

The general solution $\mathbf{X}_c$ of the homogeneous system (5) is called the **complementary function** of the nonhomogeneous system (4).

EXAMPLE 7 **General Solution–Nonhomogeneous System**

The vector $\mathbf{X}_p = \begin{pmatrix} 3t - 4 \\ -5t + 6 \end{pmatrix}$ is a particular solution of the nonhomogeneous system

$$\mathbf{X}' = \begin{pmatrix} 1 & 3 \\ 5 & 3 \end{pmatrix} \mathbf{X} + \begin{pmatrix} 12t - 11 \\ -3 \end{pmatrix} \quad \text{(11)}$$

on the interval $(-\infty, \infty)$. (Verify this.) The complementary function of (11) on the same interval, or the general solution of

$$\mathbf{X}' = \begin{pmatrix} 1 & 3 \\ 5 & 3 \end{pmatrix} \mathbf{X},$$

was seen in (10) of Example 5 to be

$$\mathbf{X}_c = c_1 \begin{pmatrix} 1 \\ -1 \end{pmatrix} e^{-2t} + c_2 \begin{pmatrix} 3 \\ 5 \end{pmatrix} e^{6t}.$$

Hence by Theorem 8.6

$$\mathbf{X} = \mathbf{X}_c + \mathbf{X}_p = c_1 \begin{pmatrix} 1 \\ -1 \end{pmatrix} e^{-2t} + c_2 \begin{pmatrix} 3 \\ 5 \end{pmatrix} e^{6t} + \begin{pmatrix} 3t - 4 \\ -5t + 6 \end{pmatrix}$$

is the general solution of (11) on $(-\infty, \infty)$. ∎

SECTION 8.1 EXERCISES

Answers to odd-numbered problems begin on page A-12.

In Problems 1–6 write the given system in matrix form.

1. $\dfrac{dx}{dt} = 3x - 5y$

$\dfrac{dy}{dt} = 4x + 8y$

2. $\dfrac{dx}{dt} = 4x - 7y$

$\dfrac{dy}{dt} = 5x$

3. $\dfrac{dx}{dt} = -3x + 4y - 9z$

$\dfrac{dy}{dt} = 6x - y$

$\dfrac{dz}{dt} = 10x + 4y + 3z$

4. $\dfrac{dx}{dt} = x - y$

$\dfrac{dy}{dt} = x + 2z$

$\dfrac{dz}{dt} = -x + z$

5. $\dfrac{dx}{dt} = x - y + z + t - 1$

$\dfrac{dy}{dt} = 2x + y - z - 3t^2$

$\dfrac{dz}{dt} = x + y + z + t^2 - t + 2$

6. $\dfrac{dx}{dt} = -3x + 4y + e^{-t} \sin 2t$

$\dfrac{dy}{dt} = 5x + 9y + 4e^{-t} \cos 2t$

In Problems 7–10 write the given system without the use of matrices.

7. $\mathbf{X}' = \begin{pmatrix} 4 & 2 \\ -1 & 3 \end{pmatrix} \mathbf{X} + \begin{pmatrix} 1 \\ -1 \end{pmatrix} e^t$

8. $\mathbf{X}' = \begin{pmatrix} 7 & 5 & -9 \\ 4 & 1 & 1 \\ 0 & -2 & 3 \end{pmatrix} \mathbf{X} + \begin{pmatrix} 0 \\ 2 \\ 1 \end{pmatrix} e^{5t} - \begin{pmatrix} 8 \\ 0 \\ 3 \end{pmatrix} e^{-2t}$

9. $\dfrac{d}{dt} \begin{pmatrix} x \\ y \\ z \end{pmatrix} = \begin{pmatrix} 1 & -1 & 2 \\ 3 & -4 & 1 \\ -2 & 5 & 6 \end{pmatrix} \begin{pmatrix} x \\ y \\ z \end{pmatrix} + \begin{pmatrix} 1 \\ 2 \\ 2 \end{pmatrix} e^{-t} - \begin{pmatrix} 3 \\ -1 \\ 1 \end{pmatrix} t$

10. $\dfrac{d}{dt} \begin{pmatrix} x \\ y \end{pmatrix} = \begin{pmatrix} 3 & -7 \\ 1 & 1 \end{pmatrix} \begin{pmatrix} x \\ y \end{pmatrix} + \begin{pmatrix} 4 \\ 8 \end{pmatrix} \sin t + \begin{pmatrix} t - 4 \\ 2t + 1 \end{pmatrix} e^{4t}$

In Problems 11–16 verify that the vector $\mathbf{X}$ is a solution of the given system.

11. $\dfrac{dx}{dt} = 3x - 4y$

$\dfrac{dy}{dt} = 4x - 7y; \quad \mathbf{X} = \begin{pmatrix} 1 \\ 2 \end{pmatrix} e^{-5t}$

12. $\dfrac{dx}{dt} = -2x + 5y$

$\dfrac{dy}{dt} = -2x + 4y; \quad \mathbf{X} = \begin{pmatrix} 5 \cos t \\ 3 \cos t - \sin t \end{pmatrix} e^t$

13. $\mathbf{X}' = \begin{pmatrix} -1 & \frac{1}{4} \\ 1 & -1 \end{pmatrix} \mathbf{X}; \quad \mathbf{X} = \begin{pmatrix} -1 \\ 2 \end{pmatrix} e^{-3t/2}$

14. $\mathbf{X}' = \begin{pmatrix} 2 & 1 \\ -1 & 0 \end{pmatrix} \mathbf{X}; \quad \mathbf{X} = \begin{pmatrix} 1 \\ 3 \end{pmatrix} e^t + \begin{pmatrix} 4 \\ -4 \end{pmatrix} t e^t$

15. $\mathbf{X}' = \begin{pmatrix} 1 & 2 & 1 \\ 6 & -1 & 0 \\ -1 & -2 & -1 \end{pmatrix} \mathbf{X}; \quad \mathbf{X} = \begin{pmatrix} 1 \\ 6 \\ -13 \end{pmatrix}$

16. $\mathbf{X}' = \begin{pmatrix} 1 & 0 & 1 \\ 1 & 1 & 0 \\ -2 & 0 & -1 \end{pmatrix} \mathbf{X}; \quad \mathbf{X} = \begin{pmatrix} \sin t \\ -\frac{1}{2} \sin t - \frac{1}{2} \cos t \\ -\sin t + \cos t \end{pmatrix}$

In Problems 17–20 the given vectors are solutions of a system $\mathbf{X}' = \mathbf{A}\mathbf{X}$. Determine whether the vectors form a fundamental set on $(-\infty, \infty)$.

17. $\mathbf{X}_1 = \begin{pmatrix} 1 \\ 1 \end{pmatrix} e^{-2t}, \quad \mathbf{X}_2 = \begin{pmatrix} 1 \\ -1 \end{pmatrix} e^{-6t}$

18. $\mathbf{X}_1 = \begin{pmatrix} 1 \\ -1 \end{pmatrix} e^t, \quad \mathbf{X}_2 = \begin{pmatrix} 2 \\ 6 \end{pmatrix} e^t + \begin{pmatrix} 8 \\ -8 \end{pmatrix} t e^t$

19. $\mathbf{X}_1 = \begin{pmatrix} 1 \\ -2 \\ 4 \end{pmatrix} + t \begin{pmatrix} 1 \\ 2 \\ 2 \end{pmatrix}, \quad \mathbf{X}_2 = \begin{pmatrix} 1 \\ -2 \\ 4 \end{pmatrix}, \quad \mathbf{X}_3 = \begin{pmatrix} 3 \\ -6 \\ 12 \end{pmatrix} + t \begin{pmatrix} 2 \\ 4 \\ 4 \end{pmatrix}$

20. $\mathbf{X}_1 = \begin{pmatrix} 1 \\ 6 \\ -13 \end{pmatrix}, \quad \mathbf{X}_2 = \begin{pmatrix} 1 \\ -2 \\ -1 \end{pmatrix} e^{-4t}, \quad \mathbf{X}_3 = \begin{pmatrix} 2 \\ 3 \\ -2 \end{pmatrix} e^{3t}$

In Problems 21–24 verify that the vector $\mathbf{X}_p$ is a particular solution of the given system.

21. $\dfrac{dx}{dt} = x + 4y + 2t - 7$

$\dfrac{dy}{dt} = 3x + 2y - 4t - 18; \quad \mathbf{X}_p = \begin{pmatrix} 2 \\ -1 \end{pmatrix} t + \begin{pmatrix} 5 \\ 1 \end{pmatrix}$

22. $\mathbf{X}' = \begin{pmatrix} 2 & 1 \\ 1 & -1 \end{pmatrix} \mathbf{X} + \begin{pmatrix} -5 \\ 2 \end{pmatrix}; \quad \mathbf{X}_p = \begin{pmatrix} 1 \\ 3 \end{pmatrix}$

23. $\mathbf{X}' = \begin{pmatrix} 2 & 1 \\ 3 & 4 \end{pmatrix} \mathbf{X} - \begin{pmatrix} 1 \\ 7 \end{pmatrix} e^t; \quad \mathbf{X}_p = \begin{pmatrix} 1 \\ 1 \end{pmatrix} e^t + \begin{pmatrix} 1 \\ -1 \end{pmatrix} te^t$

24. $\mathbf{X}' = \begin{pmatrix} 1 & 2 & 3 \\ -4 & 2 & 0 \\ -6 & 1 & 0 \end{pmatrix} \mathbf{X} + \begin{pmatrix} -1 \\ 4 \\ 3 \end{pmatrix} \sin 3t; \quad \mathbf{X}_p = \begin{pmatrix} \sin 3t \\ 0 \\ \cos 3t \end{pmatrix}$

25. Prove that the general solution of

$$\mathbf{X}' = \begin{pmatrix} 0 & 6 & 0 \\ 1 & 0 & 1 \\ 1 & 1 & 0 \end{pmatrix} \mathbf{X}$$

on the interval $(-\infty, \infty)$ is

$$\mathbf{X} = c_1 \begin{pmatrix} 6 \\ -1 \\ -5 \end{pmatrix} e^{-t} + c_2 \begin{pmatrix} -3 \\ 1 \\ 1 \end{pmatrix} e^{-2t} + c_3 \begin{pmatrix} 2 \\ 1 \\ 1 \end{pmatrix} e^{3t}.$$

26. Prove that the general solution of

$$\mathbf{X}' = \begin{pmatrix} -1 & -1 \\ -1 & 1 \end{pmatrix} \mathbf{X} + \begin{pmatrix} 1 \\ 1 \end{pmatrix} t^2 + \begin{pmatrix} 4 \\ -6 \end{pmatrix} t + \begin{pmatrix} -1 \\ 5 \end{pmatrix}$$

on the interval $(-\infty, \infty)$ is

$$\mathbf{X} = c_1 \begin{pmatrix} 1 \\ -1 - \sqrt{2} \end{pmatrix} e^{\sqrt{2}t} + c_2 \begin{pmatrix} 1 \\ -1 + \sqrt{2} \end{pmatrix} e^{-\sqrt{2}t} + \begin{pmatrix} 1 \\ 0 \end{pmatrix} t^2 + \begin{pmatrix} -2 \\ 4 \end{pmatrix} t + \begin{pmatrix} 1 \\ 0 \end{pmatrix}.$$

8.2 HOMOGENEOUS LINEAR SYSTEMS WITH CONSTANT COEFFICIENTS

■ *Characteristic equation of a square matrix* ■ *Eigenvalues of a matrix* ■ *Eigenvectors*
■ *Forms of the general solution of a homogeneous linear system with constant coefficients*

8.2.1   Distinct Real Eigenvalues

We saw in Example 5 of Section 8.1 that the general solution of the homogeneous system $\mathbf{X}' = \begin{pmatrix} 1 & 3 \\ 5 & 3 \end{pmatrix}\mathbf{X}$ is $\mathbf{X} = c_1 \begin{pmatrix} 1 \\ -1 \end{pmatrix} e^{-2t} + c_2 \begin{pmatrix} 3 \\ 5 \end{pmatrix} e^{6t}$. Since both solution vectors have the form $\mathbf{X}_i = \begin{pmatrix} k_1 \\ k_2 \end{pmatrix} e^{\lambda_i t}$, $i = 1, 2$, where k_1 and k_2 are constants, we are prompted to ask whether we can always find a solution of the form

$$\mathbf{X} = \begin{pmatrix} k_1 \\ k_2 \\ \vdots \\ k_n \end{pmatrix} e^{\lambda t} = \mathbf{K}e^{\lambda t} \tag{1}$$

for the general homogeneous linear first-order system

$$\mathbf{X}' = \mathbf{AX}, \tag{2}$$

where $\mathbf{A}$ is an $n \times n$ matrix of constants.

Eigenvalues and Eigenvectors   If (1) is to be a solution vector of (2), then $\mathbf{X}' = \mathbf{K}\lambda e^{\lambda t}$ so that the system becomes

$$\mathbf{K}\lambda e^{\lambda t} = \mathbf{AK}e^{\lambda t}.$$

After dividing out $e^{\lambda t}$ and rearranging, we obtain $\mathbf{AK} = \lambda\mathbf{K}$ or

$$(\mathbf{A} - \lambda\mathbf{I})\mathbf{K} = \mathbf{0}. \tag{3}$$

Equation (3) is equivalent to the simultaneous algebraic equations

$$\begin{aligned}
(a_{11} - \lambda)k_1 + \quad a_{12}k_2 + \cdots + \quad a_{1n}k_n &= 0 \\
a_{21}k_1 + (a_{22} - \lambda)k_2 + \cdots + \quad a_{2n}k_n &= 0 \\
\vdots \qquad\qquad \vdots \\
a_{n1}k_1 + \quad a_{n2}k_2 + \cdots + (a_{nn} - \lambda)k_n &= 0.
\end{aligned}$$

Thus to find a nontrivial solution $\mathbf{X}$ of (2) we must first find a nontrivial solution of the foregoing system; in other words, we must find a nontrivial vector $\mathbf{K}$ that satisfies (3). But in order for (3) to have nontrivial solutions we must have

$$\det(\mathbf{A} - \lambda\mathbf{I}) = 0.$$

The last equation is the **characteristic equation** of the matrix $\mathbf{A}$. In other words, $\mathbf{X} = \mathbf{K}e^{\lambda t}$ will be a solution of the system of differential equations

(2) if and only if λ is an **eigenvalue** of $\mathbf{A}$ and $\mathbf{K}$ is an **eigenvector** corresponding to λ.

When the $n \times n$ matrix $\mathbf{A}$ possesses n distinct real eigenvalues λ_1, $\lambda_2, \ldots, \lambda_n$, then a set of n linearly independent eigenvectors $\mathbf{K}_1, \mathbf{K}_2, \ldots, \mathbf{K}_n$ can always be found and

$$\mathbf{X}_1 = \mathbf{K}_1 e^{\lambda_1 t}, \quad \mathbf{X}_2 = \mathbf{K}_2 e^{\lambda_2 t}, \quad \ldots, \quad \mathbf{X}_n = \mathbf{K}_n e^{\lambda_n t}$$

is a fundamental set of solutions of (2) on $(-\infty, \infty)$.

THEOREM 8.7 **General Solution–Homogeneous Systems**

Let $\lambda_1, \lambda_2, \ldots, \lambda_n$ be n distinct real eigenvalues of the coefficient matrix $\mathbf{A}$ of the homogeneous system (2), and let $\mathbf{K}_1, \mathbf{K}_2, \ldots, \mathbf{K}_n$ be the corresponding eigenvectors. Then the **general solution** of (2) on the interval $(-\infty, \infty)$ is given by

$$\mathbf{X} = c_1 \mathbf{K}_1 e^{\lambda_1 t} + c_2 \mathbf{K}_2 e^{\lambda_2 t} + \cdots + c_n \mathbf{K}_n e^{\lambda_n t}.$$

EXAMPLE 1 **Distinct Eigenvalues**

Solve

$$\frac{dx}{dt} = 2x + 3y$$

$$\frac{dy}{dt} = 2x + y. \tag{4}$$

SOLUTION We first find the eigenvalues and eigenvectors of the matrix of coefficients.

From the characteristic equation

$$\det(\mathbf{A} - \lambda \mathbf{I}) = \begin{vmatrix} 2 - \lambda & 3 \\ 2 & 1 - \lambda \end{vmatrix} = \lambda^2 - 3\lambda - 4 = (\lambda + 1)(\lambda - 4) = 0$$

we see that the eigenvalues are $\lambda_1 = -1$ and $\lambda_2 = 4$.

Now for $\lambda_1 = -1$, (3) is equivalent to

$$3k_1 + 3k_2 = 0$$
$$2k_1 + 2k_2 = 0.$$

Thus $k_1 = -k_2$. When $k_2 = -1$, the related eigenvector is

$$\mathbf{K}_1 = \begin{pmatrix} 1 \\ -1 \end{pmatrix}.$$

For $\lambda_2 = 4$, we have
$$-2k_1 + 3k_2 = 0$$
$$2k_1 - 3k_2 = 0$$

so that $k_1 = 3k_2/2$, and therefore with $k_2 = 2$, the corresponding eigenvector is

$$\mathbf{K}_2 = \begin{pmatrix} 3 \\ 2 \end{pmatrix}.$$

Since the matrix of coefficients $\mathbf{A}$ is a 2×2 matrix and since we have found two linearly independent solutions of (4),

$$\mathbf{X}_1 = \begin{pmatrix} 1 \\ -1 \end{pmatrix} e^{-t} \quad \text{and} \quad \mathbf{X}_2 = \begin{pmatrix} 3 \\ 2 \end{pmatrix} e^{4t},$$

we conclude that the general solution of the system is

$$\mathbf{X} = c_1 \mathbf{X}_1 + c_2 \mathbf{X}_2 = c_1 \begin{pmatrix} 1 \\ -1 \end{pmatrix} e^{-t} + c_2 \begin{pmatrix} 3 \\ 2 \end{pmatrix} e^{4t}. \tag{5}$$ ∎

For the sake of review, you should keep firmly in mind that a solution of a system of first-order differential equations, when written in terms of matrices, is simply an alternative to the method that we employed in Section 4.8—namely, listing the individual functions and the relationships between the constants. If we add the vectors on the right side in (5) and equate the entries with the corresponding entries in the vector on the left, we obtain the more familiar statement

$$x(t) = c_1 e^{-t} + 3 c_2 e^{4t}$$
$$y(t) = -c_1 e^{-t} + 2 c_2 e^{4t}.$$

EXAMPLE 2   **Distinct Eigenvalues**

Solve

$$\frac{dx}{dt} = -4x + y + z$$

$$\frac{dy}{dt} = x + 5y - z \tag{6}$$

$$\frac{dz}{dt} = y - 3z.$$

SOLUTION   Using the cofactors of the third row, we find

$$\det(\mathbf{A} - \lambda \mathbf{I}) = \begin{vmatrix} -4 - \lambda & 1 & 1 \\ 1 & 5 - \lambda & -1 \\ 0 & 1 & -3 - \lambda \end{vmatrix} = -(\lambda + 3)(\lambda + 4)(\lambda - 5) = 0,$$

and so the eigenvalues are $\lambda_1 = -3$, $\lambda_2 = -4$, $\lambda_3 = 5$.

For $\lambda_1 = -3$, Gauss-Jordan elimination gives

$$(\mathbf{A} + 3\mathbf{I} | \mathbf{0}) = \begin{pmatrix} -1 & 1 & 1 & | & 0 \\ 1 & 8 & -1 & | & 0 \\ 0 & 1 & 0 & | & 0 \end{pmatrix} \xrightarrow[\text{operations}]{\text{row}} \begin{pmatrix} 1 & 0 & -1 & | & 0 \\ 0 & 1 & 0 & | & 0 \\ 0 & 0 & 0 & | & 0 \end{pmatrix}.$$

Therefore $k_1 = k_3$ and $k_2 = 0$. The choice $k_3 = 1$ gives an eigenvector and corresponding solution vector

$$\mathbf{K}_1 = \begin{pmatrix} 1 \\ 0 \\ 1 \end{pmatrix}, \quad \mathbf{X}_1 = \begin{pmatrix} 1 \\ 0 \\ 1 \end{pmatrix} e^{-3t}. \tag{7}$$

Similarly, for $\lambda_2 = -4$,

$$(\mathbf{A} + 4\mathbf{I}|\mathbf{0}) = \begin{pmatrix} 0 & 1 & 1 & | & 0 \\ 1 & 9 & -1 & | & 0 \\ 0 & 1 & 1 & | & 0 \end{pmatrix} \xrightarrow[\text{operations}]{\text{row}} \begin{pmatrix} 1 & 0 & -10 & | & 0 \\ 0 & 1 & 1 & | & 0 \\ 0 & 0 & 0 & | & 0 \end{pmatrix}$$

implies $k_1 = 10k_3$ and $k_2 = -k_3$. Choosing $k_3 = 1$, we get a second eigenvector and solution vector

$$\mathbf{K}_2 = \begin{pmatrix} 10 \\ -1 \\ 1 \end{pmatrix}, \qquad \mathbf{X}_2 = \begin{pmatrix} 10 \\ -1 \\ 1 \end{pmatrix} e^{-4t}. \tag{8}$$

Finally, when $\lambda_3 = 5$, the augmented matrices

$$(\mathbf{A} - 5\mathbf{I}|\mathbf{0}) = \begin{pmatrix} -9 & 1 & 1 & | & 0 \\ 1 & 0 & -1 & | & 0 \\ 0 & 1 & -8 & | & 0 \end{pmatrix} \xrightarrow[\text{operations}]{\text{row}} \begin{pmatrix} 1 & 0 & -1 & | & 0 \\ 0 & 1 & -8 & | & 0 \\ 0 & 0 & 0 & | & 0 \end{pmatrix}$$

yield
$$\mathbf{K}_3 = \begin{pmatrix} 1 \\ 8 \\ 1 \end{pmatrix}, \qquad \mathbf{X}_3 = \begin{pmatrix} 1 \\ 8 \\ 1 \end{pmatrix} e^{5t}. \tag{9}$$

The general solution of (6) is a linear combination of the solution vectors in (7), (8), and (9):

$$\mathbf{X} = c_1 \begin{pmatrix} 1 \\ 0 \\ 1 \end{pmatrix} e^{-3t} + c_2 \begin{pmatrix} 10 \\ -1 \\ 1 \end{pmatrix} e^{-4t} + c_3 \begin{pmatrix} 1 \\ 8 \\ 1 \end{pmatrix} e^{5t}. \qquad \blacksquare$$

Use of Computers   Software packages such as MATLAB, Mathematica, Maple, and DERIVE can be real time savers in finding eigenvalues and eigenvectors of a matrix. For example, to find the eigenvalues and eigenvectors of the matrix of coefficients in (6) using Mathematica we first input the definition of the matrix by rows:

$$\mathbf{m = \{\{-4, 1, 1\}, \{1, 5, -1\}, \{0, 1, -3\}\}.}$$

The commands

$$\mathbf{Eigenvalues[m]} \quad \text{and} \quad \mathbf{Eigenvectors[m]}$$

given in sequence yield

$$\{-4, -3, 5\} \quad \text{and} \quad \{\{10, -1, 1\}, \{1, 0, 1\}, \{1, 8, 1\}\},$$

respectively. In Mathematica eigenvalues and eigenvectors can also be obtained at the same time by using **Eigensystem[m].**

8.2.2   Repeated Eigenvalues

Of course, not all of the n eigenvalues $\lambda_1, \lambda_2, \ldots, \lambda_n$ of an $n \times n$ matrix $\mathbf{A}$ need be distinct; that is, some of the eigenvalues may be repeated. For example, the characteristic equation of the coefficient matrix in the system

$$\mathbf{X}' = \begin{pmatrix} 3 & -18 \\ 2 & -9 \end{pmatrix} \mathbf{X} \tag{10}$$

is readily shown to be $(\lambda + 3)^2 = 0$, and therefore $\lambda_1 = \lambda_2 = -3$ is a root of *multiplicity two*. For this value we find the single eigenvector

$$\mathbf{K}_1 = \begin{pmatrix} 3 \\ 1 \end{pmatrix}, \qquad \text{so} \qquad \mathbf{X}_1 = \begin{pmatrix} 3 \\ 1 \end{pmatrix} e^{-3t} \tag{11}$$

is one solution of (10). But since we are obviously interested in forming the general solution of the system, we need to pursue the question of finding a second solution.

In general, if m is a positive integer and $(\lambda - \lambda_1)^m$ is a factor of the characteristic equation while $(\lambda - \lambda_1)^{m+1}$ is not a factor, then λ_1 is said to be an **eigenvalue of multiplicity m**. The next three examples illustrate the following cases:

(*i*) For some $n \times n$ matrices $\mathbf{A}$ it may be possible to find m linearly independent eigenvectors $\mathbf{K}_1, \mathbf{K}_2, \ldots, \mathbf{K}_m$ corresponding to an eigenvalue λ_1 of multiplicity $m \le n$. In this case the general solution of the system contains the linear combination

$$c_1 \mathbf{K}_1 e^{\lambda_1 t} + c_2 \mathbf{K}_2 e^{\lambda_1 t} + \cdots + c_m \mathbf{K}_m e^{\lambda_1 t}.$$

(*ii*) If there is only one eigenvector corresponding to the eigenvalue λ_1 of multiplicity m, then m linearly independent solutions of the form

$$\mathbf{X}_1 = \mathbf{K}_{11} e^{\lambda_1 t}$$
$$\mathbf{X}_2 = \mathbf{K}_{21} t e^{\lambda_1 t} + \mathbf{K}_{22} e^{\lambda_1 t}$$
$$\vdots$$
$$\mathbf{X}_m = \mathbf{K}_{m1} \frac{t^{m-1}}{(m-1)!} e^{\lambda_1 t} + \mathbf{K}_{m2} \frac{t^{m-2}}{(m-2)!} e^{\lambda_1 t} + \cdots + \mathbf{K}_{mm} e^{\lambda_1 t},$$

where $\mathbf{K}_{ij}$ are column vectors, can always be found.

Eigenvalue of Multiplicity Two   We begin by considering eigenvalues of multiplicity two. In the first example we illustrate a matrix for which we can find two distinct eigenvectors corresponding to a double eigenvalue.

EXAMPLE 3   **Repeated Eigenvalues**

Solve $\mathbf{X}' = \begin{pmatrix} 1 & -2 & 2 \\ -2 & 1 & -2 \\ 2 & -2 & 1 \end{pmatrix} \mathbf{X}$.

SOLUTION   Expanding the determinant in the characteristic equation

$$\det(\mathbf{A} - \lambda \mathbf{I}) = \begin{vmatrix} 1 - \lambda & -2 & 2 \\ -2 & 1 - \lambda & -2 \\ 2 & -2 & 1 - \lambda \end{vmatrix} = 0$$

yields $-(\lambda + 1)^2(\lambda - 5) = 0$. We see that $\lambda_1 = \lambda_2 = -1$ and $\lambda_3 = 5$.

For $\lambda_1 = -1$, Gauss-Jordan elimination immediately gives

$$(\mathbf{A} + \mathbf{I}|\mathbf{0}) = \begin{pmatrix} 2 & -2 & 2 & | & 0 \\ -2 & 2 & -2 & | & 0 \\ 2 & -2 & 2 & | & 0 \end{pmatrix} \xrightarrow[\text{operations}]{\text{row}} \begin{pmatrix} 1 & -1 & 1 & | & 0 \\ 0 & 0 & 0 & | & 0 \\ 0 & 0 & 0 & | & 0 \end{pmatrix}.$$

The first row of the last matrix means $k_1 - k_2 + k_3 = 0$ or $k_1 = k_2 - k_3$. The choices $k_2 = 1$, $k_3 = 0$ and $k_2 = 1$, $k_3 = 1$ yield, in turn, $k_1 = 1$ and $k_1 = 0$. Thus two eigenvectors corresponding to $\lambda_1 = -1$ are

$$\mathbf{K}_1 = \begin{pmatrix} 1 \\ 1 \\ 0 \end{pmatrix} \quad \text{and} \quad \mathbf{K}_2 = \begin{pmatrix} 0 \\ 1 \\ 1 \end{pmatrix}.$$

Since neither eigenvector is a constant multiple of the other, we have found, corresponding to the same eigenvalue, two linearly independent solutions

$$\mathbf{X}_1 = \begin{pmatrix} 1 \\ 1 \\ 0 \end{pmatrix} e^{-t} \quad \text{and} \quad \mathbf{X}_2 = \begin{pmatrix} 0 \\ 1 \\ 1 \end{pmatrix} e^{-t}.$$

Last, for $\lambda_3 = 5$, the reduction

$$(\mathbf{A} - 5\mathbf{I}|\mathbf{0}) = \begin{pmatrix} -4 & -2 & 2 & | & 0 \\ -2 & -4 & -2 & | & 0 \\ 2 & -2 & -4 & | & 0 \end{pmatrix} \xrightarrow[\text{operations}]{\text{row}} \begin{pmatrix} 1 & 0 & -1 & | & 0 \\ 0 & 1 & 1 & | & 0 \\ 0 & 0 & 0 & | & 0 \end{pmatrix}$$

implies $k_1 = k_3$ and $k_2 = -k_3$. Picking $k_3 = 1$ gives $k_1 = 1$, $k_2 = -1$, and thus a third eigenvector is

$$\mathbf{K}_3 = \begin{pmatrix} 1 \\ -1 \\ 1 \end{pmatrix}.$$

We conclude that the general solution of the system is

$$\mathbf{X} = c_1 \begin{pmatrix} 1 \\ 1 \\ 0 \end{pmatrix} e^{-t} + c_2 \begin{pmatrix} 0 \\ 1 \\ 1 \end{pmatrix} e^{-t} + c_3 \begin{pmatrix} 1 \\ -1 \\ 1 \end{pmatrix} e^{5t}. \qquad \blacksquare$$

The matrix of coefficients $\mathbf{A}$ in Example 3 is a special kind of matrix known as a symmetric matrix. An $n \times n$ matrix $\mathbf{A}$ is said to be **symmetric** if its transpose $\mathbf{A}^T$ (where the rows and columns are interchanged) is the same as $\mathbf{A}$, that is, if $\mathbf{A}^T = \mathbf{A}$. It can be proved that if the matrix $\mathbf{A}$ in the system $\mathbf{X}' = \mathbf{A}\mathbf{X}$ is symmetric and has real entries, then we can always find n linearly independent eigenvectors $\mathbf{K}_1, \mathbf{K}_2, \ldots, \mathbf{K}_n$, and the general solution of such a system is as given in Theorem 8.7. As illustrated in Example 3, this result holds even when some of the eigenvalues are repeated.

Second Solution   Now suppose that λ_1 is an eigenvalue of multiplicity two and that there is only one eigenvector associated with this value. A second solution can be found of the form

$$\mathbf{X}_2 = \mathbf{K}te^{\lambda_1 t} + \mathbf{P}e^{\lambda_1 t}, \tag{12}$$

where

$$\mathbf{K} = \begin{pmatrix} k_1 \\ k_2 \\ \vdots \\ k_n \end{pmatrix} \quad \text{and} \quad \mathbf{P} = \begin{pmatrix} p_1 \\ p_2 \\ \vdots \\ p_n \end{pmatrix}.$$

To see this we substitute (12) into the system $\mathbf{X}' = \mathbf{A}\mathbf{X}$ and simplify:

$$(\mathbf{A}\mathbf{K} - \lambda_1 \mathbf{K})te^{\lambda_1 t} + (\mathbf{A}\mathbf{P} - \lambda_1 \mathbf{P} - \mathbf{K})e^{\lambda_1 t} = \mathbf{0}.$$

Since this last equation is to hold for all values of t, we must have

$$(\mathbf{A} - \lambda_1 \mathbf{I})\mathbf{K} = \mathbf{0} \tag{13}$$

and

$$(\mathbf{A} - \lambda_1 \mathbf{I})\mathbf{P} = \mathbf{K}. \tag{14}$$

Equation (13) simply states that $\mathbf{K}$ must be an eigenvector of $\mathbf{A}$ associated with λ_1. By solving (13), we find one solution $\mathbf{X}_1 = \mathbf{K}e^{\lambda_1 t}$. To find the second solution $\mathbf{X}_2$ we need only solve the additional system (14) for the vector $\mathbf{P}$.

EXAMPLE 4   Repeated Eigenvalues

Find the general solution of the system given in (10).

SOLUTION   From (11) we know that $\lambda_1 = -3$ and that one solution is $\mathbf{X}_1 = \begin{pmatrix} 3 \\ 1 \end{pmatrix} e^{-3t}$. Identifying $\mathbf{K} = \begin{pmatrix} 3 \\ 1 \end{pmatrix}$ and $\mathbf{P} = \begin{pmatrix} p_1 \\ p_2 \end{pmatrix}$, we find from (14) that we must now solve

$$(\mathbf{A} + 3\mathbf{I})\mathbf{P} = \mathbf{K} \qquad \text{or} \qquad \begin{aligned} 6p_1 - 18p_2 &= 3 \\ 2p_1 - 6p_2 &= 1. \end{aligned}$$

Since this system is obviously equivalent to one equation, we have an infinite number of choices for p_1 and p_2. For example, by choosing $p_1 = 1$ we find $p_2 = \frac{1}{6}$. However, for simplicity, we shall choose $p_1 = \frac{1}{2}$ so that $p_2 = 0$. Hence $\mathbf{P} = \begin{pmatrix} \frac{1}{2} \\ 0 \end{pmatrix}$. Thus from (12) we find

$$\mathbf{X}_2 = \begin{pmatrix} 3 \\ 1 \end{pmatrix} te^{-3t} + \begin{pmatrix} \frac{1}{2} \\ 0 \end{pmatrix} e^{-3t}.$$

The general solution of (10) is then

$$\mathbf{X} = c_1 \begin{pmatrix} 3 \\ 1 \end{pmatrix} e^{-3t} + c_2 \left[\begin{pmatrix} 3 \\ 1 \end{pmatrix} te^{-3t} + \begin{pmatrix} \frac{1}{2} \\ 0 \end{pmatrix} e^{-3t} \right]. \qquad \blacksquare$$

Eigenvalues of Multiplicity Three   When a matrix $\mathbf{A}$ has only one eigenvector associated with an eigenvalue λ_1 of multiplicity three, we can find a solution of the form (12) and a third solution of the form

$$\mathbf{X}_3 = \mathbf{K}\frac{t^2}{2}e^{\lambda_1 t} + \mathbf{P}te^{\lambda_1 t} + \mathbf{Q}e^{\lambda_1 t}, \tag{15}$$

where $\quad \mathbf{K} = \begin{pmatrix} k_1 \\ k_2 \\ \vdots \\ k_n \end{pmatrix}, \quad \mathbf{P} = \begin{pmatrix} p_1 \\ p_2 \\ \vdots \\ p_n \end{pmatrix}, \quad$ and $\quad \mathbf{Q} = \begin{pmatrix} q_1 \\ q_2 \\ \vdots \\ q_n \end{pmatrix}.$

By substituting (15) into the system $\mathbf{X}' = \mathbf{AX}$ we find that the column vectors $\mathbf{K}$, $\mathbf{P}$, and $\mathbf{Q}$ must satisfy

$$(\mathbf{A} - \lambda_1\mathbf{I})\mathbf{K} = \mathbf{0} \tag{16}$$

$$(\mathbf{A} - \lambda_1\mathbf{I})\mathbf{P} = \mathbf{K} \tag{17}$$

and $\qquad (\mathbf{A} - \lambda_1\mathbf{I})\mathbf{Q} = \mathbf{P}. \tag{18}$

Of course, the solutions of (16) and (17) can be used in forming the solutions $\mathbf{X}_1$ and $\mathbf{X}_2$.

EXAMPLE 5 Repeated Eigenvalues

Solve $\mathbf{X}' = \begin{pmatrix} 2 & 1 & 6 \\ 0 & 2 & 5 \\ 0 & 0 & 2 \end{pmatrix}\mathbf{X}.$

SOLUTION The characteristic equation $(\lambda - 2)^3 = 0$ shows that $\lambda_1 = 2$ is an eigenvalue of multiplicity three. By solving $(\mathbf{A} - 2\mathbf{I})\mathbf{K} = \mathbf{0}$ we find the single eigenvector

$$\mathbf{K} = \begin{pmatrix} 1 \\ 0 \\ 0 \end{pmatrix}.$$

We next solve the systems $(\mathbf{A} - 2\mathbf{I})\mathbf{P} = \mathbf{K}$ and $(\mathbf{A} - 2\mathbf{I})\mathbf{Q} = \mathbf{P}$ in succession and find that

$$\mathbf{P} = \begin{pmatrix} 0 \\ 1 \\ 0 \end{pmatrix} \quad \text{and} \quad \mathbf{Q} = \begin{pmatrix} 0 \\ -\frac{6}{5} \\ \frac{1}{5} \end{pmatrix}.$$

Using (12) and (15), we see that the general solution of the system is

$$\mathbf{X} = c_1\begin{pmatrix} 1 \\ 0 \\ 0 \end{pmatrix}e^{2t} + c_2\left[\begin{pmatrix} 1 \\ 0 \\ 0 \end{pmatrix}te^{2t} + \begin{pmatrix} 0 \\ 1 \\ 0 \end{pmatrix}e^{2t}\right] + c_3\left[\begin{pmatrix} 1 \\ 0 \\ 0 \end{pmatrix}\frac{t^2}{2}e^{2t} + \begin{pmatrix} 0 \\ 1 \\ 0 \end{pmatrix}te^{2t} + \begin{pmatrix} 0 \\ -\frac{6}{5} \\ \frac{1}{5} \end{pmatrix}e^{2t}\right]. \quad \blacksquare$$

> **Remark**
>
> When an eigenvalue λ_1 has multiplicity m, either we can find m linearly independent eigenvectors or the number of corresponding eigenvectors is less than m. Hence the two cases listed on page 334 are not all the possibilities under which a repeated eigenvalue can occur. It could happen, say, that a 5×5 matrix has an eigenvalue of multiplicity five and there exist three linearly independent eigenvectors. (See Problems 29 and 30 in Exercises 8.2.)

8.2.3 Complex Eigenvalues

If $\lambda_1 = \alpha + i\beta$ and $\lambda_2 = \alpha - i\beta$, $i^2 = -1$ are complex eigenvalues of the coefficient matrix $\mathbf{A}$, we can then certainly expect their corresponding eigenvectors to also have complex entries.*

For example, the characteristic equation of the system

$$\frac{dx}{dt} = 6x - y$$

$$\frac{dy}{dt} = 5x + 4y \tag{19}$$

is

$$\det(\mathbf{A} - \lambda\mathbf{I}) = \begin{vmatrix} 6 - \lambda & -1 \\ 5 & 4 - \lambda \end{vmatrix} = \lambda^2 - 10\lambda + 29 = 0.$$

From the quadratic formula we find $\lambda_1 = 5 + 2i$, $\lambda_2 = 5 - 2i$.

Now for $\lambda_1 = 5 + 2i$ we must solve

$$(1 - 2i)k_1 - k_2 = 0$$
$$5k_1 - (1 + 2i)k_2 = 0.$$

Since $k_2 = (1 - 2i)k_1$,† the choice $k_1 = 1$ gives the following eigenvector and a solution vector:

$$\mathbf{K}_1 = \begin{pmatrix} 1 \\ 1 - 2i \end{pmatrix}, \qquad \mathbf{X}_1 = \begin{pmatrix} 1 \\ 1 - 2i \end{pmatrix} e^{(5+2i)t}.$$

In like manner, for $\lambda_2 = 5 - 2i$ we find

$$\mathbf{K}_2 = \begin{pmatrix} 1 \\ 1 + 2i \end{pmatrix}, \qquad \mathbf{X}_2 = \begin{pmatrix} 1 \\ 1 + 2i \end{pmatrix} e^{(5-2i)t}.$$

We can verify by means of the Wronskian that these solution vectors are linearly independent, and so the general solution of (19) is

$$\mathbf{X} = c_1 \begin{pmatrix} 1 \\ 1 - 2i \end{pmatrix} e^{(5+2i)t} + c_2 \begin{pmatrix} 1 \\ 1 + 2i \end{pmatrix} e^{(5-2i)t}. \tag{20}$$

Note that the entries in $\mathbf{K}_2$ corresponding to λ_2 are the conjugates of the entries in $\mathbf{K}_1$ corresponding to λ_1. The conjugate of λ_1 is, of course,

---

*When the characteristic equation has real coefficients, complex eigenvalues always appear in conjugate pairs.

†Note that the second equation is simply $(1 + 2i)$ times the first.

λ_2. We write this as $\lambda_2 = \overline{\lambda}_1$ and $\mathbf{K}_2 = \overline{\mathbf{K}}_1$. We have illustrated the following general result.

THEOREM 8.8   **Solutions Corresponding to a Complex Eigenvalue**

Let $\mathbf{A}$ be the coefficient matrix having real entries of the homogeneous system (2), and let $\mathbf{K}_1$ be an eigenvector corresponding to the complex eigenvalue $\lambda_1 = \alpha + i\beta$, α and β real. Then

$$\mathbf{K}_1 e^{\lambda_1 t} \qquad \text{and} \qquad \overline{\mathbf{K}}_1 e^{\overline{\lambda}_1 t}$$

are solutions of (2).

It is desirable and relatively easy to rewrite a solution such as (20) in terms of real functions. To this end we first use Euler's formula to write

$$e^{(5+2i)t} = e^{5t} e^{2ti} = e^{5t}(\cos 2t + i \sin 2t)$$
$$e^{(5-2i)t} = e^{5t} e^{-2ti} = e^{5t}(\cos 2t - i \sin 2t).$$

Then, after we multiply complex numbers, collect terms, and replace $c_1 + c_2$ by C_1 and $(c_1 - c_2)i$ by C_2, (20) becomes

$$\mathbf{X} = C_1 \mathbf{X}_1 + C_2 \mathbf{X}_2, \tag{21}$$

where

$$\mathbf{X}_1 = \left[\begin{pmatrix} 1 \\ 1 \end{pmatrix} \cos 2t - \begin{pmatrix} 0 \\ -2 \end{pmatrix} \sin 2t \right] e^{5t}$$

and

$$\mathbf{X}_2 = \left[\begin{pmatrix} 0 \\ -2 \end{pmatrix} \cos 2t + \begin{pmatrix} 1 \\ 1 \end{pmatrix} \sin 2t \right] e^{5t}.$$

It is now important to realize that the two vectors $\mathbf{X}_1$ and $\mathbf{X}_2$ in (21) are themselves linearly independent *real* solutions of the original system. Consequently we are justified in ignoring the relationship between C_1, C_2 and c_1, c_2, and we can regard C_1 and C_2 as completely arbitrary and real. In other words, the linear combination (21) is an alternative general solution of (19).

The foregoing process can be generalized. Let $\mathbf{K}_1$ be an eigenvector of the coefficient matrix $\mathbf{A}$ (with real entries) corresponding to the complex eigenvalue $\lambda_1 = \alpha + i\beta$. Then the two solution vectors in Theorem 8.8 can be written as

$$\mathbf{K}_1 e^{\lambda_1 t} = \mathbf{K}_1 e^{\alpha t} e^{i\beta t} = \mathbf{K}_1 e^{\alpha t}(\cos \beta t + i \sin \beta t)$$
$$\overline{\mathbf{K}}_1 e^{\overline{\lambda}_1 t} = \overline{\mathbf{K}}_1 e^{\alpha t} e^{-i\beta t} = \overline{\mathbf{K}}_1 e^{\alpha t}(\cos \beta t - i \sin \beta t).$$

By the superposition principle, Theorem 8.2, the following vectors are also solutions:

$$\mathbf{X}_1 = \frac{1}{2}(\mathbf{K}_1 e^{\lambda_1 t} + \overline{\mathbf{K}}_1 e^{\overline{\lambda}_1 t}) = \frac{1}{2}(\mathbf{K}_1 + \overline{\mathbf{K}}_1)e^{\alpha t} \cos \beta t - \frac{i}{2}(-\mathbf{K}_1 + \overline{\mathbf{K}}_1)e^{\alpha t} \sin \beta t$$

$$\mathbf{X}_2 = \frac{i}{2}(-\mathbf{K}_1 e^{\lambda_1 t} + \overline{\mathbf{K}}_1 e^{\overline{\lambda}_1 t}) = \frac{i}{2}(-\mathbf{K}_1 + \overline{\mathbf{K}}_1)e^{\alpha t} \cos \beta t + \frac{1}{2}(\mathbf{K}_1 + \overline{\mathbf{K}}_1)e^{\alpha t} \sin \beta t.$$

For *any* complex number $z = a + ib$, both $\frac{1}{2}(z + \bar{z}) = a$ and $\frac{i}{2}(-z + \bar{z}) = b$ are *real* numbers. Therefore, the entries in the column vectors $\frac{1}{2}(\mathbf{K}_1 + \overline{\mathbf{K}}_1)$ and $\frac{i}{2}(-\mathbf{K}_1 + \overline{\mathbf{K}}_1)$ are real numbers. By defining

$$\mathbf{B}_1 = \frac{1}{2}(\mathbf{K}_1 + \overline{\mathbf{K}}_1) \qquad \text{and} \qquad \mathbf{B}_2 = \frac{i}{2}(-\mathbf{K}_1 + \overline{\mathbf{K}}_1) \tag{22}$$

we are led to the following theorem.

THEOREM 8.9   **Real Solutions Corresponding to a Complex Eigenvalue**

Let $\lambda_1 = \alpha + i\beta$ be a complex eigenvalue of the coefficient matrix $\mathbf{A}$ in the homogeneous system (2), and let $\mathbf{B}_1$ and $\mathbf{B}_2$ denote the column vectors defined in (22). Then

$$\mathbf{X}_1 = [\mathbf{B}_1 \cos \beta t - \mathbf{B}_2 \sin \beta t]e^{\alpha t}$$
$$\mathbf{X}_2 = [\mathbf{B}_2 \cos \beta t + \mathbf{B}_1 \sin \beta t]e^{\alpha t} \tag{23}$$

are linearly independent solutions of (2) on $(-\infty, \infty)$.

The matrices $\mathbf{B}_1$ and $\mathbf{B}_2$ in (22) are often denoted by

$$\mathbf{B}_1 = \text{Re}(\mathbf{K}_1) \qquad \text{and} \qquad \mathbf{B}_2 = \text{Im}(\mathbf{K}_1) \tag{24}$$

since these vectors are, respectively, the *real* and *imaginary* parts of the eigenvector $\mathbf{K}_1$. For example, (21) follows from (23) with

$$\mathbf{K}_1 = \begin{pmatrix} 1 \\ 1 - 2i \end{pmatrix} = \begin{pmatrix} 1 \\ 1 \end{pmatrix} + i \begin{pmatrix} 0 \\ -2 \end{pmatrix}$$

$$\mathbf{B}_1 = \text{Re}(\mathbf{K}_1) = \begin{pmatrix} 1 \\ 1 \end{pmatrix} \qquad \text{and} \qquad \mathbf{B}_2 = \text{Im}(\mathbf{K}_1) = \begin{pmatrix} 0 \\ -2 \end{pmatrix}.$$

EXAMPLE 6   **Complex Eigenvalues**

Solve $\mathbf{X}' = \begin{pmatrix} 2 & 8 \\ -1 & -2 \end{pmatrix} \mathbf{X}$.

SOLUTION   First we obtain the eigenvalues from

$$\det(\mathbf{A} - \lambda\mathbf{I}) = \begin{vmatrix} 2 - \lambda & 8 \\ -1 & -2 - \lambda \end{vmatrix} = \lambda^2 + 4 = 0.$$

Thus the eigenvalues are $\lambda_1 = 2i$ and $\lambda_2 = \overline{\lambda}_1 = -2i$. For λ_1 the system

$$(2 - 2i)k_1 + \qquad\quad 8k_2 = 0$$
$$-k_1 + (-2 - 2i)k_2 = 0$$

gives $k_1 = -(2 + 2i)k_2$. By choosing $k_2 = -1$ we get

$$\mathbf{K}_1 = \begin{pmatrix} 2 + 2i \\ -1 \end{pmatrix} = \begin{pmatrix} 2 \\ -1 \end{pmatrix} + i \begin{pmatrix} 2 \\ 0 \end{pmatrix}.$$

Now from (24) we form

$$\mathbf{B}_1 = \mathrm{Re}(\mathbf{K}_1) = \begin{pmatrix} 2 \\ -1 \end{pmatrix} \qquad \text{and} \qquad \mathbf{B}_2 = \mathrm{Im}(\mathbf{K}_1) = \begin{pmatrix} 2 \\ 0 \end{pmatrix}.$$

Since $\alpha = 0$, it follows from (23) that the general solution of the system is

$$\mathbf{X} = c_1 \left[\begin{pmatrix} 2 \\ -1 \end{pmatrix} \cos 2t - \begin{pmatrix} 2 \\ 0 \end{pmatrix} \sin 2t \right] + c_2 \left[\begin{pmatrix} 2 \\ 0 \end{pmatrix} \cos 2t + \begin{pmatrix} 2 \\ -1 \end{pmatrix} \sin 2t \right]$$

$$= c_1 \begin{pmatrix} 2 \cos 2t - 2 \sin 2t \\ -\cos 2t \end{pmatrix} + c_2 \begin{pmatrix} 2 \cos 2t + 2 \sin 2t \\ -\sin 2t \end{pmatrix}. \qquad ∎$$

SECTION 8.2 EXERCISES

Answers to odd-numbered problems begin on page A-13.

8.2.1

In Problems 1–12 find the general solution of the given system.

1. $\dfrac{dx}{dt} = x + 2y$

$\dfrac{dy}{dt} = 4x + 3y$

2. $\dfrac{dx}{dt} = 2y$

$\dfrac{dy}{dt} = 8x$

3. $\dfrac{dx}{dt} = -4x + 2y$

$\dfrac{dy}{dt} = -\dfrac{5}{2}x + 2y$

4. $\dfrac{dx}{dt} = \dfrac{1}{2}x + 9y$

$\dfrac{dy}{dt} = \dfrac{1}{2}x + 2y$

5. $\mathbf{X}' = \begin{pmatrix} 10 & -5 \\ 8 & -12 \end{pmatrix} \mathbf{X}$

6. $\mathbf{X}' = \begin{pmatrix} -6 & 2 \\ -3 & 1 \end{pmatrix} \mathbf{X}$

7. $\dfrac{dx}{dt} = x + y - z$

$\dfrac{dy}{dt} = 2y$

$\dfrac{dz}{dt} = y - z$

8. $\dfrac{dx}{dt} = 2x - 7y$

$\dfrac{dy}{dt} = 5x + 10y + 4z$

$\dfrac{dz}{dt} = 5y + 2z$

9. $\mathbf{X}' = \begin{pmatrix} -1 & 1 & 0 \\ 1 & 2 & 1 \\ 0 & 3 & -1 \end{pmatrix} \mathbf{X}$

10. $\mathbf{X}' = \begin{pmatrix} 1 & 0 & 1 \\ 0 & 1 & 0 \\ 1 & 0 & 1 \end{pmatrix} \mathbf{X}$

11. $\mathbf{X}' = \begin{pmatrix} -1 & -1 & 0 \\ \frac{3}{4} & -\frac{3}{2} & 3 \\ \frac{1}{8} & \frac{1}{4} & -\frac{1}{2} \end{pmatrix} \mathbf{X}$

12. $\mathbf{X}' = \begin{pmatrix} -1 & 4 & 2 \\ 4 & -1 & -2 \\ 0 & 0 & 6 \end{pmatrix} \mathbf{X}$

In Problems 13 and 14 solve the given system subject to the indicated initial condition.

13. $\mathbf{X}' = \begin{pmatrix} \frac{1}{2} & 0 \\ 1 & -\frac{1}{2} \end{pmatrix} \mathbf{X}, \quad \mathbf{X}(0) = \begin{pmatrix} 3 \\ 5 \end{pmatrix}$

14. $\mathbf{X}' = \begin{pmatrix} 1 & 1 & 4 \\ 0 & 2 & 0 \\ 1 & 1 & 1 \end{pmatrix} \mathbf{X}, \quad \mathbf{X}(0) = \begin{pmatrix} 1 \\ 3 \\ 0 \end{pmatrix}$

In Problems 15 and 16 use a CAS or linear algebra software as an aid in finding the general solution of the given system.

15. $\mathbf{X}' = \begin{pmatrix} 0.9 & 2.1 & 3.2 \\ 0.7 & 6.5 & 4.2 \\ 1.1 & 1.7 & 3.4 \end{pmatrix} \mathbf{X}$

16. $\mathbf{X}' = \begin{pmatrix} 1 & 0 & 2 & -1.8 & 0 \\ 0 & 5.1 & 0 & -1 & 3 \\ 1 & 2 & -3 & 0 & 0 \\ 0 & 1 & -3.1 & 4 & 0 \\ -2.8 & 0 & 0 & 1.5 & 1 \end{pmatrix} \mathbf{X}$

8.2.2

In Problems 17–26 find the general solution of the given system.

17. $\dfrac{dx}{dt} = 3x - y$

$\dfrac{dy}{dt} = 9x - 3y$

18. $\dfrac{dx}{dt} = -6x + 5y$

$\dfrac{dy}{dt} = -5x + 4y$

19. $\dfrac{dx}{dt} = -x + 3y$

$\dfrac{dy}{dt} = -3x + 5y$

20. $\dfrac{dx}{dt} = 12x - 9y$

$\dfrac{dy}{dt} = 4x$

21. $\dfrac{dx}{dt} = 3x - y - z$

$\dfrac{dy}{dt} = x + y - z$

$\dfrac{dz}{dt} = x - y + z$

22. $\dfrac{dx}{dt} = 3x + 2y + 4z$

$\dfrac{dy}{dt} = 2x + 2z$

$\dfrac{dz}{dt} = 4x + 2y + 3z$

23. $\mathbf{X}' = \begin{pmatrix} 5 & -4 & 0 \\ 1 & 0 & 2 \\ 0 & 2 & 5 \end{pmatrix} \mathbf{X}$

24. $\mathbf{X}' = \begin{pmatrix} 1 & 0 & 0 \\ 0 & 3 & 1 \\ 0 & -1 & 1 \end{pmatrix} \mathbf{X}$

25. $\mathbf{X}' = \begin{pmatrix} 1 & 0 & 0 \\ 2 & 2 & -1 \\ 0 & 1 & 0 \end{pmatrix} \mathbf{X}$

26. $\mathbf{X}' = \begin{pmatrix} 4 & 1 & 0 \\ 0 & 4 & 1 \\ 0 & 0 & 4 \end{pmatrix} \mathbf{X}$

In Problems 27 and 28 solve the given system subject to the indicated initial condition.

27. $\mathbf{X}' = \begin{pmatrix} 2 & 4 \\ -1 & 6 \end{pmatrix} \mathbf{X}, \quad \mathbf{X}(0) = \begin{pmatrix} -1 \\ 6 \end{pmatrix}$

28. $\mathbf{X}' = \begin{pmatrix} 0 & 0 & 1 \\ 0 & 1 & 0 \\ 1 & 0 & 0 \end{pmatrix} \mathbf{X}, \quad \mathbf{X}(0) = \begin{pmatrix} 1 \\ 2 \\ 5 \end{pmatrix}$

29. Show that the 5×5 matrix

$$\mathbf{A} = \begin{pmatrix} 2 & 1 & 0 & 0 & 0 \\ 0 & 2 & 0 & 0 & 0 \\ 0 & 0 & 2 & 0 & 0 \\ 0 & 0 & 0 & 2 & 1 \\ 0 & 0 & 0 & 0 & 2 \end{pmatrix}$$

has an eigenvalue λ_1 of multiplicity five. Show that three linearly independent eigenvectors corresponding to λ_1 can be found.

Discussion Problem

30. Consider the 5×5 matrix given in **Problem 29**. Solve the system $\mathbf{X}' = \mathbf{AX}$ without the aid of matrix methods, but write the general solution using matrix notation. Use the general solution as a basis for a discussion of how the system can be solved using the matrix methods of this section. Carry out your ideas.

8.2.3

In Problems 31–42 find the general solution of the given system.

31. $\dfrac{dx}{dt} = 6x - y$

$\dfrac{dy}{dt} = 5x + 2y$

32. $\dfrac{dx}{dt} = x + y$

$\dfrac{dy}{dt} = -2x - y$

33. $\dfrac{dx}{dt} = 5x + y$

$\dfrac{dy}{dt} = -2x + 3y$

34. $\dfrac{dx}{dt} = 4x + 5y$

$\dfrac{dy}{dt} = -2x + 6y$

35. $\mathbf{X}' = \begin{pmatrix} 4 & -5 \\ 5 & -4 \end{pmatrix} \mathbf{X}$

36. $\mathbf{X}' = \begin{pmatrix} 1 & -8 \\ 1 & -3 \end{pmatrix} \mathbf{X}$

37. $\dfrac{dx}{dt} = z$

$\dfrac{dy}{dt} = -z$

$\dfrac{dz}{dt} = y$

38. $\dfrac{dx}{dt} = 2x + y + 2z$

$\dfrac{dy}{dt} = 3x + 6z$

$\dfrac{dz}{dt} = -4x - 3z$

39. $\mathbf{X}' = \begin{pmatrix} 1 & -1 & 2 \\ -1 & 1 & 0 \\ -1 & 0 & 1 \end{pmatrix} \mathbf{X}$

40. $\mathbf{X}' = \begin{pmatrix} 4 & 0 & 1 \\ 0 & 6 & 0 \\ -4 & 0 & 4 \end{pmatrix} \mathbf{X}$

41. $\mathbf{X}' = \begin{pmatrix} 2 & 5 & 1 \\ -5 & -6 & 4 \\ 0 & 0 & 2 \end{pmatrix} \mathbf{X}$

42. $\mathbf{X}' = \begin{pmatrix} 2 & 4 & 4 \\ -1 & -2 & 0 \\ -1 & 0 & -2 \end{pmatrix} \mathbf{X}$

In Problems 43 and 44 solve the given system subject to the indicated initial condition.

43. $\mathbf{X}' = \begin{pmatrix} 1 & -12 & -14 \\ 1 & 2 & -3 \\ 1 & 1 & -2 \end{pmatrix} \mathbf{X}, \quad \mathbf{X}(0) = \begin{pmatrix} 4 \\ 6 \\ -7 \end{pmatrix}$

44. $\mathbf{X}' = \begin{pmatrix} 6 & -1 \\ 5 & 4 \end{pmatrix} \mathbf{X}, \quad \mathbf{X}(0) = \begin{pmatrix} -2 \\ 8 \end{pmatrix}$

8.3 VARIATION OF PARAMETERS

■ *Fundamental matrix* ■ *Finding a particular solution by variation of parameters*

Before developing a matrix version of variation of parameters for nonhomogeneous linear systems $\mathbf{X}' = \mathbf{AX} + \mathbf{F}$, we need to examine a special matrix that is formed out of the solution vectors of the corresponding homogeneous system $\mathbf{X}' = \mathbf{AX}$.

A Fundamental Matrix If $\mathbf{X}_1, \mathbf{X}_2, \ldots, \mathbf{X}_n$ is a fundamental set of solutions for the homogeneous system $\mathbf{X}' = \mathbf{AX}$ on an interval I, then its general solution on the interval is

$$\mathbf{X} = c_1\mathbf{X}_1 + c_2\mathbf{X}_2 + \cdots + c_n\mathbf{X}_n$$
$$= c_1 \begin{pmatrix} x_{11} \\ x_{21} \\ \vdots \\ x_{n1} \end{pmatrix} + c_2 \begin{pmatrix} x_{12} \\ x_{22} \\ \vdots \\ x_{n2} \end{pmatrix} + \cdots + c_n \begin{pmatrix} x_{1n} \\ x_{2n} \\ \vdots \\ x_{nn} \end{pmatrix} = \begin{pmatrix} c_1x_{11} + c_2x_{12} + \cdots + c_nx_{1n} \\ c_1x_{21} + c_2x_{22} + \cdots + c_nx_{2n} \\ \vdots \\ c_1x_{n1} + c_2x_{n2} + \cdots + c_nx_{nn} \end{pmatrix}. \quad \textbf{(1)}$$

The last matrix in (1) is recognized as the product of an $n \times n$ matrix with an $n \times 1$ matrix. In other words, the general solution (1) can be written as

$$\mathbf{X} = \mathbf{\Phi}(t)\mathbf{C}, \quad \textbf{(2)}$$

where $\mathbf{C}$ is an $n \times 1$ column vector of arbitrary constants, and the $n \times n$ matrix, whose columns consist of the entries of the solution vectors of the system $\mathbf{X}' = \mathbf{AX}$,

$$\mathbf{\Phi}(t) = \begin{pmatrix} x_{11} & x_{12} & \cdots & x_{1n} \\ x_{21} & x_{22} & \cdots & x_{2n} \\ \vdots & & & \vdots \\ x_{n1} & x_{n2} & \cdots & x_{nn} \end{pmatrix},$$

is called a **fundamental matrix** of the system on the interval.

In the discussion that follows we need to use two properties of a fundamental matrix:

- A fundamental matrix $\mathbf{\Phi}(t)$ is nonsingular.
- If $\mathbf{\Phi}(t)$ is a fundamental matrix of the system $\mathbf{X}' = \mathbf{AX}$, then

$$\mathbf{\Phi}'(t) = \mathbf{A\Phi}(t). \qquad (3)$$

A reexamination of (9) of Theorem 8.3 shows that det $\mathbf{\Phi}(t)$ is the same as the Wronskian $W(\mathbf{X}_1, \mathbf{X}_2, \ldots, \mathbf{X}_n)$. Hence the linear independence of the columns of $\mathbf{\Phi}(t)$ on the interval I guarantees that det $\mathbf{\Phi}(t) \neq 0$ for every t in the interval. Since $\mathbf{\Phi}(t)$ is nonsingular, the multiplicative inverse $\mathbf{\Phi}^{-1}(t)$ exists for every t in the interval. The result given in (3) follows immediately from the fact that every column of $\mathbf{\Phi}(t)$ is a solution vector of $\mathbf{X}' = \mathbf{AX}$.

Variation of Parameters   Analogous to the procedure in Section 4.6, we ask whether it is possible to replace the matrix of constants $\mathbf{C}$ in (2) by a column matrix of functions

$$\mathbf{U}(t) = \begin{pmatrix} u_1(t) \\ u_2(t) \\ \vdots \\ u_n(t) \end{pmatrix} \qquad \text{so that} \qquad \mathbf{X}_p = \mathbf{\Phi}(t)\mathbf{U}(t) \qquad (4)$$

is a particular solution of the nonhomogeneous system

$$\mathbf{X}' = \mathbf{AX} + \mathbf{F}(t). \qquad (5)$$

By the product rule, the derivative of the last expression in (4) is

$$\mathbf{X}_p' = \mathbf{\Phi}(t)\mathbf{U}'(t) + \mathbf{\Phi}'(t)\mathbf{U}(t). \qquad (6)$$

Note that the order of the products in (6) is very important. Since $\mathbf{U}(t)$ is a column matrix, the products $\mathbf{U}'(t)\mathbf{\Phi}(t)$ and $\mathbf{U}(t)\mathbf{\Phi}'(t)$ are not defined. Substituting (4) and (6) into (5) gives

$$\mathbf{\Phi}(t)\mathbf{U}'(t) + \mathbf{\Phi}'(t)\mathbf{U}(t) = \mathbf{A\Phi}(t)\mathbf{U}(t) + \mathbf{F}(t). \qquad (7)$$

Now if we use (3) to replace $\mathbf{\Phi}'(t)$, (7) becomes

$$\mathbf{\Phi}(t)\mathbf{U}'(t) + \mathbf{A\Phi}(t)\mathbf{U}(t) = \mathbf{A\Phi}(t)\mathbf{U}(t) + \mathbf{F}(t)$$

or

$$\mathbf{\Phi}(t)\mathbf{U}'(t) = \mathbf{F}(t). \qquad (8)$$

Multiplying both sides of equation (8) by $\mathbf{\Phi}^{-1}(t)$ gives

$$\mathbf{U}'(t) = \mathbf{\Phi}^{-1}(t)\,\mathbf{F}(t) \qquad \text{and so} \qquad \mathbf{U}(t) = \int \mathbf{\Phi}^{-1}(t)\,\mathbf{F}(t)\,dt.$$

Since $\mathbf{X}_p = \mathbf{\Phi}(t)\mathbf{U}(t)$, we conclude that a particular solution of (5) is

$$\mathbf{X}_p = \mathbf{\Phi}(t) \int \mathbf{\Phi}^{-1}(t)\mathbf{F}(t)\,dt. \qquad (9)$$

To calculate the indefinite integral of the column matrix $\mathbf{\Phi}^{-1}(t)\mathbf{F}(t)$ in (9) we integrate each entry. Thus the general solution of the system (5) is $\mathbf{X} = \mathbf{X}_c + \mathbf{X}_p$ or

$$\mathbf{X} = \mathbf{\Phi}(t)\mathbf{C} + \mathbf{\Phi}(t) \int \mathbf{\Phi}^{-1}(t)\mathbf{F}(t)\,dt. \qquad (10)$$

EXAMPLE 1   **Variation of Parameters**

Find the general solution of the nonhomogeneous system

$$\mathbf{X}' = \begin{pmatrix} -3 & 1 \\ 2 & -4 \end{pmatrix} \mathbf{X} + \begin{pmatrix} 3t \\ e^{-t} \end{pmatrix} \tag{11}$$

on the interval $(-\infty, \infty)$.

SOLUTION   We first solve the homogeneous system

$$\mathbf{X}' = \begin{pmatrix} -3 & 1 \\ 2 & -4 \end{pmatrix} \mathbf{X}. \tag{12}$$

The characteristic equation of the coefficient matrix is

$$\det(\mathbf{A} - \lambda\mathbf{I}) = \begin{vmatrix} -3 - \lambda & 1 \\ 2 & -4 - \lambda \end{vmatrix} = (\lambda + 2)(\lambda + 5) = 0,$$

so the eigenvalues are $\lambda_1 = -2$ and $\lambda_2 = -5$. By the usual method we find that the eigenvectors corresponding to λ_1 and λ_2 are, respectively,

$$\begin{pmatrix} 1 \\ 1 \end{pmatrix} \quad \text{and} \quad \begin{pmatrix} 1 \\ -2 \end{pmatrix}.$$

The solution vectors of the system (11) are then

$$\mathbf{X}_1 = \begin{pmatrix} 1 \\ 1 \end{pmatrix} e^{-2t} = \begin{pmatrix} e^{-2t} \\ e^{-2t} \end{pmatrix} \quad \text{and} \quad \mathbf{X}_2 = \begin{pmatrix} 1 \\ -2 \end{pmatrix} e^{-5t} = \begin{pmatrix} e^{-5t} \\ -2e^{-5t} \end{pmatrix}.$$

The entries in $\mathbf{X}_1$ form the first column of $\mathbf{\Phi}(t)$, and the entries in $\mathbf{X}_2$ form the second column of $\mathbf{\Phi}(t)$. Hence

$$\mathbf{\Phi}(t) = \begin{pmatrix} e^{-2t} & e^{-5t} \\ e^{-2t} & -2e^{-5t} \end{pmatrix} \quad \text{and} \quad \mathbf{\Phi}^{-1}(t) = \begin{pmatrix} \frac{2}{3}e^{2t} & \frac{1}{3}e^{2t} \\ \frac{1}{3}e^{5t} & -\frac{1}{3}e^{5t} \end{pmatrix}.$$

From (9) we obtain

$$\mathbf{X}_p = \mathbf{\Phi}(t) \int \mathbf{\Phi}^{-1}(t)\mathbf{F}(t)\, dt = \begin{pmatrix} e^{-2t} & e^{-5t} \\ e^{-2t} & -2e^{-5t} \end{pmatrix} \int \begin{pmatrix} \frac{2}{3}e^{2t} & \frac{1}{3}e^{2t} \\ \frac{1}{3}e^{5t} & -\frac{1}{3}e^{5t} \end{pmatrix} \begin{pmatrix} 3t \\ e^{-t} \end{pmatrix} dt$$

$$= \begin{pmatrix} e^{-2t} & e^{-5t} \\ e^{-2t} & -2e^{-5t} \end{pmatrix} \int \begin{pmatrix} 2te^{2t} + \frac{1}{3}e^{t} \\ te^{5t} - \frac{1}{3}e^{4t} \end{pmatrix} dt$$

$$= \begin{pmatrix} e^{-2t} & e^{-5t} \\ e^{-2t} & -2e^{-5t} \end{pmatrix} \begin{pmatrix} te^{2t} - \frac{1}{2}e^{2t} + \frac{1}{3}e^{t} \\ \frac{1}{5}te^{5t} - \frac{1}{25}e^{5t} - \frac{1}{12}e^{4t} \end{pmatrix}$$

$$= \begin{pmatrix} \frac{6}{5}t - \frac{27}{50} + \frac{1}{4}e^{-t} \\ \frac{3}{5}t - \frac{21}{50} + \frac{1}{2}e^{-t} \end{pmatrix}.$$

Hence from (10) the general solution of (11) on the interval is

$$\mathbf{X} = \begin{pmatrix} e^{-2t} & e^{-5t} \\ e^{-2t} & -2e^{-5t} \end{pmatrix} \begin{pmatrix} c_1 \\ c_2 \end{pmatrix} + \begin{pmatrix} \frac{6}{5}t - \frac{27}{50} + \frac{1}{4}e^{-t} \\ \frac{3}{5}t - \frac{21}{50} + \frac{1}{2}e^{-t} \end{pmatrix}$$

$$= c_1 \begin{pmatrix} 1 \\ 1 \end{pmatrix} e^{-2t} + c_2 \begin{pmatrix} 1 \\ -2 \end{pmatrix} e^{-5t} + \begin{pmatrix} \frac{6}{5} \\ \frac{3}{5} \end{pmatrix} t - \begin{pmatrix} \frac{27}{50} \\ \frac{21}{50} \end{pmatrix} + \begin{pmatrix} \frac{1}{4} \\ \frac{1}{2} \end{pmatrix} e^{-t}.$$

Initial-Value Problem The general solution of (5) on an interval can be written in the alternative manner

$$\mathbf{X} = \mathbf{\Phi}(t)\mathbf{C} + \mathbf{\Phi}(t)\int_{t_0}^{t}\mathbf{\Phi}^{-1}(s)\mathbf{F}(s)\,ds, \tag{13}$$

where t and t_0 are points in the interval. This last form is useful in solving (5) subject to an initial condition $\mathbf{X}(t_0) = \mathbf{X}_0$, because the limits of integration are chosen so that the particular solution vanishes at $t = t_0$. Substituting $t = t_0$ into (13) yields $\mathbf{X}_0 = \mathbf{\Phi}(t_0)\mathbf{C}$, from which we get $\mathbf{C} = \mathbf{\Phi}^{-1}(t_0)\mathbf{X}_0$. Substituting this last result into (13) gives the following solution of the initial-value problem:

$$\mathbf{X} = \mathbf{\Phi}(t)\mathbf{\Phi}^{-1}(t_0)\mathbf{X}_0 + \mathbf{\Phi}(t)\int_{t_0}^{t}\mathbf{\Phi}^{-1}(s)\mathbf{F}(s)\,ds. \tag{14}$$

SECTION 8.3 EXERCISES

Answers to odd-numbered problems begin on page A-13.

In Problems 1–20 use variation of parameters to solve the given system.

1. $\dfrac{dx}{dt} = 3x - 3y + 4$

$\dfrac{dy}{dt} = 2x - 2y - 1$

2. $\dfrac{dx}{dt} = 2x - y$

$\dfrac{dy}{dt} = 3x - 2y + 4t$

3. $\mathbf{X}' = \begin{pmatrix} 3 & -5 \\ \frac{3}{4} & -1 \end{pmatrix}\mathbf{X} + \begin{pmatrix} 1 \\ -1 \end{pmatrix}e^{t/2}$

4. $\mathbf{X}' = \begin{pmatrix} 2 & -1 \\ 4 & 2 \end{pmatrix}\mathbf{X} + \begin{pmatrix} \sin 2t \\ 2\cos 2t \end{pmatrix}e^{2t}$

5. $\mathbf{X}' = \begin{pmatrix} 0 & 2 \\ -1 & 3 \end{pmatrix}\mathbf{X} + \begin{pmatrix} 1 \\ -1 \end{pmatrix}e^{t}$

6. $\mathbf{X}' = \begin{pmatrix} 0 & 2 \\ -1 & 3 \end{pmatrix}\mathbf{X} + \begin{pmatrix} 2 \\ e^{-3t} \end{pmatrix}$

7. $\mathbf{X}' = \begin{pmatrix} 1 & 8 \\ 1 & -1 \end{pmatrix}\mathbf{X} + \begin{pmatrix} 12 \\ 12 \end{pmatrix}t$

8. $\mathbf{X}' = \begin{pmatrix} 1 & 8 \\ 1 & -1 \end{pmatrix}\mathbf{X} + \begin{pmatrix} e^{-t} \\ te^{t} \end{pmatrix}$

9. $\mathbf{X}' = \begin{pmatrix} 3 & 2 \\ -2 & -1 \end{pmatrix}\mathbf{X} + \begin{pmatrix} 2e^{-t} \\ e^{-t} \end{pmatrix}$

10. $\mathbf{X}' = \begin{pmatrix} 3 & 2 \\ -2 & -1 \end{pmatrix}\mathbf{X} + \begin{pmatrix} 1 \\ 1 \end{pmatrix}$

11. $\mathbf{X}' = \begin{pmatrix} 0 & -1 \\ 1 & 0 \end{pmatrix}\mathbf{X} + \begin{pmatrix} \sec t \\ 0 \end{pmatrix}$

12. $\mathbf{X}' = \begin{pmatrix} 1 & -1 \\ 1 & 1 \end{pmatrix}\mathbf{X} + \begin{pmatrix} 3 \\ 3 \end{pmatrix}e^{t}$

13. $\mathbf{X}' = \begin{pmatrix} 1 & -1 \\ 1 & 1 \end{pmatrix}\mathbf{X} + \begin{pmatrix} \cos t \\ \sin t \end{pmatrix}e^{t}$

14. $\mathbf{X}' = \begin{pmatrix} 2 & -2 \\ 8 & -6 \end{pmatrix}\mathbf{X} + \begin{pmatrix} 1 \\ 3 \end{pmatrix}\dfrac{e^{-2t}}{t}$

15. $\mathbf{X}' = \begin{pmatrix} 0 & 1 \\ -1 & 0 \end{pmatrix}\mathbf{X} + \begin{pmatrix} 0 \\ \sec t \tan t \end{pmatrix}$

16. $\mathbf{X}' = \begin{pmatrix} 0 & 1 \\ -1 & 0 \end{pmatrix}\mathbf{X} + \begin{pmatrix} 1 \\ \cot t \end{pmatrix}$

17. $\mathbf{X}' = \begin{pmatrix} 1 & 2 \\ -\frac{1}{2} & 1 \end{pmatrix}\mathbf{X} + \begin{pmatrix} \csc t \\ \sec t \end{pmatrix}e^{t}$

18. $\mathbf{X}' = \begin{pmatrix} 1 & -2 \\ 1 & -1 \end{pmatrix}\mathbf{X} + \begin{pmatrix} \tan t \\ 1 \end{pmatrix}$

19. $\mathbf{X}' = \begin{pmatrix} 1 & 1 & 0 \\ 1 & 1 & 0 \\ 0 & 0 & 3 \end{pmatrix} \mathbf{X} + \begin{pmatrix} e^t \\ e^{2t} \\ te^{3t} \end{pmatrix}$

20. $\mathbf{X}' = \begin{pmatrix} 3 & -1 & -1 \\ 1 & 1 & -1 \\ 1 & -1 & 1 \end{pmatrix} \mathbf{X} + \begin{pmatrix} 0 \\ t \\ 2e^t \end{pmatrix}$

In Problems 21 and 22 use (14) to solve the given system subject to the indicated initial condition.

21. $\mathbf{X}' = \begin{pmatrix} 3 & -1 \\ -1 & 3 \end{pmatrix} \mathbf{X} + \begin{pmatrix} 4e^{2t} \\ 4e^{4t} \end{pmatrix}, \quad \mathbf{X}(0) = \begin{pmatrix} 1 \\ 1 \end{pmatrix}$

22. $\mathbf{X}' = \begin{pmatrix} 1 & -1 \\ 1 & -1 \end{pmatrix} \mathbf{X} + \begin{pmatrix} 1/t \\ 1/t \end{pmatrix}, \quad \mathbf{X}(1) = \begin{pmatrix} 2 \\ -1 \end{pmatrix}$

23. The system of differential equations for the currents $i_1(t)$ and $i_2(t)$ in the electrical network shown in Figure 8.1 is

$$\frac{d}{dt} \begin{pmatrix} i_1 \\ i_2 \end{pmatrix} = \begin{pmatrix} -(R_1 + R_2)/L_2 & R_2/L_2 \\ R_2/L_1 & -R_2/L_1 \end{pmatrix} \begin{pmatrix} i_1 \\ i_2 \end{pmatrix} + \begin{pmatrix} E/L_2 \\ 0 \end{pmatrix}.$$

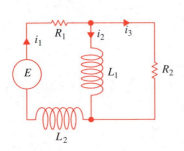

FIGURE 8.1

Solve the system if $R_1 = 8\ \Omega$, $R_2 = 3\ \Omega$, $L_1 = 1$ h, $L_2 = 1$ h, $E(t) = 100 \sin t$ V, $i_1(0) = 0$, and $i_2(0) = 0$.

24. Solving a nonhomogeneous linear system $\mathbf{X}' = \mathbf{AX} + \mathbf{F}(t)$ by variation of parameters when $\mathbf{A}$ is a 3×3 (or larger) matrix is almost an impossible task to do by hand. Consider the system

$$\mathbf{X}' = \begin{pmatrix} 2 & -2 & 2 & 1 \\ -1 & 3 & 0 & 3 \\ 0 & 0 & 4 & -2 \\ 0 & 0 & 2 & -1 \end{pmatrix} \mathbf{X} + \begin{pmatrix} te^t \\ e^{-t} \\ e^{2t} \\ 1 \end{pmatrix}.$$

(a) Use a CAS or linear algebra software to find the eigenvalues and eigenvectors of the coefficient matrix.

(b) Form a fundamental matrix $\mathbf{\Phi}(t)$ and use the computer to find $\mathbf{\Phi}^{-1}(t)$.

(c) Use the computer to carry out the computations of

$$\mathbf{\Phi}^{-1}(t)\mathbf{F}(t), \quad \int \mathbf{\Phi}^{-1}(t)\mathbf{F}(t)\ dt, \quad \mathbf{\Phi}(t) \int \mathbf{\Phi}^{-1}(t)\mathbf{F}(t)\ dt, \quad \mathbf{\Phi}(t)\mathbf{C},$$

and

$$\mathbf{\Phi}(t)\mathbf{C} + \mathbf{\Phi}(t) \int \mathbf{\Phi}^{-1}(t)\mathbf{F}(t)\ dt,$$

where $\mathbf{C}$ is a column matrix of constants c_1, c_2, c_3, and c_4.

(d) Rewrite the computer output for the general solution of the system in the form $\mathbf{X} = \mathbf{X}_c + \mathbf{X}_p$, where $\mathbf{X}_c = c_1\mathbf{X}_1 + c_2\mathbf{X}_2 + c_3\mathbf{X}_3 + c_4\mathbf{X}_4$.

8.4 MATRIX EXPONENTIAL

■ *Homogeneous systems* ■ *Power series for e^{at}* ■ *Matrix exponential* ■ *Nonhomogeneous systems*

Matrices can be utilized in an entirely different manner to solve a system of linear first-order differential equations.

Homogeneous Systems Recall that the simple linear first-order differential equation $x' = ax$, where a is a constant, has the general solution $x = ce^{at}$. It seems natural then to ask whether we can define a **matrix exponential** $e^{\mathbf{A}t}$ so that the homogeneous system $\mathbf{X}' = \mathbf{A}\mathbf{X}$, where $\mathbf{A}$ is an $n \times n$ matrix of constants, has a solution

$$\mathbf{X} = e^{\mathbf{A}t}\mathbf{C}. \tag{1}$$

Since $\mathbf{C}$ is to be an $n \times 1$ column matrix of arbitrary constants, we want $e^{\mathbf{A}t}$ to be an $n \times n$ matrix. Although the complete development of the meaning and theory of the matrix exponential would require a thorough knowledge of matrix algebra, one way of defining $e^{\mathbf{A}t}$ is inspired by the power series representation of the scalar exponential function e^{at}:

$$e^{at} = 1 + at + a^2\frac{t^2}{2!} + \cdots + a^n\frac{t^n}{n!} + \cdots = \sum_{n=0}^{\infty} a^n\frac{t^n}{n!}. \tag{2}$$

The series in (2) converges for all t. Using this series, with 1 replaced by the identity $\mathbf{I}$ and the constant a replaced by an $n \times n$ matrix $\mathbf{A}$ of constants, we arrive at a definition for the $n \times n$ matrix $e^{\mathbf{A}t}$.

DEFINITION 8.4 **Matrix Exponential**

For any $n \times n$ matrix $\mathbf{A}$,

$$e^{\mathbf{A}t} = \mathbf{I} + \mathbf{A}t + \mathbf{A}^2\frac{t^2}{2!} + \cdots + \mathbf{A}^n\frac{t^n}{n!} + \cdots = \sum_{n=0}^{\infty} \mathbf{A}^n\frac{t^n}{n!}. \tag{3}$$

It can be shown that the series given in (3) converges to an $n \times n$ matrix for every value of t. Also, $\mathbf{A}^2 = \mathbf{A}\mathbf{A}$, $\mathbf{A}^3 = \mathbf{A}(\mathbf{A}^2)$, and so on. Moreover, analogous to the differentiation property of the scalar exponential $\frac{d}{dt}e^{at} = ae^{at}$ we have

$$\frac{d}{dt}e^{\mathbf{A}t} = \mathbf{A}e^{\mathbf{A}t}. \tag{4}$$

To see this we differentiate (3) term by term:

$$\frac{d}{dt}e^{\mathbf{A}t} = \frac{d}{dt}\left[\mathbf{I} + \mathbf{A}t + \mathbf{A}^2\frac{t^2}{2!} + \cdots + \mathbf{A}^n\frac{t^n}{n!} + \cdots\right] = \mathbf{A} + \mathbf{A}^2t + \frac{1}{2!}\mathbf{A}^3t^2 + \cdots$$

$$= \mathbf{A}\left[\mathbf{I} + \mathbf{A}t + \mathbf{A}^2\frac{t^2}{2!} + \cdots\right] = \mathbf{A}e^{\mathbf{A}t}.$$

Because of (4), we can now prove that (1) is a solution of $\mathbf{X}' = \mathbf{A}\mathbf{X}$ for every $n \times 1$ vector $\mathbf{C}$ of constants:

$$\mathbf{X}' = \frac{d}{dt}e^{\mathbf{A}t}\mathbf{C} = \mathbf{A}e^{\mathbf{A}t}\mathbf{C} = \mathbf{A}(e^{\mathbf{A}t}\mathbf{C}) = \mathbf{A}\mathbf{X}.$$

$e^{\mathbf{A}t}$ Is a Fundamental Matrix

If we denote the matrix exponential $e^{\mathbf{A}t}$ by the symbol $\boldsymbol{\Psi}(t)$, then (4) is equivalent to the matrix differential equation $\boldsymbol{\Psi}'(t) = \mathbf{A}\boldsymbol{\Psi}(t)$ (see (3) of Section 8.3). In addition, it follows immediately from Definition 8.4 that $\boldsymbol{\Psi}(0) = e^{\mathbf{A}0} = \mathbf{I}$, and so det $\boldsymbol{\Psi}(0) \neq 0$. It turns out that these two properties are sufficient for us to conclude that $\boldsymbol{\Psi}(t)$ is a fundamental matrix of the system $\mathbf{X}' = \mathbf{A}\mathbf{X}$.

Nonhomogeneous Systems

Now we saw in (4) of Section 2.3 that the general solution of the single linear first-order differential equation $x' = ax + f(t)$, where a is a constant, can be expressed as

$$x = x_c + x_p = ce^{at} + e^{at}\int_{t_0}^{t}e^{-as}f(s)\,ds.$$

For a nonhomogeneous system of linear first-order differential equations it can be shown that the general solution of $\mathbf{X}' = \mathbf{A}\mathbf{X} + \mathbf{F}(t)$, where $\mathbf{A}$ is an $n \times n$ matrix of constants, is

$$\mathbf{X} = \mathbf{X}_c + \mathbf{X}_p = e^{\mathbf{A}t}\mathbf{C} + e^{\mathbf{A}t}\int_{t_0}^{t}e^{-\mathbf{A}s}\mathbf{F}(s)\,ds. \tag{5}$$

Since the matrix exponential $e^{\mathbf{A}t}$ is a fundamental matrix, it is always nonsingular and $e^{-\mathbf{A}s} = (e^{\mathbf{A}s})^{-1}$. In practice, $e^{-\mathbf{A}s}$ can be obtained from $e^{\mathbf{A}t}$ by simply replacing t by $-s$.

SECTION 8.4 EXERCISES

Answers to odd-numbered problems begin on page A-14.

In Problems 1 and 2 use (3) to compute $e^{\mathbf{A}t}$ and $e^{-\mathbf{A}t}$.

1. $\mathbf{A} = \begin{pmatrix} 1 & 0 \\ 0 & 2 \end{pmatrix}$ **2.** $\mathbf{A} = \begin{pmatrix} 0 & 1 \\ 1 & 0 \end{pmatrix}$

In Problems 3 and 4 use (3) to compute $e^{\mathbf{A}t}$.

3. $\mathbf{A} = \begin{pmatrix} 1 & 1 & 1 \\ 1 & 1 & 1 \\ -2 & -2 & -2 \end{pmatrix}$ **4.** $\mathbf{A} = \begin{pmatrix} 0 & 0 & 0 \\ 3 & 0 & 0 \\ 5 & 1 & 0 \end{pmatrix}$

In Problems 5–8 use (1) to find the general solution of the given system.

5. $\mathbf{X}' = \begin{pmatrix} 1 & 0 \\ 0 & 2 \end{pmatrix} \mathbf{X}$

6. $\mathbf{X}' = \begin{pmatrix} 0 & 1 \\ 1 & 0 \end{pmatrix} \mathbf{X}$

7. $\mathbf{X}' = \begin{pmatrix} 1 & 1 & 1 \\ 1 & 1 & 1 \\ -2 & -2 & -2 \end{pmatrix} \mathbf{X}$

8. $\mathbf{X}' = \begin{pmatrix} 0 & 0 & 0 \\ 3 & 0 & 0 \\ 5 & 1 & 0 \end{pmatrix} \mathbf{X}$

In Problems 9–12 use (5) to find the general solution of the given system.

9. $\mathbf{X}' = \begin{pmatrix} 1 & 0 \\ 0 & 2 \end{pmatrix} \mathbf{X} + \begin{pmatrix} 3 \\ -1 \end{pmatrix}$

10. $\mathbf{X}' = \begin{pmatrix} 1 & 0 \\ 0 & 2 \end{pmatrix} \mathbf{X} + \begin{pmatrix} t \\ e^{4t} \end{pmatrix}$

11. $\mathbf{X}' = \begin{pmatrix} 0 & 1 \\ 1 & 0 \end{pmatrix} \mathbf{X} + \begin{pmatrix} 1 \\ 1 \end{pmatrix}$

12. $\mathbf{X}' = \begin{pmatrix} 0 & 1 \\ 1 & 0 \end{pmatrix} \mathbf{X} + \begin{pmatrix} \cosh t \\ \sinh t \end{pmatrix}$

13. Solve the system in Problem 7 subject to the initial condition

$$\mathbf{X}(0) = \begin{pmatrix} 1 \\ -4 \\ 6 \end{pmatrix}.$$

14. Solve the system in Problem 9 subject to the initial condition

$$\mathbf{X}(0) = \begin{pmatrix} 4 \\ 3 \end{pmatrix}.$$

Let $\mathbf{P}$ denote a matrix whose columns are eigenvectors $\mathbf{K}_1, \mathbf{K}_2, \ldots, \mathbf{K}_n$ corresponding to distinct eigenvalues $\lambda_1, \lambda_2, \ldots, \lambda_n$ of an $n \times n$ matrix $\mathbf{A}$. Then it can be shown that $\mathbf{A} = \mathbf{P}\mathbf{D}\mathbf{P}^{-1}$, where $\mathbf{D}$ is defined by

$$\mathbf{D} = \begin{pmatrix} \lambda_1 & 0 & \cdots & 0 \\ 0 & \lambda_2 & \cdots & 0 \\ \vdots & & & \vdots \\ 0 & 0 & \cdots & \lambda_n \end{pmatrix}. \tag{6}$$

In Problems 15 and 16 verify the above result for the given matrix.

15. $\mathbf{A} = \begin{pmatrix} 2 & 1 \\ -3 & 6 \end{pmatrix}$

16. $\mathbf{A} = \begin{pmatrix} 2 & 1 \\ 1 & 2 \end{pmatrix}$

17. Suppose $\mathbf{A} = \mathbf{P}\mathbf{D}\mathbf{P}^{-1}$, where $\mathbf{D}$ is defined as in (6). Use (3) to show that $e^{\mathbf{A}t} = \mathbf{P}e^{\mathbf{D}t}\mathbf{P}^{-1}$.

18. Use (3) to show that

$$e^{\mathbf{D}t} = \begin{pmatrix} e^{\lambda_1 t} & 0 & \cdots & 0 \\ 0 & e^{\lambda_2 t} & \cdots & 0 \\ \vdots & & & \vdots \\ 0 & 0 & \cdots & e^{\lambda_n t} \end{pmatrix},$$

where $\mathbf{D}$ is defined as in (6).

In Problems 19 and 20 use the results of Problems 15–18 to solve the given system.

19. $\mathbf{X}' = \begin{pmatrix} 2 & 1 \\ -3 & 6 \end{pmatrix} \mathbf{X}$

20. $\mathbf{X}' = \begin{pmatrix} 2 & 1 \\ 1 & 2 \end{pmatrix} \mathbf{X}$

CHAPTER 8 REVIEW EXERCISES

Answers to odd-numbered problems begin on page A-14.

1. Verify that, on the interval $(-\infty, \infty)$, the general solution of the system

$$\mathbf{X}' = \begin{pmatrix} 4 & -2 \\ 5 & 2 \end{pmatrix} \mathbf{X}$$

   is $\quad \mathbf{X} = c_1 \begin{pmatrix} 2\cos 3t \\ \cos 3t + 3\sin 3t \end{pmatrix} e^{3t} + c_2 \begin{pmatrix} 2\sin 3t \\ \sin 3t - 3\cos 3t \end{pmatrix} e^{3t}$.

2. Verify that, on the interval $(-\infty, \infty)$, the general solution of the system

$$\frac{dx}{dt} = y$$

$$\frac{dy}{dt} = -x + 2y - 2\cos t$$

   is $\quad \mathbf{X} = c_1 \begin{pmatrix} 1 \\ 1 \end{pmatrix} e^t + c_2 \left[\begin{pmatrix} 1 \\ 1 \end{pmatrix} te^t + \begin{pmatrix} 0 \\ 1 \end{pmatrix} e^t \right] + \begin{pmatrix} \sin t \\ \cos t \end{pmatrix}$.

In Problems 3–8 use the concept of eigenvalues and eigenvectors to solve each system.

3. $\dfrac{dx}{dt} = 2x + y$

   $\dfrac{dy}{dt} = -x$

4. $\dfrac{dx}{dt} = -4x + 2y$

   $\dfrac{dy}{dt} = \quad 2x - 4y$

5. $\mathbf{X}' = \begin{pmatrix} 1 & 2 \\ -2 & 1 \end{pmatrix} \mathbf{X}$

6. $\mathbf{X}' = \begin{pmatrix} -2 & 5 \\ -2 & 4 \end{pmatrix} \mathbf{X}$

7. $\mathbf{X}' = \begin{pmatrix} 1 & 1 & 1 \\ 1 & 1 & 1 \\ 1 & 1 & 1 \end{pmatrix} \mathbf{X}$

8. $\mathbf{X}' = \begin{pmatrix} 1 & -1 & 1 \\ 0 & 1 & 3 \\ 4 & 3 & 1 \end{pmatrix} \mathbf{X}$

In Problems 9–12 use variation of parameters to solve the given system.

9. $\mathbf{X}' = \begin{pmatrix} 2 & 8 \\ 0 & 4 \end{pmatrix} \mathbf{X} + \begin{pmatrix} 2 \\ 16t \end{pmatrix}$

10. $\dfrac{dx}{dt} = x + 2y$

    $\dfrac{dy}{dt} = -\dfrac{1}{2}x + y + e^t \tan t$

11. $\mathbf{X}' = \begin{pmatrix} -1 & 1 \\ -2 & 1 \end{pmatrix} \mathbf{X} + \begin{pmatrix} 1 \\ \cot t \end{pmatrix}$

12. $\mathbf{X}' = \begin{pmatrix} 3 & 1 \\ -1 & 1 \end{pmatrix} \mathbf{X} + \begin{pmatrix} -2 \\ 1 \end{pmatrix} e^{2t}$

A Modeling Application

MODELING AN ARMS RACE

Michael Olinick
Middlebury College

The twentieth century has witnessed a number of dangerous, destabilizing, and expensive arms races. The outbreak of World War I (1914–1918) climaxed a rapid buildup of armaments among rival European powers. There was a similar mutual accumulation of conventional arms in the years just prior to World War II (1939–1945). The United States and the Soviet Union engaged in a costly nuclear arms race during the forty years of the Cold War. Stockpiling of ever more deadly weapons is common today in many parts of the world, including the Middle East and the Balkans.

The British meteorologist and educator Lewis F. Richardson (1881–1953) developed a number of mathematical models to help analyze the dynamics of such arms races. Richardson's primary model was based on *mutual fear*: a nation is spurred to increase its arms stockpile at a rate proportional to the level of armament expenditures of its rival. Richardson's model takes into account internal constraints within a nation that slow down arms buildups: the more a nation is spending on arms, the harder it is to make greater increases, because it becomes increasingly difficult to divert society's resources from basic needs such as food and housing to weapons. Richardson also built into his model other factors driving or slowing down an arms race that are independent of levels of arms expenditures.

The mathematical structure of this model is a linked system of two first-order differential equations. If x and y represent the amount of wealth being spent on arms by two nations at time t, then the model has the form

$$\frac{dx}{dt} = ay - mx + r$$

$$\frac{dy}{dt} = bx - ny + s,$$

where a, b, m, and n are positive constants and r and s are constants that can be positive or negative. The constants a and b measure mutual fear; the constants m and n represent proportionality factors for the "internal brakes" to further arms increases. Positive values for r and s correspond to underlying factors of ill will

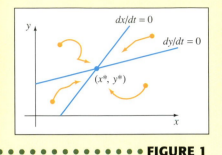

or distrust that would persist even if arms expenditures dropped to zero. Negative values for r and s indicate a contribution based on good will.

The dynamic behavior of this system of differential equations depends on the relative sizes of ab and mn together with the signs of r and s. Although the model is a relatively simple one, it allows us to consider several different long-term outcomes. It's possible that two nations might move simultaneously toward mutual disarmament, with x and y each approaching zero. A vicious cycle of unbounded increases in x and y is another possible scenario. A third eventuality is that the arms expenditures asymptotically approach a stable point (x^*, y^*) regardless of the initial level of arms expenditures. In other cases, the eventual outcome is very dependent on the starting point. Figure 1 shows one possible situation with four different initial levels, each of which leads to a "stable outcome."

FIGURE 1

Richardson's pioneering work has led to many fruitful applications of models of differential equations to problems in international relations and political science.

References

1. Richardson, Lewis F. *Arms and Insecurity: A Mathematical Study of the Causes and Origins of War*. Pittsburgh: Boxwood Press, 1960.
2. Olinick, Michael. *An Introduction to Mathematical Models in the Social and Life Sciences*. Reading, MA: Addison-Wesley, 1978.

About the Author

Michael Olinick is Professor of Mathematics and Computer Science at Middlebury College in Vermont, where he has taught since receiving his Ph.D. from the University of Wisconsin. Dr. Olinick has also held visiting positions at University College, Nairobi; the University of California, Berkeley; San Diego State University; Wesleyan University; and the University of Lancaster in England. He authored a text on mathematical modeling for Addison-Wesley and is co-author of *Calculus 6/e*, published by PWS Publishing Company.

CHAPTER

9

NUMERICAL METHODS FOR ORDINARY DIFFERENTIAL EQUATIONS

INTRODUCTION

A differential equation does not have to have a solution, and even if a solution exists we may not always be able to exhibit it in an explicit or implicit form. In many instances we may have to be content with an approximation to the solution.

If a solution of a differential equation exists, it represents a set of points in the Cartesian plane. Beginning in Section 9.2, we shall develop procedures that utilize the differential equation to obtain a sequence of distinct points whose coordinates, as shown in Figure 9.1, approximate the coordinates of the points on the actual solution curve.

Our concentration in this chapter is primarily on first-order initial-value problems: $dy/dx = f(x, y)$, $y(x_0) = y_0$. We shall see that the numerical procedures developed for first-order equations may be adapted to *systems* of first-order equations. As a consequence we can approximate solutions of higher-order initial-value problems by reducing the differential equation to a system of first-order equations. The chapter concludes with procedures for approximating solutions of linear second-order boundary-value problems.

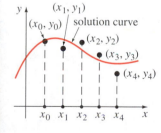

FIGURE 9.1

9.1   DIRECTION FIELDS

■ *Lineal elements* ■ *Direction field* ■ *Slope field* ■ *Lineal element field*

solution curves

$y = 2$

FIGURE 9.2

Lineal Elements   Consider the simple first-order differential equation $dy/dx = y$. Specifically, the differential equation implies that the slopes of tangent lines to the graph of a solution are given by the function $f(x, y) = y$. When $f(x, y)$ is held constant—that is, when $y = c$, where c is any real constant—we are stating that the slope of the tangents to the solution curves is the same constant value along a horizontal line. For example, for $y = 2$ let us draw a sequence of short line segments, or **lineal elements,** each having slope 2 and its midpoint on the line. As shown in Figure 9.2, the solution curves pass through this horizontal line at every point tangent to the lineal elements.

Isoclines and Direction Fields   The equation $y = c$ represents a one-parameter family of horizontal lines. In general, any member of the family $f(x, y) = c$ is called an **isocline,** which literally means a curve along which the inclination of the tangents is the same. As the parameter c is varied, we obtain a collection of isoclines on which the lineal elements are judiciously constructed. The totality of these lineal elements is variously called a **direction field, slope field,** or **lineal element field** of the differential equation $dy/dx = f(x, y)$. As we see in Figure 9.3(a), the direction field suggests the "flow pattern" for the family of solution curves of the differential equation $y' = y$. In particular, if we want a solution that passes through the point $(0, 1)$, then, as indicated in color in Figure 9.3(b), we construct a curve through this point that passes through the isoclines with the appropriate slopes.

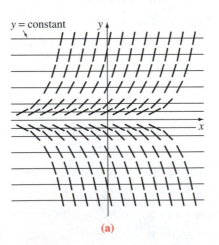

(a)

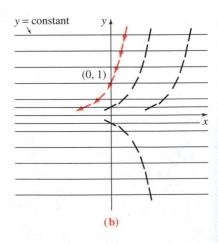

(b)

FIGURE 9.3

EXAMPLE 1   **Direction Field**

Sketch the direction field and indicate several possible members of the family of solution curves for $dy/dx = x/y$.

SOLUTION   Before sketching the direction field corresponding to the isoclines $x/y = c$ or $y = x/c$, you should examine the differential equation and see that it gives the following information.

(*i*) If a solution curve crosses the *x*-axis ($y = 0$), it does so tangent to a vertical lineal element at every point, except possibly $(0, 0)$.

(*ii*) If a solution curve crosses the *y*-axis ($x = 0$), it does so tangent to a horizontal lineal element at every point, except possibly $(0, 0)$.

(*iii*) The lineal elements corresponding to the isoclines $c = 1$ and $c = -1$ are collinear with the lines $y = x$ and $y = -x$, respectively. Indeed, $y = x$ and $y = -x$ are particular solutions of the given differential equation (verify this). Note that, *in general,* isoclines are not solutions of a differential equation.

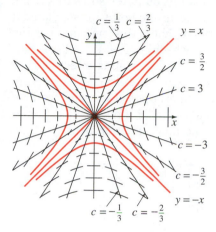

FIGURE 9.4

Figure 9.4 shows the direction field and several possible solution curves in color. Remember that on any isocline all the lineal elements are parallel. Also, the lineal elements may be drawn in such a manner as to suggest the flow of a particular curve. In other words, imagine choosing the isoclines so close together that if the lineal elements were connected we would have a polygonal curve suggestive of the shape of a smooth solution curve.   ∎

EXAMPLE 2   **Approximate Solution**

The differential equation $dy/dx = x^2 + y^2$ cannot be solved in terms of elementary functions. Use a direction field to locate an approximate solution satisfying $y(0) = 1$.

SOLUTION   The isoclines are concentric circles defined by $x^2 + y^2 = c$, $c > 0$. For $c = \frac{1}{4}$, $c = 1$, $c = \frac{9}{4}$, and $c = 4$ we obtain circles with radii $\frac{1}{2}, 1, \frac{3}{2}$, and 2, as shown in Figure 9.5(a). The lineal elements constructed on each circle have a slope corresponding to the chosen value of c. It

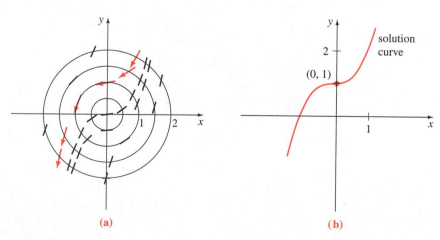

(a)

(b)

FIGURE 9.5

seems plausible from inspection of Figure 9.5(a) that an approximate solution curve passing through the point (0, 1) might have the shape given in Figure 9.5(b). ∎

Use of Computers

Sketching a direction field is straightforward but time consuming; it is one of those tasks about which an argument can be made for doing it by hand once or twice in a lifetime, but which is overall most efficiently carried out by means of computer software. Using $dy/dx = x/y$ and the direction field feature in the software supplement to this text, we obtain Figure 9.6(a). Observe that in this computer version of Figure 9.4 the lineal elements are drawn uniformly spaced on their isoclines (which themselves are not drawn). The resulting direction field suggests the flow of the solution curves even more strongly. In Figure 9.6(b), using an ODE solver, we have superimposed the approximate solution curve for the differential equation in Example 2 that passes through (0, 1) on top of its computer-generated direction field.

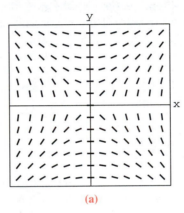

(a)

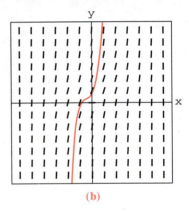

(b)

FIGURE 9.6

SECTION 9.1 EXERCISES

Answers to odd-numbered problems begin on page A-14.

In Problems 1–4 use the given computer-generated direction field to sketch several possible solution curves for the indicated differential equation.

1. $y' = xy$

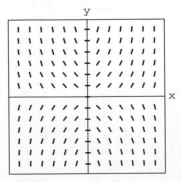

FIGURE 9.7

2. $y' = 1 - xy$

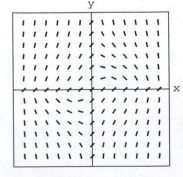

FIGURE 9.8

3. $y' = y - x$

4. $y' = \dfrac{\cos x}{\sin y}$

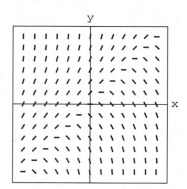

FIGURE 9.9

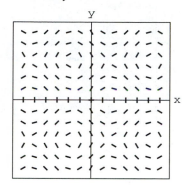

FIGURE 9.10

In Problems 5–12 sketch—or use a computer to obtain—the direction field for the given differential equation. Indicate several possible solution curves.

5. $y' = x$

6. $y' = x + y$

7. $y\dfrac{dy}{dx} = -x$

8. $\dfrac{dy}{dx} = \dfrac{1}{y}$

9. $\dfrac{dy}{dx} = 0.2x^2 + y$

10. $\dfrac{dy}{dx} = xe^y$

11. $y' = y - \cos\dfrac{\pi}{2}x$

12. $y' = 1 - \dfrac{y}{x}$

9.2 EULER METHODS

- *Euler's method* ■ *Linearization* ■ *Absolute error* ■ *Relative error*
- *Percentage relative error* ■ *Round-off error* ■ *Local truncation error*
- *Global truncation error* ■ *Improved Euler method*

Euler's Method   One of the simplest techniques for approximating solutions of the initial-value problem

$$y' = f(x, y), \qquad y(x_0) = y_0$$

is known as **Euler's method** or the **method of tangent lines.** It uses the fact that the derivative of a function $y(x)$ evaluated at a point x_0 gives the slope of the tangent line to the graph of $y(x)$ at this point. Since the initial-value problem gives the value of the derivative of the solution $y(x)$ at (x_0, y_0), we see that the slope of the tangent line to the solution curve at this point is $f(x_0, y_0)$. If we move a short distance along the tangent line, we obtain an approximation to a nearby point on the solution curve. The process is then repeated at this new point. To formalize this procedure we use the **linearization**

$$L(x) = y'(x_0)(x - x_0) + y_0 \qquad \textbf{(1)}$$

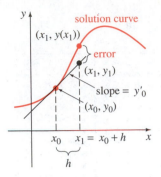

FIGURE 9.11

of $y(x)$ at $x = x_0$. The graph of this linearization is a straight line tangent to the graph of $y = y(x)$ at the point (x_0, y_0). We now let h be a positive increment on the x-axis, as shown in Figure 9.11. Then by replacing x by $x_1 = x_0 + h$ in (1) we get

$$L(x_1) = y'(x_0)(x_0 + h - x_0) + y_0 = y_0 + hy'_0$$

or

$$y_1 = y_0 + hf(x_0, y_0),$$

where $y'_0 = y'(x_0) = f(x_0, y_0)$, and $y_1 = L(x_1)$. The point (x_1, y_1) on the tangent line is an approximation to the point $(x_1, y(x_1))$ on the solution curve; that is, $L(x_1) \approx y(x_1)$ or $y_1 \approx y(x_1)$ is a *local linear approximation* of $y(x)$ at x_1. Of course the accuracy of the approximation depends heavily on the size of the increment h. Usually we must choose this **step size** to be "reasonably small." If we now repeat this process, identifying the new starting point (x_1, y_1) with (x_0, y_0) in the above discussion, we obtain the approximation

$$y(x_2) = y(x_0 + 2h) = y(x_1 + h) \approx y_2 = y_1 + hf(x_1, y_1).$$

In general it follows that

$$y_{n+1} = y_n + hf(x_n, y_n), \tag{2}$$

where $x_n = x_0 + nh$.

To illustrate Euler's method we use the iteration scheme (2) on a differential equation for which we know the explicit solution; in this way we can compare the estimated values y_n with the true values $y(x_n)$.

EXAMPLE 1   **Euler's Method**

Consider the initial-value problem

$$y' = 0.2xy, \qquad y(1) = 1.$$

Use Euler's method to obtain an approximation to $y(1.5)$ using first $h = 0.1$ and then $h = 0.05$.

SOLUTION   We first identify $f(x, y) = 0.2xy$ so that (2) becomes

$$y_{n+1} = y_n + h(0.2x_n y_n).$$

Then for $h = 0.1$ we find

$$y_1 = y_0 + (0.1)(0.2x_0 y_0) = 1 + (0.1)[0.2(1)(1)] = 1.02,$$

which is an estimate of the value of $y(1.1)$. However, if we use $h = 0.05$, it takes *two* iterations to reach $x = 1.1$. We have

$$y_1 = 1 + (0.05)[0.2(1)(1)] = 1.01$$
$$y_2 = 1.01 + (0.05)[0.2(1.05)(1.01)] = 1.020605.$$

Here we note that $y_1 \approx y(1.05)$ and $y_2 \approx y(1.1)$. The remaining calculations are summarized in Tables 9.1 and 9.2. Each entry is rounded to four decimal places.

TABLE 9.1 Euler's Method with $h = 0.1$

x_n	y_n	True value	Abs. error	% Rel. error
1.00	1.0000	1.0000	0.0000	0.00
1.10	1.0200	1.0212	0.0012	0.12
1.20	1.0424	1.0450	0.0025	0.24
1.30	1.0675	1.0714	0.0040	0.37
1.40	1.0952	1.1008	0.0055	0.50
1.50	1.1259	1.1331	0.0073	0.64

TABLE 9.2 Euler's Method with $h = 0.05$

x_n	y_n	True value	Abs. error	% Rel. error
1.00	1.0000	1.0000	0.0000	0.00
1.05	1.0100	1.0103	0.0003	0.03
1.10	1.0206	1.0212	0.0006	0.06
1.15	1.0318	1.0328	0.0009	0.09
1.20	1.0437	1.0450	0.0013	0.12
1.25	1.0562	1.0579	0.0016	0.16
1.30	1.0694	1.0714	0.0020	0.19
1.35	1.0833	1.0857	0.0024	0.22
1.40	1.0980	1.1008	0.0028	0.25
1.45	1.1133	1.1166	0.0032	0.29
1.50	1.1295	1.1331	0.0037	0.32

In Example 1 the true values were calculated from the known solution $y = e^{0.1(x^2-1)}$. Also, the **absolute error** is defined to be

$$|true\ value - approximation|.$$

The **relative error** and the **percentage relative error** are, respectively,

$$\frac{|true\ value - approximation|}{|true\ value|}$$

and

$$\frac{|true\ value - approximation|}{|true\ value|} \times 100 = \frac{absolute\ error}{|true\ value|} \times 100.$$

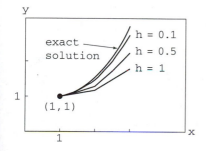

FIGURE 9.12

Computer software enables us to examine approximations to the graph of the solution $y(x)$ of an initial-value problem by plotting straight lines through the points (x_n, y_n) generated by Euler's method. In Figure 9.12 we have compared, on the interval $[1, 3]$, the graph of the exact solution of the initial-value problem in Example 1 with the graphs obtained from Euler's method using the step sizes $h = 1$, $h = 0.5$, and $h = 0.1$. It is apparent from the figure that the approximation improves as the step size decreases.

Although we see that the percentage relative error in Tables 9.1 and 9.2 is growing, it does not appear to be that bad. But you should not be deceived by Example 1 and Figure 9.12. Watch what happens in the next example when we simply change the coefficient 0.2 of the differential equation in Example 1 to the number 2.

EXAMPLE 2 **Comparison of Exact and Approximate Values**

Use Euler's method to obtain the approximate value of $y(1.5)$ for the solution of

$$y' = 2xy, \qquad y(1) = 1.$$

SOLUTION You should verify that the exact, or analytic, solution is now $y = e^{x^2-1}$. Proceeding as in Example 1, we obtain the results shown in Tables 9.3 and 9.4.

TABLE 9.3   Euler's Method with $h = 0.1$

x_n	y_n	True value	Abs. error	% Rel. error
1.00	1.0000	1.0000	0.0000	0.00
1.10	1.2000	1.2337	0.0337	2.73
1.20	1.4640	1.5527	0.0887	5.71
1.30	1.8154	1.9937	0.1784	8.95
1.40	2.2874	2.6117	0.3244	12.42
1.50	2.9278	3.4904	0.5625	16.12

TABLE 9.4   Euler's Method with $h = 0.05$

x_n	y_n	True value	Abs. error	% Rel. error
1.00	1.0000	1.0000	0.0000	0.00
1.05	1.1000	1.1079	0.0079	0.72
1.10	1.2155	1.2337	0.0182	1.47
1.15	1.3492	1.3806	0.0314	2.27
1.20	1.5044	1.5527	0.0483	3.11
1.25	1.6849	1.7551	0.0702	4.00
1.30	1.8955	1.9937	0.0982	4.93
1.35	2.1419	2.2762	0.1343	5.90
1.40	2.4311	2.6117	0.1806	6.92
1.45	2.7714	3.0117	0.2403	7.98
1.50	3.1733	3.4904	0.3171	9.08

In this case, with a step size $h = 0.1$, a 16% relative error in the calculation of the approximation to $y(1.5)$ is totally unacceptable. At the expense of doubling the number of calculations, some improvement in accuracy is obtained by halving the step size to $h = 0.05$.   ∎

Errors in Numerical Methods   In choosing and using a numerical method for the solution of an initial-value problem, we must be aware of the various sources of errors. For some kinds of computation the accumulation of errors might reduce the accuracy of an approximation to the point of making the computation useless. On the other hand, depending on the use to which a numerical solution may be put, extreme accuracy may not be worth the added expense and complication.

One source of error always present in calculations is **round-off error.** This error results from the fact that any calculator or computer can represent numbers using only a finite number of digits. Suppose, for the sake of illustration, that we have a calculator that uses base 10 arithmetic and carries four digits, so that $\frac{1}{3}$ is represented in the calculator as 0.3333 and $\frac{1}{9}$ is represented as 0.1111. If we use this calculator to compute $(x^2 - \frac{1}{9})/(x - \frac{1}{3})$ for $x = 0.3334$, we obtain

$$\frac{(0.3334)^2 - 0.1111}{0.3334 - 0.3333} = \frac{0.1112 - 0.1111}{0.3334 - 0.3333} = 1.$$

With the help of a little algebra, however, we see that

$$\frac{x^2 - 1/9}{x - 1/3} = \frac{(x - 1/3)(x + 1/3)}{x - 1/3} = x + \frac{1}{3},$$

so that when $x = 0.3334$, $(x^2 - \frac{1}{9})/(x - \frac{1}{3}) \approx 0.3334 + 0.3333 = 0.6667$. This example shows that the effects of round-off error can be quite serious unless some care is taken. One way to reduce the effect of round-off error is to minimize the number of calculations. Another technique on a computer is to use double-precision arithmetic to check the results. In general, round-off error is unpredictable and difficult to analyze, and we will neglect it in the error analysis that follows. We will concentrate on investigating the error introduced by using a formula or algorithm to approximate the values of the solution.

Truncation Errors for Euler's Method When iterating Euler's formula

$$y_{n+1} = y_n + hf(x_n, y_n)$$

we obtain a sequence of values $y_1, y_2, y_3, \ldots$. Usually the value y_1 will not agree with $y(x_1)$, the actual solution evaluated at x_1, because the algorithm gives only a straight-line approximation to the solution. See Figure 9.11. The error is called the **local truncation error, formula error,** or **discretization error.** It occurs at each step; that is, if we assume that y_n is accurate, then y_{n+1} will contain local truncation error.

To derive a formula for the local truncation error for Euler's method we use Taylor's formula with remainder. If a function $y(x)$ possesses $k + 1$ derivatives that are continuous on an open interval containing a and x, then

$$y(x) = y(a) + y'(a)\frac{(x-a)}{1!} + \cdots + y^{(k)}(a)\frac{(x-a)^k}{k!} + y^{(k+1)}(c)\frac{(x-a)^{k+1}}{(k+1)!}, \quad \textbf{(3)}$$

where c is some point between a and x. Setting $k = 1$, $a = x_n$, and $x = x_{n+1} = x_n + h$, we get

$$y(x_{n+1}) = y(x_n) + y'(x_n)\frac{h}{1!} + y''(c)\frac{h^2}{2!}$$

or

$$y(x_{n+1}) = \underbrace{y_n + hf(x_n, y_n)}_{y_{n+1}} + y''(c)\frac{h^2}{2!}.$$

Euler's method is this formula without the last term; hence the local truncation error in y_{n+1} is

$$y''(c)\frac{h^2}{2!}, \qquad \text{where} \qquad x_n < c < x_{n+1}.$$

Unfortunately, the value of c is usually unknown (it exists theoretically) and so the exact error cannot be calculated, but an upper bound on the absolute value of the error is

$$M\frac{h^2}{2}, \qquad \text{where} \qquad M = \max_{x_n < x < x_{n+1}} |y''(x)|.$$

In discussing errors arising from the use of numerical methods it is helpful to use the notation $O(h^n)$. To define this concept we let $e(h)$ denote the error in a numerical calculation depending on h. Then $e(h)$ is said to be of order h^n, denoted by $O(h^n)$, if there exist a constant C and a positive integer n such that $|e(h)| \leq Ch^n$ for h sufficiently small. Thus the local truncation error for Euler's method is $O(h^2)$. We note that, in general, if a numerical method has order h^n and h is halved, the new error is approximately $C(h/2)^n = Ch^n/2^n$; that is, the error is reduced by a factor of $1/2^n$.

EXAMPLE 3 **Bound for Local Truncation Errors**

Find a bound for the local truncation errors for Euler's method applied to

$$y' = 2xy, \qquad y(1) = 1.$$

SOLUTION   This differential equation was studied in Example 2, and its analytic solution is $y(x) = e^{x^2 - 1}$.

The local truncation error is

$$y''(c)\frac{h^2}{2} = (2 + 4c^2)e^{(c^2 - 1)}\frac{h^2}{2},$$

where c is between x_n and $x_n + h$. In particular, for $h = 0.1$ we can get an upper bound on the local truncation error for y_1 by replacing c by 1.1:

$$[2 + (4)(1.1)^2]e^{((1.1)^2 - 1)}\frac{(0.1)^2}{2} = 0.0422.$$

From Table 9.3 we see that the error after the first step is 0.0337, less than the value given by the bound.

Similarly, we can get a bound for the local truncation error for any of the five steps given in Table 9.3 by replacing c by 1.5 (this value of c gives the largest value of $y''(c)$ for any of the steps and may be too generous for the first few steps). Doing this gives

$$[2 + (4)(1.5)^2]e^{((1.5)^2 - 1)}\frac{(0.1)^2}{2} = 0.1920 \tag{4}$$

as an upper bound for the local truncation error in each step.   ∎

Note in Example 3 that if h is halved to 0.05, then the error bound is 0.0480, about one-fourth as much as shown in (4). This is expected because the local truncation error for Euler's method is $O(h^2)$.

In the above analysis we assumed that the value of y_n was exact in the calculation of y_{n+1}, but it is not because it contains local truncation errors from previous steps. The total error in y_{n+1} is an accumulation of the errors in each of the previous steps. This total error is called the **global truncation error.** A complete analysis of the global truncation error is beyond the scope of this text, but it can be shown that the global truncation error for Euler's method is $O(h)$.

We expect that, for Euler's method, if the step size is halved the error will be approximately halved as well. This is borne out in Example 2, where the absolute error at $x = 1.50$ with $h = 0.1$ is 0.5625 and with $h = 0.05$ is 0.3171, approximately half as large. See Tables 9.3 and 9.4.

In general it can be shown that if a method for the numerical solution of a differential equation has local truncation error $O(h^{\alpha+1})$, then the global truncation error is $O(h^\alpha)$.

Improved Euler Method   Although Euler's formula is attractive for its simplicity, it is seldom used in serious calculations. In the remainder of this section and in subsequent sections we study methods that give significantly greater accuracy than Euler's method.

The formula

$$y_{n+1} = y_n + h\frac{f(x_n, y_n) + f(x_{n+1}, y_{n+1}^*)}{2},$$

where

$$y_{n+1}^* = y_n + hf(x_n, y_n), \tag{5}$$

is known as the **improved Euler formula** or **Heun's formula.** Euler's formula is used to obtain the initial estimate y_{n+1}^*. The values $f(x_n, y_n)$ and $f(x_{n+1}, y_{n+1}^*)$ are approximations to the slopes of the solution curve at $(x_n, y(x_n))$ and $(x_{n+1}, y(x_{n+1}))$, and consequently the quotient

$$\frac{f(x_n, y_n) + f(x_{n+1}, y_{n+1}^*)}{2}$$

can be interpreted as an average slope on the interval from x_n to x_{n+1}. The value of y_{n+1} is then computed in a manner similar to that used in Euler's method, except that an average slope over the interval is used instead of the slope at $(x_n, y(x_n))$. We say that the value of y_{n+1}^* *predicts* a value of $y(x_n)$, whereas

$$y_{n+1} = y_n + h\frac{f(x_n, y_n) + f(x_{n+1}, y_{n+1}^*)}{2}$$

corrects this estimate.

EXAMPLE 4 **Improved Euler Method**

Use the improved Euler formula to obtain the approximate value of $y(1.5)$ for the solution of the initial-value problem in Example 2. Compare the results for $h = 0.1$ and $h = 0.05$.

SOLUTION   For $n = 0$ and $h = 0.1$ we first compute

$$y_1^* = y_0 + (0.1)(2x_0y_0) = 1.2.$$

Then from (5)

$$y_1 = y_0 + (0.1)\frac{2x_0y_0 + 2x_1y_1^*}{2} = 1 + (0.1)\frac{2(1)(1) + 2(1.1)(1.2)}{2} = 1.232.$$

The comparative values of the calculations for $h = 0.1$ and $h = 0.5$ are given in Tables 9.5 and 9.6, respectively.

TABLE 9.5   Improved Euler Method with $h = 0.1$

x_n	y_n	True value	Abs. error	% Rel. error
1.00	1.0000	1.0000	0.0000	0.00
1.10	1.2320	1.2337	0.0017	0.14
1.20	1.5479	1.5527	0.0048	0.31
1.30	1.9832	1.9937	0.0106	0.53
1.40	2.5908	2.6117	0.0209	0.80
1.50	3.4509	3.4904	0.0394	1.13

TABLE 9.6   Improved Euler Method with $h = 0.05$

x_n	y_n	True value	Abs. error	% Rel. error
1.00	1.0000	1.0000	0.0000	0.00
1.05	1.1077	1.1079	0.0002	0.02
1.10	1.2332	1.2337	0.0004	0.04
1.15	1.3798	1.3806	0.0008	0.06
1.20	1.5514	1.5527	0.0013	0.08
1.25	1.7531	1.7551	0.0020	0.11
1.30	1.9909	1.9937	0.0029	0.14
1.35	2.2721	2.2762	0.0041	0.18
1.40	2.6060	2.6117	0.0057	0.22
1.45	3.0038	3.0117	0.0079	0.26
1.50	3.4795	3.4904	0.0108	0.31

A brief word of caution is in order here. We cannot compute all the values of y_n^* first and then substitute these values into the first formula of (5). In other words, we cannot use the data in Table 9.3 to help construct the values in Table 9.5. Why not?

Truncation Errors for the Improved Euler Method   The local truncation error for the improved Euler method is $O(h^3)$. The derivation of this result is similar to the derivation of the local truncation error for Euler's method and is left for you. See Problem 16. Since the local truncation for the improved Euler method is $O(h^3)$, the global truncation error is $O(h^2)$. This can be seen in Example 4; when the step size is halved from $h = 0.1$ to $h = 0.05$, the absolute error at $x = 1.50$ is reduced from 0.0394 to 0.0108, a reduction of approximately $(\frac{1}{2})^2 = \frac{1}{4}$.

ODE Solvers   As already noted, Euler's method uses local linear approximations to generate the sequence of points (x_n, y_n). If we connect these points with straight line segments, we obtain a polygonal curve that approximates the actual solution curve. This is illustrated in Figure 9.12, using step sizes $h = 1$, $h = 0.5$, and $h = 0.1$. It is apparent from the figure that the approximation improves as the step size decreases. For h sufficiently small, the polygonal curve will appear to be smooth and will, it is hoped, be close to the actual solution curve. This polygonal approximation is well suited to computers, and a program that does this is sometimes referred to as an **ODE solver.** Frequently an ODE solver is part of a larger, more versatile software package.

SECTION 9.2 EXERCISES

Answers to odd-numbered problems begin on page A-15.

1. Consider the initial-value problem

$$y' = (x + y - 1)^2, \qquad y(0) = 2.$$

  (a) Solve the initial-value problem in terms of elementary functions. [*Hint:* Let $u = x + y - 1$.]

  (b) Use Euler's formula with $h = 0.1$ and $h = 0.05$ to obtain approximate values of the solution of the initial-value problem at $x = 0.5$. Compare the approximate values with the exact values computed using the solution from part (a).

2. Repeat the calculations of Problem 1(b) using the improved Euler formula.

Given the initial-value problems in Problems 3–12, use Euler's formula to obtain a four-decimal approximation to the indicated value. First use $h = 0.1$ and then use $h = 0.05$.

3. $y' = 2x - 3y + 1, y(1) = 5$; $y(1.5)$

4. $y' = 4x - 2y, y(0) = 2$; $y(0.5)$

5. $y' = 1 + y^2, y(0) = 0$; $y(0.5)$

6. $y' = x^2 + y^2, y(0) = 1$; $y(0.5)$

7. $y' = e^{-y}$, $y(0) = 0$; $y(0.5)$

8. $y' = x + y^2$, $y(0) = 0$; $y(0.5)$

9. $y' = (x - y)^2$, $y(0) = 0.5$; $y(0.5)$

10. $y' = xy + \sqrt{y}$, $y(0) = 1$; $y(0.5)$

11. $y' = xy^2 - \dfrac{y}{x}$, $y(1) = 1$; $y(1.5)$

12. $y' = y - y^2$, $y(0) = 0.5$; $y(0.5)$

13. As parts (a)–(e) of this problem, repeat the calculations of Problems 3, 5, 7, 9, and 11 using the improved Euler formula.

14. As parts (a)–(e) of this problem, repeat the calculations of Problems 4, 6, 8, 10, and 12 using the improved Euler formula.

15. Although it may not be obvious from the differential equation, its solution could "behave badly" near a point x at which we wish to approximate $y(x)$. Numerical procedures may give widely differing results near this point. Let $y(x)$ be the solution of the initial-value problem

$$y' = x^2 + y^3, \qquad y(1) = 1.$$

(a) Use an ODE solver to obtain a graph of the solution on the interval $[1, 1.4]$.

(b) Using the step size $h = 0.1$, compare the results obtained from Euler's formula with the results from the improved Euler formula in the approximation of $y(1.4)$.

16. In this problem we show that the local truncation error for the improved Euler method is $O(h^3)$.

(a) Use Taylor's formula with remainder to show that

$$y''(x_n) = \frac{y'(x_{n+1}) - y'(x_n)}{h} - \frac{1}{2} hy'''(c).$$

[*Hint*: Set $k = 2$ and differentiate the Taylor polynomial (3).]

(b) Use the Taylor polynomial with $k = 2$ to show that

$$y(x_{n+1}) = y(x_n) + h\frac{y'(x_n) + y'(x_{n+1})}{2} + O(h^3).$$

17. Consider the initial-value problem $y' = 2y$, $y(0) = 1$. The analytic solution is $y(x) = e^{2x}$.

(a) Approximate $y(0.1)$ using one step and Euler's method.

(b) Find a bound for the local truncation error in y_1.

(c) Compare the actual error in y_1 with your error bound.

(d) Approximate $y(0.1)$ using two steps and Euler's method.

(e) Verify that the global truncation error for Euler's method is $O(h)$ by comparing the errors in parts (a) and (d).

18. Repeat Problem 17 using the improved Euler method. Its global truncation error is $O(h^2)$.

19. Repeat Problem 17 using the initial-value problem $y' = -2y + x$, $y(0) = 1$. The analytic solution is $y(x) = \frac{1}{2}x - \frac{1}{4} + \frac{5}{4}e^{-2x}$.

20. Repeat Problem 19 using the improved Euler method. Its global truncation error is $O(h^2)$.

21. Consider the initial-value problem $y' = 2x - 3y + 1$, $y(1) = 5$. The analytic solution is $y(x) = \frac{1}{9} + \frac{2}{3}x + \frac{38}{9}e^{-3(x-1)}$.
    (a) Find a formula involving c and h for the local truncation error in the nth step if Euler's method is used.
    (b) Find a bound for the local truncation error in each step if $h = 0.1$ is used to approximate $y(1.5)$.
    (c) Approximate $y(1.5)$ using $h = 0.1$ and $h = 0.05$ with Euler's method. See Problem 3.
    (d) Calculate the errors in part (c) and verify that the global truncation error of Euler's method is $O(h)$.

22. Repeat Problem 21 using the improved Euler method, which has global truncation error $O(h^2)$. See Problem 13(a). You may need to keep more than four decimal places to see the effect of reducing the order of error.

23. Repeat Problem 21 for the initial-value problem $y' = e^{-y}$, $y(0) = 0$. The analytic solution is $y(x) = \ln(x + 1)$. Approximate $y(0.5)$. See Problem 7.

24. Repeat Problem 23 using the improved Euler method, which has global truncation error $O(h^2)$. See Problem 13(a). You may need to keep more than four decimal places to see the effect of reducing the order of error.

9.3   RUNGE-KUTTA METHODS

- *First-order Runge-Kutta method* ■ *Second-order Runge-Kutta method*
- *Fourth-order Runge-Kutta method* ■ *Truncation errors* ■ *Adaptive methods*

Probably one of the most popular as well as most accurate numerical procedures used in obtaining approximate solutions to the initial-value problem $y' = f(x, y)$, $y(x_0) = y_0$ is the **fourth-order Runge-Kutta method.** As the name suggests, there are Runge-Kutta methods of different orders. These methods are derived using the Taylor series expansion with remainder for $y(x_n + h)$:

$$y(x_{n+1}) = y(x_n + h) = y(x_n) + hy'(x_n) + \frac{h^2}{2!}y''(x_n) + \frac{h^3}{3!}y'''(x_n) + \cdots + \frac{h^{k+1}}{(k+1)!}y^{(k+1)}(c),$$

where c is a number between x_n and $x_n + h$. In the case when $k = 1$ and the remainder $\frac{h^2}{2}y''(c)$ is small, we obtain the familiar iteration formula

$$y_{n+1} = y_n + hy'_n = y_n + hf(x_n, y_n).$$

In other words, the basic Euler method is a **first-order Runge-Kutta procedure.**

We consider now the **second-order Runge-Kutta procedure.** This consists of finding constants a, b, α, and β so that the formula

$$y_{n+1} = y_n + ak_1 + bk_2, \tag{1}$$

where
$$k_1 = hf(x_n, y_n)$$
$$k_2 = hf(x_n + \alpha h, y_n + \beta k_1),$$

agrees with a Taylor polynomial of degree 2. It can be shown that this can be done whenever the constants satisfy

$$a + b = 1, \quad b\alpha = \frac{1}{2}, \quad \text{and} \quad b\beta = \frac{1}{2}. \tag{2}$$

This is a system of three equations in four unknowns and has infinitely many solutions. Observe that when $a = b = \frac{1}{2}$ and $\alpha = \beta = 1$, (1) reduces to the improved Euler formula. Since the formula agrees with a Taylor polynomial of degree 2, the local truncation error for this method is $O(h^3)$ and the global truncation error is $O(h^2)$.

Notice that the sum $ak_1 + bk_2$ in (1) is a weighted average of k_1 and k_2 since $a + b = 1$. The numbers k_1 and k_2 are multiples of approximations to the slope of the solution curve $y(x)$ at two different points in the interval from x_n to x_{n+1}.

Fourth-Order Runge-Kutta Formula
The **fourth-order Runge-Kutta procedure** consists of finding appropriate constants so that the formula

$$y_{n+1} = y_n + ak_1 + bk_2 + ck_3 + dk_4,$$

where
$$k_1 = hf(x_n, y_n)$$
$$k_2 = hf(x_n + \alpha_1 h, y_n + \beta_1 k_1)$$
$$k_3 = hf(x_n + \alpha_2 h, y_n + \beta_2 k_1 + \beta_3 k_2)$$
$$k_4 = hf(x_n + \alpha_3 h, y_n + \beta_4 k_1 + \beta_5 k_2 + \beta_6 k_3),$$

agrees with a Taylor polynomial of degree 4. This results in 11 equations in 13 unknowns. The most commonly used set of values for the constants yields the following result:

$$y_{n+1} = y_n + \frac{1}{6}(k_1 + 2k_2 + 2k_3 + k_4),$$

$$k_1 = hf(x_n, y_n)$$
$$k_2 = hf(x_n + \tfrac{1}{2}h, y_n + \tfrac{1}{2}k_1)$$
$$k_3 = hf(x_n + \tfrac{1}{2}h, y_n + \tfrac{1}{2}k_2) \tag{3}$$
$$k_4 = hf(x_n + h, y_n + k_3).$$

You are advised to look carefully at the formulas in (3); note that k_2 depends on k_1, k_3 depends on k_2, and k_4 depends on k_3. Also k_2 and k_3 involve approximations to the slope at the midpoint of the interval between x_n and x_{n+1}.

EXAMPLE 1　**Runge-Kutta Method**

Use the Runge-Kutta method with $h = 0.1$ to obtain an approximation to $y(1.5)$ for the solution of

$$y' = 2xy, \quad y(1) = 1.$$

SOLUTION   For the sake of illustration let us compute the case when $n = 0$. From (3) we find

$$k_1 = (0.1)f(x_0, y_0) = (0.1)(2x_0 y_0) = 0.2$$

$$k_2 = (0.1)f(x_0 + \tfrac{1}{2}(0.1), y_0 + \tfrac{1}{2}(0.2))$$

$$= (0.1)2\left(x_0 + \frac{1}{2}(0.1)\right)\left(y_0 + \frac{1}{2}(0.2)\right) = 0.231$$

$$k_3 = (0.1)f(x_0 + \tfrac{1}{2}(0.1), y_0 + \tfrac{1}{2}(0.231))$$

$$= (0.1)2\left(x_0 + \frac{1}{2}(0.1)\right)\left(y_0 + \frac{1}{2}(0.231)\right) = 0.234255$$

$$k_4 = (0.1)f(x_0 + 0.1, y_0 + 0.234255)$$

$$= (0.1)2(x_0 + 0.1)(y_0 + 0.234255) = 0.2715361$$

and therefore

$$y_1 = y_0 + \frac{1}{6}(k_1 + 2k_2 + 2k_3 + k_4)$$

$$= 1 + \frac{1}{6}(0.2 + 2(0.231) + 2(0.234255) + 0.2715361) = 1.23367435.$$

The remaining calculations are summarized in Table 9.7, whose entries are rounded to four decimal places.

TABLE 9.7   Runge-Kutta Method with $h = 0.1$

x_n	y_n	True value	Abs. error	% Rel. error
1.00	1.0000	1.0000	0.0000	0.00
1.10	1.2337	1.2337	0.0000	0.00
1.20	1.5527	1.5527	0.0000	0.00
1.30	1.9937	1.9937	0.0000	0.00
1.40	2.6116	2.6117	0.0001	0.00
1.50	3.4902	3.4904	0.0001	0.00

Inspection of Table 9.7 shows why the fourth-order Runge-Kutta method is so popular. If four-decimal-place accuracy is all that we desire, there is no need to use a smaller step size. Table 9.8 compares the results of applying the Euler, improved Euler, and fourth-order Runge-Kutta methods to the initial-value problem $y' = 2xy$, $y(1) = 1$. (See Examples 2 and 3 in Section 9.2.)

Truncation Errors for the Runge-Kutta Method   Since the first equation in (3) agrees with a Taylor polynomial of degree 4, the local truncation error for this method is

$$y^{(5)}(c)\frac{h^5}{5!} \quad \text{or} \quad O(h^5),$$

and the global truncation error is thus $O(h^4)$. It is now obvious why this is called the *fourth-order* Runge-Kutta method.

TABLE 9.8 $y' = 2xy$, $y(1) = 1$

	Comparison of numerical methods with $h = 0.1$					Comparison of numerical methods with $h = 0.05$			
x_n	Euler	Improved Euler	Runge-Kutta	True value	x_n	Euler	Improved Euler	Runge-Kutta	True value
1.00	1.0000	1.0000	1.0000	1.0000	1.00	1.0000	1.0000	1.0000	1.0000
1.10	1.2000	1.2320	1.2337	1.2337	1.05	1.1000	1.1077	1.1079	1.1079
1.20	1.4640	1.5479	1.5527	1.5527	1.10	1.2155	1.2332	1.2337	1.2337
1.30	1.8154	1.9832	1.9937	1.9937	1.15	1.3492	1.3798	1.3806	1.3806
1.40	2.2874	2.5908	2.6116	2.6117	1.20	1.5044	1.5514	1.5527	1.5527
1.50	2.9278	3.4509	3.4902	3.4904	1.25	1.6849	1.7531	1.7551	1.7551
					1.30	1.8955	1.9909	1.9937	1.9937
					1.35	2.1419	2.2721	2.2762	2.2762
					1.40	2.4311	2.6060	2.6117	2.6117
					1.45	2.7714	3.0038	3.0117	3.0117
					1.50	3.1733	3.4795	3.4903	3.4904

EXAMPLE 2   **Bound for Local/Global Truncation Errors**

Analyze the local and global truncation errors for the fourth-order Runge-Kutta method applied to $y' = 2xy$, $y(1) = 1$.

SOLUTION   By differentiating the known solution $y(x) = e^{x^2 - 1}$ we get

$$y^{(5)}(c) \frac{h^5}{5!} = (120c + 160c^3 + 32c^5)e^{c^2 - 1} \frac{h^5}{5!}. \tag{4}$$

Thus with $c = 1.5$, (4) yields a bound of 0.00028 on the local truncation error for each of the five steps when $h = 0.1$. Note that in Table 9.7 the actual error in y_1 is much less than this bound.

Table 9.9 gives the approximations to the solution of the initial-value problem at $x = 1.5$ that are obtained from the fourth-order Runge-Kutta method. By computing the value of the exact solution at $x = 1.5$ we can find the error in these approximations. Because the method is so accurate, many decimal places must be used in the numerical solution to see the effect of halving the step size. Note that when h is halved, from $h = 0.1$ to $h = 0.05$, the error is divided by a factor of about $2^4 = 16$, as expected.

TABLE 9.9 Runge-Kutta Method

h	Approximation	Error
0.1	3.49021064	$1.323210889 \times 10^{-4}$
0.05	3.49033382	$9.137760898 \times 10^{-6}$

Adaptive Methods   We have seen that the accuracy of a numerical method can be improved by decreasing the step size h. Of course, this enhanced accuracy is usually obtained at a cost—namely, increased computation time and greater possibility of round-off error. In general, over the interval of approximation there may be subintervals where a relatively large step size suffices and other subintervals where a smaller step size is necessary in order to keep the truncation error within a desired limit. Numerical methods that use a variable step size are called **adaptive methods.** One of the more popular of these methods for approximating solutions of differential equations is called the **Runge-Kutta-Fehlberg algorithm.**

SECTION 9.3 EXERCISES

Answers to odd-numbered problems begin on page A-17.

1. Use the fourth-order Runge-Kutta method with $h = 0.1$ to obtain a four-decimal-place approximation to the solution of the initial-value problem

$$y' = (x + y - 1)^2, \qquad y(0) = 2$$

at $x = 0.5$. Compare the approximate values with the exact values obtained in Problem 1 in Exercises 9.2.

2. Solve the equations in (2) using the assumption $a = \frac{1}{4}$. Use the resulting second-order Runge-Kutta method to obtain a four-decimal-place approximation to the solution of the initial-value problem

$$y' = (x + y - 1)^2, \qquad y(0) = 2$$

at $x = 0.5$. Compare the approximate values with the values obtained in Problem 2 in Exercises 9.2.

Given the initial-value problems in Problems 3–12, use the Runge-Kutta method with $h = 0.1$ to obtain a four-decimal-place approximation to the indicated value.

3. $y' = 2x - 3y + 1, y(1) = 5; \quad y(1.5)$

4. $y' = 4x - 2y, y(0) = 2; \quad y(0.5)$

5. $y' = 1 + y^2, y(0) = 0; \quad y(0.5)$

6. $y' = x^2 + y^2, y(0) = 1; \quad y(0.5)$

7. $y' = e^{-y}, y(0) = 0; \quad y(0.5)$

8. $y' = x + y^2, y(0) = 0; \quad y(0.5)$

9. $y' = (x - y)^2, y(0) = 0.5; \quad y(0.5)$

10. $y' = xy + \sqrt{y}, y(0) = 1; \quad y(0.5)$

11. $y' = xy^2 - \dfrac{y}{x}, y(1) = 1; \quad y(1.5)$

12. $y' = y - y^2, y(0) = 0.5; \quad y(0.5)$

13. If air resistance is proportional to the square of the instantaneous velocity, then the velocity v of a mass m dropped from a height h is

determined from

$$m\frac{dv}{dt} = mg - kv^2, \quad k > 0.$$

Let $v(0) = 0$, $k = 0.125$, $m = 5$ slugs, and $g = 32$ ft/s^2.
  (a) Use the Runge-Kutta method with $h = 1$ to find an approximation to the velocity of the falling mass at $t = 5$ s.
  (b) Use an ODE solver to graph the solution of the initial-value problem.
  (c) Use separation of variables to solve the initial-value problem and find the true value $v(5)$.

14. A mathematical model for the area A (in cm^2) that a colony of bacteria (*B. dendroides*) occupies is given by

$$\frac{dA}{dt} = A(2.128 - 0.0432A).*$$

Suppose that the initial area is 0.24 cm^2.
  (a) Use the Runge-Kutta method with $h = 0.5$ to complete the following table.

t (days)	1	2	3	4	5
A(observed)	2.78	13.53	36.30	47.50	49.40
A (approximated)					

  (b) Use an ODE solver to graph the solution of the initial-value problem. Estimate the values $A(1)$, $A(2)$, $A(3)$, $A(4)$, and $A(5)$ from the graph.
  (c) Use separation of variables to solve the initial-value problem and compute the values $A(1)$, $A(2)$, $A(3)$, $A(4)$, and $A(5)$.

15. Consider the initial-value problem

$$y' = x^2 + y^3, \qquad y(1) = 1.$$

(See Problem 15 in Exercises 9.2.)
  (a) Compare the results obtained from using the Runge-Kutta formula over the interval $[1, 1.4]$ with step sizes $h = 0.1$ and $h = 0.05$.
  (b) Use an ODE solver to obtain a graph of the solution on the interval $[1, 1.4]$.

16. Consider the initial-value problem $y' = 2y$, $y(0) = 1$. The analytic solution is $y(x) = e^{2x}$.
  (a) Approximate $y(0.1)$ using one step and the fourth-order Runge-Kutta method.
  (b) Find a bound for the local truncation error in y_1.
  (c) Compare the actual error in y_1 with your error bound.

*See V. A. Kostitzin, *Mathematical Biology* (London: Harrap, 1939).

    (d) Approximate $y(0.1)$ using two steps and the fourth-order Runge-Kutta method.

    (e) Verify that the global truncation error for the fourth-order Runge-Kutta method is $O(h^4)$ by comparing the errors in parts (a) and (d).

17. Repeat Problem 16 using the initial-value problem $y' = -2y + x$, $y(0) = 1$. The analytic solution is $y(x) = \frac{1}{2}x - \frac{1}{4} + \frac{5}{4}e^{-2x}$.

18. Consider the initial-value problem $y' = 2x - 3y + 1$, $y(1) = 5$. The analytic solution is $y(x) = \frac{1}{9} + \frac{2}{3}x + \frac{38}{9}e^{-3(x-1)}$.

    (a) Find a formula involving c and h for the local truncation error in the nth step if the fourth-order Runge-Kutta method is used.

    (b) Find a bound for the local truncation error in each step if $h = 0.1$ is used to approximate $y(1.5)$.

    (c) Approximate $y(1.5)$ using the fourth-order Runge-Kutta method with $h = 0.1$ and $h = 0.05$. See Problem 3. You will need to carry more than six decimal places to see the effect of reducing the step size.

19. Repeat Problem 18 for the initial-value problem $y' = e^{-y}$, $y(0) = 0$. The analytic solution is $y(x) = \ln(x + 1)$. Approximate $y(0.5)$. See Problem 7.

20. The Runge-Kutta method for solving an initial-value problem over an interval $[a, b]$ results in a finite set of points that are supposed to approximate points on the graph of the exact solution. In order to expand this set of discrete points to an approximate solution defined at all points on the interval $[a, b]$ we can use an **interpolating function.** This is a function, supported by most computer algebra systems, that agrees with the given data exactly and assumes a smooth transition between data points. These interpolating functions may be polynomials or sets of polynomials joined together smoothly. In Mathematica the command **y=Interpolation[data]** can be used to obtain an interpolating function through the points **data** $= \{\{x_0, y_0\}, \{x_1, y_1\}, \dots, \{x_n, y_n\}\}$. The interpolating function **y[x]** can now be treated like any other function built into the computer algebra system.

    (a) Find the exact solution of the initial-value problem $y' = -y + 10 \sin 3x$, $y(0) = 0$ on the interval $[0, 2]$. Graph this solution and find its positive roots.

    (b) Use the fourth-order Runge-Kutta method with $h = 0.1$ to approximate a solution of the initial-value problem in part (a). Obtain an interpolating function and graph it. Find the positive roots of the interpolating function on the interval $[0, 2]$.

Discussion Problem

21. A count of the number of evaluations of the function f used in solving the initial-value problem $y' = f(x, y)$, $y(x_0) = y_0$ is used as a measure of the computational complexity of a numerical method. Determine the number of evaluations of f required for each step of the Euler, improved Euler, and Runge-Kutta methods. By considering some specific examples, compare the accuracy of these methods when used with comparable computational complexities.

9.4 MULTISTEP METHODS

- *Single-step methods* ■ *Multistep methods* ■ *Predictor-corrector methods*
- *Adams-Bashforth/Adams-Moulton method* ■ *Stability of numerical methods*

The Euler and Runge-Kutta methods discussed in the preceding sections are examples of **single-step** methods. In these methods each successive value y_{n+1} is computed based only on information about the immediately preceding value y_n. A **multistep** or **continuing method,** on the other hand, uses the values from several previously computed steps to obtain the value of y_{n+1}. There are a large number of multistep formulas that can be applied to obtain approximations to solutions of differential equations. Since it is not our intention to survey the vast field of numerical procedures, we will consider only one such method here. This, like the improved Euler formula, is a **predictor-corrector method.** That is, one formula is used to predict a value y_{n+1}^*, which in turn is used to obtain a corrected value y_{n+1}.

Adams-Bashforth/Adams-Moulton Method One of the most popular multistep methods is the fourth-order **Adams-Bashforth/Adams-Moulton method.** The predictor in this method is the Adams-Bashforth formula

$$y_{n+1}^* = y_n + \frac{h}{24}(55y_n' - 59y_{n-1}' + 37y_{n-2}' - 9y_{n-3}'), \tag{1}$$

$$y_n' = f(x_n, y_n)$$
$$y_{n-1}' = f(x_{n-1}, y_{n-1})$$
$$y_{n-2}' = f(x_{n-2}, y_{n-2})$$
$$y_{n-3}' = f(x_{n-3}, y_{n-3})$$

for $n \geq 3$. The value of y_{n+1}^* is then substituted into the Adams-Moulton corrector

$$y_{n+1} = y_n + \frac{h}{24}(9y_{n+1}' + 19y_n' - 5y_{n-1}' + y_{n-2}'), \tag{2}$$
$$y_{n+1}' = f(x_{n+1}, y_{n+1}^*).$$

Notice that formula (1) requires that we know the values of y_0, y_1, y_2, and y_3 in order to obtain y_4. The value of y_0 is, of course, the given initial condition. Since the local truncation error of the Adams-Bashforth/Adams-Moulton method is $O(h^5)$, the values of y_1, y_2, and y_3 are generally computed by a method with the same error property, such as the fourth-order Runge-Kutta formula.

EXAMPLE 1 **Adams-Bashforth/Adams-Moulton Method**

Use the Adams-Bashforth/Adams-Moulton method with $h = 0.2$ to obtain an approximation to $y(0.8)$ for the solution of

$$y' = x + y - 1, \qquad y(0) = 1.$$

SOLUTION   With a step size of $h = 0.2$, $y(0.8)$ will be approximated by y_4. To get started we use the Runge-Kutta method with $x_0 = 0$, $y_0 = 1$, and $h = 0.2$ to obtain

$$y_1 = 1.02140000, \qquad y_2 = 1.09181796, \qquad y_3 = 1.22210646.$$

Now with the identifications $x_0 = 0$, $x_1 = 0.2$, $x_2 = 0.4$, $x_3 = 0.6$, and $f(x, y) = x + y - 1$, we find

$$y_0' = f(x_0, y_0) = (0) + (1) - 1 = 0$$

$$y_1' = f(x_1, y_1) = (0.2) + (1.02140000) - 1 = 0.22140000$$

$$y_2' = f(x_2, y_2) = (0.4) + (1.09181796) - 1 = 0.49181796$$

$$y_3' = f(x_3, y_3) = (0.6) + (1.22210646) - 1 = 0.82210646.$$

With the foregoing values the predictor (1) then gives

$$y_4^* = y_3 + \frac{0.2}{24}(55y_3' - 59y_2' + 37y_1' - 9y_0') = 1.42535975.$$

To use the corrector (2) we first need

$$y_4' = f(x_4, y_4^*) = 0.8 + 1.42535975 - 1 = 1.22535975.$$

Finally, (2) yields

$$y_4 = y_3 + \frac{0.2}{24}(9y_4' + 19y_3' - 5y_2' + y_1') = 1.42552788. \qquad \blacksquare$$

You should verify that the exact value of $y(0.8)$ in Example 1 is $y(0.8) = 1.42554093$.

Stability of Numerical Methods

An important consideration in using numerical methods to approximate the solution of an initial-value problem is the stability of the method. Simply stated, a numerical method is **stable** if small changes in the initial condition result in only small changes in the computed solution. A numerical method is said to be **unstable** if it is not stable. The reason that stability considerations are important is that in each step after the first step of a numerical technique we are essentially starting over again with a new initial-value problem, where the initial condition is the approximate solution value computed in the preceding step. Because of the presence of round-off error, this value will almost certainly vary at least slightly from the true value of the solution. Besides round-off error, another common source of error occurs in the initial condition itself; in physical applications the data are often obtained by imprecise measurements.

One possible method for detecting instability in the numerical solution of a specific initial-value problem is to compare the approximate solutions obtained when decreasing step sizes are used. If the numerical method is unstable, the error may actually increase with smaller step sizes. Another way of checking stability is to observe what happens to solutions when the initial condition is slightly perturbed (for example, change $y(0) = 1$ to $y(0) = 0.999$).

For a more detailed and precise discussion of stability consult a numerical analysis text. In general, all of the methods we have discussed in this chapter have good stability characteristics.

Advantages/Disadvantages of Multistep Methods   Many considerations enter into the choice of a method to solve a differential equation numerically. Single-step methods, particularly the Runge-Kutta method, are often chosen because of their accuracy and the fact that they are easy to program. However, a major drawback is that the right-hand side of the differential equation must be evaluated many times at each step. For instance, the fourth-order Runge-Kutta method requires four function evaluations for each step. See Problem 21 in Exercises 9.3. On the other hand, if the function evaluations in the previous step have been calculated and stored, a multistep method requires only one new function evaluation for each step. This can lead to great savings in time and expense.

As an example, to solve $y' = f(x, y)$, $y(x_0) = y_0$ numerically using n steps by the fourth-order Runge-Kutta method requires $4n$ function evaluations. The Adams-Bashforth multistep method requires 16 function evaluations for the Runge-Kutta fourth-order starter and $n - 4$ for the n Adams-Bashforth steps, giving a total of $n + 12$ function evaluations for this method. In general the Adams-Bashforth multistep method requires slightly more than a quarter of the number of function evaluations required for the fourth-order Runge-Kutta method. If the evaluation of $f(x, y)$ is complicated, the multistep method will be more efficient.

Another issue involved with multistep methods is how many times the Adams-Moulton corrector formula should be repeated in each step. Each time the corrector is used, another function evaluation is done, and so the accuracy is increased at the expense of losing an advantage of the multistep method. In practice, the corrector is calculated once, and if the value of y_{n+1} is changed by a large amount, the entire problem is restarted using a smaller step size. This is often the basis of the variable step size methods, whose discussion is beyond the scope of this text.

SECTION 9.4 EXERCISES

Answers to odd-numbered problems begin on page A-18.

1. Find the exact solution of the initial-value problem in Example 1. Compare the exact values of $y(0.2)$, $y(0.4)$, $y(0.6)$, and $y(0.8)$ with the approximations y_1, y_2, y_3, and y_4.

2. Write a computer program for the Adams-Bashforth/Adams-Moulton method.

In Problems 3 and 4 use the Adams-Bashforth/Adams-Moulton method to approximate $y(0.8)$, where $y(x)$ is the solution of the given initial-value problem. Use $h = 0.2$ and the Runge-Kutta method to compute y_1, y_2, and y_3.

3. $y' = 2x - 3y + 1$, $y(0) = 1$ 4. $y' = 4x - 2y$, $y(0) = 2$

In Problems 5–8 use the Adams-Bashforth/Adams-Moulton method to approximate $y(1.0)$, where $y(x)$ is the solution of the given initial-value problem. Use $h = 0.2$ and $h = 0.1$ and the Runge-Kutta method to compute y_1, y_2, and y_3.

5. $y' = 1 + y^2$, $y(0) = 0$ **6.** $y' = y + \cos x$, $y(0) = 1$

7. $y' = (x - y)^2$, $y(0) = 0$ **8.** $y' = xy + \sqrt{y}$, $y(0) = 1$

9.5 HIGHER-ORDER EQUATIONS AND SYSTEMS

- *Second-order initial-value problem as a system* ■ *Systems of DEs reduced to first-order systems*
- *Numerical methods applied to systems*

Second-Order Initial-Value Problems In Sections 9.2–9.4 we focused on numerical techniques that could be applied to obtain an approximation to the solution of a first-order initial-value problem $y' = f(x, y), y(x_0) = y_0$. To approximate the solution of a second-order initial-value problem

$$y'' = f(x, y, y'), \qquad y(x_0) = y_0, \quad y'(x_0) = y_1 \tag{1}$$

we reduce the differential equation to a system of two first-order equations. When we let $y' = u$, the initial-value problem in (1) becomes

$$y' = u$$
$$u' = f(x, y, u) \tag{2}$$
$$y(x_0) = y_0, \quad u(x_0) = y_1.$$

We can now solve this system numerically by adapting the techniques discussed in Sections 9.2–9.4 to the system. We do this by simply applying a particular method to each equation in the system. For example, **Euler's method** applied to the system (2) would be

$$y_{n+1} = y_n + hu_n$$
$$u_{n+1} = u_n + hf(x_n, y_n, u_n). \tag{3}$$

EXAMPLE 1 **Euler's Method**

Use Euler's method to obtain the approximate value of $y(0.2)$, where $y(x)$ is the solution of the initial-value problem

$$y'' + xy' + y = 0, \qquad y(0) = 1, \quad y'(0) = 2.$$

SOLUTION In terms of the substitution $y' = u$, the equation is equivalent to the system

$$y' = u$$
$$u' = -xu - y.$$

Thus from (3) we obtain

$$y_{n+1} = y_n + hu_n$$
$$u_{n+1} = u_n + h[-x_n u_n - y_n].$$

Using the step size $h = 0.1$ and $y_0 = 1$, $u_0 = 2$, we find

$$y_1 = y_0 + (0.1)u_0 = 1 + (0.1)2 = 1.2$$
$$u_1 = u_0 + (0.1)[-x_0 u_0 - y_0] = 2 + (0.1)[-(0)(2) - 1] = 1.9$$
$$y_2 = y_1 + (0.1)u_1 = 1.2 + (0.1)(1.9) = 1.39$$
$$u_2 = u_1 + (0.1)[-x_1 u_1 - y_1] = 1.9 + (0.1)[-(0.1)(1.9) - 1.2] = 1.761.$$

In other words, $y(0.2) \approx 1.39$ and $y'(0.2) \approx 1.761$. ∎

In general, we can reduce every nth-order differential equation $y^{(n)} = f(x, y, y', \ldots, y^{(n-1)})$ to a system of n first-order equations using the substitutions $y = u_1$, $y' = u_2$, $y'' = u_3$, $\ldots$, $y^{(n-1)} = u_n$.

Systems Reduced to First-Order Systems   Using a procedure similar to that just discussed, we can often reduce a system of higher-order differential equations to a system of first-order equations by first solving for the highest-order derivative of each dependent variable and then making appropriate substitutions for the lower-order derivatives.

EXAMPLE 2   **A System Rewritten as a First-Order System**

Write
$$x'' - x' + 5x + 2y'' = e^t$$
$$-2x + y'' + 2y = 3t^2$$

as a system of first-order differential equations.

SOLUTION   Write the system as
$$x'' + 2y'' = e^t - 5x + x'$$
$$y'' = 3t^2 + 2x - 2y$$

and then eliminate y'' by multiplying the second equation by 2 and subtracting. This gives

$$x'' = -9x + 4y + x' + e^t - 6t^2.$$

Since the second equation of the system already expresses the highest-order derivative of y in terms of the remaining functions, we are now in a position to introduce new variables. If we let $x' = u$ and $y' = v$, the expressions for x'' and y'' become, respectively,

$$u' = x'' = -9x + 4y + u + e^t - 6t^2$$
$$v' = y'' = 2x - 2y + 3t^2.$$

The original system can then be written in the form
$$x' = u$$
$$y' = v$$
$$u' = -9x + 4y + u + e^t - 6t^2$$
$$v' = 2x - 2y + 3t^2.$$
∎

It may not always be possible to carry out the reductions illustrated in Example 2.

Numerical Solution of a System The solution of a system of the form

$$\frac{dx_1}{dt} = f_1(t, x_1, x_2, \ldots, x_n)$$

$$\frac{dx_2}{dt} = f_2(t, x_1, x_2, \ldots, x_n)$$

$$\vdots \qquad \qquad \vdots$$

$$\frac{dx_n}{dt} = f_n(t, x_1, x_2, \ldots, x_n)$$

can be approximated by a version of the Euler, Runge-Kutta, or Adams-Bashforth/Adams-Moulton method adapted to the system. For example, the **fourth-order Runge-Kutta method** applied to the system

$$x' = f(t, x, y)$$
$$y' = g(t, x, y) \tag{4}$$
$$x(t_0) = x_0, \quad y(t_0) = y_0$$

looks like this:

$$x_{n+1} = x_n + \frac{1}{6}(m_1 + 2m_2 + 2m_3 + m_4)$$

$$\tag{5}$$

$$y_{n+1} = y_n + \frac{1}{6}(k_1 + 2k_2 + 2k_3 + k_4),$$

where

$$\begin{aligned}
m_1 &= hf(t_n, x_n, y_n) & k_1 &= hg(t_n, x_n, y_n) \\
m_2 &= hf(t_n + \tfrac{1}{2}h, x_n + \tfrac{1}{2}m_1, y_n + \tfrac{1}{2}k_1) & k_2 &= hg(t_n + \tfrac{1}{2}h, x_n + \tfrac{1}{2}m_1, y_n + \tfrac{1}{2}k_1) \\
m_3 &= hf(t_n + \tfrac{1}{2}h, x_n + \tfrac{1}{2}m_2, y_n + \tfrac{1}{2}k_2) & k_3 &= hg(t_n + \tfrac{1}{2}h, x_n + \tfrac{1}{2}m_2, y_n + \tfrac{1}{2}k_2) \\
m_4 &= hf(t_n + h, x_n + m_3, y_n + k_3) & k_4 &= hg(t_n + h, x_n + m_3, y_n + k_3).
\end{aligned} \tag{6}$$

EXAMPLE 3   **Runge-Kutta Method**

Consider the initial-value problem

$$x' = 2x + 4y$$
$$y' = -x + 6y$$
$$x(0) = -1, \quad y(0) = 6.$$

Use the fourth-order Runge-Kutta method to approximate $x(0.6)$ and $y(0.6)$. Compare the results for $h = 0.2$ and $h = 0.1$.

SOLUTION   We illustrate the computations of x_1 and y_1 with the step size $h = 0.2$. With the identifications $f(t, x, y) = 2x + 4y$, $g(t, x, y) = -x + 6y$, $t_0 = 0$, $x_0 = -1$, and $y_0 = 6$, we see from (6) that

$$m_1 = hf(t_0, x_0, y_0) = 0.2f(0, -1, 6) = 0.2[2(-1) + 4(6)] = 4.4000$$
$$k_1 = hg(t_0, x_0, y_0) = 0.2g(0, -1, 6) = 0.2[-1(-1) + 6(6)] = 7.4000$$
$$m_2 = hf(t_0 + \tfrac{1}{2}h, x_0 + \tfrac{1}{2}m_1, y_0 + \tfrac{1}{2}k_1) = 0.2f(0.1, 1.2, 9.7) = 8.2400$$

$$k_2 = hg(t_0 + \tfrac{1}{2}h, x_0 + \tfrac{1}{2}m_1, y_0 + \tfrac{1}{2}k_1) = 0.2\,g(0.1, 1.2, 9.7) = 11.4000$$
$$m_3 = hf(t_0 + \tfrac{1}{2}h, x_0 + \tfrac{1}{2}m_2, y_0 + \tfrac{1}{2}k_2) = 0.2\,f(0.1, 3.12, 11.7) = 10.6080$$
$$k_3 = hg(t_0 + \tfrac{1}{2}h, x_0 + \tfrac{1}{2}m_2, y_0 + \tfrac{1}{2}k_2) = 0.2\,g(0.1, 3.12, 11.7) = 13.4160$$
$$m_4 = hf(t_0 + h, x_0 + m_3, y_0 + k_3) = 0.2\,f(0.2, 8, 20.216) = 19.3760$$
$$k_4 = hg(t_0 + h, x_0 + m_3, y_0 + k_3) = 0.2g(0.2, 8, 20.216) = 21.3776.$$

Therefore from (5) we get

$$x_1 = x_0 + \frac{1}{6}(m_1 + 2m_2 + 2m_3 + m_4)$$

$$= -1 + \frac{1}{6}(4.4 + 2(8.24) + 2(10.608) + 19.3760) = 9.2453$$

$$y_1 = y_0 + \frac{1}{6}(k_1 + 2k_2 + 2k_3 + k_4)$$

$$= 6 + \frac{1}{6}(7.4 + 2(11.4) + 2(13.416) + 21.3776) = 19.0683,$$

where, as usual, the computed values are rounded to four decimal places. These numbers give us the approximations $x_1 \approx x(0.2)$ and $y_1 \approx y(0.2)$. The subsequent values, obtained with the aid of a computer, are summarized in Tables 9.10 and 9.11.

TABLE 9.10 Runge-Kutta Method with $h = 0.2$

m_1	m_2	m_3	m_4	k_1	k_2	k_3	k_4	t_n	x_n	y_n
								0.00	−1.0000	6.0000
4.4000	8.2400	10.6080	19.3760	7.4000	11.4000	13.4160	21.3776	0.20	9.2453	19.0683
18.9527	31.1564	37.8870	63.6848	21.0329	31.7573	36.9716	57.8214	0.40	46.0327	55.1203
62.5093	97.7863	116.0063	187.3669	56.9378	84.8495	98.0688	151.4191	0.60	158.9430	150.8192

TABLE 9.11 Runge-Kutta Method with $h = 0.1$

m_1	m_2	m_3	m_4	k_1	k_2	k_3	k_4	t_n	x_n	y_n
								0.00	−1.0000	6.0000
2.2000	3.1600	3.4560	4.8720	3.7000	4.7000	4.9520	6.3256	0.10	2.3840	10.8883
4.8321	6.5742	7.0778	9.5870	6.2946	7.9413	8.3482	10.5957	0.20	9.3379	19.1332
9.5208	12.5821	13.4258	17.7609	10.5461	13.2339	13.8872	17.5358	0.30	22.5541	32.8539
17.6524	22.9090	24.3055	31.6554	17.4569	21.8114	22.8549	28.7393	0.40	46.5103	55.4420
31.4788	40.3496	42.6387	54.9202	28.6141	35.6245	37.2840	46.7207	0.50	88.5729	93.3006
54.6348	69.4029	73.1247	93.4107	46.5231	57.7482	60.3774	75.4370	0.60	160.7563	152.0025

You should verify that the solution of the initial-value problem in Example 3 is given by $x(t) = (26t - 1)e^{4t}$, $y(t) = (13t + 6)e^{4t}$. From these equations we see that the exact values are $x(0.6) = 160.9384$ and $y(0.6) = 152.1198$.

In conclusion, we state **Euler's method** for the general system (4) as

$$x_{n+1} = x_n + hf(t_n, x_n, y_n)$$
$$y_{n+1} = y_n + hg(t_n, x_n, y_n).$$

Answers to odd-numbered problems begin on page A-19.

1. Use Euler's method to approximate $y(0.2)$, where $y(x)$ is the solution of the initial-value problem

$$y'' - 4y' + 4y = 0, \qquad y(0) = -2, \quad y'(0) = 1.$$

Use $h = 0.1$. Find the exact solution of the problem, and compare the exact value of $y(0.2)$ with y_2.

2. Use Euler's method to approximate $y(1.2)$, where $y(x)$ is the solution of the initial-value problem

$$x^2 y'' - 2xy' + 2y = 0, \qquad y(1) = 4, \quad y'(1) = 9,$$

where $x > 0$. Use $h = 0.1$. Find the exact solution of the problem, and compare the exact value of $y(1.2)$ with y_2.

3. Repeat Problem 1 using the Runge-Kutta method with $h = 0.2$ and $h = 0.1$.

4. Repeat Problem 2 using the Runge-Kutta method with $h = 0.2$ and $h = 0.1$.

5. Use the Runge-Kutta method to obtain the approximate value of $y(0.2)$, where $y(x)$ is a solution of the initial-value problem

$$y'' - 2y' + 2y = e^t \cos t, \qquad y(0) = 1, \quad y'(0) = 2.$$

Use $h = 0.2$ and $h = 0.1$.

6. When $E = 100$ V, $R = 10\ \Omega$, and $L = 1$ h, the system of differential equations for the currents $i_1(t)$ and $i_3(t)$ in the electrical network given in Figure 9.13 is

$$\frac{di_1}{dt} = -20i_1 + 10i_3 + 100$$

$$\frac{di_3}{dt} = 10i_1 - 20i_3,$$

where $i_1(0) = 0$ and $i_3(0) = 0$. Use the Runge-Kutta method to approximate $i_1(t)$ and $i_3(t)$ at $t = 0.1, 0.2, 0.3, 0.4,$ and 0.5. Use $h = 0.1$.

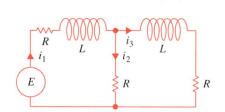

FIGURE 9.13

In Problems 7–12 use the Runge-Kutta method to approximate $x(0.2)$ and $y(0.2)$. Compare the results for $h = 0.2$ and $h = 0.1$.

7. $x' = 2x - y$
$y' = x$
$x(0) = 6, y(0) = 2$

8. $x' = x + 2y$
$y' = 4x + 3y$
$x(0) = 1, y(0) = 1$

9. $x' = -y + t$
$y' = x - t$
$x(0) = -3, y(0) = 5$

10. $x' = 6x + y + 6t$
$y' = 4x + 3y - 10t + 4$
$x(0) = 0.5, y(0) = 0.2$

11. $x' + 4x - y' = 7t$
$x' + y' - 2y = 3t$
$x(0) = 1, y(0) = -2$

12. $\quad\quad x'' + y' = 4t$
$-x'' + y' + y = 6t^2 + 10$
$x(0) = 3, y(0) = -1$

Discussion Problem

13. Recall from Section 5.3 that the nonlinear differential equation

$$\frac{d^2\theta}{dt^2} + \frac{g}{l}\sin\theta = 0$$

is a model for the motion of a simple pendulum of length l. For small values of θ we saw that a linearization of this differential equation is

$$\frac{d^2\theta}{dt^2} + \frac{g}{l}\theta = 0.$$

(a) Discuss: For what "small values of θ" is the linear differential equation a good approximation to the nonlinear differential equation?

(b) Find the exact solution of the linear differential equation subject to $\theta(0) = \theta_0$, $\theta'(0) = -1$.

(c) Use the fourth-order Runge-Kutta method on the interval $[0, 3]$ with $h = 0.1$ to approximate the solution of the nonlinear equation with initial conditions $\theta(0) = \theta_0$, $\theta'(0) = -1$ for various values of θ_0. Discuss: For what values of θ_0 is the solution of the linear initial-value problem a good approximation to the numerical solution of the nonlinear initial-value problem? Does this agree with your conjecture in part (a)?

9.6 SECOND-ORDER BOUNDARY-VALUE PROBLEMS

■ *Difference quotients* ■ *Finite differences* ■ *Forward difference* ■ *Backward difference*
■ *Central difference* ■ *Interior mesh points* ■ *Finite difference equation* ■ *Shooting method*

In Sections 9.2–9.4 we considered techniques that yield an approximation to a solution of a first-order initial-value problem $y' = f(x, y)$, $y(x_0) = y_0$. In addition, we saw in Section 9.5 that we can adapt the approximation techniques to a second-order *initial-value problem* $y'' = f(x, y, y')$, $y(x_0) = y_0, y'(x_0) = y_1$ by reducing the second-order differential equation to a system of first-order equations. In this section we are going to examine a method for approximating a solution to a second-order *boundary-value problem* $y'' = f(x, y, y')$, $y(a) = \alpha$, $y(b) = \beta$. We note at the outset that this method does not require reducing the second-order differential equation to a system of equations.

Finite Difference Approximations   The Taylor series expansion of a function $y(x)$ centered at a point a is

$$y(x) = y(a) + y'(a)\frac{x-a}{1!} + y''(a)\frac{(x-a)^2}{2!} + y'''(a)\frac{(x-a)^3}{3!} + \cdots.$$

If we set $h = x - a$, then the preceding line is the same as

$$y(a + h) = y(a) + y'(a)\frac{h}{1!} + y''(a)\frac{h^2}{2!} + y'''(a)\frac{h^3}{3!} + \cdots.$$

For the subsequent discussion it is convenient then to rewrite this last expression in two alternative forms:

$$y(x + h) = y(x) + y'(x)h + y''(x)\frac{h^2}{2} + y'''(x)\frac{h^3}{6} + \cdots \tag{1}$$

and

$$y(x - h) = y(x) - y'(x)h + y''(x)\frac{h^2}{2} - y'''(x)\frac{h^3}{6} + \cdots. \tag{2}$$

If h is small, we can ignore terms involving h^4, h^5, ... since these values are negligible. Indeed, if we ignore all terms involving h^2 and higher, then (1) and (2) yield, in turn, the following approximations for the first derivative $y'(x)$:

$$y'(x) \approx \frac{1}{h}[y(x + h) - y(x)] \tag{3}$$

$$y'(x) \approx \frac{1}{h}[y(x) - y(x - h)]. \tag{4}$$

Subtracting (1) and (2) also gives

$$y'(x) \approx \frac{1}{2h}[y(x + h) - y(x - h)]. \tag{5}$$

On the other hand, if we ignore terms involving h^3 and higher, then by adding (1) and (2) we obtain an approximation for the second derivative $y''(x)$:

$$y''(x) \approx \frac{1}{h^2}[y(x + h) - 2y(x) + y(x - h)]. \tag{6}$$

The right sides of (3), (4), (5), and (6) are called **difference quotients.** The expressions

$$y(x + h) - y(x) \qquad\qquad y(x) - y(x - h)$$
$$y(x + h) - y(x - h) \qquad \text{and} \qquad y(x + h) - 2y(x) + y(x - h)$$

are called **finite differences.** Specifically, $y(x + h) - y(x)$ is called a **forward difference,** $y(x) - y(x - h)$ is a **backward difference,** and both $y(x + h) - y(x - h)$ and $y(x + h) - 2y(x) + y(x - h)$ are called **central differences.** The results given in (5) and (6) are referred to as **central difference approximations** for the derivatives y' and y''.

Finite Difference Equation   Consider now a linear second-order boundary-value problem

$$y'' + P(x)y' + Q(x)y = f(x), \qquad y(a) = \alpha, \quad y(b) = \beta. \tag{7}$$

Suppose $a = x_0 < x_1 < x_2 < \cdots < x_{n-1} < x_n = b$ represents a regular partition of the interval $[a, b]$; that is, $x_i = a + ih$, where $i = 0, 1, 2, \ldots, n$ and $h = (b - a)/n$. The points

$$x_1 = a + h, \quad x_2 = a + 2h, \quad \ldots, \quad x_{n-1} = a + (n - 1)h$$

are called **interior mesh points** of the interval $[a, b]$. If we let

$$y_i = y(x_i), \qquad P_i = P(x_i), \qquad Q_i = Q(x_i), \qquad \text{and} \qquad f_i = f(x_i)$$

and if y'' and y' in (7) are replaced by the central difference approximations (5) and (6), we get

$$\frac{y_{i+1} - 2y_i + y_{i-1}}{h^2} + P_i \frac{y_{i+1} - y_{i-1}}{2h} + Q_i y_i = f_i$$

or, after simplifying,

$$\left(1 + \frac{h}{2}P_i\right)y_{i+1} + (-2 + h^2 Q_i)y_i + \left(1 - \frac{h}{2}P_i\right)y_{i-1} = h^2 f_i. \qquad \textbf{(8)}$$

The last equation, known as a **finite difference equation,** is an approximation to the differential equation. It enables us to approximate the solution $y(x)$ of (7) at the interior mesh points $x_1, x_2, \ldots, x_{n-1}$ of the interval $[a, b]$. By letting i take on the values $1, 2, \ldots, n - 1$ in (8), we obtain $n - 1$ equations in the $n - 1$ unknowns $y_1, y_2, \ldots, y_{n-1}$. Bear in mind that we know y_0 and y_n since these are the prescribed boundary conditions $y_0 = y(x_0) = y(a) = \alpha$ and $y_n = y(x_n) = y(b) = \beta$.

In Example 1 we consider a boundary-value problem for which we can compare the approximate values found with the exact values of an explicit solution.

EXAMPLE 1   Using the Finite Difference Method

Use the difference equation (8) with $n = 4$ to approximate the solution of the boundary-value problem

$$y'' - 4y = 0, \qquad y(0) = 0, \quad y(1) = 5.$$

SOLUTION   To use (8) we identify $P(x) = 0$, $Q(x) = -4$, $f(x) = 0$, and $h = (1 - 0)/4 = \frac{1}{4}$. Hence the difference equation is

$$y_{i+1} - 2.25y_i + y_{i-1} = 0. \qquad \textbf{(9)}$$

Now the interior points are $x_1 = 0 + \frac{1}{4}$, $x_2 = 0 + \frac{2}{4}$, $x_3 = 0 + \frac{3}{4}$, and so for $i = 1, 2$, and 3, (9) yields the following system for the corresponding y_1, y_2, and y_3:

$$y_2 - 2.25y_1 + y_0 = 0$$
$$y_3 - 2.25y_2 + y_1 = 0$$
$$y_4 - 2.25y_3 + y_2 = 0.$$

With the boundary conditions $y_0 = 0$ and $y_4 = 5$, the foregoing system becomes

$$-2.25y_1 + \qquad y_2 \qquad\qquad = 0$$
$$y_1 - 2.25y_2 + \qquad y_3 = 0$$
$$y_2 - 2.25y_3 = -5.$$

Solving the system gives $y_1 = 0.7256$, $y_2 = 1.6327$, and $y_3 = 2.9479$.

Now the general solution of the given differential equation is $y = c_1 \cosh 2x + c_2 \sinh 2x$. The condition $y(0) = 0$ implies $c_1 = 0$. The other boundary condition gives c_2. In this way we see that an explicit solution of the boundary-value problem is $y(x) = (5 \sinh 2x)/\sinh 2$.

Thus the exact values (rounded to four decimal places) of this solution at the interior points are as follows: $y(0.25) = 0.7184$, $y(0.5) = 1.6201$, and $y(0.75) = 2.9354$.   ∎

The accuracy of the approximations in Example 1 can be improved by using a smaller value of h. Of course, the trade-off here is that a smaller value of h necessitates solving a larger system of equations. It is left as an exercise to show that with $h = \frac{1}{8}$, approximations to $y(0.25)$, $y(0.5)$, and $y(0.75)$ are 0.7202, 1.6233, and 2.9386, respectively. See Problem 11 in Exercises 9.6.

EXAMPLE 2   Using the Finite Difference Method

Use the difference equation (8) with $n = 10$ to approximate the solution of

$$y'' + 3y' + 2y = 4x^2, \quad y(1) = 1, \quad y(2) = 6.$$

SOLUTION   In this case we identify $P(x) = 3$, $Q(x) = 2$, $f(x) = 4x^2$, and $h = (2 - 1)/10 = 0.1$, and so (8) becomes

$$1.15y_{i+1} - 1.98y_i + 0.85y_{i-1} = 0.04x_i^2. \tag{10}$$

Now the interior points are $x_1 = 1.1$, $x_2 = 1.2$, $x_3 = 1.3$, $x_4 = 1.4$, $x_5 = 1.5$, $x_6 = 1.6$, $x_7 = 1.7$, $x_8 = 1.8$, and $x_9 = 1.9$. For $i = 1, 2, \ldots, 9$ and $y_0 = 1$, $y_{10} = 6$, (10) gives a system of nine equations and nine unknowns:

$$
\begin{aligned}
1.15y_2 - 1.98y_1 &= -0.8016 \\
1.15y_3 - 1.98y_2 + 0.85y_1 &= 0.0576 \\
1.15y_4 - 1.98y_3 + 0.85y_2 &= 0.0676 \\
1.15y_5 - 1.98y_4 + 0.85y_3 &= 0.0784 \\
1.15y_6 - 1.98y_5 + 0.85y_4 &= 0.0900 \\
1.15y_7 - 1.98y_6 + 0.85y_5 &= 0.1024 \\
1.15y_8 - 1.98y_7 + 0.85y_6 &= 0.1156 \\
1.15y_9 - 1.98y_8 + 0.85y_7 &= 0.1296 \\
- 1.98y_9 + 0.85y_8 &= -6.7556.
\end{aligned}
$$

We can solve this large system using Gaussian elimination or, with relative ease, by means of a computer algebra system such as Mathematica. The result is found to be $y_1 = 2.4047$, $y_2 = 3.4432$, $y_3 = 4.2010$, $y_4 = 4.7469$, $y_5 = 5.1359$, $y_6 = 5.4124$, $y_7 = 5.6117$, $y_8 = 5.7620$, and $y_9 = 5.8855$.   ∎

Shooting Method   Another way of approximating a solution of a boundary-value problem $y'' = f(x, y, y')$, $y(a) = \alpha$, $y(b) = \beta$ is called the **shooting method.** The starting point in this method is the replacement of the boundary-value problem by an initial-value problem

$$y'' = f(x, y, y'), \quad y(a) = \alpha, \quad y'(a) = m_1. \tag{11}$$

The number m_1 in (11) is simply a guess for the unknown slope of the solution curve at the known point $(a, y(a))$. We then apply one of the step-by-step numerical techniques to the second-order equation in (11) to find an approximation β_1 for the value of $y(b)$. If β_1 agrees with the given value $y(b) = \beta$ to some preassigned tolerance, we stop; otherwise the calculations are repeated, starting with a different guess $y'(a) = m_2$ to obtain a second approximation β_2 for $y(b)$. This method can be continued in a trial-and-error manner or the subsequent slopes $m_3, m_4, \ldots$ can be adjusted in some systematic way; linear interpolation is particularly successful when the differential equation in (11) is linear. The procedure is analogous to shooting (the "aim" is the choice of the initial slope) at a target until the bulls-eye $y(b)$ is hit.

Of course, underlying the use of these numerical methods is the assumption, which we know is not always warranted, that a solution of the boundary-value problem exists.

Remark

The approximation method using finite differences can be extended to boundary-value problems in which the first derivative is specified at a boundary—for example, a problem such as $y'' = f(x, y, y')$, $y'(a) = \alpha$, $y(b) = \beta$. See Problem 13 in Exercises 9.6.

SECTION 9.6 EXERCISES

Answers to odd-numbered problems begin on page A-19.

In Problems 1–10 use the finite difference method and the indicated value of n to approximate the solution of the given boundary-value problem.

1. $y'' + 9y = 0$, $y(0) = 4$, $y(2) = 1$; $n = 4$
2. $y'' - y = x^2$, $y(0) = 0$, $y(1) = 0$; $n = 4$
3. $y'' + 2y' + y = 5x$, $y(0) = 0$, $y(1) = 0$; $n = 5$
4. $y'' - 10y' + 25y = 1$, $y(0) = 1$, $y(1) = 0$; $n = 5$
5. $y'' - 4y' + 4y = (x + 1)e^{2x}$, $y(0) = 3$, $y(1) = 0$; $n = 6$
6. $y'' + 5y' = 4\sqrt{x}$, $y(1) = 1$, $y(2) = -1$; $n = 6$
7. $x^2y'' + 3xy' + 3y = 0$, $y(1) = 5$, $y(2) = 0$; $n = 8$
8. $x^2y'' - xy' + y = \ln x$, $y(1) = 0$, $y(2) = -2$; $n = 8$
9. $y'' + (1 - x)y' + xy = x$, $y(0) = 0$, $y(1) = 2$; $n = 10$
10. $y'' + xy' + y = x$, $y(0) = 1$, $y(1) = 0$; $n = 10$
11. Rework Example 1 using $n = 8$.
12. The electrostatic potential u between two concentric spheres of radius $r = 1$ and $r = 4$ is determined from

$$\frac{d^2u}{dr^2} + \frac{2}{r}\frac{du}{dr} = 0, \qquad u(1) = 50, \quad u(4) = 100.$$

Use the method of this section with $n = 6$ to approximate the solution of this boundary-value problem.

13. Consider the boundary-value problem $y'' + xy = 0$, $y'(0) = 1$, $y(1) = -1$.

(a) Find the difference equation corresponding to the differential equation. Show that for $i = 0, 1, 2, \ldots, n - 1$ the difference equation yields n equations in $n + 1$ unknowns $y_{-1}, y_0, y_1, y_2, \ldots, y_{n-1}$. Here y_{-1} and y_0 are unknowns since y_{-1} represents an approximation to y at the exterior point $x = -h$ and y_0 is not specified at $x = 0$.

(b) Use the central difference approximation (5) to show that $y_1 - y_{-1} = 2h$. Use this equation to eliminate y_{-1} from the system in part (a).

(c) Use $n = 5$ and the system of equations found in parts (a) and (b) to approximate the solution of the original boundary-value problem.

14. Consider the boundary-value problem $y'' = y' - \sin(xy)$, $y(0) = 1$, $y(1) = 1.5$. Use the shooting method to approximate the solution of this problem. (The actual approximation can be obtained using a numerical technique—say, the fourth-order Runge-Kutta method with $h = 0.1$; or, even better, if you have access to a CAS such as Mathematica or Maple, the **NDSolve** function can be used.)

CHAPTER 9 REVIEW EXERCISES

Answers to odd-numbered problems begin on page A-19.

In Problems 1 and 2 sketch the direction field for the given differential equation. Indicate several possible solution curves.

1. $y \, dx - x \, dy = 0$
2. $y' = 2x - y$

In Problems 3–6 construct a table comparing the indicated values of $y(x)$ using the Euler, improved Euler, and Runge-Kutta methods. Compute to four rounded decimal places. Use $h = 0.1$ and $h = 0.05$.

3. $y' = 2 \ln xy$, $y(1) = 2$;
   $y(1.1)$, $y(1.2)$, $y(1.3)$, $y(1.4)$, $y(1.5)$

4. $y' = \sin x^2 + \cos y^2$, $y(0) = 0$;
   $y(0.1)$, $y(0.2)$, $y(0.3)$, $y(0.4)$, $y(0.5)$

5. $y' = \sqrt{x + y}$, $y(0.5) = 0.5$;
   $y(0.6)$, $y(0.7)$, $y(0.8)$, $y(0.9)$, $y(1.0)$

6. $y' = xy + y^2$, $y(1) = 1$;
   $y(1.1)$, $y(1.2)$, $y(1.3)$, $y(1.4)$, $y(1.5)$

7. Use Euler's method to obtain the approximate value of $y(0.2)$, where $y(x)$ is the solution of the initial-value problem

$$y'' - (2x + 1)y = 1, \quad y(0) = 3, \quad y'(0) = 1.$$

First use one step with $h = 0.2$, and then repeat the calculations using $h = 0.1$.

8. Use the Adams-Bashforth/Adams-Moulton method to approximate the value of $y(0.4)$, where $y(x)$ is the solution of

$$y' = 4x - 2y, \qquad y(0) = 2.$$

Use the Runge-Kutta formula and $h = 0.1$ to obtain the values of y_1, y_2, and y_3.

9. Use Euler's method and $h = 0.1$ to approximate the values of $x(0.2)$, $y(0.2)$, where $x(t)$ and $y(t)$ are solutions of

$$x' = x + y$$
$$y' = x - y,$$
$$x(0) = 1, \quad y(0) = 2.$$

10. Use the finite difference method with $n = 10$ to approximate the solution of the boundary-value problem

$$y'' + 6.55(1 + x)y = 1, \qquad y(0) = 0, \quad y(1) = 0.$$

APPENDIX

I

GAMMA FUNCTION

Euler's integral definition of the **gamma function** is

$$\Gamma(x) = \int_0^\infty t^{x-1} e^{-t}\, dt. \tag{1}$$

Convergence of the integral requires that $x - 1 > -1$ or $x > 0$. The recurrence relation

$$\Gamma(x + 1) = x\Gamma(x), \tag{2}$$

which we saw in Section 6.4, can be obtained from (1) with integration by parts. Now when $x = 1$, $\Gamma(1) = \int_0^\infty e^{-t}\, dt = 1$, and thus (2) gives

$$\Gamma(2) = 1\Gamma(1) = 1$$
$$\Gamma(3) = 2\Gamma(2) = 2 \cdot 1$$
$$\Gamma(4) = 3\Gamma(3) = 3 \cdot 2 \cdot 1$$

and so on. In this manner it is seen that when n is a positive integer,

$$\Gamma(n + 1) = n!.$$

For this reason the gamma function is often called the **generalized factorial function.**

Although the integral form (1) does not converge for $x < 0$, it can be shown by means of alternative definitions that the gamma function is defined for all real and complex numbers *except $x = -n, n = 0, 1, 2, \ldots$.* As a consequence, (2) is actually valid for $x \neq -n$. Considered as a function of a real variable x, the graph of $\Gamma(x)$ is as given in Figure I.1. Observe that the nonpositive integers correspond to vertical asymptotes of the graph.

In Problems 27–33 of Exercises 6.4 we utilized the fact that $\Gamma(\frac{1}{2}) = \sqrt{\pi}$. This result can be derived from (1) by setting $x = \frac{1}{2}$:

$$\Gamma\left(\frac{1}{2}\right) = \int_0^\infty t^{-1/2} e^{-t}\, dt. \tag{3}$$

When we let $t = u^2$, (3) can be written as $\Gamma(\frac{1}{2}) = 2\int_0^\infty e^{-u^2}\, du$. But

$$\int_0^\infty e^{-u^2}\, du = \int_0^\infty e^{-v^2}\, dv,$$

and so

$$\left[\Gamma\left(\frac{1}{2}\right)\right]^2 = \left(2\int_0^\infty e^{-u^2}\, du\right)\left(2\int_0^\infty e^{-v^2}\, dv\right)$$

$$= 4\int_0^\infty \int_0^\infty e^{-(u^2+v^2)}\, du\, dv.$$

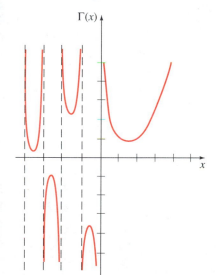

$\Gamma(x)$

FIGURE I.1

Switching to polar coordinates $u = r \cos \theta$, $v = r \sin \theta$ enables us to evaluate the double integral:

$$4 \int_0^\infty \int_0^\infty e^{-(u^2+v^2)} \, du \, dv = 4 \int_0^{\pi/2} \int_0^\infty e^{-r^2} r \, dr \, d\theta = \pi.$$

Hence
$$\left[\Gamma\left(\frac{1}{2}\right) \right]^2 = \pi \qquad \text{or} \qquad \Gamma\left(\frac{1}{2}\right) = \sqrt{\pi}. \tag{4}$$

EXAMPLE 1   **Value of $\Gamma(-\frac{1}{2})$**

Evaluate $\Gamma(-\frac{1}{2})$.

SOLUTION   In view of (2) and (4) it follows that, with $x = -\frac{1}{2}$,

$$\Gamma\left(\frac{1}{2}\right) = -\frac{1}{2} \Gamma\left(-\frac{1}{2}\right).$$

Therefore
$$\Gamma\left(-\frac{1}{2}\right) = -2 \Gamma\left(\frac{1}{2}\right) = -2\sqrt{\pi}. \qquad \blacksquare$$

APPENDIX I EXERCISES

Answers to odd-numbered problems begin on page A-20.

1. Evaluate.
   (a) $\Gamma(5)$ (b) $\Gamma(7)$ (c) $\Gamma(-\frac{3}{2})$ (d) $\Gamma(-\frac{5}{2})$

2. Use (1) and the fact that $\Gamma(\frac{6}{5}) = 0.92$ to evaluate $\int_0^\infty x^5 e^{-x^5} \, dx$.
   [*Hint:* Let $t = x^5$.]

3. Use (1) and the fact that $\Gamma(\frac{5}{3}) = 0.89$ to evaluate $\int_0^\infty x^4 e^{-x^3} \, dx$.

4. Evaluate $\int_0^1 x^3 \left(\ln \frac{1}{x} \right)^3 \, dx$. [*Hint:* Let $t = -\ln x$.]

5. Use the fact that $\Gamma(x) > \int_0^1 t^{x-1} e^{-t} \, dt$ to show that $\Gamma(x)$ is unbounded as $x \to 0^+$.

6. Use (1) to derive (2) for $x > 0$.

INTRODUCTION TO MATRICES

II.1 Basic Definitions and Theory

DEFINITION II.1   Matrix

A **matrix A** is any rectangular array of numbers or functions:

$$\mathbf{A} = \begin{pmatrix} a_{11} & a_{12} & \cdots & a_{1n} \\ a_{21} & a_{22} & \cdots & a_{2n} \\ \vdots & & & \vdots \\ a_{m1} & a_{m2} & \cdots & a_{mn} \end{pmatrix}. \tag{1}$$

If a matrix has m rows and n columns, we say that its **size** is m by n (written $m \times n$). An $n \times n$ matrix is called a **square** matrix of order n.

The element, or entry, in the ith row and jth column of an $m \times n$ matrix $\mathbf{A}$ is written a_{ij}. An $m \times n$ matrix $\mathbf{A}$ is then abbreviated as $\mathbf{A} = (a_{ij})_{m \times n}$ or simply $\mathbf{A} = (a_{ij})$. A 1×1 matrix is simply one constant or function.

DEFINITION II.2   Equality of Matrices

Two $m \times n$ matrices $\mathbf{A}$ and $\mathbf{B}$ are **equal** if $a_{ij} = b_{ij}$ for each i and j.

DEFINITION II.3   Column Matrix

A **column matrix X** is any matrix having n rows and one column:

$$\mathbf{X} = \begin{pmatrix} b_{11} \\ b_{21} \\ \vdots \\ b_{n1} \end{pmatrix} = (b_{i1})_{n \times 1}.$$

A column matrix is also called a **column vector** or simply a **vector**.

DEFINITION II.4   **Multiples of Matrices**

A **multiple** of a matrix **A** is defined to be

$$k\mathbf{A} = \begin{pmatrix} ka_{11} & ka_{12} & \cdots & ka_{1n} \\ ka_{21} & ka_{22} & \cdots & ka_{2n} \\ \vdots & & & \vdots \\ ka_{m1} & ka_{m2} & \cdots & ka_{mn} \end{pmatrix} = (ka_{ij})_{m \times n},$$

where k is a constant or a function.

EXAMPLE 1   **Multiples of Matrices**

(a) $5\begin{pmatrix} 2 & -3 \\ 4 & -1 \\ \frac{1}{5} & 6 \end{pmatrix} = \begin{pmatrix} 10 & -15 \\ 20 & -5 \\ 1 & 30 \end{pmatrix}$ (b) $e^t\begin{pmatrix} 1 \\ -2 \\ 4 \end{pmatrix} = \begin{pmatrix} e^t \\ -2e^t \\ 4e^t \end{pmatrix}$

We note in passing that for any matrix **A** the product $k\mathbf{A}$ is the same as $\mathbf{A}k$. For example,

$$e^{-3t}\begin{pmatrix} 2 \\ 5 \end{pmatrix} = \begin{pmatrix} 2e^{-3t} \\ 5e^{-3t} \end{pmatrix} = \begin{pmatrix} 2 \\ 5 \end{pmatrix}e^{-3t}.$$

DEFINITION II.5   **Addition of Matrices**

The **sum** of two $m \times n$ matrices **A** and **B** is defined to be the matrix

$$\mathbf{A} + \mathbf{B} = (a_{ij} + b_{ij})_{m \times n}.$$

In other words, when adding two matrices of the same size, we add the corresponding elements.

EXAMPLE 2   **Matrix Addition**

The sum of $\mathbf{A} = \begin{pmatrix} 2 & -1 & 3 \\ 0 & 4 & 6 \\ -6 & 10 & -5 \end{pmatrix}$ and $\mathbf{B} = \begin{pmatrix} 4 & 7 & -8 \\ 9 & 3 & 5 \\ 1 & -1 & 2 \end{pmatrix}$ is

$$\mathbf{A} + \mathbf{B} = \begin{pmatrix} 2+4 & -1+7 & 3+(-8) \\ 0+9 & 4+3 & 6+5 \\ -6+1 & 10+(-1) & -5+2 \end{pmatrix} = \begin{pmatrix} 6 & 6 & -5 \\ 9 & 7 & 11 \\ -5 & 9 & -3 \end{pmatrix}.$$

EXAMPLE 3 **A Matrix Written as a Sum of Column Matrices**

The single matrix $\begin{pmatrix} 3t^2 - 2e^t \\ t^2 + 7t \\ 5t \end{pmatrix}$ can be written as the sum of three col-

umn vectors:

$$\begin{pmatrix} 3t^2 - 2e^t \\ t^2 + 7t \\ 5t \end{pmatrix} = \begin{pmatrix} 3t^2 \\ t^2 \\ 0 \end{pmatrix} + \begin{pmatrix} 0 \\ 7t \\ 5t \end{pmatrix} + \begin{pmatrix} -2e^t \\ 0 \\ 0 \end{pmatrix} = \begin{pmatrix} 3 \\ 1 \\ 0 \end{pmatrix} t^2 + \begin{pmatrix} 0 \\ 7 \\ 5 \end{pmatrix} t + \begin{pmatrix} -2 \\ 0 \\ 0 \end{pmatrix} e^t.$$ ∎

The **difference** of two $m \times n$ matrices is defined in the usual manner: $\mathbf{A} - \mathbf{B} = \mathbf{A} + (-\mathbf{B})$, where $-\mathbf{B} = (-1)\mathbf{B}$.

DEFINITION II.6 **Multiplication of Matrices**

Let $\mathbf{A}$ be a matrix having m rows and n columns and $\mathbf{B}$ be a matrix having n rows and p columns. We define the **product AB** to be the $m \times p$ matrix

$$\mathbf{AB} = \begin{pmatrix} a_{11} & a_{12} & \cdots & a_{1n} \\ a_{21} & a_{22} & \cdots & a_{2n} \\ \vdots & & & \vdots \\ a_{m1} & a_{m2} & \cdots & a_{mn} \end{pmatrix} \begin{pmatrix} b_{11} & b_{12} & \cdots & b_{1p} \\ b_{21} & b_{22} & \cdots & b_{2p} \\ \vdots & & & \vdots \\ b_{n1} & b_{n2} & \cdots & b_{np} \end{pmatrix}$$

$$= \begin{pmatrix} a_{11}b_{11} + a_{12}b_{21} + \cdots + a_{1n}b_{n1} & \cdots & a_{11}b_{1p} + a_{12}b_{2p} + \cdots + a_{1n}b_{np} \\ a_{21}b_{11} + a_{22}b_{21} + \cdots + a_{2n}b_{n1} & \cdots & a_{21}b_{1p} + a_{22}b_{2p} + \cdots + a_{2n}b_{np} \\ \vdots & & \vdots \\ a_{m1}b_{11} + a_{m2}b_{21} + \cdots + a_{mn}b_{n1} & \cdots & a_{m1}b_{1p} + a_{m2}b_{2p} + \cdots + a_{mn}b_{np} \end{pmatrix}$$

$$= \left(\sum_{k=1}^{n} a_{ik}b_{kj} \right)_{m \times p}.$$

Note carefully in Definition II.6 that the product $\mathbf{AB} = \mathbf{C}$ is defined only when the number of columns in the matrix $\mathbf{A}$ is the same as the number of rows in $\mathbf{B}$. The size of the product can be determined from

$$\mathbf{A}_{m \times n} \mathbf{B}_{n \times p} = \mathbf{C}_{m \times p}.$$

Also, you might recognize that the entries in, say, the ith row of the final matrix $\mathbf{AB}$ are formed by using the component definition of the inner, or dot, product of the ith row of $\mathbf{A}$ with each of the columns of $\mathbf{B}$.

EXAMPLE 4   **Multiplication of Matrices**

(a) For $\mathbf{A} = \begin{pmatrix} 4 & 7 \\ 3 & 5 \end{pmatrix}$ and $\mathbf{B} = \begin{pmatrix} 9 & -2 \\ 6 & 8 \end{pmatrix}$,

$$\mathbf{AB} = \begin{pmatrix} 4 \cdot 9 + 7 \cdot 6 & 4 \cdot (-2) + 7 \cdot 8 \\ 3 \cdot 9 + 5 \cdot 6 & 3 \cdot (-2) + 5 \cdot 8 \end{pmatrix} = \begin{pmatrix} 78 & 48 \\ 57 & 34 \end{pmatrix}.$$

(b) For $\mathbf{A} = \begin{pmatrix} 5 & 8 \\ 1 & 0 \\ 2 & 7 \end{pmatrix}$ and $\mathbf{B} = \begin{pmatrix} -4 & -3 \\ 2 & 0 \end{pmatrix}$,

$$\mathbf{AB} = \begin{pmatrix} 5 \cdot (-4) + 8 \cdot 2 & 5 \cdot (-3) + 8 \cdot 0 \\ 1 \cdot (-4) + 0 \cdot 2 & 1 \cdot (-3) + 0 \cdot 0 \\ 2 \cdot (-4) + 7 \cdot 2 & 2 \cdot (-3) + 7 \cdot 0 \end{pmatrix} = \begin{pmatrix} -4 & -15 \\ -4 & -3 \\ 6 & -6 \end{pmatrix}. \quad \blacksquare$$

In general, *matrix multiplication is not commutative*; that is, $\mathbf{AB} \neq \mathbf{BA}$. Observe in part (a) of Example 4 that $\mathbf{BA} = \begin{pmatrix} 30 & 53 \\ 48 & 82 \end{pmatrix}$, whereas in part (b) the product $\mathbf{BA}$ is not defined since Definition II.6 requires that the first matrix (in this case $\mathbf{B}$) have the same number of columns as the second matrix has rows.

We are particularly interested in the product of a square matrix and a column vector.

EXAMPLE 5   **Multiplication of Matrices**

(a) $\begin{pmatrix} 2 & -1 & 3 \\ 0 & 4 & 5 \\ 1 & -7 & 9 \end{pmatrix} \begin{pmatrix} -3 \\ 6 \\ 4 \end{pmatrix} = \begin{pmatrix} 2 \cdot (-3) + (-1) \cdot 6 + 3 \cdot 4 \\ 0 \cdot (-3) + 4 \cdot 6 + 5 \cdot 4 \\ 1 \cdot (-3) + (-7) \cdot 6 + 9 \cdot 4 \end{pmatrix} = \begin{pmatrix} 0 \\ 44 \\ -9 \end{pmatrix}$

(b) $\begin{pmatrix} -4 & 2 \\ 3 & 8 \end{pmatrix} \begin{pmatrix} x \\ y \end{pmatrix} = \begin{pmatrix} -4x + 2y \\ 3x + 8y \end{pmatrix}$ $\quad \blacksquare$

Multiplicative Identity   For a given positive integer n, the $n \times n$ matrix

$$\mathbf{I} = \begin{pmatrix} 1 & 0 & 0 & \cdots & 0 \\ 0 & 1 & 0 & \cdots & 0 \\ \vdots & & & & \vdots \\ 0 & 0 & 0 & \cdots & 1 \end{pmatrix}$$

is called the **multiplicative identity matrix**. It follows from Definition II.6 that for any $n \times n$ matrix $\mathbf{A}$,

$$\mathbf{AI} = \mathbf{IA} = \mathbf{A}.$$

Also, it is readily verified that if $\mathbf{X}$ is an $n \times 1$ column matrix, then $\mathbf{IX} = \mathbf{X}$.

Zero Matrix A matrix consisting of all zero entries is called a **zero matrix** and is denoted by **0**. For example,

$$\mathbf{0} = \begin{pmatrix} 0 \\ 0 \end{pmatrix}, \qquad \mathbf{0} = \begin{pmatrix} 0 & 0 \\ 0 & 0 \end{pmatrix}, \qquad \mathbf{0} = \begin{pmatrix} 0 & 0 \\ 0 & 0 \\ 0 & 0 \end{pmatrix},$$

and so on. If **A** and **0** are $m \times n$ matrices, then

$$\mathbf{A} + \mathbf{0} = \mathbf{0} + \mathbf{A} = \mathbf{A}.$$

Associative Law Although we shall not prove it, matrix multiplication is **associative.** If **A** is an $m \times p$ matrix, **B** a $p \times r$ matrix, and **C** an $r \times n$ matrix, then

$$\mathbf{A}(\mathbf{BC}) = (\mathbf{AB})\mathbf{C}$$

is an $m \times n$ matrix.

Distributive Law If all products are defined, multiplication is **distributive** over addition:

$$\mathbf{A}(\mathbf{B} + \mathbf{C}) = \mathbf{AB} + \mathbf{AC} \qquad \text{and} \qquad (\mathbf{B} + \mathbf{C})\mathbf{A} = \mathbf{BA} + \mathbf{CA}.$$

Determinant of a Matrix Associated with every *square* matrix **A** of constants is a number called the **determinant of the matrix,** which is denoted by det **A**.

EXAMPLE 6 **Determinant of a Square Matrix**

For $\mathbf{A} = \begin{pmatrix} 3 & 6 & 2 \\ 2 & 5 & 1 \\ -1 & 2 & 4 \end{pmatrix}$ we expand det **A** by cofactors of the first row:

$$\det \mathbf{A} = \begin{vmatrix} 3 & 6 & 2 \\ 2 & 5 & 1 \\ -1 & 2 & 4 \end{vmatrix} = 3 \begin{vmatrix} 5 & 1 \\ 2 & 4 \end{vmatrix} - 6 \begin{vmatrix} 2 & 1 \\ -1 & 4 \end{vmatrix} + 2 \begin{vmatrix} 2 & 5 \\ -1 & 2 \end{vmatrix}$$

$$= 3(20 - 2) - 6(8 + 1) + 2(4 + 5) = 18. \qquad \blacksquare$$

It can be proved that a determinant det **A** can be expanded by cofactors using any row or column. If det **A** has a row (or a column) containing many zero entries, then wisdom dictates that we expand the determinant by that row (or column).

DEFINITION II.7   **Transpose of a Matrix**

The **transpose** of the $m \times n$ matrix (1) is the $n \times m$ matrix $\mathbf{A}^T$ given by

$$\mathbf{A}^T = \begin{pmatrix} a_{11} & a_{21} & \cdots & a_{m1} \\ a_{12} & a_{22} & \cdots & a_{m2} \\ \vdots & \vdots & & \vdots \\ a_{1n} & a_{2n} & \cdots & a_{mn} \end{pmatrix}.$$

In other words, the rows of a matrix $\mathbf{A}$ become the columns of its transpose $\mathbf{A}^T$.

EXAMPLE 7   **Transpose of a Matrix**

(a) The transpose of $\mathbf{A} = \begin{pmatrix} 3 & 6 & 2 \\ 2 & 5 & 1 \\ -1 & 2 & 4 \end{pmatrix}$ is $\mathbf{A}^T = \begin{pmatrix} 3 & 2 & -1 \\ 6 & 5 & 2 \\ 2 & 1 & 4 \end{pmatrix}$.

(b) If $\mathbf{X} = \begin{pmatrix} 5 \\ 0 \\ 3 \end{pmatrix}$, then $\mathbf{X}^T = (5 \quad 0 \quad 3)$.

DEFINITION II.8   **Multiplicative Inverse of a Matrix**

Let $\mathbf{A}$ be an $n \times n$ matrix. If there exists an $n \times n$ matrix $\mathbf{B}$ such that

$$\mathbf{AB} = \mathbf{BA} = \mathbf{I},$$

where $\mathbf{I}$ is the multiplicative identity, then $\mathbf{B}$ is said to be the **multiplicative inverse of $\mathbf{A}$** and is denoted by $\mathbf{B} = \mathbf{A}^{-1}$.

DEFINITION II.9   **Nonsingular/Singular Matrices**

Let $\mathbf{A}$ be an $n \times n$ matrix. If $\det \mathbf{A} \neq 0$, then $\mathbf{A}$ is said to be **nonsingular.** If $\det \mathbf{A} = 0$, then $\mathbf{A}$ is said to be **singular.**

The following theorem gives a necessary and sufficient condition for a square matrix to have a multiplicative inverse.

> **THEOREM II.1** **Nonsingularity Implies A Has an Inverse**
>
> An $n \times n$ matrix $\mathbf{A}$ has a multiplicative inverse $\mathbf{A}^{-1}$ if and only if $\mathbf{A}$ is nonsingular.

The following theorem gives one way of finding the multiplicative inverse for a nonsingular matrix.

> **THEOREM II.2** **A Formula for the Inverse of a Matrix**
>
> Let $\mathbf{A}$ be an $n \times n$ nonsingular matrix and let $C_{ij} = (-1)^{i+j} M_{ij}$, where M_{ij} is the determinant of the $(n-1) \times (n-1)$ matrix obtained by deleting the ith row and jth column from $\mathbf{A}$. Then
>
> $$\mathbf{A}^{-1} = \frac{1}{\det \mathbf{A}} (C_{ij})^T. \tag{2}$$

Each C_{ij} in Theorem II.2 is simply the **cofactor** (signed minor) of the corresponding entry a_{ij} in $\mathbf{A}$. Note that the transpose is utilized in formula (2).

For future reference we observe in the case of a 2×2 nonsingular matrix

$$\mathbf{A} = \begin{pmatrix} a_{11} & a_{12} \\ a_{21} & a_{22} \end{pmatrix}$$

that $C_{11} = a_{22}$, $C_{12} = -a_{21}$, $C_{21} = -a_{12}$, and $C_{22} = a_{11}$. Thus

$$\mathbf{A}^{-1} = \frac{1}{\det \mathbf{A}} \begin{pmatrix} a_{22} & -a_{21} \\ -a_{12} & a_{11} \end{pmatrix}^T = \frac{1}{\det \mathbf{A}} \begin{pmatrix} a_{22} & -a_{12} \\ -a_{21} & a_{11} \end{pmatrix}. \tag{3}$$

For a 3×3 nonsingular matrix

$$\mathbf{A} = \begin{pmatrix} a_{11} & a_{12} & a_{13} \\ a_{21} & a_{22} & a_{23} \\ a_{31} & a_{32} & a_{33} \end{pmatrix},$$

$$C_{11} = \begin{vmatrix} a_{22} & a_{23} \\ a_{32} & a_{33} \end{vmatrix}, \qquad C_{12} = - \begin{vmatrix} a_{21} & a_{23} \\ a_{31} & a_{33} \end{vmatrix}, \qquad C_{13} = \begin{vmatrix} a_{21} & a_{22} \\ a_{31} & a_{32} \end{vmatrix},$$

and so on. Carrying out the transposition gives

$$\mathbf{A}^{-1} = \frac{1}{\det \mathbf{A}} \begin{pmatrix} C_{11} & C_{21} & C_{31} \\ C_{12} & C_{22} & C_{32} \\ C_{13} & C_{23} & C_{33} \end{pmatrix}. \tag{4}$$

EXAMPLE 8   **Inverse of a 2 × 2 Matrix**

Find the multiplicative inverse for $\mathbf{A} = \begin{pmatrix} 1 & 4 \\ 2 & 10 \end{pmatrix}$.

SOLUTION   Since det $\mathbf{A} = 10 - 8 = 2 \neq 0$, $\mathbf{A}$ is nonsingular. It follows from Theorem II.1 that $\mathbf{A}^{-1}$ exists. From (3) we find

$$\mathbf{A}^{-1} = \frac{1}{2}\begin{pmatrix} 10 & -4 \\ -2 & 1 \end{pmatrix} = \begin{pmatrix} 5 & -2 \\ -1 & \frac{1}{2} \end{pmatrix}.$$ ∎

Not every square matrix has a multiplicative inverse. The matrix $\mathbf{A} = \begin{pmatrix} 2 & 2 \\ 3 & 3 \end{pmatrix}$ is singular since det $\mathbf{A} = 0$. Hence $\mathbf{A}^{-1}$ does not exist.

EXAMPLE 9   **Inverse of a 3 × 3 Matrix**

Find the multiplicative inverse for $\mathbf{A} = \begin{pmatrix} 2 & 2 & 0 \\ -2 & 1 & 1 \\ 3 & 0 & 1 \end{pmatrix}$.

SOLUTION   Since det $\mathbf{A} = 12 \neq 0$, the given matrix is nonsingular. The cofactors corresponding to the entries in each row of det $\mathbf{A}$ are

$$C_{11} = \begin{vmatrix} 1 & 1 \\ 0 & 1 \end{vmatrix} = 1 \qquad C_{12} = -\begin{vmatrix} -2 & 1 \\ 3 & 1 \end{vmatrix} = 5 \qquad C_{13} = \begin{vmatrix} -2 & 1 \\ 3 & 0 \end{vmatrix} = -3$$

$$C_{21} = -\begin{vmatrix} 2 & 0 \\ 0 & 1 \end{vmatrix} = -2 \qquad C_{22} = \begin{vmatrix} 2 & 0 \\ 3 & 1 \end{vmatrix} = 2 \qquad C_{23} = -\begin{vmatrix} 2 & 2 \\ 3 & 0 \end{vmatrix} = 6$$

$$C_{31} = \begin{vmatrix} 2 & 0 \\ 1 & 1 \end{vmatrix} = 2 \qquad C_{32} = -\begin{vmatrix} 2 & 0 \\ -2 & 1 \end{vmatrix} = -2 \qquad C_{33} = \begin{vmatrix} 2 & 2 \\ -2 & 1 \end{vmatrix} = 6.$$

It follows from (4) that

$$\mathbf{A}^{-1} = \frac{1}{12}\begin{pmatrix} 1 & -2 & 2 \\ 5 & 2 & -2 \\ -3 & 6 & 6 \end{pmatrix} = \begin{pmatrix} \frac{1}{12} & -\frac{1}{6} & \frac{1}{6} \\ \frac{5}{12} & \frac{1}{6} & -\frac{1}{6} \\ -\frac{1}{4} & \frac{1}{2} & \frac{1}{2} \end{pmatrix}.$$

You are urged to verify that $\mathbf{A}^{-1}\mathbf{A} = \mathbf{A}\mathbf{A}^{-1} = \mathbf{I}$.   ∎

Formula (2) presents obvious difficulties for nonsingular matrices larger than 3×3. For example, to apply (2) to a 4×4 matrix we would have to calculate *sixteen* 3×3 determinants.* In the case of a large matrix there are more efficient ways of finding $\mathbf{A}^{-1}$. The curious reader is referred to any text in linear algebra.

Since our goal is to apply the concept of a matrix to systems of linear first-order differential equations, we need the following definitions.

---

*Strictly speaking, a determinant is a number, but it is sometimes convenient to refer to a determinant as if it were an array.

DEFINITION II.10   **Derivative of a Matrix of Functions**

If $\mathbf{A}(t) = (a_{ij}(t))_{m \times n}$ is a matrix whose entries are functions differentiable on a common interval, then

$$\frac{d\mathbf{A}}{dt} = \left(\frac{d}{dt} a_{ij} \right)_{m \times n}.$$

DEFINITION II.11   **Integral of a Matrix of Functions**

If $\mathbf{A}(t) = (a_{ij}(t))_{m \times n}$ is a matrix whose entries are functions continuous on a common interval containing t and t_0, then

$$\int_{t_0}^{t} \mathbf{A}(s) \, ds = \left(\int_{t_0}^{t} a_{ij}(s) \, ds \right)_{m \times n}.$$

To differentiate (integrate) a matrix of functions we simply differentiate (integrate) each entry. The derivative of a matrix is also denoted by $\mathbf{A}'(t)$.

EXAMPLE 10   **Derivative/Integral of a Matrix**

If

$$\mathbf{X}(t) = \begin{pmatrix} \sin 2t \\ e^{3t} \\ 8t - 1 \end{pmatrix}, \quad \text{then} \quad \mathbf{X}'(t) = \begin{pmatrix} \dfrac{d}{dt} \sin 2t \\[4pt] \dfrac{d}{dt} e^{3t} \\[4pt] \dfrac{d}{dt} (8t - 1) \end{pmatrix} = \begin{pmatrix} 2 \cos 2t \\ 3e^{3t} \\ 8 \end{pmatrix}$$

and

$$\int_{0}^{t} \mathbf{X}(s) \, ds = \begin{pmatrix} \int_{0}^{t} \sin 2s \, ds \\[4pt] \int_{0}^{t} e^{3s} \, ds \\[4pt] \int_{0}^{t} (8s - 1) \, ds \end{pmatrix} = \begin{pmatrix} -\frac{1}{2} \cos 2t + \frac{1}{2} \\[4pt] \frac{1}{3} e^{3t} - \frac{1}{3} \\[4pt] 4t^2 - t \end{pmatrix}. \qquad \blacksquare$$

II.2   Gaussian and Gauss-Jordan Elimination

Matrices are an invaluable aid in solving algebraic systems of n linear equations in n unknowns

$$
\begin{aligned}
a_{11}x_1 + a_{12}x_2 + \cdots + a_{1n}x_n &= b_1 \\
a_{21}x_1 + a_{22}x_2 + \cdots + a_{2n}x_n &= b_2 \\
&\;\;\vdots \qquad\qquad \vdots \\
a_{n1}x_1 + a_{n2}x_2 + \cdots + a_{nn}x_n &= b_n.
\end{aligned}
\tag{5}
$$

If $\mathbf{A}$ denotes the matrix of coefficients in (5), we know that Cramer's rule could be used to solve the system whenever $\det \mathbf{A} \neq 0$. However, that rule requires a herculean effort if $\mathbf{A}$ is larger than 3×3. The procedure that we shall now consider has the distinct advantage of being not only an efficient way of handling large systems but also a means of solving consistent systems (5) in which $\det \mathbf{A} = 0$ and a means of solving m linear equations in n unknowns.

DEFINITION II.12   **Augmented Matrix**

The **augmented matrix** of the system (5) is the $n \times (n + 1)$ matrix

$$
\begin{pmatrix}
a_{11} & a_{12} & \cdots & a_{1n} & \bigm| & b_1 \\
a_{21} & a_{22} & \cdots & a_{2n} & \bigm| & b_2 \\
\vdots & & & & \bigm| & \vdots \\
a_{n1} & a_{n2} & \cdots & a_{nn} & \bigm| & b_n
\end{pmatrix}.
$$

If $\mathbf{B}$ is the column matrix of the b_i, $i = 1, 2, \ldots, n$, the augmented matrix of (5) is denoted by $(\mathbf{A} \mid \mathbf{B})$.

Elementary Row Operations   Recall from algebra that we can transform an algebraic system of equations into an equivalent system (that is, one having the same solution) by multiplying an equation by a nonzero constant, interchanging the positions of any two equations in a system, and adding a nonzero constant multiple of an equation to another equation. These operations on equations in a system are, in turn, equivalent to **elementary row operations** on an augmented matrix:

(i) Multiply a row by a nonzero constant.
(ii) Interchange any two rows.
(iii) Add a nonzero constant multiple of one row to any other row.

Elimination Methods   To solve a system such as (5) using an augmented matrix we use either **Gaussian elimination** or the **Gauss-Jordan elimination method.** In the former method we carry out a succession of elementary row operations until we arrive at an augmented matrix in **row-echelon form:**

(i) The first nonzero entry in a nonzero row is 1.
(ii) In consecutive nonzero rows, the first entry 1 in the lower row appears to the right of the first 1 in the higher row.
(iii) Rows consisting of all 0's are at the bottom of the matrix.

In the Gauss-Jordan method the row operations are continued until we obtain an augmented matrix that is in **reduced row-echelon form.** A reduced row-echelon matrix has the same three properties listed above in addition to the following one:

(iv) A column containing a first entry 1 has 0's everywhere else.

EXAMPLE 11 **Row-Echelon/Reduced Row-Echelon Form**

(a) The augmented matrices

$$\begin{pmatrix} 1 & 5 & 0 & | & 2 \\ 0 & 1 & 0 & | & -1 \\ 0 & 0 & 0 & | & 0 \end{pmatrix} \quad \text{and} \quad \begin{pmatrix} 0 & 0 & 1 & -6 & 2 & | & 2 \\ 0 & 0 & 0 & 0 & 1 & | & 4 \end{pmatrix}$$

are in row-echelon form. You should verify that the three criteria are satisfied.

(b) The augmented matrices

$$\begin{pmatrix} 1 & 0 & 0 & | & 7 \\ 0 & 1 & 0 & | & -1 \\ 0 & 0 & 0 & | & 0 \end{pmatrix} \quad \text{and} \quad \begin{pmatrix} 0 & 0 & 1 & -6 & 0 & | & -6 \\ 0 & 0 & 0 & 0 & 1 & | & 4 \end{pmatrix}$$

are in reduced row-echelon form. Note that the remaining entries in the columns containing a leading entry 1 are all 0's. ∎

Note that in Gaussian elimination we stop once we have obtained *an* augmented matrix in row-echelon form. In other words, by using different sequences of row operations we may arrive at different row-echelon forms. This method then requires the use of back-substitution. In Gauss-Jordan elimination we stop when we have obtained *the* augmented matrix in reduced row-echelon form. Any sequence of row operations will lead to the same augmented matrix in reduced row-echelon form. This method does not require back-substitution; the solution of the system will be apparent by inspection of the final matrix. In terms of the equations of the original system, our goal in both methods is simply to make the coefficient of x_1 in the first equation* equal to 1 and then use multiples of that equation to eliminate x_1 from other equations. The process is repeated on the other variables.

To keep track of the row operations on an augmented matrix, we utilize the following notation:

Symbol	Meaning
R_{ij}	Interchange rows i and j
cR_i	Multiply the ith row by the nonzero constant c
$cR_i + R_j$	Multiply the ith row by c and add to the jth row

EXAMPLE 12 **Solution by Elimination**

Solve
$$2x_1 + 6x_2 + x_3 = 7$$
$$x_1 + 2x_2 - x_3 = -1$$
$$5x_1 + 7x_2 - 4x_3 = 9$$

using (a) Gaussian elimination and (b) Gauss-Jordan elimination.

---

*We can always interchange equations so that the first equation contains the variable x_1.

SOLUTION   (a) Using row operations on the augmented matrix of the system, we obtain

$$\begin{pmatrix} 2 & 6 & 1 & | & 7 \\ 1 & 2 & -1 & | & -1 \\ 5 & 7 & -4 & | & 9 \end{pmatrix} \xrightarrow{R_{12}} \begin{pmatrix} 1 & 2 & -1 & | & -1 \\ 2 & 6 & 1 & | & 7 \\ 5 & 7 & -4 & | & 9 \end{pmatrix} \xrightarrow[-5R_1 + R_3]{-2R_1 + R_2} \begin{pmatrix} 1 & 2 & -1 & | & -1 \\ 0 & 2 & 3 & | & 9 \\ 0 & -3 & 1 & | & 14 \end{pmatrix}$$

$$\xrightarrow{\frac{1}{2}R_2} \begin{pmatrix} 1 & 2 & -1 & | & -1 \\ 0 & 1 & \frac{3}{2} & | & \frac{9}{2} \\ 0 & -3 & 1 & | & 14 \end{pmatrix} \xrightarrow{3R_2 + R_3} \begin{pmatrix} 1 & 2 & -1 & | & -1 \\ 0 & 1 & \frac{3}{2} & | & \frac{9}{2} \\ 0 & 0 & \frac{11}{2} & | & \frac{55}{2} \end{pmatrix} \xrightarrow{\frac{2}{11}R_3} \begin{pmatrix} 1 & 2 & -1 & | & -1 \\ 0 & 1 & \frac{3}{2} & | & \frac{9}{2} \\ 0 & 0 & 1 & | & 5 \end{pmatrix}.$$

The last matrix is in row-echelon form and represents the system

$$x_1 + 2x_2 - x_3 = -1$$
$$x_2 + \frac{3}{2}x_3 = \frac{9}{2}$$
$$x_3 = 5.$$

Substituting $x_3 = 5$ into the second equation then gives $x_2 = -3$. Substituting both these values back into the first equation finally yields $x_1 = 10$.

(b) We start with the last matrix above. Since the first entries in the second and third rows are 1's, we must, in turn, make the remaining entries in the second and third columns 0's:

$$\begin{pmatrix} 1 & 2 & -1 & | & -1 \\ 0 & 1 & \frac{3}{2} & | & \frac{9}{2} \\ 0 & 0 & 1 & | & 5 \end{pmatrix} \xrightarrow{-2R_2 + R_1} \begin{pmatrix} 1 & 0 & -4 & | & -10 \\ 0 & 1 & \frac{3}{2} & | & \frac{9}{2} \\ 0 & 0 & 1 & | & 5 \end{pmatrix} \xrightarrow[-\frac{3}{2}R_3 + R_2]{4R_3 + R_1} \begin{pmatrix} 1 & 0 & 0 & | & 10 \\ 0 & 1 & 0 & | & -3 \\ 0 & 0 & 1 & | & 5 \end{pmatrix}.$$

The last matrix is now in reduced row-echelon form. Because of what the matrix means in terms of equations, it is evident that the solution of the system is $x_1 = 10$, $x_2 = -3$, $x_3 = 5$. ■

EXAMPLE 13   **Gauss-Jordan Elimination**

Solve
$$x + 3y - 2z = -7$$
$$4x + y + 3z = 5$$
$$2x - 5y + 7z = 19.$$

SOLUTION   We solve the system using Gauss-Jordan elimination:

$$\begin{pmatrix} 1 & 3 & -2 & | & -7 \\ 4 & 1 & 3 & | & 5 \\ 2 & -5 & 7 & | & 19 \end{pmatrix} \xrightarrow[-2R_1 + R_3]{-4R_1 + R_2} \begin{pmatrix} 1 & 3 & -2 & | & -7 \\ 0 & -11 & 11 & | & 33 \\ 0 & -11 & 11 & | & 33 \end{pmatrix}$$

$$\xrightarrow[-\frac{1}{11}R_3]{-\frac{1}{11}R_2} \begin{pmatrix} 1 & 3 & -2 & | & -7 \\ 0 & 1 & -1 & | & -3 \\ 0 & 1 & -1 & | & -3 \end{pmatrix} \xrightarrow[-R_2 + R_3]{-3R_2 + R_1} \begin{pmatrix} 1 & 0 & 1 & | & 2 \\ 0 & 1 & -1 & | & -3 \\ 0 & 0 & 0 & | & 0 \end{pmatrix}.$$

In this case the last matrix in reduced row-echelon form implies that the original system of three equations in three unknowns is really equivalent to two equations in three unknowns. Since only z is common to both equations (the nonzero rows), we can assign its values arbitrarily. If we let $z = t$, where t represents any real number, then we see that the system has infinitely many solutions: $x = 2 - t$, $y = -3 + t$, $z = t$. Geometrically, these equations are the parametric equations for the line of intersection of the planes $x + 0y + z = 2$ and $0x + y - z = -3$. ∎

II.3 The Eigenvalue Problem

Gauss-Jordan elimination can be used to find the **eigenvectors** of a square matrix.

DEFINITION II.13 **Eigenvalues and Eigenvectors**

Let **A** be an $n \times n$ matrix. A number λ is said to be an **eigenvalue** of **A** if there exists a *nonzero* solution vector **K** of the linear system

$$\mathbf{AK} = \lambda\mathbf{K}. \tag{6}$$

The solution vector **K** is said to be an **eigenvector** corresponding to the eigenvalue λ.

The word *eigenvalue* is a combination of German and English terms adapted from the German word *eigenwert,* which, translated literally, is "proper value." Eigenvalues and eigenvectors are also called **characteristic values** and **characteristic vectors,** respectively.

EXAMPLE 14 **Eigenvector of a Matrix**

Verify that $\mathbf{K} = \begin{pmatrix} 1 \\ -1 \\ 1 \end{pmatrix}$ is an eigenvector of the matrix

$$\mathbf{A} = \begin{pmatrix} 0 & -1 & -3 \\ 2 & 3 & 3 \\ -2 & 1 & 1 \end{pmatrix}.$$

SOLUTION By carrying out the multiplication **AK** we see that

$$\mathbf{AK} = \begin{pmatrix} 0 & -1 & -3 \\ 2 & 3 & 3 \\ -2 & 1 & 1 \end{pmatrix} \begin{pmatrix} 1 \\ -1 \\ 1 \end{pmatrix} = \begin{pmatrix} -2 \\ 2 \\ -2 \end{pmatrix} = (-2) \begin{pmatrix} 1 \\ -1 \\ 1 \end{pmatrix} = \overset{\text{eigenvalue}}{\underset{\downarrow}{(-2)}}\mathbf{K}.$$

We see from the preceding line and Definition II.13 that $\lambda = -2$ is an eigenvalue of **A**. ∎

Using properties of matrix algebra, we can write (6) in the alternative form

$$(\mathbf{A} - \lambda\mathbf{I})\mathbf{K} = \mathbf{0}, \tag{7}$$

where $\mathbf{I}$ is the multiplicative identity. If we let

$$\mathbf{K} = \begin{pmatrix} k_1 \\ k_2 \\ \vdots \\ k_n \end{pmatrix},$$

then (7) is the same as

$$\begin{array}{rcl}
(a_{11} - \lambda)k_1 + & a_{12}k_2 + \cdots + & a_{1n}k_n = 0 \\
a_{21}k_1 + (a_{22} - \lambda)k_2 + \cdots + & a_{2n}k_n = 0 \\
\vdots & \vdots \\
a_{n1}k_1 + & a_{n2}k_2 + \cdots + (a_{nn} - \lambda)k_n = 0.
\end{array} \tag{8}$$

Although an obvious solution of (8) is $k_1 = 0, k_2 = 0, \ldots, k_n = 0$, we are seeking only nontrivial solutions. It is known that a homogeneous system of n linear equations in n unknowns (that is, $b_i = 0, i = 1, 2, \ldots, n$ in (5)) has a nontrivial solution if and only if the determinant of the coefficient matrix is equal to zero. Thus to find a nonzero solution $\mathbf{K}$ for (7) we must have

$$\det(\mathbf{A} - \lambda\mathbf{I}) = 0. \tag{9}$$

Inspection of (8) shows that the expansion of $\det(\mathbf{A} - \lambda\mathbf{I})$ by cofactors results in an nth-degree polynomial in λ. The equation (9) is called the **characteristic equation** of $\mathbf{A}$. Thus *the eigenvalues of $\mathbf{A}$ are the roots of the characteristic equation.* To find an eigenvector corresponding to an eigenvalue λ we simply solve the system of equations $(\mathbf{A} - \lambda\mathbf{I})\mathbf{K} = \mathbf{0}$ by applying Gauss-Jordan elimination to the augmented matrix $(\mathbf{A} - \lambda\mathbf{I}|\mathbf{0})$.

EXAMPLE 15   Eigenvalues/Eigenvectors

Find the eigenvalues and eigenvectors of $\mathbf{A} = \begin{pmatrix} 1 & 2 & 1 \\ 6 & -1 & 0 \\ -1 & -2 & -1 \end{pmatrix}$.

SOLUTION   To expand the determinant in the characteristic equation we use the cofactors of the second row:

$$\det(\mathbf{A} - \lambda\mathbf{I}) = \begin{vmatrix} 1 - \lambda & 2 & 1 \\ 6 & -1 - \lambda & 0 \\ -1 & -2 & -1 - \lambda \end{vmatrix} = -\lambda^3 - \lambda^2 + 12\lambda = 0.$$

From $-\lambda^3 - \lambda^2 + 12\lambda = -\lambda(\lambda + 4)(\lambda - 3) = 0$ we see that the eigenvalues are $\lambda_1 = 0$, $\lambda_2 = -4$, and $\lambda_3 = 3$. To find the eigenvectors we must now reduce $(\mathbf{A} - \lambda\mathbf{I}|\mathbf{0})$ three times corresponding to the three distinct eigenvalues.

For $\lambda_1 = 0$ we have

$$(\mathbf{A} - 0\mathbf{I}|\mathbf{0}) = \begin{pmatrix} 1 & 2 & 1 & | & 0 \\ 6 & -1 & 0 & | & 0 \\ -1 & -2 & -1 & | & 0 \end{pmatrix} \xrightarrow[R_1 + R_3]{-6R_1 + R_2} \begin{pmatrix} 1 & 2 & 1 & | & 0 \\ 0 & -13 & -6 & | & 0 \\ 0 & 0 & 0 & | & 0 \end{pmatrix}$$

$$\xrightarrow{-\frac{1}{13}R_2} \begin{pmatrix} 1 & 2 & 1 & | & 0 \\ 0 & 1 & \frac{6}{13} & | & 0 \\ 0 & 0 & 0 & | & 0 \end{pmatrix} \xrightarrow{-2R_2 + R_1} \begin{pmatrix} 1 & 0 & \frac{1}{13} & | & 0 \\ 0 & 1 & \frac{6}{13} & | & 0 \\ 0 & 0 & 0 & | & 0 \end{pmatrix}.$$

Thus we see that $k_1 = -\frac{1}{13}k_3$ and $k_2 = -\frac{6}{13}k_3$. Choosing $k_3 = -13$, we get the eigenvector*

$$\mathbf{K}_1 = \begin{pmatrix} 1 \\ 6 \\ -13 \end{pmatrix}.$$

For $\lambda_2 = -4$,

$$(\mathbf{A} + 4\mathbf{I}|\mathbf{0}) = \begin{pmatrix} 5 & 2 & 1 & | & 0 \\ 6 & 3 & 0 & | & 0 \\ -1 & -2 & 3 & | & 0 \end{pmatrix} \xrightarrow[R_{31}]{-R_3} \begin{pmatrix} 1 & 2 & -3 & | & 0 \\ 6 & 3 & 0 & | & 0 \\ 5 & 2 & 1 & | & 0 \end{pmatrix}$$

$$\xrightarrow[-5R_1 + R_3]{-6R_1 + R_2} \begin{pmatrix} 1 & 2 & -3 & | & 0 \\ 0 & -9 & 18 & | & 0 \\ 0 & -8 & 16 & | & 0 \end{pmatrix} \xrightarrow[-\frac{1}{8}R_3]{-\frac{1}{9}R_2} \begin{pmatrix} 1 & 2 & -3 & | & 0 \\ 0 & 1 & -2 & | & 0 \\ 0 & 1 & -2 & | & 0 \end{pmatrix} \xrightarrow[-R_2 + R_3]{-2R_2 + R_1} \begin{pmatrix} 1 & 0 & 1 & | & 0 \\ 0 & 1 & -2 & | & 0 \\ 0 & 0 & 0 & | & 0 \end{pmatrix}$$

implies $k_1 = -k_3$ and $k_2 = 2k_3$. Choosing $k_3 = 1$ then yields the second eigenvector

$$\mathbf{K}_2 = \begin{pmatrix} -1 \\ 2 \\ 1 \end{pmatrix}.$$

Finally, for $\lambda_3 = 3$ Gauss-Jordan elimination gives

$$(\mathbf{A} - 3\mathbf{I}|\mathbf{0}) = \begin{pmatrix} -2 & 2 & 1 & | & 0 \\ 6 & -4 & 0 & | & 0 \\ -1 & -2 & -4 & | & 0 \end{pmatrix} \xrightarrow[\text{operations}]{\text{row}} \begin{pmatrix} 1 & 0 & 1 & | & 0 \\ 0 & 1 & \frac{3}{2} & | & 0 \\ 0 & 0 & 0 & | & 0 \end{pmatrix},$$

and so $k_1 = -k_3$ and $k_2 = -\frac{3}{2}k_3$. The choice of $k_3 = -2$ leads to the third eigenvector:

$$\mathbf{K}_3 = \begin{pmatrix} 2 \\ 3 \\ -2 \end{pmatrix}. \qquad \blacksquare$$

When an $n \times n$ matrix $\mathbf{A}$ possesses n distinct eigenvalues $\lambda_1, \lambda_2, \ldots, \lambda_n$, it can be proved that a set of n linearly independent[†] eigenvectors

---

*Of course k_3 could be chosen as any nonzero number. In other words, a nonzero constant multiple of an eigenvector is also an eigenvector.

[†]Linear independence of column vectors is defined in exactly the same manner as for functions.

$K_1, K_2, \ldots, K_n$ can be found. However, when the characteristic equation has repeated roots, it may not be possible to find n linearly independent eigenvectors for A.

EXAMPLE 16   **Eigenvalues/Eigenvectors**

Find the eigenvalues and eigenvectors of $A = \begin{pmatrix} 3 & 4 \\ -1 & 7 \end{pmatrix}$.

SOLUTION   From the characteristic equation

$$\det(A - \lambda I) = \begin{vmatrix} 3 - \lambda & 4 \\ -1 & 7 - \lambda \end{vmatrix} = (\lambda - 5)^2 = 0$$

we see that $\lambda_1 = \lambda_2 = 5$ is an eigenvalue of multiplicity two. In the case of a 2×2 matrix there is no need to use Gauss-Jordan elimination. To find the eigenvector(s) corresponding to $\lambda_1 = 5$ we resort to the system $(A - 5I|0)$ in its equivalent form

$$-2k_1 + 4k_2 = 0$$
$$-k_1 + 2k_2 = 0.$$

It is apparent from this system that $k_1 = 2k_2$. Thus if we choose $k_2 = 1$, we find the single eigenvector

$$K_1 = \begin{pmatrix} 2 \\ 1 \end{pmatrix}.$$

EXAMPLE 17   **Eigenvalues/Eigenvectors**

Find the eigenvalues and eigenvectors of $A = \begin{pmatrix} 9 & 1 & 1 \\ 1 & 9 & 1 \\ 1 & 1 & 9 \end{pmatrix}$.

SOLUTION   The characteristic equation

$$\det(A - \lambda I) = \begin{vmatrix} 9 - \lambda & 1 & 1 \\ 1 & 9 - \lambda & 1 \\ 1 & 1 & 9 - \lambda \end{vmatrix} = -(\lambda - 11)(\lambda - 8)^2 = 0$$

shows that $\lambda_1 = 11$ and that $\lambda_2 = \lambda_3 = 8$ is an eigenvalue of multiplicity two.

For $\lambda_1 = 11$ Gauss-Jordan elimination gives

$$(A - 11I|0) = \begin{pmatrix} -2 & 1 & 1 & | & 0 \\ 1 & -2 & 1 & | & 0 \\ 1 & 1 & -2 & | & 0 \end{pmatrix} \xrightarrow[\text{operations}]{\text{row}} \begin{pmatrix} 1 & 0 & -1 & | & 0 \\ 0 & 1 & -1 & | & 0 \\ 0 & 0 & 0 & | & 0 \end{pmatrix}.$$

Hence $k_1 = k_3$ and $k_2 = k_3$. If $k_3 = 1$, then

$$\mathbf{K}_1 = \begin{pmatrix} 1 \\ 1 \\ 1 \end{pmatrix}.$$

Now for $\lambda_2 = 8$ we have

$$(\mathbf{A} - 8\mathbf{I}|\mathbf{0}) = \begin{pmatrix} 1 & 1 & 1 & | & 0 \\ 1 & 1 & 1 & | & 0 \\ 1 & 1 & 1 & | & 0 \end{pmatrix} \xrightarrow[\text{operations}]{\text{row}} \begin{pmatrix} 1 & 1 & 1 & | & 0 \\ 0 & 0 & 0 & | & 0 \\ 0 & 0 & 0 & | & 0 \end{pmatrix}.$$

In the equation $k_1 + k_2 + k_3 = 0$ we are free to select two of the variables arbitrarily. Choosing, on the one hand, $k_2 = 1$, $k_3 = 0$ and, on the other, $k_2 = 0$, $k_3 = 1$, we obtain two linearly independent eigenvectors

$$\mathbf{K}_2 = \begin{pmatrix} -1 \\ 1 \\ 0 \end{pmatrix} \quad \text{and} \quad \mathbf{K}_3 = \begin{pmatrix} -1 \\ 0 \\ 1 \end{pmatrix}. \qquad \blacksquare$$

APPENDIX II EXERCISES

Answers to odd-numbered problems begin on page A-20.

II.1

1. If $\mathbf{A} = \begin{pmatrix} 4 & 5 \\ -6 & 9 \end{pmatrix}$ and $\mathbf{B} = \begin{pmatrix} -2 & 6 \\ 8 & -10 \end{pmatrix}$, find

   (a) $\mathbf{A} + \mathbf{B}$ (b) $\mathbf{B} - \mathbf{A}$ (c) $2\mathbf{A} + 3\mathbf{B}$

2. If $\mathbf{A} = \begin{pmatrix} -2 & 0 \\ 4 & 1 \\ 7 & 3 \end{pmatrix}$ and $\mathbf{B} = \begin{pmatrix} 3 & -1 \\ 0 & 2 \\ -4 & -2 \end{pmatrix}$, find

   (a) $\mathbf{A} - \mathbf{B}$ (b) $\mathbf{B} - \mathbf{A}$ (c) $2(\mathbf{A} + \mathbf{B})$

3. If $\mathbf{A} = \begin{pmatrix} 2 & -3 \\ -5 & 4 \end{pmatrix}$ and $\mathbf{B} = \begin{pmatrix} -1 & 6 \\ 3 & 2 \end{pmatrix}$, find

   (a) $\mathbf{AB}$ (b) $\mathbf{BA}$ (c) $\mathbf{A}^2 = \mathbf{AA}$ (d) $\mathbf{B}^2 = \mathbf{BB}$

4. If $\mathbf{A} = \begin{pmatrix} 1 & 4 \\ 5 & 10 \\ 8 & 12 \end{pmatrix}$ and $\mathbf{B} = \begin{pmatrix} -4 & 6 & -3 \\ 1 & -3 & 2 \end{pmatrix}$, find

   (a) $\mathbf{AB}$ (b) $\mathbf{BA}$

5. If $\mathbf{A} = \begin{pmatrix} 1 & -2 \\ -2 & 4 \end{pmatrix}$, $\mathbf{B} = \begin{pmatrix} 6 & 3 \\ 2 & 1 \end{pmatrix}$, and $\mathbf{C} = \begin{pmatrix} 0 & 2 \\ 3 & 4 \end{pmatrix}$, find

   (a) $\mathbf{BC}$ (b) $\mathbf{A}(\mathbf{BC})$ (c) $\mathbf{C}(\mathbf{BA})$ (d) $\mathbf{A}(\mathbf{B} + \mathbf{C})$

6. If $\mathbf{A} = (5 \quad -6 \quad 7)$, $\mathbf{B} = \begin{pmatrix} 3 \\ 4 \\ -1 \end{pmatrix}$, and $\mathbf{C} = \begin{pmatrix} 1 & 2 & 4 \\ 0 & 1 & -1 \\ 3 & 2 & 1 \end{pmatrix}$, find

   (a) $\mathbf{AB}$ (b) $\mathbf{BA}$ (c) $(\mathbf{BA})\mathbf{C}$ (d) $(\mathbf{AB})\mathbf{C}$

7. If $\mathbf{A} = \begin{pmatrix} 4 \\ 8 \\ -10 \end{pmatrix}$ and $\mathbf{B} = (2 \quad 4 \quad 5)$, find

   (a) $\mathbf{A}^T\mathbf{A}$ (b) $\mathbf{B}^T\mathbf{B}$ (c) $\mathbf{A} + \mathbf{B}^T$

8. If $\mathbf{A} = \begin{pmatrix} 1 & 2 \\ 2 & 4 \end{pmatrix}$ and $\mathbf{B} = \begin{pmatrix} -2 & 3 \\ 5 & 7 \end{pmatrix}$, find

   (a) $\mathbf{A} + \mathbf{B}^T$ (b) $2\mathbf{A}^T - \mathbf{B}^T$ (c) $\mathbf{A}^T(\mathbf{A} - \mathbf{B})$

9. If $\mathbf{A} = \begin{pmatrix} 3 & 4 \\ 8 & 1 \end{pmatrix}$ and $\mathbf{B} = \begin{pmatrix} 5 & 10 \\ -2 & -5 \end{pmatrix}$, find

   (a) $(\mathbf{AB})^T$ (b) $\mathbf{B}^T\mathbf{A}^T$

10. If $\mathbf{A} = \begin{pmatrix} 5 & 9 \\ -4 & 6 \end{pmatrix}$ and $\mathbf{B} = \begin{pmatrix} -3 & 11 \\ -7 & 2 \end{pmatrix}$, find

    (a) $\mathbf{A}^T + \mathbf{B}^T$ (b) $(\mathbf{A} + \mathbf{B})^T$

In Problems 11–14 write the given sum as a single column matrix.

11. $4\begin{pmatrix} -1 \\ 2 \end{pmatrix} - 2\begin{pmatrix} 2 \\ 8 \end{pmatrix} + 3\begin{pmatrix} -2 \\ 3 \end{pmatrix}$

12. $3t\begin{pmatrix} 2 \\ t \\ -1 \end{pmatrix} + (t - 1)\begin{pmatrix} -1 \\ -t \\ 3 \end{pmatrix} - 2\begin{pmatrix} 3t \\ 4 \\ -5t \end{pmatrix}$

13. $\begin{pmatrix} 2 & -3 \\ 1 & 4 \end{pmatrix}\begin{pmatrix} -2 \\ 5 \end{pmatrix} - \begin{pmatrix} -1 & 6 \\ -2 & 3 \end{pmatrix}\begin{pmatrix} -7 \\ 2 \end{pmatrix}$

14. $\begin{pmatrix} 1 & -3 & 4 \\ 2 & 5 & -1 \\ 0 & -4 & -2 \end{pmatrix}\begin{pmatrix} t \\ 2t - 1 \\ -t \end{pmatrix} + \begin{pmatrix} -t \\ 1 \\ 4 \end{pmatrix} - \begin{pmatrix} 2 \\ 8 \\ -6 \end{pmatrix}$

In Problems 15–22 determine whether the given matrix is singular or nonsingular. If nonsingular, find $\mathbf{A}^{-1}$.

15. $\mathbf{A} = \begin{pmatrix} -3 & 6 \\ -2 & 4 \end{pmatrix}$ 16. $\mathbf{A} = \begin{pmatrix} 2 & 5 \\ 1 & 4 \end{pmatrix}$

17. $\mathbf{A} = \begin{pmatrix} 4 & 8 \\ -3 & -5 \end{pmatrix}$ 18. $\mathbf{A} = \begin{pmatrix} 7 & 10 \\ 2 & 2 \end{pmatrix}$

19. $\mathbf{A} = \begin{pmatrix} 2 & 1 & 0 \\ -1 & 2 & 1 \\ 1 & 2 & 1 \end{pmatrix}$ 20. $\mathbf{A} = \begin{pmatrix} 3 & 2 & 1 \\ 4 & 1 & 0 \\ -2 & 5 & -1 \end{pmatrix}$

21. $\mathbf{A} = \begin{pmatrix} 2 & 1 & 1 \\ 1 & -2 & -3 \\ 3 & 2 & 4 \end{pmatrix}$ 22. $\mathbf{A} = \begin{pmatrix} 4 & 1 & -1 \\ 6 & 2 & -3 \\ -2 & -1 & 2 \end{pmatrix}$

In Problems 23 and 24 show that the given matrix is nonsingular for every real value of t. Find $\mathbf{A}^{-1}(t)$.

23. $\mathbf{A}(t) = \begin{pmatrix} 2e^{-t} & e^{4t} \\ 4e^{-t} & 3e^{4t} \end{pmatrix}$

24. $\mathbf{A}(t) = \begin{pmatrix} 2e^t \sin t & -2e^t \cos t \\ e^t \cos t & e^t \sin t \end{pmatrix}$

In Problems 25–28 find $d\mathbf{X}/dt$.

25. $\mathbf{X} = \begin{pmatrix} 5e^{-t} \\ 2e^{-t} \\ -7e^{-t} \end{pmatrix}$

26. $\mathbf{X} = \begin{pmatrix} \frac{1}{2}\sin 2t - 4\cos 2t \\ -3\sin 2t + 5\cos 2t \end{pmatrix}$

27. $\mathbf{X} = 2\begin{pmatrix} 1 \\ -1 \end{pmatrix}e^{2t} + 4\begin{pmatrix} 2 \\ 1 \end{pmatrix}e^{-3t}$

28. $\mathbf{X} = \begin{pmatrix} 5te^{2t} \\ t\sin 3t \end{pmatrix}$

29. Let $\mathbf{A}(t) = \begin{pmatrix} e^{4t} & \cos \pi t \\ 2t & 3t^2 - 1 \end{pmatrix}$. Find

(a) $\dfrac{d\mathbf{A}}{dt}$ (b) $\displaystyle\int_0^2 \mathbf{A}(t)\, dt$ (c) $\displaystyle\int_0^t \mathbf{A}(s)\, ds$

30. Let $\mathbf{A}(t) = \begin{pmatrix} \dfrac{1}{t^2 + 1} & 3t \\ t^2 & t \end{pmatrix}$ and $\mathbf{B}(t) = \begin{pmatrix} 6t & 2 \\ 1/t & 4t \end{pmatrix}$. Find

(a) $\dfrac{d\mathbf{A}}{dt}$ (b) $\dfrac{d\mathbf{B}}{dt}$ (c) $\displaystyle\int_0^1 \mathbf{A}(t)\, dt$ (d) $\displaystyle\int_1^2 \mathbf{B}(t)\, dt$

(e) $\mathbf{A}(t)\mathbf{B}(t)$ (f) $\dfrac{d}{dt}\mathbf{A}(t)\mathbf{B}(t)$ (g) $\displaystyle\int_1^t \mathbf{A}(s)\mathbf{B}(s)\, ds$

II.2

In Problems 31–38 solve the given system of equations by either Gaussian elimination or Gauss-Jordan elimination.

31. $\begin{aligned} x + y - 2z &= 14 \\ 2x - y + z &= 0 \\ 6x + 3y + 4z &= 1 \end{aligned}$

32. $\begin{aligned} 5x - 2y + 4z &= 10 \\ x + y + z &= 9 \\ 4x - 3y + 3z &= 1 \end{aligned}$

33. $\begin{aligned} y + z &= -5 \\ 5x + 4y - 16z &= -10 \\ x - y - 5z &= 7 \end{aligned}$

34. $\begin{aligned} 3x + y + z &= 4 \\ 4x + 2y - z &= 7 \\ x + y - 3z &= 6 \end{aligned}$

35. $\begin{aligned} 2x + y + z &= 4 \\ 10x - 2y + 2z &= -1 \\ 6x - 2y + 4z &= 8 \end{aligned}$

36. $\begin{aligned} x + 2z &= 8 \\ x + 2y - 2z &= 4 \\ 2x + 5y - 6z &= 6 \end{aligned}$

37. $\begin{aligned} x_1 + x_2 - x_3 - x_4 &= -1 \\ x_1 + x_2 + x_3 + x_4 &= 3 \\ x_1 - x_2 + x_3 - x_4 &= 3 \\ 4x_1 + x_2 - 2x_3 + x_4 &= 0 \end{aligned}$

38. $\begin{aligned} 2x_1 + x_2 + x_3 &= 0 \\ x_1 + 3x_2 + x_3 &= 0 \\ 7x_1 + x_2 + 3x_3 &= 0 \end{aligned}$

In Problems 39 and 40 use Gauss-Jordan elimination to demonstrate that the given system of equations has no solution.

39. $\begin{aligned} x + 2y + 4z &= 2 \\ 2x + 4y + 3z &= 1 \\ x + 2y - z &= 7 \end{aligned}$

40. $\begin{aligned} x_1 + x_2 - x_3 + 3x_4 &= 1 \\ x_2 - x_3 - 4x_4 &= 0 \\ x_1 + 2x_2 - 2x_3 - x_4 &= 6 \\ 4x_1 + 7x_2 - 7x_3 &= 9 \end{aligned}$

II.3

In Problems 41–48 find the eigenvalues and eigenvectors of the given matrix.

41. $\begin{pmatrix} -1 & 2 \\ -7 & 8 \end{pmatrix}$

42. $\begin{pmatrix} 2 & 1 \\ 2 & 1 \end{pmatrix}$

43. $\begin{pmatrix} -8 & -1 \\ 16 & 0 \end{pmatrix}$

44. $\begin{pmatrix} 1 & 1 \\ \frac{1}{4} & 1 \end{pmatrix}$

45. $\begin{pmatrix} 5 & -1 & 0 \\ 0 & -5 & 9 \\ 5 & -1 & 0 \end{pmatrix}$

46. $\begin{pmatrix} 3 & 0 & 0 \\ 0 & 2 & 0 \\ 4 & 0 & 1 \end{pmatrix}$

47. $\begin{pmatrix} 0 & 4 & 0 \\ -1 & -4 & 0 \\ 0 & 0 & -2 \end{pmatrix}$

48. $\begin{pmatrix} 1 & 6 & 0 \\ 0 & 2 & 1 \\ 0 & 1 & 2 \end{pmatrix}$

In Problems 49 and 50 show that the given matrix has complex eigenvalues. Find the eigenvectors of the matrix.

49. $\begin{pmatrix} -1 & 2 \\ -5 & 1 \end{pmatrix}$

50. $\begin{pmatrix} 2 & -1 & 0 \\ 5 & 2 & 4 \\ 0 & 1 & 2 \end{pmatrix}$

51. If $\mathbf{A}(t)$ is a 2×2 matrix of differentiable functions and $\mathbf{X}(t)$ is a 2×1 column matrix of differentiable functions, prove the product rule

$$\frac{d}{dt}[\mathbf{A}(t)\mathbf{X}(t)] = \mathbf{A}(t)\mathbf{X}'(t) + \mathbf{A}'(t)\mathbf{X}(t).$$

52. Derive formula (3). [*Hint:* Find a matrix $\mathbf{B} = \begin{pmatrix} b_{11} & b_{12} \\ b_{21} & b_{22} \end{pmatrix}$ for which $\mathbf{AB} = \mathbf{I}$. Solve for b_{11}, b_{12}, b_{21}, and b_{22}. Then show that $\mathbf{BA} = \mathbf{I}$.]

53. If $\mathbf{A}$ is nonsingular and $\mathbf{AB} = \mathbf{AC}$, show that $\mathbf{B} = \mathbf{C}$.

54. If $\mathbf{A}$ and $\mathbf{B}$ are nonsingular, show that $(\mathbf{AB})^{-1} = \mathbf{B}^{-1}\mathbf{A}^{-1}$.

55. Let $\mathbf{A}$ and $\mathbf{B}$ be $n \times n$ matrices. In general, is $(\mathbf{A} + \mathbf{B})^2 = \mathbf{A}^2 + 2\mathbf{AB} + \mathbf{B}^2$?

III

LAPLACE TRANSFORMS

$f(t)$	$\mathscr{L}\{f(t)\} = F(s)$
1. 1	$\dfrac{1}{s}$
2. t	$\dfrac{1}{s^2}$
3. t^n	$\dfrac{n!}{s^{n+1}}$, $\quad n$ a positive integer
4. $t^{-1/2}$	$\sqrt{\dfrac{\pi}{s}}$
5. $t^{1/2}$	$\dfrac{\sqrt{\pi}}{2s^{3/2}}$
6. t^{α}	$\dfrac{\Gamma(\alpha + 1)}{s^{\alpha+1}}$, $\quad \alpha > -1$
7. $\sin kt$	$\dfrac{k}{s^2 + k^2}$
8. $\cos kt$	$\dfrac{s}{s^2 + k^2}$
9. $\sin^2 kt$	$\dfrac{2k^2}{s(s^2 + 4k^2)}$
10. $\cos^2 kt$	$\dfrac{s^2 + 2k^2}{s(s^2 + 4k^2)}$
11. e^{at}	$\dfrac{1}{s - a}$
12. $\sinh kt$	$\dfrac{k}{s^2 - k^2}$
13. $\cosh kt$	$\dfrac{s}{s^2 - k^2}$
14. $\sinh^2 kt$	$\dfrac{2k^2}{s(s^2 - 4k^2)}$
15. $\cosh^2 kt$	$\dfrac{s^2 - 2k^2}{s(s^2 - 4k^2)}$
16. te^{at}	$\dfrac{1}{(s - a)^2}$
17. $t^n e^{at}$	$\dfrac{n!}{(s - a)^{n+1}}$, $\quad n$ a positive integer

$f(t)$	$\mathscr{L}\{f(t)\} = F(s)$
18. $e^{at} \sin kt$	$\dfrac{k}{(s-a)^2 + k^2}$
19. $e^{at} \cos kt$	$\dfrac{s-a}{(s-a)^2 + k^2}$
20. $e^{at} \sinh kt$	$\dfrac{k}{(s-a)^2 - k^2}$
21. $e^{at} \cosh kt$	$\dfrac{s-a}{(s-a)^2 - k^2}$
22. $t \sin kt$	$\dfrac{2ks}{(s^2 + k^2)^2}$
23. $t \cos kt$	$\dfrac{s^2 - k^2}{(s^2 + k^2)^2}$
24. $\sin kt + kt \cos kt$	$\dfrac{2ks^2}{(s^2 + k^2)^2}$
25. $\sin kt - kt \cos kt$	$\dfrac{2k^3}{(s^2 + k^2)^2}$
26. $t \sinh kt$	$\dfrac{2ks}{(s^2 - k^2)^2}$
27. $t \cosh kt$	$\dfrac{s^2 + k^2}{(s^2 - k^2)^2}$
28. $\dfrac{e^{at} - e^{bt}}{a - b}$	$\dfrac{1}{(s-a)(s-b)}$
29. $\dfrac{ae^{at} - be^{bt}}{a - b}$	$\dfrac{s}{(s-a)(s-b)}$
30. $1 - \cos kt$	$\dfrac{k^2}{s(s^2 + k^2)}$
31. $kt - \sin kt$	$\dfrac{k^3}{s^2(s^2 + k^2)}$
32. $\dfrac{a \sin bt - b \sin at}{ab(a^2 - b^2)}$	$\dfrac{1}{(s^2 + a^2)(s^2 + b^2)}$
33. $\dfrac{\cos bt - \cos at}{a^2 - b^2}$	$\dfrac{s}{(s^2 + a^2)(s^2 + b^2)}$
34. $\sin kt \sinh kt$	$\dfrac{2k^2s}{s^4 + 4k^4}$
35. $\sin kt \cosh kt$	$\dfrac{k(s^2 + 2k^2)}{s^4 + 4k^4}$
36. $\cos kt \sinh kt$	$\dfrac{k(s^2 - 2k^2)}{s^4 + 4k^4}$
37. $\cos kt \cosh kt$	$\dfrac{s^3}{s^4 + 4k^4}$
38. $J_0(kt)$	$\dfrac{1}{\sqrt{s^2 + k^2}}$

$f(t)$	$\mathscr{L}\{f(t)\} = F(s)$
39. $\dfrac{e^{bt} - e^{at}}{t}$	$\ln \dfrac{s - a}{s - b}$
40. $\dfrac{2(1 - \cos kt)}{t}$	$\ln \dfrac{s^2 + k^2}{s^2}$
41. $\dfrac{2(1 - \cosh kt)}{t}$	$\ln \dfrac{s^2 - k^2}{s^2}$
42. $\dfrac{\sin at}{t}$	$\arctan \left(\dfrac{a}{s} \right)$
43. $\dfrac{\sin at \cos bt}{t}$	$\dfrac{1}{2} \arctan \dfrac{a + b}{s} + \dfrac{1}{2} \arctan \dfrac{a - b}{s}$
44. $\delta(t)$	1
45. $\delta(t - t_0)$	e^{-st_0}
46. $e^{at} f(t)$	$F(s - a)$
47. $f(t - a)\,\mathscr{U}(t - a)$	$e^{-as} F(s)$
48. $\mathscr{U}(t - a)$	$\dfrac{e^{-as}}{s}$
49. $f^{(n)}(t)$	$s^n F(s) - s^{(n-1)} f(0) - \cdots - f^{(n-1)}(0)$
50. $t^n f(t)$	$(-1)^n \dfrac{d^n}{ds^n} F(s)$
51. $\displaystyle\int_0^t f(\tau) g(t - \tau)\, d\tau$	$F(s) G(s)$

ANSWERS TO
ODD-NUMBERED PROBLEMS

SECTION 1.1 EXERCISES, page 8

1. linear, second-order **3.** nonlinear, first-order
5. linear, fourth-order **7.** nonlinear, second-order
9. linear, third-order **43.** $y = -1$
45. $m = 2$ and $m = 3$ **47.** $m = \dfrac{1 \pm \sqrt{5}}{2}$

SECTION 1.2 EXERCISES, page 15

1. half-planes defined by either $y > 0$ or $y < 0$
3. half-planes defined by either $x > 0$ or $x < 0$
5. the regions defined by either $y > 2$, $y < -2$, or $-2 < y < 2$
7. any region not containing $(0, 0)$
9. the entire xy-plane **11.** $y = 0$, $y = x^3$ **13.** yes
15. no
17. (a) $y = cx$
   (b) any rectangular region not touching the y-axis
   (c) No, the function is not differentiable at $x = 0$.
19. (c) $(-\infty, \infty)$; $(-\infty, \frac{1}{2})$; $(-\infty, -\frac{4}{3})$; $(-\infty, -2)$; $(-2, \infty)$; $(-\frac{1}{2}, \infty)$; $(0, \infty)$
21. $y = 1/(1 - 4e^{-x})$ **23.** $y = \frac{3}{2}e^x - \frac{1}{2}e^{-x}$ **25.** $y = 5e^{-x-1}$

SECTION 1.3 EXERCISES, page 25

1. $\dfrac{dP}{dt} = kP + r$ **3.** $\dfrac{dx}{dt} + kx = r$, $k > 0$
5. $\dfrac{dA}{dt} = -\dfrac{A}{100}$ **7.** $\dfrac{dh}{dt} = -\dfrac{c\pi}{450}\sqrt{h}$
9. $L\dfrac{di}{dt} + Ri = E(t)$ **11.** $\dfrac{dA}{dt} = k(M - A)$, $k > 0$
13. $\dfrac{dy}{dx} = -\dfrac{y}{\sqrt{s^2 - y^2}}$ **15.** $m\dfrac{d^2x}{dt^2} = -kx$
17. $\dfrac{dy}{dx} = \dfrac{-x + \sqrt{x^2 + y^2}}{y}$

CHAPTER 1 REVIEW EXERCISES, page 28

1. the regions defined by $x^2 + y^2 > 25$ and $x^2 + y^2 < 25$
3. false **5.** ordinary, first-order, nonlinear
7. partial, second-order **13.** $y = x^2$ **15.** $y = \dfrac{x^2}{2}$
17. $y = 0$, $y = e^x$ **19.** $y = 0$, $y = \cos x$, $y = \sin x$
21. $x < 0$ or $x > 1$ **23.** $\dfrac{dh}{dt} = -\dfrac{25\sqrt{2g}}{16\pi}h^{-3/2}$
25. (a) $k = gR^2$ (b) $\dfrac{d^2r}{dt^2} - \dfrac{gR^2}{r^2} = 0$
   (c) $v\dfrac{dv}{dr} - \dfrac{gR^2}{r^2} = 0$

SECTION 2.1 EXERCISES, page 35

1. $y = -\frac{1}{5}\cos 5x + c$ **3.** $y = \frac{1}{3}e^{-3x} + c$
5. $y = x + 5\ln|x + 1| + c$ **7.** $y = cx^4$
9. $y^{-2} = 2x^{-1} + c$ **11.** $-3 + 3x\ln|x| = xy^3 + cx$
13. $-3e^{-2y} = 2e^{3x} + c$ **15.** $2 + y^2 = c(4 + x^2)$
17. $y^2 = x - \ln|x + 1| + c$
19. $\dfrac{x^3}{3}\ln x - \dfrac{1}{9}x^3 = \dfrac{1}{2}y^2 + 2y + \ln|y| + c$
21. $S = ce^{kr}$ **23.** $\dfrac{P}{1 - P} = ce^t$ or $P = \dfrac{ce^t}{1 + ce^t}$
25. $4\cos y = 2x + \sin 2x + c$
27. $-2\cos x + e^y + ye^{-y} + e^{-y} = c$
29. $(e^x + 1)^{-2} + 2(e^y + 1)^{-1} = c$
31. $(y + 1)^{-1} + \ln|y + 1| = \dfrac{1}{2}\ln\left|\dfrac{x + 1}{x - 1}\right| + c$
33. $y - 5\ln|y + 3| = x - 5\ln|x + 4| + c$
   or $\left(\dfrac{y + 3}{x + 4}\right)^5 = c_1 e^{y-x}$
35. $-\cot y = \cos x + c$ **37.** $y = \sin\left(\dfrac{x^2}{2} + c\right)$
39. $-y^{-1} = \tan^{-1}(e^x) + c$ **41.** $(1 + \cos x)(1 + e^y) = 4$
43. $\sqrt{y^2 + 1} = 2x^2 + \sqrt{2}$ **45.** $x = \tan\left(4y - \dfrac{3\pi}{4}\right)$
47. $xy = e^{-(1+1/x)}$
49. (a) $y = 3\dfrac{1 - e^{6x}}{1 + e^{6x}}$ (b) $y = 3$ (c) $y = 3\dfrac{2 - e^{6x-2}}{2 + e^{6x-2}}$
51. $y = 1$ **53.** $y = 1$ **55.** $y = 1 + \dfrac{1}{10}\tan\dfrac{x}{10}$

SECTION 2.2 EXERCISES, page 42

1. $x^2 - x + \frac{3}{2}y^2 + 7y = c$ **3.** $\frac{5}{2}x^2 + 4xy - 2y^4 = c$
5. $x^2y^2 - 3x + 4y = c$ **7.** not exact
9. $xy^3 + y^2\cos x - \frac{1}{2}x^2 = c$ **11.** not exact
13. $xy - 2xe^x + 2e^x - 2x^3 = c$
15. $x + y + xy - 3\ln|xy| = c$ **17.** $x^3y^3 - \tan^{-1}3x = c$
19. $-\ln|\cos x| + \cos x \sin y = c$
21. $y - 2x^2y - y^2 - x^4 = c$
23. $x^4y - 5x^3 - xy + y^3 = c$
25. $\frac{1}{3}x^3 + x^2y + xy^2 - y = \frac{4}{3}$
27. $4xy + x^2 - 5x + 3y^2 - y = 8$
29. $y^2\sin x - x^3y - x^2 + y\ln y - y = 0$
31. $k = 10$ **33.** $k = 1$
35. $M(x, y) = ye^{xy} + y^2 - \dfrac{y}{x^2} + h(x)$
37. $3x^2y^3 + y^4 = c$ **39.** $x^2y^2\cos x = c$
41. $x^2y^2 + x^3 = c$

SECTION 2.3 EXERCISES, page 51

1. $y = ce^{5x}$, $-\infty < x < \infty$
3. $y = \frac{1}{3} + ce^{-4x}$, $-\infty < x < \infty$

A-1

5. $y = \frac{1}{4}e^{3x} + ce^{-x}, \; -\infty < x < \infty$

7. $y = \frac{1}{3} + ce^{-x^3}, \; -\infty < x < \infty$

9. $y = x^{-1} \ln x + cx^{-1}, \; 0 < x < \infty$

11. $x = -\frac{4}{5}y^2 + cy^{-1/2}, \; 0 < y < \infty$

13. $y = -\cos x + \dfrac{\sin x}{x} + \dfrac{c}{x}, \; 0 < x < \infty$

15. $y = \dfrac{c}{e^x + 1}, \; -\infty < x < \infty$

17. $y = \sin x + c \cos x, \; -\frac{\pi}{2} < x < \frac{\pi}{2}$

19. $y = \frac{1}{7}x^3 - \frac{1}{5}x + cx^{-4}, \; 0 < x < \infty$

21. $y = \dfrac{1}{2x^2}e^x + \dfrac{c}{x^2}e^{-x}, \; 0 < x < \infty$

23. $y = \sec x + c \csc x, \; 0 < x < \frac{\pi}{2}$

25. $x = \dfrac{1}{2}e^y - \dfrac{1}{2y}e^y + \dfrac{1}{4y^2}e^y + \dfrac{c}{y^2}e^{-y}, \; 0 < y < \infty$

27. $y = e^{-3x} + \dfrac{c}{x}e^{-3x}, \; 0 < x < \infty$

29. $x = 2y^6 + cy^4, \; 0 < y < \infty$

31. $y = e^{-x}\ln(e^x + e^{-x}) + ce^{-x}, \; -\infty < x < \infty$

33. $x = \dfrac{1}{y} + \dfrac{c}{y}e^{-y^2}, \; 0 < y < \infty$

35. $(\sec \theta + \tan \theta)r = \theta - \cos \theta + c, \; -\frac{\pi}{2} < \theta < \frac{\pi}{2}$

37. $y = \frac{5}{3}(x + 2)^{-1} + c(x + 2)^{-4}, \; -2 < x < \infty$

39. $y = 10 + ce^{-\sinh x}, \; -\infty < x < \infty$

41. $y = 4 - 2e^{-5x}, \; -\infty < x < \infty$

43. $i(t) = \dfrac{E}{R} + \left(i_0 - \dfrac{E}{R}\right)e^{-Rt/L}, \; -\infty < t < \infty$

45. $y = \sin x \cos x - \cos x, \; -\frac{\pi}{2} < x < \frac{\pi}{2}$

47. $T(t) = 50 + 150e^{kt}, \; -\infty < t < \infty$

49. $(x + 1)y = x \ln x - x + 21, \; 0 < x < \infty$

51. $y = \begin{cases} \frac{1}{2}(1 - e^{-2x}), & 0 \le x \le 3 \\ \frac{1}{2}(e^6 - 1)e^{-2x}, & x > 3 \end{cases}$

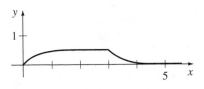

53. $y = \begin{cases} \frac{1}{2} + \frac{3}{2}e^{-x^2}, & 0 \le x < 1 \\ \left(\frac{1}{2}e + \frac{3}{2}\right)e^{-x^2}, & x \ge 1 \end{cases}$

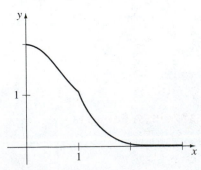

55. $y = \dfrac{10}{x^2}[\text{Si}(x) - \text{Si}(1)]$

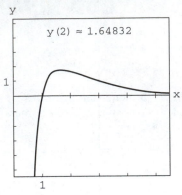

$y(2) \approx 1.64832$

57. $y = e^{x^2 - 1} + \dfrac{\sqrt{\pi}}{2}e^{x^2}[\text{erf}(x) - \text{erf}(1)]$

SECTION 2.4 EXERCISES, page 57

1. $x \ln|x| + y = cx$

3. $(x - y)\ln|x - y| = y + c(x - y)$

5. $x + y \ln|x| = cy$ **7.** $\ln(x^2 + y^2) + 2 \tan^{-1}\left(\dfrac{y}{x}\right) = c$

9. $4x = y(\ln|y| - c)^2$ **11.** $y^3 + 3x^3 \ln|x| = 8x^3$

13. $\ln|x| = e^{y/x} - 1$ **15.** $y^3 = 1 + cx^{-3}$

17. $y^{-3} = x + \frac{1}{3} + ce^{3x}$ **19.** $e^{x/y} = cx$

21. $y^{-3} = -\frac{9}{5}x^{-1} + \frac{49}{5}x^{-6}$

23. $y = -x - 1 + \tan(x + c)$

25. $2y - 2x + \sin 2(x + y) = c$

27. $4(y - 2x + 3) = (x + c)^2$

29. $-\cot(x + y) + \csc(x + y) = x + \sqrt{2} - 1$

CHAPTER 2 REVIEW EXERCISES, page 59

1. homogeneous, exact, linear in y

3. separable, exact, linear in y **5.** separable

7. linear in x **9.** Bernoulli

11. separable, homogeneous, exact, linear in x and in y

13. homogeneous **15.** $2x + \sin 2x = 2 \ln(y^2 + 1) + c$

17. $(6x + 1)y^3 = -3x^3 + c$ **19.** $Q = \dfrac{c}{t} + \dfrac{t^4}{25}(5 \ln t - 1)$

21. $2y^2 \ln y - y^2 = 4te^t - 4e^t - 1$

23. $y = \frac{1}{4} - 320(x^2 + 4)^{-4}$ **25.** $e^x = 2e^{2y} - e^{2y+x}$

SECTION 3.1 EXERCISES, page 68

1. 7.9 y; 10 y **3.** 760 **5.** 11 h **7.** 136.5 h

9. $I(15) = 0.00098I_0$; approximately 0.1% of I_0

11. 15,600 y **13.** $T(1) = 36.67°$; approximately 3.06'

15. $i(t) = \frac{3}{5} - \frac{3}{5}e^{-500t}$; $i \to \frac{3}{5}$ as $t \to \infty$

17. $q(t) = \frac{1}{100} - \frac{1}{100}e^{-50t}$; $i(t) = \frac{1}{2}e^{-50t}$

19. $i(t) = \begin{cases} 60 - 60e^{-t/10}, & 0 \le t \le 20 \\ 60(e^2 - 1)e^{-t/10}, & t > 20 \end{cases}$

21. $A(t) = 200 - 170e^{-t/50}$

23. $A(t) = 1000 - 1000e^{-t/100}$ **25.** 64.38 lb

27. (a) $v(t) = \dfrac{mg}{k} + \left(v_0 - \dfrac{mg}{k}\right)e^{-kt/m}$

(b) $v \to \dfrac{mg}{k}$ as $t \to \infty$

(c) $s(t) = \dfrac{mg}{k}t - \dfrac{m}{k}\left(v_0 - \dfrac{mg}{k}\right)e^{-kt/m}$

$\qquad + \dfrac{m}{k}\left(v_0 - \dfrac{mg}{k}\right) + s_0$

29. (a) $P(t) = P_0 e^{(k_1 - k_2)t}$

(b) $k_1 > k_2$, births surpass deaths so population increases.

$k_1 = k_2$, a constant population since the number of births equals the number of deaths.

$k_1 < k_2$, deaths surpass births so population decreases.

31. $A = \dfrac{k_1 M}{k_1 + k_2} + ce^{-(k_1 + k_2)t}$

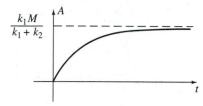

$\displaystyle\lim_{t \to \infty} A(t) = \dfrac{k_1 M}{k_1 + k_2}$

If $k_2 > 0$, the material will never be completely memorized.

33. (a) Let $t = 0$ correspond to 1790 so that $P(0) = 3.929$. The growth constant k in the solution $P(t) = 3.929e^{kt}$ depends on which census population is used. For example, when $t = 10$, $P(10) = 5.308$ gives $k = 0.030$. Thus $P(t) = 3.929e^{0.030t}$.

SECTION 3.2 EXERCISES, page 78

1. 1834; 2000   **3.** 1,000,000; 5.29 mo

5. (a) The result in (7) can be obtained by separation of variables.

(b) $c = \dfrac{a}{b} - \ln P_0$

7. 29.3 g; $X \to 60$ as $t \to \infty$; 0 g of A and 30 g of B

9. For $\alpha \neq \beta$, $\dfrac{1}{\alpha - \beta} \ln\left|\dfrac{\alpha - X}{\beta - X}\right| = kt + c$

For $\alpha = \beta$, $X = \alpha - \dfrac{1}{kt + c}$

11. $2h^{1/2} = -\frac{1}{25}t + 2\sqrt{20}$; $t = 50\sqrt{20}$ s

13. To evaluate the indefinite integral of the left side of

$$\frac{\sqrt{100 - y^2}}{y}\,dy = -dx$$

we use the substitution $y = 10\cos\theta$. It follows that

$$x = 10\ln\left(\frac{10 + \sqrt{100 - y^2}}{y}\right) - \sqrt{100 - y^2}.$$

15. (a) $v(t) = \sqrt{\dfrac{mg}{k}}\tanh\left(\sqrt{\dfrac{mg}{k}}\,t + c_1\right)$,

$\qquad$ where $c_1 = \tanh^{-1}\sqrt{\dfrac{k}{mg}}\,v_0$.

(b) $\sqrt{\dfrac{mg}{k}}$

(c) $s(t) = \dfrac{m}{k}\ln\cosh\left(\sqrt{\dfrac{kg}{m}}\,t + c_1\right) + c_2$,

$\qquad$ where $c_2 = s_0 - \ln\cosh c_1$

17. (a) $P(t) = \dfrac{4(P_0 - 1) - (P_0 - 4)e^{-3t}}{(P_0 - 1) - (P_0 - 4)e^{-3t}}$

(b) When $P_0 > 4$ or $1 < P_0 < 4$, $\displaystyle\lim_{t \to \infty} P(t) = 4$.
When $0 < P_0 < 1$, $P(t) \to 0$ for a finite value of time t.

(c) $P(t) = 0$ for $0 < P_0 < 1$

$\qquad$ when $t = \dfrac{1}{3}\ln\left(\dfrac{P_0 - 4}{4P_0 - 4}\right)$.

19. $y^3 = 3x + c$

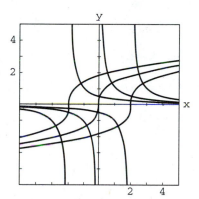

21. (b) The curve is $y^2 = 2c_1 x + c_1^2 = 2c_1\left(x + \dfrac{c_1}{2}\right)$.

SECTION 3.3 EXERCISES, page 87

1. $x(t) = x_0 e^{-\lambda_1 t}$

$y(t) = \dfrac{x_0 \lambda_1}{\lambda_2 - \lambda_1}(e^{-\lambda_1 t} - e^{-\lambda_2 t})$

$z(t) = x_0\left(1 - \dfrac{\lambda_2}{\lambda_2 - \lambda_1}e^{-\lambda_1 t} + \dfrac{\lambda_1}{\lambda_2 - \lambda_1}e^{-\lambda_2 t}\right)$

3. 5, 20, 147 days. The time when $y(t)$ and $z(t)$ are the same makes sense because most of A and half of B are gone, so half of C should have been formed.

5. $\dfrac{dx_1}{dt} = 6 - \dfrac{2}{25}x_1 + \dfrac{1}{50}x_2$

$\dfrac{dx_2}{dt} = \dfrac{2}{25}x_1 - \dfrac{2}{25}x_2$

7. $\dfrac{dx_1}{dt} = 3\dfrac{x_2}{100 - t} - 2\dfrac{x_1}{100 + t}$

$\dfrac{dx_2}{dt} = 2\dfrac{x_1}{100 + t} - 3\dfrac{x_2}{100 - t}$

9.

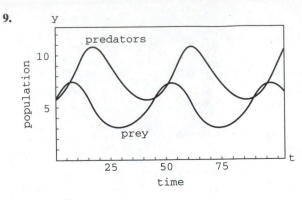

Populations are first equal at about $t = 5.6$. Periods are about 45.

11. In all cases $x(t) \to 6$ and $y(t) \to 8$ as $t \to \infty$.

13. $L_1 \dfrac{di_2}{dt} + (R_1 + R_2)i_2 + R_1 i_3 = E(t)$

$L_2 \dfrac{di_3}{dt} + R_1 i_2 + (R_1 + R_3)i_3 = E(t)$

15. $i(0) = i_0,\ s(0) = n - i_0,\ r(0) = 0$

CHAPTER 3 REVIEW EXERCISES, page 91

1. $P(45) = 8.99$ billion   **3.** $E(t) = E_0 e^{-(t-t_1)/RC}$

5. **(a)** $T(t) = \dfrac{T_2 + BT_1}{1 + B} + \dfrac{T_1 - T_2}{1 + B} e^{k(1+B)t}$

(b) $\dfrac{T_2 + BT_1}{1 + B}$

(c) $\dfrac{T_2 + BT_1}{1 + B}$

7. $x(\theta) = k\theta - \dfrac{k}{2}\sin 2\theta + c,\ y(\theta) = k\sin^2\theta$

9. $x(t) = \dfrac{\alpha c_1 e^{\alpha k_1 t}}{1 + c_1 e^{\alpha k_1 t}},\ y(t) = c_2(1 + c_1 e^{\alpha k_1 t})^{k_2/k_1}$

SECTION 4.1 EXERCISES, page 106

1. $y = \frac{1}{2}e^x - \frac{1}{2}e^{-x}$ **3.** $y = \frac{3}{5}e^{4x} + \frac{2}{5}e^{-x}$

5. $y = 3x - 4x \ln x$ **7.** $y = 0,\ y = x^2$

9. **(a)** $y = e^x \cos x - e^x \sin x$

(b) no solution

(c) $y = e^x \cos x + e^{-\pi/2}e^x \sin x$

(d) $y = c_2 e^x \sin x$, where c_2 is arbitrary

11. $(-\infty, 2)$ **15.** dependent   **17.** dependent

19. dependent   **21.** independent

23. The functions satisfy the differential equation and are linearly independent on the interval since $W(e^{-3x}, e^{4x}) = 7e^x \neq 0;\ y = c_1 e^{-3x} + c_2 e^{4x}$.

25. The functions satisfy the differential equation and are linearly independent on the interval since $W(e^x \cos 2x, e^x \sin 2x) = 2e^{2x} \neq 0;$ $y = c_1 e^x \cos 2x + c_2 e^x \sin 2x$.

27. The functions satisfy the differential equation and are linearly independent on the interval since $W(x^3, x^4) = x^6 \neq 0;\ y = c_1 x^3 + c_2 x^4$.

29. The functions satisfy the differential equation and are linearly independent on the interval since

$W(x, x^{-2}, x^{-2} \ln x) = 9x^{-6} \neq 0;$ $y = c_1 x + c_2 x^{-2} + c_3 x^{-2} \ln x$.

33. e^{2x} and e^{5x} form a fundamental set of solutions of the homogeneous equation; $6e^x$ is a particular solution of the nonhomogeneous equation.

35. e^{2x} and xe^{2x} form a fundamental set of solutions of the homogeneous equation; $x^2 e^{2x} + x - 2$ is a particular solution of the nonhomogeneous equation.

37. $y_p = x^2 + 3x + 3e^{2x};\ y_p = -2x^2 - 6x - \frac{1}{3}e^{2x}$

SECTION 4.2 EXERCISES, page 112

1. $y_2 = e^{-5x}$ **3.** $y_2 = xe^{2x}$ **5.** $y_2 = \sin 4x$

7. $y_2 = \sinh x$ **9.** $y_2 = xe^{2x/3}$ **11.** $y_2 = x^4 \ln|x|$

13. $y_2 = 1$ **15.** $y_2 = x^2 + x + 2$ **17.** $y_2 = x \cos(\ln x)$

19. $y_2 = x$ **21.** $y_2 = x \ln x$ **23.** $y_2 = x^3$

25. $y_2 = e^{2x},\ y_p = -\frac{1}{2}$ **27.** $y_2 = e^{2x},\ y_p = \frac{5}{2}e^{3x}$

SECTION 4.3 EXERCISES, page 119

1. $y = c_1 + c_2 e^{-x/4}$ **3.** $y = c_1 e^{-6x} + c_2 e^{6x}$

5. $y = c_1 \cos 3x + c_2 \sin 3x$ **7.** $y = c_1 e^{3x} + c_2 e^{-2x}$

9. $y = c_1 e^{-4x} + c_2 x e^{-4x}$

11. $y = c_1 e^{(-3+\sqrt{29})x/2} + c_2 e^{(-3-\sqrt{29})x/2}$

13. $y = c_1 e^{2x/3} + c_2 e^{-x/4}$ **15.** $y = e^{2x}(c_1 \cos x + c_2 \sin x)$

17. $y = e^{-x/3}\left(c_1 \cos \dfrac{\sqrt{2}}{3}x + c_2 \sin \dfrac{\sqrt{2}}{3}x\right)$

19. $y = c_1 + c_2 e^{-x} + c_3 e^{5x}$

21. $y = c_1 e^x + e^{-x/2}\left(c_2 \cos \dfrac{\sqrt{3}}{2}x + c_3 \sin \dfrac{\sqrt{3}}{2}x\right)$

23. $y = c_1 e^{-x} + c_2 e^{3x} + c_3 x e^{3x}$

25. $y = c_1 e^x + e^{-x}(c_2 \cos x + c_3 \sin x)$

27. $y = c_1 e^{-x} + c_2 x e^{-x} + c_3 x^2 e^{-x}$

29. $y = c_1 + c_2 x + e^{-x/2}\left(c_3 \cos \dfrac{\sqrt{3}}{2}x + c_4 \sin \dfrac{\sqrt{3}}{2}x\right)$

31. $y = c_1 \cos \dfrac{\sqrt{3}}{2}x + c_2 \sin \dfrac{\sqrt{3}}{2}x$

$\qquad + c_3 x \cos \dfrac{\sqrt{3}}{2}x + c_4 x \sin \dfrac{\sqrt{3}}{2}x$

33. $y = c_1 + c_2 e^{-2x} + c_3 e^{2x} + c_4 \cos 2x + c_5 \sin 2x$

35. $y = c_1 e^x + c_2 x e^x + c_3 e^{-x} + c_4 x e^{-x} + c_5 e^{-5x}$

37. $y = 2 \cos 4x - \frac{1}{2}\sin 4x$ **39.** $y = -\frac{3}{4}e^{-5x} + \frac{3}{4}e^{-x}$

41. $y = -e^{x/2}\cos \dfrac{x}{2} + e^{x/2}\sin \dfrac{x}{2}$ **43.** $y = 0$

45. $y = e^{2(x-1)} - e^{x-1}$ **47.** $y = \frac{5}{36} - \frac{5}{36}e^{-6x} + \frac{1}{6}xe^{-6x}$

49. $y = -\dfrac{1}{6}e^{2x} + \dfrac{1}{6}e^{-x}\cos \sqrt{3}\,x - \dfrac{\sqrt{3}}{6}e^{-x}\sin \sqrt{3}\,x$

51. $y = 2 - 2e^x + 2xe^x - \frac{1}{2}x^2 e^x$ **53.** $y = e^{5x} - xe^{5x}$

55. $y = -2 \cos x$

57. $y = c_1 e^{-0.270534x} + c_2 e^{0.658675x} + c_3 e^{5.61186x}$

59. $y = c_1 e^{-1.74806x} + c_2 e^{0.501219x} + c_3 e^{0.62342x}\cos(0.588965x)$

$\qquad + c_4 e^{0.62342x}\sin(0.588965x)$

SECTION 4.4 EXERCISES, page 130

1. $y = c_1 e^{-x} + c_2 e^{-2x} + 3$

3. $y = c_1 e^{5x} + c_2 x e^{5x} + \frac{6}{5}x + \frac{3}{5}$

5. $y = c_1 e^{-2x} + c_2 x e^{-2x} + x^2 - 4x + \frac{7}{2}$

7. $y = c_1 \cos \sqrt{3}\, x + c_2 \sin \sqrt{3}\, x + (-4x^2 + 4x - \frac{4}{3})e^{3x}$

9. $y = c_1 + c_2 e^x + 3x$

11. $y = c_1 e^{x/2} + c_2 x e^{x/2} + 12 + \frac{1}{2}x^2 e^{x/2}$

13. $y = c_1 \cos 2x + c_2 \sin 2x - \frac{3}{4}x \cos 2x$

15. $y = c_1 \cos x + c_2 \sin x - \frac{1}{2}x^2 \cos x + \frac{1}{2}x \sin x$

17. $y = c_1 e^x \cos 2x + c_2 e^x \sin 2x + \frac{1}{4}x e^x \sin 2x$

19. $y = c_1 e^{-x} + c_2 x e^{-x} - \frac{1}{2}\cos x + \frac{12}{25}\sin 2x - \frac{9}{25}\cos 2x$

21. $y = c_1 + c_2 x + c_3 e^{6x} - \frac{1}{4}x^2 - \frac{6}{37}\cos x + \frac{1}{37}\sin x$

23. $y = c_1 e^x + c_2 x e^x + c_3 x^2 e^x - x - 3 - \frac{2}{3}x^3 e^x$

25. $y = c_1 \cos x + c_2 \sin x + c_3 x \cos x$
$\qquad + c_4 x \sin x + x^2 - 2x - 3$

27. $y = \sqrt{2} \sin 2x - \frac{1}{2}$

29. $y = -200 + 200 e^{-x/5} - 3x^2 + 30x$

31. $y = -10 e^{-2x} \cos x + 9 e^{-2x} \sin x + 7 e^{-4x}$

33. $x = \dfrac{F_0}{2\omega^2} \sin \omega t - \dfrac{F_0}{2\omega} t \cos \omega t$

35. $y = 11 - 11 e^x + 9x e^x + 2x - 12 x^2 e^x + \frac{1}{2}e^{5x}$

37. $y = 6 \cos x - 6(\cot 1) \sin x + x^2 - 1$

39. $y = \begin{cases} \cos 2x + \frac{5}{6}\sin 2x + \frac{1}{3}\sin x, & 0 \le x \le \pi/2 \\ \frac{2}{3}\cos 2x + \frac{5}{6}\sin 2x, & x > \pi/2 \end{cases}$

SECTION 4.5 EXERCISES, page 139

1. $(3D - 2)(3D + 2)y = \sin x$

3. $(D - 6)(D + 2)y = x - 6$ **5.** $D(D + 5)^2 y = e^x$

7. $(D - 1)(D - 2)(D + 5)y = x e^{-x}$

9. $D(D + 2)(D^2 - 2D + 4)y = 4$ **15.** D^4 **17.** $D(D - 2)$

19. $D^2 + 4$ **21.** $D^3(D^2 + 16)$ **23.** $(D + 1)(D - 1)^3$

25. $D(D^2 - 2D + 5)$ **27.** $1, x, x^2, x^3, x^4$ **29.** $e^{6x}, e^{-3x/2}$

31. $\cos \sqrt{5}\, x, \sin \sqrt{5}\, x$ **33.** $1, e^{5x}, x e^{5x}$

35. $y = c_1 e^{-3x} + c_2 e^{3x} - 6$ **37.** $y = c_1 + c_2 e^{-x} + 3x$

39. $y = c_1 e^{-2x} + c_2 x e^{-2x} + \frac{1}{2}x + 1$

41. $y = c_1 + c_2 x + c_3 e^{-x} + \frac{2}{3}x^4 - \frac{8}{3}x^3 + 8x^2$

43. $y = c_1 e^{-3x} + c_2 e^{4x} + \frac{1}{7}x e^{4x}$

45. $y = c_1 e^{-x} + c_2 e^{3x} - e^x + 3$

47. $y = c_1 \cos 5x + c_2 \sin 5x + \frac{1}{4}\sin x$

49. $y = c_1 e^{-3x} + c_2 x e^{-3x} - \frac{1}{49}x e^{4x} + \frac{2}{343}e^{4x}$

51. $y = c_1 e^{-x} + c_2 e^x + \frac{1}{6}x^3 e^x - \frac{1}{4}x^2 e^x + \frac{1}{4}x e^x - 5$

53. $y = e^x(c_1 \cos 2x + c_2 \sin 2x) + \frac{1}{3}e^x \sin x$

55. $y = c_1 \cos 5x + c_2 \sin 5x - 2x \cos 5x$

57. $y = e^{-x/2}\left(c_1 \cos \dfrac{\sqrt{3}}{2}x + c_2 \sin \dfrac{\sqrt{3}}{2}x \right)$
$\qquad + \sin x + 2 \cos x - x \cos x$

59. $y = c_1 + c_2 x + c_3 e^{-8x} + \frac{11}{256}x^2 + \frac{7}{32}x^3 - \frac{1}{16}x^4$

61. $y = c_1 e^x + c_2 x e^x + c_3 x^2 e^x + \frac{1}{6}x^3 e^x + x - 13$

63. $y = c_1 + c_2 x + c_3 e^x + c_4 x e^x + \frac{1}{2}x^2 e^x + \frac{1}{2}x^2$

65. $y = \frac{5}{8}e^{-8x} + \frac{5}{8}e^{8x} - \frac{1}{4}$

67. $y = -\frac{41}{125} + \frac{41}{125}e^{5x} - \frac{1}{10}x^2 + \frac{9}{25}x$

69. $y = -\pi \cos x - \frac{11}{3}\sin x - \frac{8}{3}\cos 2x + 2x \cos x$

71. $y = 2 e^{2x} \cos 2x - \frac{3}{64}e^{2x} \sin 2x + \frac{1}{8}x^3 + \frac{3}{16}x^2 + \frac{3}{32}x$

SECTION 4.6 EXERCISES, page 146

1. $y = c_1 \cos x + c_2 \sin x + x \sin x$
$\qquad + \cos x \ln|\cos x|; \; (-\pi/2, \pi/2)$

3. $y = c_1 \cos x + c_2 \sin x + \frac{1}{2}\sin x - \frac{1}{2}x \cos x$
$\qquad = c_1 \cos x + c_3 \sin x - \frac{1}{2}x \cos x; \; (-\infty, \infty)$

5. $y = c_1 \cos x + c_2 \sin x + \frac{1}{2} - \frac{1}{6}\cos 2x; \; (-\infty, \infty)$

7. $y = c_1 e^x + c_2 e^{-x} + \frac{1}{4}x e^x - \frac{1}{4}x e^{-x}$
$\qquad = c_1 e^x + c_2 e^{-x} + \frac{1}{2}x \sinh x; \; (-\infty, \infty)$

9. $y = c_1 e^{2x} + c_2 e^{-2x}$
$\qquad + \dfrac{1}{4}\left(e^{2x} \ln|x| - e^{-2x} \displaystyle\int_{x_0}^{x} \dfrac{e^{4t}}{t}\, dt \right), x_0 > 0; \; (0, \infty)$

11. $y = c_1 e^{-x} + c_2 e^{-2x} + (e^{-x} + e^{-2x}) \ln(1 + e^x); \; (-\infty, \infty)$

13. $y = c_1 e^{-2x} + c_2 e^{-x} - e^{-2x} \sin e^x; \; (-\infty, \infty)$

15. $y = c_1 e^x + c_2 x e^x - \frac{1}{2}e^x \ln(1 + x^2) + x e^x \tan^{-1} x;$
$\qquad (-\infty, \infty)$

17. $y = c_1 e^{-x} + c_2 x e^{-x} + \frac{1}{2}x^2 e^{-x} \ln x - \frac{3}{4}x^2 e^{-x}; \; (0, \infty)$

19. $y = c_1 e^x \cos 3x + c_2 e^x \sin x$
$\qquad - \frac{1}{27}e^x \cos 3x \ln|\sec 3x + \tan 3x|; \; (-\pi/6, \pi/6)$

21. $y = c_1 + c_2 \cos x + c_3 \sin x - \ln|\cos x|$
$\qquad - \sin x \ln|\sec x + \tan x|; \; (-\pi/2, \pi/2)$

23. $y = c_1 e^x + c_2 e^{2x} + c_3 e^{-x} + \frac{1}{8}e^{3x}; \; (-\infty, \infty)$

25. $y = \frac{1}{4}e^{-x/2} + \frac{3}{4}e^{x/2} + \frac{1}{8}x^2 e^{x/2} - \frac{1}{4}x e^{x/2}$

27. $y = \frac{4}{9}e^{-4x} + \frac{25}{36}e^{2x} - \frac{1}{4}e^{-2x} + \frac{1}{9}e^{-x}$

29. $y = c_1 x^{-1/2} \cos x + c_2 x^{-1/2} \sin x + x^{-1/2}$

SECTION 4.7 EXERCISES, page 152

1. $y = c_1 x^{-1} + c_2 x^2$ **3.** $y = c_1 + c_2 \ln x$

5. $y = c_1 \cos(2 \ln x) + c_2 \sin(2 \ln x)$

7. $y = c_1 x^{(2-\sqrt{6})} + c_2 x^{(2+\sqrt{6})}$

9. $y_1 = c_1 \cos(\frac{1}{5} \ln x) + c_2 \sin(\frac{1}{5} \ln x)$

11. $y = c_1 x^{-2} + c_2 x^{-2} \ln x$

13. $y = x[c_1 \cos(\ln x) + c_2 \sin(\ln x)]$

15. $y = x^{-1/2}\left[c_1 \cos\left(\dfrac{\sqrt{3}}{6} \ln x \right) + c_2 \sin\left(\dfrac{\sqrt{3}}{6} \ln x \right) \right]$

17. $y = c_1 x^3 + c_2 \cos(\sqrt{2} \ln x) + c_3 \sin(\sqrt{2} \ln x)$

19. $y = c_1 x^{-1} + c_2 x^2 + c_3 x^4$

21. $y = c_1 + c_2 x + c_3 x^2 + c_4 x^{-3}$

23. $y = 2 - 2x^{-2}$ **25.** $y = \cos(\ln x) + 2 \sin(\ln x)$

27. $y = 2(-x)^{1/2} - 5(-x)^{1/2} \ln(-x)$

29. $y = c_1 + c_2 \ln x + \dfrac{x^2}{4}$

31. $y = c_1 x^{-1/2} + c_2 x^{-1} + \frac{1}{15}x^2 - \frac{1}{6}x$

33. $y = c_1 x + c_2 x \ln x + x(\ln x)^2$

35. $y = c_1 x^{-1} + c_2 x^{-8} + \frac{1}{30}x^2$

37. $y = x^2[c_1 \cos(3 \ln x) + c_2 \sin(3 \ln x)] + \frac{4}{13} + \frac{3}{10}x$

39. $y = c_1 x^2 + c_2 x^{-10} - \frac{1}{7}x^{-3}$

SECTION 4.8 EXERCISES, page 160

1. $x = c_1 e^t + c_2 t e^t$
$\quad y = (c_1 - c_2)e^t + c_2 t e^t$

3. $x = c_1 \cos t + c_2 \sin t + t + 1$
$\quad y = c_1 \sin t - c_2 \cos t + t - 1$

5. $x = \frac{1}{2}c_1 \sin t + \frac{1}{2}c_2 \cos t - 2c_3 \sin \sqrt{6}t - 2c_4 \cos \sqrt{6}t$
$\quad y = c_1 \sin t + c_2 \cos t + c_3 \sin \sqrt{6}t + c_4 \cos \sqrt{6}t$

7. $x = c_1 e^{2t} + c_2 e^{-2t} + c_3 \sin 2t + c_4 \cos 2t + \frac{1}{5}e^t$
$\quad y = c_1 e^{2t} + c_2 e^{-2t} - c_3 \sin 2t - c_4 \cos 2t - \frac{1}{5}e^t$

9. $x = c_1 - c_2 \cos t + c_3 \sin t + \frac{17}{15}e^{3t}$
$\quad y = c_1 + c_2 \sin t + c_3 \cos t - \frac{4}{15}e^{3t}$

11. $x = c_1 e^t + c_2 e^{-t/2} \cos \dfrac{\sqrt{3}}{2} t + c_3 e^{-t/2} \sin \dfrac{\sqrt{3}}{2} t$

$y = \left(-\dfrac{3}{2} c_2 - \dfrac{\sqrt{3}}{2} c_3 \right) e^{-t/2} \cos \dfrac{\sqrt{3}}{2} t$

$\quad + \left(\dfrac{\sqrt{3}}{2} c_2 - \dfrac{3}{2} c_3 \right) e^{-t/2} \sin \dfrac{\sqrt{3}}{2} t$

13. $x = c_1 e^{4t} + \frac{4}{3} e^t$

$y = -\frac{3}{4} c_1 e^{4t} + c_2 + 5e^t$

15. $x = c_1 + c_2 t + c_3 e^t + c_4 e^{-t} - \frac{1}{2} t^2$

$y = (c_1 - c_2 + 2) + (c_2 + 1)t + c_4 e^{-t} - \frac{1}{2} t^2$

17. $x = c_1 e^t + c_2 e^{-t/2} \sin \dfrac{\sqrt{3}}{2} t + c_3 e^{-t/2} \cos \dfrac{\sqrt{3}}{2} t$

$y = c_1 e^t + \left(-\dfrac{1}{2} c_2 - \dfrac{\sqrt{3}}{2} c_3 \right) e^{-t/2} \sin \dfrac{\sqrt{3}}{2} t$

$\quad + \left(\dfrac{\sqrt{3}}{2} c_2 - \dfrac{1}{2} c_3 \right) e^{-t/2} \cos \dfrac{\sqrt{3}}{2} t$

$z = c_1 e^t + \left(-\dfrac{1}{2} c_2 + \dfrac{\sqrt{3}}{2} c_3 \right) e^{-t/2} \sin \dfrac{\sqrt{3}}{2} t$

$\quad + \left(-\dfrac{\sqrt{3}}{2} c_2 - \dfrac{1}{2} c_3 \right) e^{-t/2} \cos \dfrac{\sqrt{3}}{2} t$

19. $x = -6c_1 e^{-t} - 3c_2 e^{-2t} + 2c_3 e^{3t}$

$y = c_1 e^{-t} + c_2 e^{-2t} + c_3 e^{3t}$

$z = 5c_1 e^{-t} + c_2 e^{-2t} + c_3 e^{3t}$

21. $x = -c_1 e^{-t} + c_2 + \frac{1}{3} t^3 - 2t^2 + 5t$

$y = c_1 e^{-t} + 2t^2 - 5t + 5$

23. $x = e^{-3t+3} - te^{-3t+3}$

$y = -e^{-3t+3} + 2te^{-3t+3}$

25. $m \dfrac{d^2 x}{dt^2} = 0$

$m \dfrac{d^2 y}{dt^2} = -mg$

$x = c_1 t + c_2$

$y = -\frac{1}{2} g t^2 + c_3 t + c_4$

SECTION 4.9 EXERCISES, page 165

3. $y = \ln|\cos(c_1 - x)| + c_2$

5. $y = \dfrac{1}{c_1^2} \ln|c_1 x + 1| - \dfrac{1}{c_1} x + c_2$

7. $\frac{1}{3} y^3 - c_1 y = x + c_2$

9. $y = \tan\left(\dfrac{\pi}{4} - \dfrac{x}{2} \right), \ -\dfrac{\pi}{2} < x < \dfrac{3\pi}{2}$

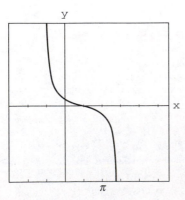

11. $y = -\dfrac{1}{c_1} \sqrt{1 - c_1^2 x^2} + c_2$

13. $y = 1 + x + \frac{1}{2} x^2 + \frac{1}{2} x^3 + \frac{1}{6} x^4 + \frac{1}{10} x^5 + \cdots$

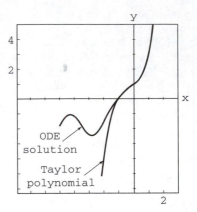

15. $y = 1 + x - \frac{1}{2} x^2 + \frac{2}{3} x^3 - \frac{1}{4} x^4 + \frac{7}{60} x^5 + \cdots$

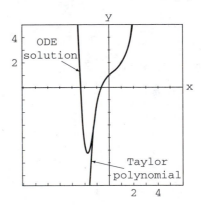

17. $y = -\sqrt{1 - x^2}$

CHAPTER 4 REVIEW EXERCISES, page 167

1. $y = 0$

3. false. The functions $f_1(x) = 0$ and $f_2(x) = e^x$ are linearly dependent on $(-\infty, \infty)$, but f_2 is not a constant multiple of f_1.

5. $(-\infty, 0); (0, \infty)$ **7.** false   **9.** $y_p = A + Bxe^x$

11. $y_2 = \sin 2x$ **13.** $y = c_1 e^{(1+\sqrt{3})x} + c_2 e^{(1-\sqrt{3})x}$

15. $y = c_1 + c_2 e^{-5x} + c_3 x e^{-5x}$

17. $y = c_1 e^{-x/3} + e^{-3x/2} \left(c_2 \cos \dfrac{\sqrt{7}}{2} x + c_3 \sin \dfrac{\sqrt{7}}{2} x \right)$

19. $y = c_1 x^{-1/3} + c_2 x^{1/2}$

21. $y = e^{3x/2} \left(c_1 \cos \dfrac{\sqrt{11}}{2} x + c_2 \sin \dfrac{\sqrt{11}}{2} x \right)$

$\quad + \frac{4}{5} x^3 + \frac{36}{25} x^2 + \frac{46}{125} x - \frac{222}{625}$

23. $y = c_1 + c_2 e^{2x} + c_3 e^{3x} + \frac{1}{5} \sin x - \frac{1}{5} \cos x + \frac{4}{3} x$

25. $y = e^{x-\pi} \cos x$ **27.** $y = x^2 + 4$

29. $y = e^x (c_1 \cos x + c_2 \sin x) - e^x \cos x \ln|\sec x + \tan x|$

31. $y = c_1 x^2 + c_2 x^3 + x^4 - x^2 \ln x$

33. $y = \frac{2}{5} e^{x/2} - \frac{2}{5} e^{3x} + x e^{3x} - 4$

35. $x = -c_1 e^t - \frac{3}{2} c_2 e^{2t} + \frac{5}{2}$

$y = c_1 e^t + c_2 e^{2t} - 3$

37. $x = c_1 e^t + c_2 e^{5t} + te^t$
$y = -c_1 e^t + 3c_2 e^{5t} - te^t + 2e^t$

SECTION 5.1 EXERCISES, page 184

1. $\dfrac{\sqrt{2}\,\pi}{8}$ **3.** $x(t) = -\frac{1}{4}\cos 4\sqrt{6}\,t$

5. **(a)** $x\left(\dfrac{\pi}{12}\right) = -\dfrac{1}{4};\ x\left(\dfrac{\pi}{8}\right) = -\dfrac{1}{2};\ x\left(\dfrac{\pi}{6}\right) = -\dfrac{1}{4};$

$x\left(\dfrac{\pi}{4}\right) = \dfrac{1}{2};\ x\left(\dfrac{9\pi}{32}\right) = \dfrac{\sqrt{2}}{4}$

   (b) 4 ft/s; downward

   (c) $t = \dfrac{(2n+1)\pi}{16}, n = 0, 1, 2, \ldots$

7. **(a)** the 20-kg mass
   (b) the 20-kg mass; the 50-kg mass
   (c) $t = n\pi, n = 0, 1, 2, \ldots$; at the equilibrium position;
   the 50-kg mass is moving upward whereas the 20-kg
   mass is moving upward when n is even and downward when n is odd.

9. $x(t) = \dfrac{1}{2}\cos 2t + \dfrac{3}{4}\sin 2t = \dfrac{\sqrt{13}}{4}\sin(2t + 0.5880)$

11. **(a)** $x(t) = -\frac{2}{3}\cos 10t + \frac{1}{2}\sin 10t = \frac{5}{6}\sin(10t - 0.927)$

   (b) $\dfrac{5}{6}$ ft; $\dfrac{\pi}{5}$

   (c) 15 cycles

   (d) 0.721 s

   (e) $\dfrac{(2n+1)\pi}{20} + 0.0927, n = 0, 1, 2, \ldots$

   (f) $x(3) = -0.597$ ft

   (g) $x'(3) = -5.814$ ft/s

   (h) $x''(3) = 59.702$ ft/s^2

   (i) $\pm 8\frac{1}{3}$ ft/s

   (j) $0.1451 + \dfrac{n\pi}{5}; 0.3545 + \dfrac{n\pi}{5}, n = 0, 1, 2, \ldots$

   (k) $0.3545 + \dfrac{n\pi}{5}, n = 0, 1, 2, \ldots$

13. 120 lb/ft; $x(t) = \dfrac{\sqrt{3}}{12}\sin 8\sqrt{3}\,t$

17. **(a)** above   **(b)** heading upward
19. **(a)** below   **(b)** heading upward
21. $\frac{1}{4}$ s; $\frac{1}{2}$ s, $x(\frac{1}{2}) = e^{-2}$; that is, the weight is approximately
   0.14 ft below the equilibrium position.
23. **(a)** $x(t) = \frac{4}{3}e^{-2t} - \frac{1}{3}e^{-8t}$
   (b) $x(t) = -\frac{2}{3}e^{-2t} + \frac{5}{3}e^{-8t}$
25. **(a)** $x(t) = e^{-2t}(-\cos 4t - \frac{1}{2}\sin 4t)$

   (b) $x(t) = \dfrac{\sqrt{5}}{2}\,e^{-2t}\sin(4t + 4.249)$

   (c) $t = 1.294$ s
27. **(a)** $\beta > \frac{5}{2}$ **(b)** $\beta = \frac{5}{2}$ **(c)** $0 < \beta < \frac{5}{2}$

29. $x(t) = e^{-t/2}\left(-\dfrac{4}{3}\cos\dfrac{\sqrt{47}}{2}t - \dfrac{64}{3\sqrt{47}}\sin\dfrac{\sqrt{47}}{2}t\right)$

$+ \dfrac{10}{3}(\cos 3t + \sin 3t)$

31. $x(t) = \frac{1}{4}e^{-4t} + te^{-4t} - \frac{1}{4}\cos 4t$

33. $x(t) = -\frac{1}{2}\cos 4t + \frac{9}{4}\sin 4t + \frac{1}{2}e^{-2t}\cos 4t$
$- 2e^{-2t}\sin 4t$

35. **(a)** $m\dfrac{d^2x}{dt^2} = -k(x - h) - \beta\dfrac{dx}{dt}$ or

$\dfrac{d^2x}{dt^2} + 2\lambda\dfrac{dx}{dt} + \omega^2 x = \omega^2 h(t),$

where $2\lambda = \beta/m$ and $\omega^2 = k/m$

   (b) $x(t) = e^{-2t}(-\frac{56}{13}\cos 2t - \frac{72}{13}\sin 2t) + \frac{56}{13}\cos t + \frac{32}{13}\sin t$
37. $x(t) = -\cos 2t - \frac{1}{8}\sin 2t + \frac{3}{4}t\sin 2t + \frac{5}{4}t\cos 2t$

39. **(b)** $\dfrac{F_0}{2\omega}t\sin\omega t$

45. 4.568 C; 0.0509 s
47. $q(t) = 10 - 10e^{-3t}(\cos 3t + \sin 3t)$
$i(t) = 60e^{-3t}\sin 3t; 10.432$ C
49. $q_p = \frac{100}{13}\sin t + \frac{150}{13}\cos t$
$i_p = \frac{100}{13}\cos t - \frac{150}{13}\sin t$
53. $q(t) = -\frac{1}{2}e^{-10t}(\cos 10t + \sin 10t) + \frac{3}{2}; \frac{3}{2}$ C

57. $q(t) = \left(q_0 - \dfrac{E_0 C}{1 - \gamma^2 LC}\right)\cos\dfrac{t}{\sqrt{LC}}$

$+ \sqrt{LC}i_0\sin\dfrac{t}{\sqrt{LC}} + \dfrac{E_0 C}{1 - \gamma^2 LC}\cos\gamma t$

$i(t) = i_0\cos\dfrac{t}{\sqrt{LC}}$

$- \dfrac{1}{\sqrt{LC}}\left(q_0 - \dfrac{E_0 C}{1 - \gamma^2 LC}\right)\sin\dfrac{t}{\sqrt{LC}}$

$- \dfrac{E_0 C\gamma}{1 - \gamma^2 LC}\sin\gamma t$

SECTION 5.2 EXERCISES, page 197

1. **(a)** $y(x) = \dfrac{w_0}{24EI}(6L^2x^2 - 4Lx^3 + x^4)$

   (b)

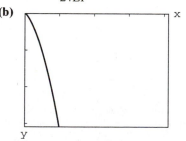

3. **(a)** $y(x) = \dfrac{w_0}{48EI}(3L^2x^2 - 5Lx^3 + 2x^4)$

   (b)

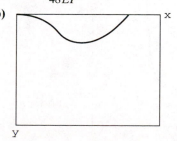

5. **(a)** $y_{\max} = \dfrac{w_0 L^4}{8EI}$

   (b) $\frac{1}{16}$th of the maximum deflection in part (a)

7. $y(x) = -\dfrac{w_0 EI}{P^2} \cosh \sqrt{\dfrac{P}{EI}}\, x$

$+ \left(\dfrac{w_0 EI}{P^2} \sinh \sqrt{\dfrac{P}{EI}}\, L - \dfrac{w_0 L \sqrt{EI}}{P\sqrt{P}} \right) \dfrac{\sinh \sqrt{\dfrac{P}{EI}}\, x}{\cosh \sqrt{\dfrac{P}{EI}}\, L}$

$+ \dfrac{w_0}{2P} x^2 + \dfrac{w_0 EI}{P^2}$

9. $\lambda = n^2$, $n = 1, 2, 3, \ldots$; $y = \sin nx$

11. $\lambda = \dfrac{(2n-1)^2 \pi^2}{4L^2}$, $n = 1, 2, 3, \ldots$; $y = \cos \dfrac{(2n-1)\pi x}{2L}$

13. $\lambda = n^2$, $n = 0, 1, 2, \ldots$; $y = \cos nx$

15. $\lambda = \dfrac{n^2 \pi^2}{25}$, $n = 1, 2, 3, \ldots$; $y = e^{-x} \sin \dfrac{n\pi x}{5}$

17. $\lambda = \dfrac{n\pi}{L}$, $n = 1, 2, 3, \ldots$; $y = \sin \dfrac{n\pi x}{L}$

19. $\lambda = n^2$, $n = 1, 2, 3, \ldots$; $y = \sin(n \ln x)$

21. $\lambda = 0$; $y = 1$

$\lambda = \dfrac{n^2 \pi^2}{4}$, $n = 1, 2, 3, \ldots$; $y = \cos\left(\dfrac{n\pi}{2} \ln x \right)$

25. $\omega_n = \dfrac{n\pi \sqrt{T}}{L\sqrt{\rho}}$, $n = 1, 2, 3, \ldots$; $y = \sin \dfrac{n\pi x}{L}$

27. $u(r) = \left(\dfrac{u_0 - u_1}{b - a} \right) \dfrac{ab}{r} + \dfrac{u_1 b - u_0 a}{b - a}$

SECTION 5.3 EXERCISES, page 206

1. x

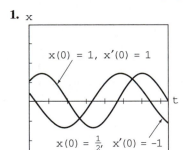

For the first IVP the period T is approximately 6; for the second IVP the period T is approximately 6.3.

3. x

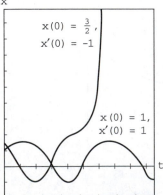

For the first IVP the period T is approximately 6; the solution of the second IVP appears not to be periodic.

A-8

5. $|x_1| \approx 1.2$

7. $\dfrac{d^2x}{dt^2} + x = 0$

9. **(a)** Expect $x \to 0$ as $t \to \infty$.
(b) x

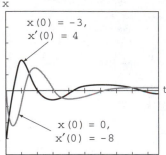

11. For k_1 very small the effect of nonlinearity is reduced and the system is close to pure resonance.

13. For $\lambda = 2$ and $\omega = 1$ the motion corresponds to the overdamped case. For $\lambda = \frac{1}{2}$ and $\omega = 1$ the motion corresponds to the underdamped case.

15. **(a)** $xy'' = r\sqrt{1 + (y')^2}$. When $t = 0$, $x = a$, $y = 0$, and $dy/dx = 0$.

(b) When $r \neq 1$,

$$y(x) = \dfrac{a}{2}\left[\dfrac{1}{1+r}\left(\dfrac{x}{a} \right)^{1+r} - \dfrac{1}{1-r}\left(\dfrac{x}{a} \right)^{1-r} \right] + \dfrac{ar}{1 - r^2}.$$

When $r = 1$, $y(x) = \dfrac{1}{2}\left[\dfrac{1}{2a}(x^2 - a^2) + \dfrac{1}{a} \ln \dfrac{a}{x} \right]$.

(c) The paths intersect when $r < 1$.

19. **(a)** 0.666404   **(b)** 3.84411, 7.0218

CHAPTER 5 REVIEW EXERCISES, page 210

1. 8 ft   **3.** $\frac{5}{4}$ m

5. false; there could be an impressed force driving the system.

7. overdamped   **9.** $\frac{9}{2}$ lb/ft

11. $x(t) = -\frac{2}{3}e^{-2t} + \frac{1}{3}e^{-4t}$ **13.** $0 < m \leq 2$

15. $\gamma = \dfrac{8\sqrt{3}}{3}$

17. $x(t) = e^{-4t}\left(\dfrac{26}{17} \cos 2\sqrt{2}t + \dfrac{28\sqrt{2}}{17} \sin 2\sqrt{2}t \right) + \dfrac{8}{17} e^{-t}$

19. **(a)** $q(t) = -\frac{1}{150} \sin 100t + \frac{1}{75} \sin 50t$
(b) $i(t) = -\frac{2}{3} \cos 100t + \frac{2}{3} \cos 50t$

(c) $t = \dfrac{n\pi}{50}$, $n = 0, 1, 2, \ldots$

SECTION 6.1 EXERCISES, page 221

1. $(-1, 1]$ **3.** $[-\frac{1}{2}, \frac{1}{2})$ **5.** $[2, 4]$ **7.** $(-5, 15)$ **9.** $\{0\}$

11. $x + x^2 + \frac{1}{3}x^3 - \frac{1}{30}x^5 + \cdots$

13. $x - \frac{2}{3}x^3 + \frac{2}{15}x^5 - \frac{4}{315}x^7 + \cdots$

15. $y = ce^{-x}$; $y = c_0 \displaystyle\sum_{n=0}^{\infty} \dfrac{(-1)^n}{n!} x^n$

17. $y = ce^{x^3/3}$; $y = c_0 \displaystyle\sum_{n=0}^{\infty} \dfrac{1}{n!}\left(\dfrac{x^3}{3} \right)^n$

19. $y = \dfrac{c}{1-x}$; $y = c_0 \displaystyle\sum_{n=0}^{\infty} x^n$

21. $y = C_1 \cos x + C_2 \sin x$

$y = c_0 \displaystyle\sum_{n=0}^{\infty} \dfrac{(-1)^n}{(2n)!} x^{2n} + c_1 \displaystyle\sum_{n=0}^{\infty} \dfrac{(-1)^n}{(2n+1)!} x^{2n+1}$

23. $y = C_1 + C_2 e^x$

$y = c_0 + c_1 \displaystyle\sum_{n=1}^{\infty} \dfrac{x^n}{n!} = c_0 - c_1 + c_1 \displaystyle\sum_{n=0}^{\infty} \dfrac{x^n}{n!}$

$= c_0 - c_1 + c_1 e^x$

SECTION 6.2 EXERCISES, page 229

1. $y_1(x) = c_0 \left[1 + \dfrac{1}{3 \cdot 2} x^3 + \dfrac{1}{6 \cdot 5 \cdot 3 \cdot 2} x^6 \right.$

$\left. + \dfrac{1}{9 \cdot 8 \cdot 6 \cdot 5 \cdot 3 \cdot 2} x^9 + \cdots \right]$

$y_2(x) = c_1 \left[x + \dfrac{1}{4 \cdot 3} x^4 + \dfrac{1}{7 \cdot 6 \cdot 4 \cdot 3} x^7 \right.$

$\left. + \dfrac{1}{10 \cdot 9 \cdot 7 \cdot 6 \cdot 4 \cdot 3} x^{10} + \cdots \right]$

3. $y_1(x) = c_0 \left[1 - \dfrac{1}{2!} x^2 - \dfrac{3}{4!} x^4 - \dfrac{21}{6!} x^6 - \cdots \right]$

$y_2(x) = c_1 \left[x + \dfrac{1}{3!} x^3 + \dfrac{5}{5!} x^5 + \dfrac{45}{7!} x^7 + \cdots \right]$

5. $y_1(x) = c_0 \left[1 - \dfrac{1}{3!} x^3 + \dfrac{4^2}{6!} x^6 - \dfrac{7^2 \cdot 4^2}{9!} x^9 + \cdots \right]$

$y_2(x) = c_1 \left[x - \dfrac{2^2}{4!} x^4 + \dfrac{5^2 \cdot 2^2}{7!} x^7 - \dfrac{8^2 \cdot 5^2 \cdot 2^2}{10!} x^{10} + \cdots \right]$

7. $y_1(x) = c_0$; $y_2(x) = c_1 \displaystyle\sum_{n=1}^{\infty} \dfrac{1}{n} x^n$

9. $y_1(x) = c_0 \displaystyle\sum_{n=0}^{\infty} x^{2n}$; $y_2(x) = c_1 \displaystyle\sum_{n=0}^{\infty} x^{2n+1}$

11. $y_1(x) = c_0 \left[1 + \dfrac{1}{4} x^2 - \dfrac{7}{4 \cdot 4!} x^4 + \dfrac{23 \cdot 7}{8 \cdot 6!} x^6 - \cdots \right]$

$y_2(x) = c_1 \left[x - \dfrac{1}{6} x^3 + \dfrac{14}{2 \cdot 5!} x^5 - \dfrac{34 \cdot 14}{4 \cdot 7!} x^7 - \cdots \right]$

13. $y_1(x) = c_0 [1 + \frac{1}{2} x^2 + \frac{1}{6} x^3 + \frac{1}{6} x^4 + \cdots]$

$y_2(x) = c_1 [x + \frac{1}{2} x^2 + \frac{1}{2} x^3 + \frac{1}{4} x^4 + \cdots]$

15. $y(x) = -2 \left[1 + \dfrac{1}{2!} x^2 + \dfrac{1}{3!} x^3 + \dfrac{1}{4!} x^4 + \cdots \right] + 6x$

$= 8x - 2e^x$

17. $y(x) = 3 - 12x^2 + 4x^4$

19. $y_1(x) = c_0 [1 - \frac{1}{6} x^3 + \frac{1}{120} x^5 + \cdots]$

$y_2(x) = c_1 [x - \frac{1}{12} x^4 + \frac{1}{180} x^6 + \cdots]$

21. $y_1(x) = c_0 [1 - \frac{1}{2} x^2 + \frac{1}{6} x^3 - \frac{1}{40} x^5 + \cdots]$

$y_2(x) = c_1 [x - \frac{1}{6} x^3 + \frac{1}{12} x^4 - \frac{1}{60} x^5 + \cdots]$

23. $y_1(x) = c_0 \left[1 + \dfrac{1}{3!} x^3 + \dfrac{4}{6!} x^6 + \dfrac{7 \cdot 4}{9!} x^9 + \cdots \right]$

$+ c_1 \left[x + \dfrac{2}{4!} x^4 + \dfrac{5 \cdot 2}{7!} x^7 + \dfrac{8 \cdot 5 \cdot 2}{10!} x^{10} + \cdots \right]$

$+ \dfrac{1}{2!} x^2 + \dfrac{3}{5!} x^5 + \dfrac{6 \cdot 3}{8!} x^8 + \dfrac{9 \cdot 6 \cdot 3}{11!} x^{11} + \cdots$

SECTION 6.3 EXERCISES, page 242

1. $x = 0$, irregular singular point

3. $x = -3$, regular singular point; $x = 3$, irregular singular point

5. $x = 0, 2i, -2i$, regular singular points

7. $x = -3, 2$, regular singular points

9. $x = 0$, irregular singular point; $x = -5, 5, 2$, regular singular points

11. $r_1 = \frac{3}{2}, r_2 = 0$

$y(x) = C_1 x^{3/2} \left[1 - \dfrac{2}{5} x + \dfrac{2^2}{7 \cdot 5 \cdot 2} x^2 \right.$

$\left. - \dfrac{2^3}{9 \cdot 7 \cdot 5 \cdot 3!} x^3 + \cdots \right]$

$+ C_2 \left[1 + 2x - 2x^2 + \dfrac{2^3}{3 \cdot 3!} x^3 - \cdots \right]$

13. $r_1 = \frac{7}{8}, r_2 = 0$

$y(x) = C_1 x^{7/8} \left[1 - \dfrac{2}{15} x + \dfrac{2^2}{23 \cdot 15 \cdot 2} x^2 \right.$

$\left. - \dfrac{2^3}{31 \cdot 23 \cdot 15 \cdot 3!} x^3 + \cdots \right]$

$+ C_2 \left[1 - 2x + \dfrac{2^2}{9 \cdot 2} x^2 - \dfrac{2^3}{17 \cdot 9 \cdot 3!} x^3 + \cdots \right]$

15. $r_1 = \frac{1}{3}, r_2 = 0$

$y(x) = C_1 x^{1/3} \left[1 + \dfrac{1}{3} x + \dfrac{1}{3^2 \cdot 2} x^2 + \dfrac{1}{3^3 \cdot 3!} x^3 + \cdots \right]$

$+ C_2 \left[1 + \dfrac{1}{2} x + \dfrac{1}{5 \cdot 2} x^2 + \dfrac{1}{8 \cdot 5 \cdot 2} x^3 + \cdots \right]$

17. $r_1 = \frac{5}{2}, r_2 = 0$

$y(x) = C_1 x^{5/2} \left[1 + \dfrac{2 \cdot 2}{7} x + \dfrac{2^2 \cdot 3}{9 \cdot 7} x^2 \right.$

$\left. + \dfrac{2^3 \cdot 4}{11 \cdot 9 \cdot 7} x^3 + \cdots \right]$

$+ C_2 \left[1 + \dfrac{1}{3} x - \dfrac{1}{6} x^2 - \dfrac{1}{6} x^3 - \cdots \right]$

19. $r_1 = \frac{2}{3}, r_2 = \frac{1}{3}$

$y(x) = C_1 x^{2/3} [1 - \frac{1}{2} x + \frac{5}{28} x^2 - \frac{1}{21} x^3 + \cdots]$

$+ C_2 x^{1/3} [1 - \frac{1}{2} x + \frac{1}{5} x^2 - \frac{7}{120} x^3 + \cdots]$

21. $r_1 = 1, r_2 = -\frac{1}{2}$

$y(x) = C_1 x \left[1 + \dfrac{1}{5} x + \dfrac{1}{5 \cdot 7} x^2 + \dfrac{1}{5 \cdot 7 \cdot 9} x^3 + \cdots \right]$

$+ C_2 x^{-1/2} \left[1 + \dfrac{1}{2} x + \dfrac{1}{2 \cdot 4} x^2 + \dfrac{1}{2 \cdot 4 \cdot 6} x^3 + \cdots \right]$

23. $r_1 = 0, r_2 = -1$

$y(x) = C_1 x^{-1} \displaystyle\sum_{n=0}^{\infty} \dfrac{1}{(2n)!} x^{2n} + C_2 x^{-1} \displaystyle\sum_{n=0}^{\infty} \dfrac{1}{(2n+1)!} x^{2n+1}$

$= \dfrac{1}{x} [C_1 \cosh x + C_2 \sinh x]$

25. $r_1 = 4, r_2 = 0$

$y(x) = C_1 \left[1 + \dfrac{2}{3} x + \dfrac{1}{3} x^2 \right] + C_2 \displaystyle\sum_{n=0}^{\infty} (n+1) x^{n+4}$

27. $r_1 = r_2 = 0$

$$y(x) = C_1 y_1(x) + C_2 \left[y_1(x) \ln x \right.$$
$$\left. + y_1(x) \left(-x + \frac{1}{4} x^2 - \frac{1}{3 \cdot 3!} x^3 + \frac{1}{4 \cdot 4!} x^4 - \cdots \right) \right],$$

where $y_1(x) = \sum_{n=0}^{\infty} \frac{1}{n!} x^n = e^x$

29. $r_1 = r_2 = 0$

$$y(x) = C_1 y_1(x) + C_2 [y_1(x) \ln x$$
$$+ y_1(x)(2x + \tfrac{5}{4} x^2 + \tfrac{23}{27} x^3 + \cdots)],$$

where $y_1(x) = \sum_{n=0}^{\infty} \frac{(-1)^n}{(n!)^{2n}} x^n$

SECTION 6.4 EXERCISES, page 253

1. $y = c_1 J_{1/3}(x) + c_2 J_{-1/3}(x)$
3. $y = c_1 J_{5/2}(x) + c_2 J_{-5/2}(x)$
5. $y = c_1 J_0(x) + c_2 Y_0(x)$
7. $y = c_1 J_2(3x) + c_2 Y_2(3x)$
9. $y = c_1 x^{-1/2} J_{1/2}(\lambda x) + c_2 x^{-1/2} J_{-1/2}(\lambda x)$
13. From Problem 10, $y = x^{1/2} J_{1/2}(x)$; from Problem 11, $y = x^{1/2} J_{-1/2}(x)$.
15. From Problem 10, $y = x^{-1} J_{-1}(x)$; from Problem 11, $y = x^{-1} J_1(x)$. Since $J_{-1}(x) = -J_1(x)$, no new solution results.
17. From Problem 12 with $\lambda = 1$ and $\nu = \pm \frac{3}{2}$, $y = \sqrt{x} \, J_{3/2}(x)$ and $y = \sqrt{x} J_{-3/2}(x)$.
27. $J_{-1/2}(x) = \sqrt{\dfrac{2}{\pi x}} \cos x$
29. $J_{-3/2}(x) = \sqrt{\dfrac{2}{\pi x}} \left[-\sin x - \dfrac{\cos x}{x} \right]$
31. $J_{-5/2}(x) = \sqrt{\dfrac{2}{\pi x}} \left[\dfrac{3}{x} \sin x + \left(\dfrac{3}{x^2} - 1 \right) \cos x \right]$
33. $J_{-7/2}(x) = \sqrt{\dfrac{2}{\pi x}} \left[\left(1 - \dfrac{15}{x^2} \right) \sin x + \left(\dfrac{6}{x} - \dfrac{15}{x^3} \right) \cos x \right]$
35. $y = c_1 I_\nu(x) + c_2 I_{-\nu}(x)$, $\nu \neq$ integer
43. (a) $x(t) = -0.809264 x^{1/2} J_{1/3}(\frac{1}{3} x^{3/2})$
$+ 0.782397 x^{1/2} J_{-1/3}(\frac{1}{3} x^{3/2})$
45. (a) $P_6(x) = \frac{1}{16}(231 x^6 - 315 x^4 + 105 x^2 - 5)$
$P_7(x) = \frac{1}{16}(429 x^7 - 693 x^5 + 315 x^3 - 35x)$
(b) $P_6(x)$ satisfies $(1 - x^2) y'' - 2xy' + 42y = 0$.
$P_7(x)$ satisfies $(1 - x^2) y'' - 2xy' + 56y = 0$.

CHAPTER 6 REVIEW EXERCISES, page 257

1. The singular points are $x = 0$, $x = -1 + \sqrt{3} i$, $x = -1 - \sqrt{3} i$; all other finite values of x, real or complex, are ordinary points.
3. $x = 0$, regular singular point; $x = 5$, irregular singular point
5. $x = -3, 3$, regular singular points; $x = 0$, irregular singular point
7. $|x| < \infty$
9. $y_1(x) = c_0 \left[1 + \frac{1}{2} x^2 + \frac{1}{2 \cdot 4} x^4 + \cdots \right]$
$y_2(x) = c_1 \left[x + \frac{1}{3} x^3 + \frac{1}{3 \cdot 5} x^5 + \cdots \right]$

11. $y_1(x) = c_0 [1 + \frac{3}{2} x^2 + \frac{1}{2} x^3 + \frac{5}{8} x^4 + \cdots]$
$y_2(x) = c_1 [x + \frac{1}{2} x^3 + \frac{1}{4} x^4 + \cdots]$
13. $y = 3 \left[1 - x^2 + \frac{1}{3} x^4 - \frac{1}{3 \cdot 5} x^6 + \cdots \right]$
$- 2 \left[x - \frac{1}{2} x^3 + \frac{1}{2 \cdot 4} x^5 - \frac{1}{2 \cdot 4 \cdot 6} x^7 + \cdots \right]$
15. $r_1 = 1, r_2 = -\frac{1}{2}$
$$y(x) = C_1 x \left[1 + \frac{1}{5} x + \frac{1}{7 \cdot 5 \cdot 2} x^2 \right.$$
$$+ \frac{1}{9 \cdot 7 \cdot 5 \cdot 3 \cdot 2} x^3 + \cdots \bigg]$$
$$+ C_2 x^{-1/2} \left[1 - x - \frac{1}{2} x^2 - \frac{1}{3^2 \cdot 2} x^3 - \cdots \right]$$
17. $r_1 = 3, r_2 = 0$
$$y_1(x) = C_3 \left[x^3 + \frac{5}{4} x^4 + \frac{11}{8} x^5 + \cdots \right]$$
$$y(x) = C_1 y_1(x) + C_2 \left[-\frac{1}{36} y_1(x) \ln x \right.$$
$$\left. + y_1(x) \left(-\frac{1}{3} \frac{1}{x^3} + \frac{1}{4} \frac{1}{x^2} + \frac{1}{16} \frac{1}{x} + \cdots \right) \right]$$
19. $r_1 = r_2 = 0$; $y(x) = C_1 e^x + C_2 e^x \ln x$

SECTION 7.1 EXERCISES, page 265

1. $\dfrac{2}{s} e^{-s} - \dfrac{1}{s}$ **3.** $\dfrac{1}{s^2} - \dfrac{1}{s^2} e^{-s}$ **5.** $\dfrac{1 + e^{-s\pi}}{s^2 + 1}$
7. $\dfrac{e^{-s}}{s} + \dfrac{e^{-s}}{s^2}$ **9.** $\dfrac{1}{s} - \dfrac{1}{s^2} + \dfrac{e^{-s}}{s^2}$ **11.** $\dfrac{e^7}{s - 1}$
13. $\dfrac{1}{(s - 4)^2}$ **15.** $\dfrac{1}{s^2 + 2s + 2}$ **17.** $\dfrac{s^2 - 1}{(s^2 + 1)^2}$
19. $\dfrac{48}{s^5}$ **21.** $\dfrac{4}{s^2} - \dfrac{10}{s}$ **23.** $\dfrac{2}{s^3} + \dfrac{6}{s^2} - \dfrac{3}{s}$
25. $\dfrac{6}{s^4} + \dfrac{6}{s^3} + \dfrac{3}{s^2} + \dfrac{1}{s}$ **27.** $\dfrac{1}{s} + \dfrac{1}{s - 4}$
29. $\dfrac{1}{s} + \dfrac{2}{s - 2} + \dfrac{1}{s - 4}$ **31.** $\dfrac{8}{s^3} - \dfrac{15}{s^2 + 9}$
33. Use $\sinh kt = \dfrac{e^{kt} - e^{-kt}}{2}$ to show that
$$\mathcal{L}\{\sinh kt\} = \frac{k}{s^2 - k^2}.$$
35. $\dfrac{1}{2(s - 2)} - \dfrac{1}{2s}$ **37.** $\dfrac{2}{s^2 + 16}$ **41.** $\dfrac{\frac{1}{2} \Gamma(\frac{1}{2})}{s^{3/2}} = \dfrac{\sqrt{\pi}}{2 s^{3/2}}$

SECTION 7.2 EXERCISES, page 272

1. $\frac{1}{2} t^2$ **3.** $t - 2t^4$ **5.** $1 + 3t + \frac{3}{2} t^2 + \frac{1}{6} t^3$
7. $t - 1 + e^{2t}$ **9.** $\frac{1}{4} e^{-t/4}$ **11.** $\frac{5}{7} \sin 7t$
13. $\cos \dfrac{t}{2}$ **15.** $\frac{1}{4} \sinh 4t$ **17.** $2 \cos 3t - 2 \sin 3t$
19. $\frac{1}{3} - \frac{1}{3} e^{-3t}$ **21.** $\frac{3}{4} e^{-3t} + \frac{1}{4} e^t$
23. $0.3 e^{0.1t} + 0.6 e^{-0.2t}$ **25.** $\frac{1}{2} e^{2t} - e^{3t} + \frac{1}{2} e^{6t}$
27. $-\frac{1}{3} e^{-t} + \frac{8}{15} e^{2t} - \frac{1}{5} e^{-3t}$ **29.** $\frac{1}{4} t - \frac{1}{8} \sin 2t$
31. $-\frac{1}{4} e^{-2t} + \frac{1}{4} \cos 2t + \frac{1}{4} \sin 2t$ **33.** $\frac{1}{3} \sin t - \frac{1}{6} \sin 2t$

SECTION 7.3 EXERCISES, page 281

1. $\dfrac{1}{(s-10)^2}$ **3.** $\dfrac{6}{(s+2)^4}$ **5.** $\dfrac{3}{(s-1)^2+9}$

7. $\dfrac{3}{(s-5)^2-9}$ **9.** $\dfrac{1}{(s-2)^2}+\dfrac{2}{(s-3)^2}+\dfrac{1}{(s-4)^2}$

11. $\dfrac{1}{2}\left[\dfrac{1}{s+1}-\dfrac{s+1}{(s+1)^2+4}\right]$ **13.** $\frac{1}{2}t^2e^{-2t}$

15. $e^{3t}\sin t$ **17.** $e^{-2t}\cos t-2e^{-2t}\sin t$

19. $e^{-t}-te^{-t}$ **21.** $5-t-5e^{-t}-4te^{-t}-\frac{3}{2}t^2e^{-t}$

23. $\dfrac{e^{-s}}{s^2}$ **25.** $\dfrac{e^{-2s}}{s^2}+2\dfrac{e^{-2s}}{s}$ **27.** $\dfrac{s}{s^2+4}e^{-\pi s}$

29. $\dfrac{6e^{-s}}{(s-1)^4}$ **31.** $\frac{1}{2}(t-2)^2\,\mathcal{U}(t-2)$

33. $-\sin t\,\mathcal{U}(t-\pi)$ **35.** $\mathcal{U}(t-1)-e^{-(t-1)}\,\mathcal{U}(t-1)$

37. $\dfrac{s^2-4}{(s^2+4)^2}$ **39.** $\dfrac{6s^2+2}{(s^2-1)^3}$ **41.** $\dfrac{12s-24}{[(s-2)^2+36]^2}$

43. $\frac{1}{2}t\sin t$ **45.** (c)   **47.** (f)   **49.** (a)

51. $f(t)=2-4\,\mathcal{U}(t-3)$; $\mathcal{L}\{f(t)\}=\dfrac{2}{s}-\dfrac{4}{s}e^{-3s}$

53. $f(t)=t^2\,\mathcal{U}(t-1)$
$=(t-1)^2\,\mathcal{U}(t-1)+2(t-1)\,\mathcal{U}(t-1)+\mathcal{U}(t-1)$
$\mathcal{L}\{f(t)\}=2\dfrac{e^{-s}}{s^3}+2\dfrac{e^{-s}}{s^2}+\dfrac{e^{-s}}{s}$

55. $f(t)=t-t\,\mathcal{U}(t-2)$
$=t-(t-2)\,\mathcal{U}(t-2)-2\,\mathcal{U}(t-2)$
$\mathcal{L}\{f(t)\}=\dfrac{1}{s^2}-\dfrac{e^{-2s}}{s^2}-2\dfrac{e^{-2s}}{s}$

57. $f(t)=\mathcal{U}(t-a)-\mathcal{U}(t-b)$; $\mathcal{L}\{f(t)\}=\dfrac{e^{-as}}{s}-\dfrac{e^{-bs}}{s}$

59.

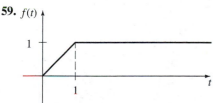

61. $\dfrac{e^{-t}-e^{3t}}{t}$ **63.** $e^{-2s}\left[\dfrac{2}{s^3}+\dfrac{1}{s^2}-\dfrac{2}{s}\right]$

SECTION 7.4 EXERCISES, page 289

1. Since $f'(t)=e^t$, $f(0)=1$, it follows from (1) that $\mathcal{L}\{e^t\}=s\mathcal{L}\{e^t\}-1$. Solving gives $\mathcal{L}\{e^t\}=1/(s-1)$.

3. $(s^2+3s)Y(s)-s-2$ **5.** $Y(s)=\dfrac{2s-1}{(s-1)^2}$

7. $\dfrac{1}{s(s-1)}$ **9.** $\dfrac{s+1}{s[(s+1)^2+1]}$ **11.** $\dfrac{1}{s^2(s-1)}$

13. $\dfrac{3s^2+1}{s^2(s^2+1)^2}$ **15.** $\dfrac{6}{s^5}$ **17.** $\dfrac{48}{s^8}$

19. $\dfrac{s-1}{(s+1)[(s-1)^2+1]}$ **21.** $\int_0^t f(\tau)e^{-5(t-\tau)}\,d\tau$

23. $1-e^{-t}$ **25.** $-\frac{1}{3}e^{-t}+\frac{1}{3}e^{2t}$ **27.** $\frac{1}{4}t\sin 2t$

31. $\dfrac{(1-e^{-as})^2}{s(1-e^{-2as})}=\dfrac{1-e^{-as}}{s(1+e^{-as})}$ **33.** $\dfrac{a}{s}\left(\dfrac{1}{bs}-\dfrac{1}{e^{bs}-1}\right)$

35. $\dfrac{\coth(\pi s/2)}{s^2+1}$ **37.** $\dfrac{1}{s^2+1}$

SECTION 7.5 EXERCISES, page 300

1. $y=-1+e^t$ **3.** $y=te^{-4t}+2e^{-4t}$

5. $y=\frac{4}{3}e^{-t}-\frac{1}{3}e^{-4t}$ **7.** $y=\frac{1}{9}t+\frac{2}{27}-\frac{2}{27}e^{3t}+\frac{10}{9}te^{3t}$

9. $y=\frac{1}{20}t^5e^{2t}$ **11.** $y=\cos t-\frac{1}{2}\sin t-\frac{1}{2}t\cos t$

13. $y=\frac{1}{2}-\frac{1}{2}e^t\cos t+\frac{1}{2}e^t\sin t$

15. $y=-\frac{8}{9}e^{-t/2}+\frac{1}{9}e^{-2t}+\frac{5}{18}e^t+\frac{1}{2}e^{-t}$

17. $y=\cos t$ **19.** $y=[5-5e^{-(t-1)}]\,\mathcal{U}(t-1)$

21. $y=-\frac{1}{4}+\frac{1}{2}t+\frac{1}{4}e^{-2t}-\frac{1}{4}\mathcal{U}(t-1)$
$-\frac{1}{2}(t-1)\,\mathcal{U}(t-1)$
$+\frac{1}{4}e^{-2(t-1)}\,\mathcal{U}(t-1)$

23. $y=\cos 2t-\frac{1}{6}\sin 2(t-2\pi)\,\mathcal{U}(t-2\pi)$
$+\frac{1}{3}\sin(t-2\pi)\,\mathcal{U}(t-2\pi)$

25. $y=\sin t+[1-\cos(t-\pi)]\,\mathcal{U}(t-\pi)$
$-[1-\cos(t-2\pi)]\,\mathcal{U}(t-2\pi)$

27. $y=(e+1)te^{-t}+(e-1)e^{-t}$ **29.** $f(t)=\sin t$

31. $f(t)=-\frac{1}{8}e^{-t}+\frac{1}{8}e^t+\frac{3}{4}te^t+\frac{1}{4}t^2e^t$ **33.** $f(t)=e^{-t}$

35. $f(t)=\frac{3}{8}e^{2t}+\frac{1}{8}e^{-2t}+\frac{1}{2}\cos 2t+\frac{1}{4}\sin 2t$

37. $y=\sin t-\frac{1}{2}t\sin t$

39. $i(t)=20{,}000[te^{-100t}-(t-1)e^{-100(t-1)}\,\mathcal{U}(t-1)]$

41. $q(t)=\dfrac{E_0 C}{1-kRC}(e^{-kt}-e^{-t/RC})$;
$q(t)=\dfrac{E_0}{R}te^{-t/RC}$

43. $q(t)=\frac{2}{5}\mathcal{U}(t-3)-\frac{2}{5}e^{-5(t-3)}\,\mathcal{U}(t-3)$

45. (a) $i(t)=\dfrac{1}{101}e^{-10t}-\dfrac{1}{101}\cos t+\dfrac{10}{101}\sin t$
$-\dfrac{10}{101}e^{-10(t-3\pi/2)}\,\mathcal{U}\!\left(t-\dfrac{3\pi}{2}\right)$
$+\dfrac{10}{101}\cos\!\left(t-\dfrac{3\pi}{2}\right)\mathcal{U}\!\left(t-\dfrac{3\pi}{2}\right)$
$+\dfrac{1}{101}\sin\!\left(t-\dfrac{3\pi}{2}\right)\mathcal{U}\!\left(t-\dfrac{3\pi}{2}\right)$

(b) $i_{\max}\approx 0.1$ at $t\approx 1.6$
$i_{\min}\approx -0.1$ at $t\approx 4.7$

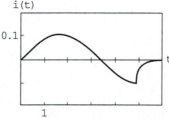

47. $i(t)=\dfrac{t}{R}+\dfrac{L}{R^2}(e^{-Rt/L}-1)$
$+\dfrac{1}{R}\sum_{n=1}^{\infty}(e^{-R(t-n)/L}-1)\,\mathcal{U}(t-n)$

For $0\le t<2$,

$i(t)=\begin{cases}\dfrac{t}{R}+\dfrac{L}{R^2}(e^{-Rt/L}-1), & 0\le t<1\\[2mm]\dfrac{t}{R}+\dfrac{L}{R^2}(e^{-Rt/L}-1)\\[1mm]\quad +\dfrac{1}{R}(e^{-R(t-1)/L}-1), & 1\le t<2\end{cases}$

49. $q(t) = \frac{3}{5}e^{-10t} + 6te^{-10t} - \frac{3}{5}\cos 10t$

$i(t) = -60te^{-10t} + 6\sin 10t$

Steady-state current is $6\sin 10t$.

51. $q(t) = \dfrac{E_0}{L(k^2 + 1/LC)}\left[e^{-kt} - \cos\dfrac{t}{\sqrt{LC}}\right]$

$\qquad + \dfrac{kE_0\sqrt{C/L}}{k^2 + 1/LC}\sin\dfrac{t}{\sqrt{LC}}$

53. $x(t) = -\dfrac{3}{2}e^{-7t/2}\cos\dfrac{\sqrt{15}}{2}t - \dfrac{7\sqrt{15}}{10}e^{-7t/2}\sin\dfrac{\sqrt{15}}{2}t$

55. $y(x) = \dfrac{w_0 L^2}{16EI}x^2 - \dfrac{w_0 L}{12EI}x^3 + \dfrac{w_0}{24EI}x^4$

$\qquad - \dfrac{w_0}{24EI}\left(x - \dfrac{L}{2}\right)^4 \mathcal{U}\left(x - \dfrac{L}{2}\right)$

57. $y(x) = \dfrac{w_0 L^2}{48EI}x^2 - \dfrac{w_0 L}{24EI}x^3$

$\qquad + \dfrac{w_0}{60EIL}\left[\dfrac{5L}{2}x^4 - x^5 + \left(x - \dfrac{L}{2}\right)^5 \mathcal{U}\left(x - \dfrac{L}{2}\right)\right]$

SECTION 7.6 EXERCISES, page 309

1. $y = e^{3(t-2)}\mathcal{U}(t - 2)$ **3.** $y = \sin t + \sin t\,\mathcal{U}(t - 2\pi)$

5. $y = -\cos t\,\mathcal{U}\left(t - \dfrac{\pi}{2}\right) + \cos t\,\mathcal{U}\left(t - \dfrac{3\pi}{2}\right)$

7. $y = \frac{1}{2} - \frac{1}{2}e^{-2t} + [\frac{1}{2} - \frac{1}{2}e^{-2(t-1)}]\mathcal{U}(t - 1)$

9. $y = e^{-2(t-2\pi)}\sin t\,\mathcal{U}(t - 2\pi)$

11. $y = e^{-2t}\cos 3t + \frac{2}{3}e^{-2t}\sin 3t$

$\qquad + \frac{1}{3}e^{-2(t-\pi)}\sin 3(t - \pi)\,\mathcal{U}(t - \pi)$

$\qquad + \frac{1}{3}e^{-2(t-3\pi)}\sin 3(t - 3\pi)\,\mathcal{U}(t - 3\pi)$

13. $y(x) = \begin{cases} \dfrac{P_0}{EI}\left(\dfrac{L}{4}x^2 - \dfrac{1}{6}x^3\right), & 0 \le x < \dfrac{L}{2} \\[3mm] \dfrac{P_0 L^2}{4EI}\left(\dfrac{1}{2}x - \dfrac{L}{12}\right), & \dfrac{L}{2} \le x \le L \end{cases}$

SECTION 7.7 EXERCISES, page 313

1. $x = -\frac{1}{3}e^{-2t} + \frac{1}{3}e^{t}$ **3.** $x = -\cos 3t - \frac{5}{3}\sin 3t$

$\quad\, y = \frac{1}{3}e^{-2t} + \frac{2}{3}e^{t}$ $\qquad\quad y = 2\cos 3t - \frac{7}{3}\sin 3t$

5. $x = -2e^{3t} + \frac{5}{2}e^{2t} - \frac{1}{2}$ **7.** $x = -\frac{1}{2}t - \frac{3}{4}\sqrt{2}\sin\sqrt{2}t$

$\quad\, y = \frac{8}{3}e^{3t} - \frac{5}{2}e^{2t} - \frac{1}{6}$ $\qquad\, y = -\frac{1}{2}t + \frac{3}{4}\sqrt{2}\sin\sqrt{2}t$

9. $x = 8 + \dfrac{2}{3!}t^3 + \dfrac{1}{4!}t^4$

$\quad\, y = -\dfrac{2}{3!}t^3 + \dfrac{1}{4!}t^4$

11. $x = \frac{1}{2}t^2 + t + 1 - e^{-t}$

$\quad\,\, y = -\frac{1}{3} + \frac{1}{3}e^{-t} + \frac{1}{3}te^{-t}$

13. $x_1 = \dfrac{1}{5}\sin t + \dfrac{2\sqrt{6}}{15}\sin\sqrt{6}t + \dfrac{2}{5}\cos t - \dfrac{2}{5}\cos\sqrt{6}t$

$\quad\,\, x_2 = \dfrac{2}{5}\sin t - \dfrac{\sqrt{6}}{15}\sin\sqrt{6}t + \dfrac{4}{5}\cos t + \dfrac{1}{5}\cos\sqrt{6}t$

15. **(b)** $i_2 = \frac{100}{9} - \frac{100}{9}e^{-900t}$

$\qquad\quad i_3 = \frac{80}{9} - \frac{80}{9}e^{-900t}$

$\quad\,$ **(c)** $i_1 = 20 - 20e^{-900t}$

17. $i_2 = -\frac{20}{13}e^{-2t} + \frac{375}{1469}e^{-15t} + \frac{145}{113}\cos t + \frac{85}{113}\sin t$

$\quad\,\, i_3 = \frac{30}{13}e^{-2t} + \frac{250}{1469}e^{-15t} - \frac{280}{113}\cos t + \frac{810}{113}\sin t$

19. $i_1 = \dfrac{6}{5} - \dfrac{6}{5}e^{-100t}\cosh 50\sqrt{2}t - \dfrac{9\sqrt{2}}{10}e^{-100t}\sinh 50\sqrt{2}t$

$\quad\,\, i_2 = \dfrac{6}{5} - \dfrac{6}{5}e^{-100t}\cosh 50\sqrt{2}t - \dfrac{6\sqrt{2}}{5}e^{-100t}\sinh 50\sqrt{2}t$

21. $\theta_1 = \dfrac{1}{4}\cos\dfrac{2}{\sqrt{3}}t + \dfrac{3}{4}\cos 2t$

$\quad\,\, \theta_2 = \dfrac{1}{2}\cos\dfrac{2}{\sqrt{3}}t - \dfrac{3}{2}\cos 2t$

CHAPTER 7 REVIEW EXERCISES, page 316

1. $\dfrac{1}{s^2} - \dfrac{2}{s^2}e^{-s}$ **3.** false **5.** true **7.** $\dfrac{1}{s + 7}$

9. $\dfrac{2}{s^2 + 4}$ **11.** $\dfrac{4s}{(s^2 + 4)^2}$ **13.** $\frac{1}{6}t^5$ **15.** $\frac{1}{2}t^2 e^{5t}$

17. $e^{5t}\cos 2t + \frac{5}{2}e^{5t}\sin 2t$

19. $\cos\pi(t - 1)\,\mathcal{U}(t - 1) + \sin\pi(t - 1)\,\mathcal{U}(t - 1)$

21. -5 **23.** $e^{-k(s-a)}F(s - a)$

25. **(a)** $f(t) = t - (t - 1)\mathcal{U}(t - 1) - \mathcal{U}(t - 4)$

$\quad\,$ **(b)** $\mathscr{L}\{f(t)\} = \dfrac{1}{s^2} - \dfrac{1}{s^2}e^{-s} - \dfrac{1}{s}e^{-4s}$

$\quad\,$ **(c)** $\mathscr{L}\{e^t f(t)\} = \dfrac{1}{(s - 1)^2} - \dfrac{1}{(s - 1)^2}e^{-(s-1)}$

$\qquad\qquad\qquad\qquad - \dfrac{1}{s - 1}e^{-4(s-1)}$

27. **(a)** $f(t) = 2 + (t - 2)\mathcal{U}(t - 2)$

$\quad\,$ **(b)** $\mathscr{L}\{f(t)\} = \dfrac{2}{s} + \dfrac{1}{s^2}e^{-2s}$

$\quad\,$ **(c)** $\mathscr{L}\{e^t f(t)\} = \dfrac{2}{s - 1} + \dfrac{1}{(s - 1)^2}e^{-2(s-1)}$

29. $y = 5te^t + \frac{1}{2}t^2 e^t$

31. $y = 5\mathcal{U}(t - \pi) - 5e^{2(t-\pi)}\cos\sqrt{2}(t - \pi)\,\mathcal{U}(t - \pi)$

$\qquad + 5\sqrt{2}e^{2(t-\pi)}\sin\sqrt{2}(t - \pi)\,\mathcal{U}(t - \pi)$

33. $y = -\frac{2}{125} - \frac{2}{25}t - \frac{1}{5}t^2 + \frac{127}{125}e^{5t}$

$\qquad - [-\frac{37}{125} - \frac{12}{25}(t - 1) - \frac{1}{5}(t - 1)^2$

$\qquad\quad + \frac{37}{125}e^{5(t-1)}]\mathcal{U}(t - 1)$

35. $y = 1 + t + \frac{1}{2}t^2$

37. $x = -\frac{1}{4} + \frac{9}{8}e^{-2t} + \frac{1}{8}e^{2t}$

$\quad\,\, y = t + \frac{9}{4}e^{-2t} - \frac{1}{4}e^{2t}$

39. $i(t) = -9 + 2t + 9e^{-t/5}$

41. $y(x) = \dfrac{w_0}{12EIL}\left[-\dfrac{1}{5}x^5 + \dfrac{L}{2}x^4 - \dfrac{L^2}{2}x^3 + \dfrac{L^3}{4}x^2\right.$

$\qquad\qquad\qquad\left. + \dfrac{1}{5}\left(x - \dfrac{L}{2}\right)^5 \mathcal{U}\left(x - \dfrac{L}{2}\right)\right]$

SECTION 8.1 EXERCISES, page 327

1. $\mathbf{X}' = \begin{pmatrix} 3 & -5 \\ 4 & 8 \end{pmatrix}\mathbf{X}$, where $\mathbf{X} = \begin{pmatrix} x \\ y \end{pmatrix}$

3. $\mathbf{X}' = \begin{pmatrix} -3 & 4 & -9 \\ 6 & -1 & 0 \\ 10 & 4 & 3 \end{pmatrix}\mathbf{X}$, where $\mathbf{X} = \begin{pmatrix} x \\ y \\ z \end{pmatrix}$

5. $\mathbf{X}' = \begin{pmatrix} 1 & -1 & 1 \\ 2 & 1 & -1 \\ 1 & 1 & 1 \end{pmatrix}\mathbf{X} + \begin{pmatrix} 0 \\ -3t^2 \\ t^2 \end{pmatrix} + \begin{pmatrix} t \\ 0 \\ -t \end{pmatrix} + \begin{pmatrix} -1 \\ 0 \\ 2 \end{pmatrix}$,

$\quad\,$ where $\mathbf{X} = \begin{pmatrix} x \\ y \\ z \end{pmatrix}$

7. $\dfrac{dx}{dt} = 4x + 2y + e^t$

$\dfrac{dy}{dt} = -x + 3y - e^t$

9. $\dfrac{dx}{dt} = x - y + 2z + e^{-t} - 3t$

$\dfrac{dy}{dt} = 3x - 4y + z + 2e^{-t} + t$

$\dfrac{dz}{dt} = -2x + 5y + 6z + 2e^{-t} - t$

17. Yes; $W(\mathbf{X}_1, \mathbf{X}_2) = -2e^{-8t} \neq 0$ implies that $\mathbf{X}_1$ and $\mathbf{X}_2$ are linearly independent on $(-\infty, \infty)$.

19. No; $W(\mathbf{X}_1, \mathbf{X}_2, \mathbf{X}_3) = 0$ for every t. The solution vectors are linearly dependent on $(-\infty, \infty)$. Note that $\mathbf{X}_3 = 2\mathbf{X}_1 + \mathbf{X}_2$.

SECTION 8.2 EXERCISES, page 341

1. $\mathbf{X} = c_1 \begin{pmatrix} 1 \\ 2 \end{pmatrix} e^{5t} + c_2 \begin{pmatrix} 1 \\ -1 \end{pmatrix} e^{-t}$

3. $\mathbf{X} = c_1 \begin{pmatrix} 2 \\ 1 \end{pmatrix} e^{-3t} + c_2 \begin{pmatrix} 2 \\ 5 \end{pmatrix} e^{t}$

5. $\mathbf{X} = c_1 \begin{pmatrix} 5 \\ 2 \end{pmatrix} e^{8t} + c_2 \begin{pmatrix} 1 \\ 4 \end{pmatrix} e^{-10t}$

7. $\mathbf{X} = c_1 \begin{pmatrix} 1 \\ 0 \\ 0 \end{pmatrix} e^{t} + c_2 \begin{pmatrix} 2 \\ 3 \\ 1 \end{pmatrix} e^{2t} + c_3 \begin{pmatrix} 1 \\ 0 \\ 2 \end{pmatrix} e^{-t}$

9. $\mathbf{X} = c_1 \begin{pmatrix} -1 \\ 0 \\ 1 \end{pmatrix} e^{-t} + c_2 \begin{pmatrix} 1 \\ 4 \\ 3 \end{pmatrix} e^{3t} + c_3 \begin{pmatrix} 1 \\ -1 \\ 3 \end{pmatrix} e^{-2t}$

11. $\mathbf{X} = c_1 \begin{pmatrix} 4 \\ 0 \\ -1 \end{pmatrix} e^{-t} + c_2 \begin{pmatrix} -12 \\ 6 \\ 5 \end{pmatrix} e^{-t/2} + c_3 \begin{pmatrix} 4 \\ 2 \\ -1 \end{pmatrix} e^{-3t/2}$

13. $\mathbf{X} = 3 \begin{pmatrix} 1 \\ 1 \end{pmatrix} e^{t/2} + 2 \begin{pmatrix} 0 \\ 1 \end{pmatrix} e^{-t/2}$

15. $\mathbf{X} = c_1 \begin{pmatrix} 0.382175 \\ 0.851161 \\ 0.359815 \end{pmatrix} e^{8.58979t} + c_2 \begin{pmatrix} 0.405188 \\ -0.676043 \\ 0.615458 \end{pmatrix} e^{2.25684t}$

$+ c_3 \begin{pmatrix} -0.923562 \\ -0.132174 \\ 0.35995 \end{pmatrix} e^{-0.0466321t}$

17. $\mathbf{X} = c_1 \begin{pmatrix} 1 \\ 3 \end{pmatrix} + c_2 \left[\begin{pmatrix} 1 \\ 3 \end{pmatrix} t + \begin{pmatrix} \frac{1}{4} \\ -\frac{1}{4} \end{pmatrix} \right]$

19. $\mathbf{X} = c_1 \begin{pmatrix} 1 \\ 1 \end{pmatrix} e^{2t} + c_2 \left[\begin{pmatrix} 1 \\ 1 \end{pmatrix} te^{2t} + \begin{pmatrix} -\frac{1}{3} \\ 0 \end{pmatrix} e^{2t} \right]$

21. $\mathbf{X} = c_1 \begin{pmatrix} 1 \\ 1 \\ 1 \end{pmatrix} e^{t} + c_2 \begin{pmatrix} 1 \\ 1 \\ 0 \end{pmatrix} e^{2t} + c_3 \begin{pmatrix} 1 \\ 0 \\ 1 \end{pmatrix} e^{2t}$

23. $\mathbf{X} = c_1 \begin{pmatrix} -4 \\ -5 \\ 2 \end{pmatrix} + c_2 \begin{pmatrix} 2 \\ 0 \\ -1 \end{pmatrix} e^{5t}$

$+ c_3 \left[\begin{pmatrix} 2 \\ 0 \\ -1 \end{pmatrix} te^{5t} + \begin{pmatrix} -\frac{1}{2} \\ -\frac{1}{2} \\ -1 \end{pmatrix} e^{5t} \right]$

25. $\mathbf{X} = c_1 \begin{pmatrix} 0 \\ 1 \\ 1 \end{pmatrix} e^{t} + c_2 \left[\begin{pmatrix} 0 \\ 1 \\ 1 \end{pmatrix} te^{t} + \begin{pmatrix} 0 \\ 1 \\ 0 \end{pmatrix} e^{t} \right]$

$+ c_3 \left[\begin{pmatrix} 0 \\ 1 \\ 1 \end{pmatrix} \frac{t^2}{2} e^{t} + \begin{pmatrix} 0 \\ 1 \\ 0 \end{pmatrix} te^{t} + \begin{pmatrix} \frac{1}{2} \\ 0 \\ 0 \end{pmatrix} e^{t} \right]$

27. $\mathbf{X} = -7 \begin{pmatrix} 2 \\ 1 \end{pmatrix} e^{4t} + 13 \begin{pmatrix} 2t + 1 \\ t + 1 \end{pmatrix} e^{4t}$

29. Corresponding to the eigenvalue $\lambda_1 = 2$ of multiplicity five, eigenvectors are

$$\mathbf{K}_1 = \begin{pmatrix} 1 \\ 0 \\ 0 \\ 0 \\ 0 \end{pmatrix}, \mathbf{K}_2 = \begin{pmatrix} 0 \\ 0 \\ 1 \\ 0 \\ 0 \end{pmatrix}, \mathbf{K}_3 = \begin{pmatrix} 0 \\ 0 \\ 0 \\ 1 \\ 0 \end{pmatrix}.$$

31. $\mathbf{X} = c_1 \begin{pmatrix} \cos t \\ 2\cos t + \sin t \end{pmatrix} e^{4t} + c_2 \begin{pmatrix} \sin t \\ 2\sin t - \cos t \end{pmatrix} e^{4t}$

33. $\mathbf{X} = c_1 \begin{pmatrix} \cos t \\ -\cos t - \sin t \end{pmatrix} e^{4t} + c_2 \begin{pmatrix} \sin t \\ -\sin t + \cos t \end{pmatrix} e^{4t}$

35. $\mathbf{X} = c_1 \begin{pmatrix} 5\cos 3t \\ 4\cos 3t + 3\sin 3t \end{pmatrix} + c_2 \begin{pmatrix} 5\sin 3t \\ 4\sin 3t - 3\cos 3t \end{pmatrix}$

37. $\mathbf{X} = c_1 \begin{pmatrix} 1 \\ 0 \\ 0 \end{pmatrix} + c_2 \begin{pmatrix} -\cos t \\ \cos t \\ \sin t \end{pmatrix} + c_3 \begin{pmatrix} \sin t \\ -\sin t \\ \cos t \end{pmatrix}$

39. $\mathbf{X} = c_1 \begin{pmatrix} 0 \\ 2 \\ 1 \end{pmatrix} e^{t} + c_2 \begin{pmatrix} \sin t \\ \cos t \\ \cos t \end{pmatrix} e^{t} + c_3 \begin{pmatrix} \cos t \\ -\sin t \\ -\sin t \end{pmatrix} e^{t}$

41. $\mathbf{X} = \begin{pmatrix} 28 \\ -5 \\ 25 \end{pmatrix} e^{2t} + c_2 \begin{pmatrix} 5\cos 3t \\ -4\cos 3t - 3\sin 3t \\ 0 \end{pmatrix} e^{-2t}$

$+ c_3 \begin{pmatrix} 5\sin 3t \\ -4\sin 3t + 3\cos 3t \\ 0 \end{pmatrix} e^{-2t}$

43. $\mathbf{X} = -\begin{pmatrix} 25 \\ -7 \\ 6 \end{pmatrix} e^{t} - \begin{pmatrix} \cos 5t - 5\sin 5t \\ \cos 5t \\ \cos 5t \end{pmatrix}$

$+ 6 \begin{pmatrix} 5\cos 5t + \sin 5t \\ \sin 5t \\ \sin 5t \end{pmatrix}$

SECTION 8.3 EXERCISES, page 347

1. $\mathbf{X} = c_1 \begin{pmatrix} 1 \\ 1 \end{pmatrix} + c_2 \begin{pmatrix} 3 \\ 2 \end{pmatrix} e^{t} - \begin{pmatrix} 11 \\ 11 \end{pmatrix} t - \begin{pmatrix} 15 \\ 10 \end{pmatrix}$

3. $\mathbf{X} = c_1 \begin{pmatrix} 2 \\ 1 \end{pmatrix} e^{t/2} + c_2 \begin{pmatrix} 10 \\ 3 \end{pmatrix} e^{3t/2} - \begin{pmatrix} \frac{13}{2} \\ \frac{13}{4} \end{pmatrix} te^{t/2} - \begin{pmatrix} \frac{15}{2} \\ \frac{9}{4} \end{pmatrix} e^{t/2}$

5. $\mathbf{X} = c_1 \begin{pmatrix} 2 \\ 1 \end{pmatrix} e^{t} + c_2 \begin{pmatrix} 1 \\ 1 \end{pmatrix} e^{2t} + \begin{pmatrix} 3 \\ 3 \end{pmatrix} e^{t} + \begin{pmatrix} 4 \\ 2 \end{pmatrix} te^{t}$

7. $\mathbf{X} = c_1 \begin{pmatrix} 4 \\ 1 \end{pmatrix} e^{3t} + c_2 \begin{pmatrix} -2 \\ 1 \end{pmatrix} e^{-3t} + \begin{pmatrix} -12 \\ 0 \end{pmatrix} t - \begin{pmatrix} \frac{4}{3} \\ \frac{4}{3} \end{pmatrix}$

9. $\mathbf{X} = c_1 \begin{pmatrix} 1 \\ -1 \end{pmatrix} e^{t} + c_2 \begin{pmatrix} -t \\ \frac{1}{2} - t \end{pmatrix} e^{t} + \begin{pmatrix} \frac{1}{2} \\ -2 \end{pmatrix} e^{-t}$

11. $\mathbf{X} = c_1 \begin{pmatrix} \cos t \\ \sin t \end{pmatrix} + c_2 \begin{pmatrix} \sin t \\ -\cos t \end{pmatrix}$
$+ \begin{pmatrix} \cos t \\ \sin t \end{pmatrix} t + \begin{pmatrix} -\sin t \\ \cos t \end{pmatrix} \ln|\cos t|$

13. $\mathbf{X} = c_1 \begin{pmatrix} \cos t \\ \sin t \end{pmatrix} e^{t} + c_2 \begin{pmatrix} \sin t \\ -\cos t \end{pmatrix} e^{t} + \begin{pmatrix} \cos t \\ \sin t \end{pmatrix} te^{t}$

15. $\mathbf{X} = c_1 \begin{pmatrix} \cos t \\ -\sin t \end{pmatrix} + c_2 \begin{pmatrix} \sin t \\ \cos t \end{pmatrix} + \begin{pmatrix} \cos t \\ -\sin t \end{pmatrix} t$
$+ \begin{pmatrix} -\sin t \\ \sin t \tan t \end{pmatrix} - \begin{pmatrix} \sin t \\ \cos t \end{pmatrix} \ln|\cos t|$

17. $\mathbf{X} = c_1 \begin{pmatrix} 2 \sin t \\ \cos t \end{pmatrix} e^{t} + c_2 \begin{pmatrix} 2 \cos t \\ -\sin t \end{pmatrix} e^{t} + \begin{pmatrix} 3 \sin t \\ \frac{3}{2} \cos t \end{pmatrix} te^{t}$
$+ \begin{pmatrix} \cos t \\ -\frac{1}{2} \sin t \end{pmatrix} e^{t} \ln|\sin t| + \begin{pmatrix} 2 \cos t \\ -\sin t \end{pmatrix} e^{t} \ln|\cos t|$

19. $\mathbf{X} = c_1 \begin{pmatrix} 1 \\ -1 \\ 0 \end{pmatrix} + c_2 \begin{pmatrix} 1 \\ 1 \\ 1 \end{pmatrix} e^{2t} + c_3 \begin{pmatrix} 0 \\ 0 \\ 1 \end{pmatrix} e^{3t}$
$+ \begin{pmatrix} -\frac{1}{4} e^{2t} + \frac{1}{2} te^{2t} \\ -e^{t} + \frac{1}{4} e^{2t} + \frac{1}{2} te^{2t} \\ \frac{1}{2} t^2 e^{3t} \end{pmatrix}$

21. $\mathbf{X} = \begin{pmatrix} 2 \\ 2 \end{pmatrix} te^{2t} + \begin{pmatrix} -1 \\ 1 \end{pmatrix} e^{2t} + \begin{pmatrix} -2 \\ 2 \end{pmatrix} te^{4t} + \begin{pmatrix} 2 \\ 0 \end{pmatrix} e^{4t}$

23. $\begin{pmatrix} i_1 \\ i_2 \end{pmatrix} = 2 \begin{pmatrix} 1 \\ 3 \end{pmatrix} e^{-2t} + \frac{6}{29} \begin{pmatrix} 3 \\ -1 \end{pmatrix} e^{-12t}$
$+ \begin{pmatrix} \frac{332}{29} \\ \frac{276}{29} \end{pmatrix} \sin t - \begin{pmatrix} \frac{76}{29} \\ \frac{168}{29} \end{pmatrix} \cos t$

SECTION 8.4 EXERCISES, page 350

1. $e^{\mathbf{A}t} = \begin{pmatrix} e^{t} & 0 \\ 0 & e^{2t} \end{pmatrix}; e^{-\mathbf{A}t} = \begin{pmatrix} e^{-t} & 0 \\ 0 & e^{-2t} \end{pmatrix}$

3. $e^{\mathbf{A}t} = \begin{pmatrix} t+1 & t & t \\ t & t+1 & t \\ -2t & -2t & -2t+1 \end{pmatrix}$

5. $\mathbf{X} = c_1 \begin{pmatrix} 1 \\ 0 \end{pmatrix} e^{t} + c_2 \begin{pmatrix} 0 \\ 1 \end{pmatrix} e^{2t}$

7. $\mathbf{X} = c_1 \begin{pmatrix} t+1 \\ t \\ -2t \end{pmatrix} + c_2 \begin{pmatrix} t \\ t+1 \\ -2t \end{pmatrix} + c_3 \begin{pmatrix} t \\ t \\ -2t+1 \end{pmatrix}$

9. $\mathbf{X} = c_3 \begin{pmatrix} 1 \\ 0 \end{pmatrix} e^{t} + c_4 \begin{pmatrix} 0 \\ 1 \end{pmatrix} e^{2t} + \begin{pmatrix} -3 \\ \frac{1}{2} \end{pmatrix}$

11. $\mathbf{X} = c_1 \begin{pmatrix} \cosh t \\ \sinh t \end{pmatrix} + c_2 \begin{pmatrix} \sinh t \\ \cosh t \end{pmatrix} - \begin{pmatrix} 1 \\ 1 \end{pmatrix}$

13. $\mathbf{X} = \begin{pmatrix} t+1 \\ t \\ -2t \end{pmatrix} - 4 \begin{pmatrix} t \\ t+1 \\ -2t \end{pmatrix} + 6 \begin{pmatrix} t \\ t \\ -2t+1 \end{pmatrix}$

19. $\mathbf{X} = \begin{pmatrix} \frac{3}{2} e^{3t} - \frac{1}{2} e^{5t} & -\frac{1}{2} e^{3t} + \frac{1}{2} e^{5t} \\ \frac{3}{2} e^{3t} - \frac{3}{2} e^{5t} & -\frac{1}{2} e^{3t} + \frac{3}{2} e^{5t} \end{pmatrix} \begin{pmatrix} c_1 \\ c_2 \end{pmatrix}$

CHAPTER 8 REVIEW EXERCISES, page 352

3. $\mathbf{X} = c_1 \begin{pmatrix} 1 \\ -1 \end{pmatrix} e^{t} + c_2 \left[\begin{pmatrix} 1 \\ -1 \end{pmatrix} te^{t} + \begin{pmatrix} 0 \\ 1 \end{pmatrix} e^{t} \right]$

5. $\mathbf{X} = c_1 \begin{pmatrix} \cos 2t \\ -\sin 2t \end{pmatrix} e^{t} + c_2 \begin{pmatrix} \sin 2t \\ \cos 2t \end{pmatrix} e^{t}$

7. $\mathbf{X} = c_1 \begin{pmatrix} -1 \\ 1 \\ 0 \end{pmatrix} + c_2 \begin{pmatrix} -1 \\ 0 \\ 1 \end{pmatrix} + c_3 \begin{pmatrix} 1 \\ 1 \\ 1 \end{pmatrix} e^{3t}$

9. $\mathbf{X} = c_1 \begin{pmatrix} 1 \\ 0 \end{pmatrix} e^{2t} + c_2 \begin{pmatrix} 4 \\ 1 \end{pmatrix} e^{4t} + \begin{pmatrix} 16 \\ -4 \end{pmatrix} t + \begin{pmatrix} 11 \\ -1 \end{pmatrix}$

11. $\mathbf{X} = c_1 \begin{pmatrix} \cos t \\ \cos t - \sin t \end{pmatrix} + c_2 \begin{pmatrix} \sin t \\ \sin t + \cos t \end{pmatrix} - \begin{pmatrix} 1 \\ 1 \end{pmatrix}$
$+ \begin{pmatrix} \sin t \\ \sin t + \cos t \end{pmatrix} \ln|\csc t - \cot t|$

SECTION 9.1 EXERCISES, page 356

1.

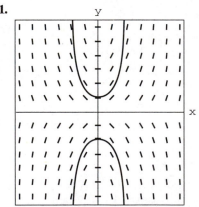

3.

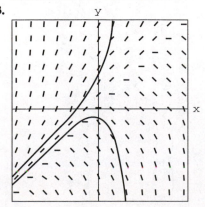

5.

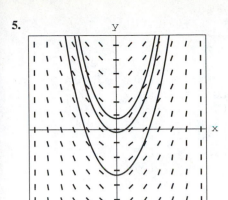

7.

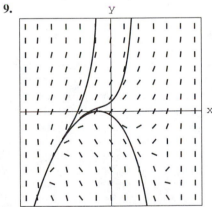

9.

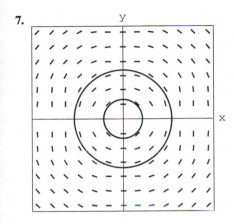

11.

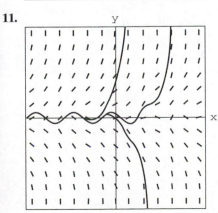

1. (a) $y = 1 - x + \tan\left(x + \dfrac{\pi}{4}\right)$

(b) $h = 0.1$

x_n	y_n	True value
0.00	2.0000	2.0000
0.10	2.1000	2.1230
0.20	2.2440	2.3085
0.30	2.4525	2.5958
0.40	2.7596	3.0650
0.50	3.2261	3.9082

$h = 0.05$

x_n	y_n	True value
0.00	2.0000	2.0000
0.05	2.0500	2.0554
0.10	2.1105	2.1230
0.15	2.1838	2.2061
0.20	2.2727	2.3085
0.25	2.3812	2.4358
0.30	2.5142	2.5958
0.35	2.6788	2.7997
0.40	2.8845	3.0650
0.45	3.1455	3.4189
0.50	3.4823	3.9082

3. $h = 0.1$

x_n	y_n
1.00	5.0000
1.10	3.8000
1.20	2.9800
1.30	2.4260
1.40	2.0582
1.50	1.8207

$h = 0.05$

x_n	y_n
1.00	5.0000
1.05	4.4000
1.10	3.8950
1.15	3.4707
1.20	3.1151
1.25	2.8179
1.30	2.5702
1.35	2.3647
1.40	2.1950
1.45	2.0557
1.50	1.9424

5.

$h = 0.1$			$h = 0.05$	
x_n	y_n		x_n	y_n
0.00	0.0000		0.00	0.0000
0.10	0.1000		0.05	0.0500
0.20	0.2010		0.10	0.1001
0.30	0.3050		0.15	0.1506
0.40	0.4143		0.20	0.2018
0.50	0.5315		0.25	0.2538
			0.30	0.3070
			0.35	0.3617
			0.40	0.4183
			0.45	0.4770
			0.50	0.5384

7.

$h = 0.1$			$h = 0.05$	
x_n	y_n		x_n	y_n
0.00	0.0000		0.00	0.0000
0.10	0.1000		0.05	0.0500
0.20	0.1905		0.10	0.0976
0.30	0.2731		0.15	0.1429
0.40	0.3492		0.20	0.1863
0.50	0.4198		0.25	0.2278
			0.30	0.2676
			0.35	0.3058
			0.40	0.3427
			0.45	0.3782
			0.50	0.4124

9.

$h = 0.1$			$h = 0.05$	
x_n	y_n		x_n	y_n
0.00	0.5000		0.00	0.5000
0.10	0.5250		0.05	0.5125
0.20	0.5431		0.10	0.5232
0.30	0.5548		0.15	0.5322
0.40	0.5613		0.20	0.5395
0.50	0.5639		0.25	0.5452
			0.30	0.5496
			0.35	0.5527
			0.40	0.5547
			0.45	0.5559
			0.50	0.5565

11.

$h = 0.1$			$h = 0.05$	
x_n	y_n		x_n	y_n
1.00	1.0000		1.00	1.0000
1.10	1.0000		1.05	1.0000
1.20	1.0191		1.10	1.0049
1.30	1.0588		1.15	1.0147
1.40	1.1231		1.20	1.0298
1.50	1.2194		1.25	1.0506
			1.30	1.0775
			1.35	1.1115
			1.40	1.1538
			1.45	1.2057
			1.50	1.2696

13. (a)

$h = 0.1$			$h = 0.05$	
x_n	y_n		x_n	y_n
1.00	5.0000		1.00	5.0000
1.10	3.9900		1.05	4.4475
1.20	3.2545		1.10	3.9763
1.30	2.7236		1.15	3.5751
1.40	2.3451		1.20	3.2342
1.50	2.0801		1.25	2.9452
			1.30	2.7009
			1.35	2.4952
			1.40	2.3226
			1.45	2.1786
			1.50	2.0592

(b)

$h = 0.1$			$h = 0.05$	
x_n	y_n		x_n	y_n
0.00	0.0000		0.00	0.0000
0.10	0.1005		0.05	0.0501
0.20	0.2030		0.10	0.1004
0.30	0.3098		0.15	0.1512
0.40	0.4234		0.20	0.2028
0.50	0.5470		0.25	0.2554
			0.30	0.3095
			0.35	0.3652
			0.40	0.4230
			0.45	0.4832
			0.50	0.5465

(c)

$h = 0.1$			$h = 0.05$	
x_n	y_n		x_n	y_n
0.00	0.0000		0.00	0.0000
0.10	0.0952		0.05	0.0488
0.20	0.1822		0.10	0.0953
0.30	0.2622		0.15	0.1397
0.40	0.3363		0.20	0.1823
0.50	0.4053		0.25	0.2231
			0.30	0.2623
			0.35	0.3001
			0.40	0.3364
			0.45	0.3715
			0.50	0.4054

(d)

$h = 0.1$			$h = 0.05$	
x_n	y_n		x_n	y_n
0.00	5.0000		0.00	0.5000
0.10	0.5215		0.05	0.5116
0.20	0.5362		0.10	0.5214
0.30	0.5449		0.15	0.5294
0.40	0.5490		0.20	0.5359
0.50	0.5503		0.25	0.5408
			0.30	0.5444
			0.35	0.5469
			0.40	0.5484
			0.45	0.5492
			0.50	0.5495

(e)

$h = 0.1$			$h = 0.05$	
x_n	y_n		x_n	y_n
1.00	1.0000		1.00	1.0000
1.10	1.0095		1.05	1.0024
1.20	1.0404		1.10	1.0100
1.30	1.0967		1.15	1.0228
1.40	1.1866		1.20	1.0414
1.50	1.3260		1.25	1.0663
			1.30	1.0984
			1.35	1.1389
			1.40	1.1895
			1.45	1.2526
			1.50	1.3315

15. (a) The appearance of the graph will depend on the ODE solver used. The following graph was obtained using Mathematica on the interval $[1, 1.3556]$.

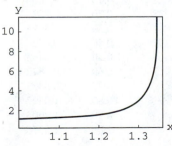

(b)

x_n	Euler	Improved Euler
1.0	1.0000	1.0000
1.1	1.2000	1.2469
1.2	1.4938	1.6668
1.3	1.9711	2.6427
1.4	2.9060	8.7989

17. (a) $y_1 = 1.2$

(b) $y''(c)\dfrac{h^2}{2} = 4e^{2c}\dfrac{(0.1)^2}{2} = 0.02e^{2c} \le 0.02e^{0.2} = 0.0244$

(c) Actual value is $y(0.1) = 1.2214$. Error is 0.0214.

(d) If $h = 0.05$, $y_2 = 1.21$.

(e) Error with $h = 0.1$ is 0.0214. Error with $h = 0.05$ is 0.0114.

19. (a) $y_1 = 0.8$

(b) $y''(c)\dfrac{h^2}{2} = 5e^{-2c}\dfrac{(0.1)^2}{2} = 0.025e^{-2c} \le 0.025$ for

$0 \le c \le 0.1$.

(c) Actual value is $y(0.1) = 0.8234$. Error is 0.0234.

(d) If $h = 0.05$, $y_2 = 0.8125$.

(e) Error with $h = 0.1$ is 0.0234. Error with $h = 0.05$ is 0.0109.

21. (a) Error is $19h^2e^{-3(c-1)}$.

(b) $y''(c)\dfrac{h^2}{2} \le 19(0.1)^2(1) = 0.19$

(c) If $h = 0.1$, $y_5 = 1.8207$. If $h = 0.05$, $y_{10} = 1.9424$

(d) Error with $h = 0.1$ is 0.2325. Error with $h = 0.05$ is 0.1109.

23. (a) Error is $\dfrac{1}{(c+1)^2}\dfrac{h^2}{2}$.

(b) $\left| y''(c)\dfrac{h^2}{2} \right| \le (1)\dfrac{(0.1)^2}{2} = 0.005$

(c) If $h = 0.1$, $y_5 = 0.4198$. If $h = 0.05$, $y_{10} = 0.4124$

(d) Error with $h = 0.1$ is 0.0143. Error with $h = 0.05$ is 0.0069.

SECTION 9.3 EXERCISES, page 370

1.

x_n	y_n	True value
0.00	2.0000	2.0000
0.10	2.1230	2.1230
0.20	2.3085	2.3085
0.30	2.5958	2.5958
0.40	3.0649	3.0650
0.50	3.9078	3.9082

3.

x_n	y_n
1.00	5.0000
1.10	3.9724
1.20	3.2284
1.30	2.6945
1.40	2.3163
1.50	2.0533

5.

x_n	y_n
0.00	0.0000
0.10	0.1003
0.20	0.2027
0.30	0.3093
0.40	0.4228
0.50	0.5463

7.

x_n	y_n
0.00	0.0000
0.10	0.0953
0.20	0.1823
0.30	0.2624
0.40	0.3365
0.50	0.4055

9.

x_n	y_n
0.00	0.5000
0.10	0.5213
0.20	0.5358
0.30	0.5443
0.40	0.5482
0.50	0.5493

11.

x_n	y_n
1.00	1.0000
1.10	1.0101
1.20	1.0417
1.30	1.0989
1.40	1.1905
1.50	1.3333

13. (a) $v(5) = 35.7678$

(b)

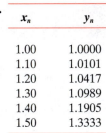

(c) $v(t) = \sqrt{\dfrac{mg}{k}} \tanh \sqrt{\dfrac{kg}{m}}\, t; \; v(5) = 35.7678$

15. (a) $h = 0.1$ $h = 0.05$

x_n	y_n	x_n	y_n
1.00	1.0000	1.00	1.0000
1.10	1.2511	1.05	1.1112
1.20	1.6934	1.10	1.2511
1.30	2.9425	1.15	1.4348
1.40	903.0282	1.20	1.6934
		1.25	2.1047
		1.30	2.9560
		1.35	7.8981
		1.40	1.1E + 15

(b) The appearance of the graph will depend on the ODE solver used. The following graph was ob-

tained using Mathematica on the interval [1, 1.3556].

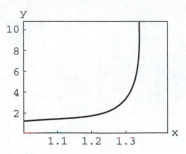

17. (a) $y_1 = 0.82341667$

(b) $y^{(5)}(c)\dfrac{h^5}{5!} = 40e^{-2c}\dfrac{h^5}{5!} \le 40e^{2(0)}\dfrac{(0.1)^5}{5!}$
$= 3.333 \times 10^{-6}$

(c) Actual value is $y(0.1) = 0.8234134413$. Error is $3.225 \times 10^{-6} \le 3.333 \times 10^{-6}$.

(d) If $h = 0.05$, $y_2 = 0.82341363$.

(e) Error with $h = 0.1$ is 3.225×10^{-6}. Error with $h = 0.05$ is 1.854×10^{-7}.

19. (a) $y^{(5)}(c)\dfrac{h^5}{5!} = \dfrac{24}{(c+1)^5}\dfrac{h^5}{5!}$

(b) $\dfrac{24}{(c+1)^5}\dfrac{h^5}{5!} \le 24\dfrac{(0.1)^5}{5!} = 2.0000 \times 10^{-6}$

(c) From calculation with $h = 0.1$, $y_5 = 0.40546517$. From calculation with $h = 0.05$, $y_{10} = 0.40546511$.

SECTION 9.4 EXERCISES, page 375

1. $y(x) = -x + e^x$; $y(0.2) = 1.0214$, $y(0.4) = 1.0918$, $y(0.6) = 1.2221$, $y(0.8) = 1.4255$

3.

x_n	y_n
0.00	1.0000
0.20	0.7328
0.40	0.6461
0.60	0.6585
0.80	0.7232

5.

x_n	y_n	x_n	y_n
0.00	0.0000	0.00	0.0000
0.20	0.2027	0.10	0.1003
0.40	0.4228	0.20	0.2027
0.60	0.6841	0.30	0.3093
0.80	1.0297	0.40	0.4228
1.00	1.5569	0.50	0.5463
		0.60	0.6842
		0.70	0.8423
		0.80	1.0297
		0.90	1.2603
		1.00	1.5576

7.

x_n	y_n
0.00	0.0000
0.20	0.0026
0.40	0.0201
0.60	0.0630
0.80	0.1360
1.00	0.2385

x_n	y_n
0.00	0.0000
0.10	0.0003
0.20	0.0026
0.30	0.0087
0.40	0.0200
0.50	0.0379
0.60	0.0629
0.70	0.0956
0.80	0.1360
0.90	0.1837
1.00	0.2384

SECTION 9.5 EXERCISES, page 380

1. $y(x) = -2e^{2x} + 5xe^{2x}$; $y(0.2) = -1.4918$,
  $y_2 = -1.6800$
3. $y_1 = -1.4928$, $y_2 = -1.4919$
5. $y_1 = 1.4640$, $y_2 = 1.4640$
7. $x_1 = 8.3055$, $y_1 = 3.4199$;
  $x_2 = 8.3055$, $y_2 = 3.4199$
9. $x_1 = -3.9123$, $y_1 = 4.2857$;
  $x_2 = -3.9123$, $y_2 = 4.2857$
11. $x_1 = 0.4179$, $y_1 = -2.1824$;
  $x_2 = 0.4173$, $y_2 = -2.1821$

SECTION 9.6 EXERCISES, page 385

1. $y_1 = -5.6774$, $y_2 = -2.5807$, $y_3 = 6.3226$
3. $y_1 = -0.2259$, $y_2 = -0.3356$, $y_3 = -0.3308$, $y_4 = -0.2167$
5. $y_1 = 3.3751$, $y_2 = 3.6306$, $y_3 = 3.6448$, $y_4 = 3.2355$,
  $y_5 = 2.1411$
7. $y_1 = 3.8842$, $y_2 = 2.9640$, $y_3 = 2.2064$, $y_4 = 1.5826$,
  $y_5 = 1.0681$, $y_6 = 0.6430$, $y_7 = 0.2913$
9. $y_1 = 0.2660$, $y_2 = 0.5097$, $y_3 = 0.7357$, $y_4 = 0.9471$,
  $y_5 = 1.1465$, $y_6 = 1.3353$, $y_7 = 1.5149$, $y_8 = 1.6855$,
  $y_9 = 1.8474$
11. $y_1 = 0.3492$, $y_2 = 0.7202$, $y_3 = 1.1363$, $y_4 = 1.6233$,
  $y_5 = 2.2118$, $y_6 = 2.9386$, $y_7 = 3.8490$
13. (c) $y_0 = -2.2755$, $y_1 = -2.0755$,
  $y_2 = -1.8589$, $y_3 = -1.6126$, $y_4 = -1.3275$

CHAPTER 9 REVIEW EXERCISES, page 386

1. All isoclines $y = cx$ are solutions of the differential equation.

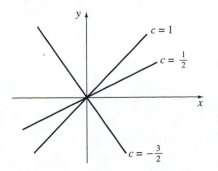

3. Comparison of Numerical Methods with $h = 0.1$

x_n	Euler	Improved Euler	Runge-Kutta
1.00	2.0000	2.0000	2.0000
1.10	2.1386	2.1549	2.1556
1.20	2.3097	2.3439	2.3454
1.30	2.5136	2.5672	2.5695
1.40	2.7504	2.8246	2.8278
1.50	3.0201	3.1157	3.1197

Comparison of Numerical Methods with $h = 0.05$

x_n	Euler	Improved Euler	Runge-Kutta
1.00	2.0000	2.0000	2.0000
1.05	2.0693	2.0735	2.0736
1.10	2.1469	2.1554	2.1556
1.15	2.2329	2.2459	2.2462
1.20	2.3272	2.3450	2.3454
1.25	2.4299	3.4527	2.4532
1.30	2.5410	2.5689	2.5695
1.35	2.6604	2.6937	2.6944
1.40	2.7883	2.8269	2.8278
1.45	2.9245	2.9686	2.9696
1.50	3.0690	3.1187	3.1197

5. Comparison of Numerical Methods with $h = 0.1$

x_n	Euler	Improved Euler	Runge-Kutta
0.50	0.5000	0.5000	0.5000
0.60	0.6000	0.6048	0.6049
0.70	0.7095	0.7191	0.7194
0.80	0.8283	0.8427	0.8431
0.90	0.9559	0.9752	0.9757
1.00	1.0921	1.1163	1.1169

Comparison of Numerical Methods with $h = 0.05$

x_n	Euler	Improved Euler	Runge-Kutta
0.50	0.5000	0.5000	0.5000
0.55	0.5500	0.5512	0.5512
0.60	0.6024	0.6049	0.6049
0.65	0.6573	0.6610	0.6610
0.70	0.7144	0.7194	0.7194
0.75	0.7739	0.7802	0.7801
0.80	0.8356	0.8431	0.8431
0.85	0.8996	0.9083	0.9083
0.90	0.9657	0.9757	0.9757
0.95	1.0340	1.0453	1.0452
1.00	1.1044	1.1170	1.1169

7. $h = 0.2$: $y(0.2) \approx 3.2$; $h = 0.1$: $y(0.2) \approx 3.23$
9. $x(0.2) \approx 1.62$, $y(0.2) \approx 1.84$

APPENDIX I EXERCISES, page APP-2

1. (a) 24 **(b)** 720 **(c)** $\dfrac{4\sqrt{\pi}}{3}$ **(d)** $-\dfrac{8\sqrt{\pi}}{15}$

3. 0.297

APPENDIX II EXERCISES, page APP-19

1. (a) $\begin{pmatrix} 2 & 11 \\ 2 & -1 \end{pmatrix}$ **(b)** $\begin{pmatrix} -6 & 1 \\ 14 & -19 \end{pmatrix}$

(c) $\begin{pmatrix} 2 & 28 \\ 12 & -12 \end{pmatrix}$

3. (a) $\begin{pmatrix} -11 & 6 \\ 17 & -22 \end{pmatrix}$ **(b)** $\begin{pmatrix} -32 & 27 \\ -4 & -1 \end{pmatrix}$

(c) $\begin{pmatrix} 19 & -18 \\ -30 & 31 \end{pmatrix}$ **(d)** $\begin{pmatrix} 19 & 6 \\ 3 & 22 \end{pmatrix}$

5. (a) $\begin{pmatrix} 9 & 24 \\ 3 & 8 \end{pmatrix}$ **(b)** $\begin{pmatrix} 3 & 8 \\ -6 & -16 \end{pmatrix}$

(c) $\begin{pmatrix} 0 & 0 \\ 0 & 0 \end{pmatrix}$ **(d)** $\begin{pmatrix} -4 & -5 \\ 8 & 10 \end{pmatrix}$

7. (a) 180 **(b)** $\begin{pmatrix} 4 & 8 & 10 \\ 8 & 16 & 20 \\ 10 & 20 & 25 \end{pmatrix}$ **(c)** $\begin{pmatrix} 6 \\ 12 \\ -5 \end{pmatrix}$

9. (a) $\begin{pmatrix} 7 & 38 \\ 10 & 75 \end{pmatrix}$ **(b)** $\begin{pmatrix} 7 & 38 \\ 10 & 75 \end{pmatrix}$

11. $\begin{pmatrix} -14 \\ 1 \end{pmatrix}$ **13.** $\begin{pmatrix} -38 \\ -2 \end{pmatrix}$ **15.** singular

17. nonsingular; $\mathbf{A}^{-1} = \dfrac{1}{4}\begin{pmatrix} -5 & -8 \\ 3 & 4 \end{pmatrix}$

19. nonsingular; $\mathbf{A}^{-1} = \dfrac{1}{2}\begin{pmatrix} 0 & -1 & 1 \\ 2 & 2 & -2 \\ -4 & -3 & 5 \end{pmatrix}$

21. nonsingular; $\mathbf{A}^{-1} = -\dfrac{1}{9}\begin{pmatrix} -2 & -2 & -1 \\ -13 & 5 & 7 \\ 8 & -1 & -5 \end{pmatrix}$

23. $\mathbf{A}^{-1}(t) = \dfrac{1}{2e^{3t}}\begin{pmatrix} 3e^{4t} & -e^{4t} \\ -4e^{-t} & 2e^{-t} \end{pmatrix}$

25. $\dfrac{d\mathbf{X}}{dt} = \begin{pmatrix} -5e^{-t} \\ -2e^{-t} \\ 7e^{-t} \end{pmatrix}$

27. $\dfrac{d\mathbf{X}}{dt} = 4\begin{pmatrix} 1 \\ -1 \end{pmatrix}e^{2t} - 12\begin{pmatrix} 2 \\ 1 \end{pmatrix}e^{-3t}$

29. (a) $\begin{pmatrix} 4e^{4t} & -\pi\sin\pi t \\ 2 & 6t \end{pmatrix}$ **(b)** $\begin{pmatrix} \frac{1}{4}e^8 - \frac{1}{4} & 0 \\ 4 & 6 \end{pmatrix}$

(c) $\begin{pmatrix} \frac{1}{4}e^{4t} - \frac{1}{4} & (1/\pi)\sin\pi t \\ t^2 & t^3 - t \end{pmatrix}$

31. $x = 3$, $y = 1$, $z = -5$

33. $x = 2 + 4t$, $y = -5 - t$, $z = t$

35. $x = -\frac{1}{2}$, $y = \frac{3}{2}$, $z = \frac{7}{2}$

37. $x_1 = 1$, $x_2 = 0$, $x_3 = 2$, $x_4 = 0$

41. $\lambda_1 = 6$, $\lambda_2 = 1$, $\mathbf{K}_1 = \begin{pmatrix} 2 \\ 7 \end{pmatrix}$, $\mathbf{K}_2 = \begin{pmatrix} 1 \\ 1 \end{pmatrix}$

43. $\lambda_1 = \lambda_2 = -4$, $\mathbf{K}_1 = \begin{pmatrix} 1 \\ -4 \end{pmatrix}$

45. $\lambda_1 = 0$, $\lambda_2 = 4$, $\lambda_3 = -4$,

$\mathbf{K}_1 = \begin{pmatrix} 9 \\ 45 \\ 25 \end{pmatrix}$, $\mathbf{K}_2 = \begin{pmatrix} 1 \\ 1 \\ 1 \end{pmatrix}$, $\mathbf{K}_3 = \begin{pmatrix} 1 \\ 9 \\ 1 \end{pmatrix}$

47. $\lambda_1 = \lambda_2 = \lambda_3 = -2$,

$\mathbf{K}_1 = \begin{pmatrix} 2 \\ -1 \\ 0 \end{pmatrix}$, $\mathbf{K}_2 = \begin{pmatrix} 0 \\ 0 \\ 1 \end{pmatrix}$

49. $\lambda_1 = 3i$, $\lambda_2 = -3i$,

$\mathbf{K}_1 = \begin{pmatrix} 1 - 3i \\ 5 \end{pmatrix}$, $\mathbf{K}_2 = \begin{pmatrix} 1 + 3i \\ 5 \end{pmatrix}$

INDEX

TABLE OF INTEGRALS

1. $\displaystyle\int u\,dv = uv - \int v\,du$

2. $\displaystyle\int u^n\,du = \frac{1}{n+1}u^{n+1} + C, n \neq -1$

3. $\displaystyle\int \frac{du}{u} = \ln|u| + C$

4. $\displaystyle\int e^u\,du = e^u + C$

5. $\displaystyle\int a^u\,du = \frac{1}{\ln a}a^u + C$

6. $\displaystyle\int \sin u\,du = -\cos u + C$

7. $\displaystyle\int \cos u\,du = \sin u + C$

8. $\displaystyle\int \sec^2 u\,du = \tan u + C$

9. $\displaystyle\int \csc^2 u\,du = -\cot u + C$

10. $\displaystyle\int \sec u \tan u\,du = \sec u + C$

11. $\displaystyle\int \csc u \cot u\,du = -\csc u + C$

12. $\displaystyle\int \tan u\,du = -\ln|\cos u| + C$

13. $\displaystyle\int \cot u\,du = \ln|\sin u| + C$

14. $\displaystyle\int \sec u\,du = \ln|\sec u + \tan u| + C$

15. $\displaystyle\int \csc u\,du = \ln|\csc u - \cot u| + C$

16. $\displaystyle\int \frac{du}{\sqrt{a^2 - u^2}} = \sin^{-1}\frac{u}{a} + C$

17. $\displaystyle\int \frac{du}{a^2 + u^2} = \frac{1}{a}\tan^{-1}\frac{u}{a} + C$

18. $\displaystyle\int \frac{du}{u\sqrt{u^2 - a^2}} = \frac{1}{a}\sec^{-1}\frac{u}{a} + C$

19. $\displaystyle\int \frac{du}{a^2 - u^2} = \frac{1}{2a}\ln\left|\frac{u+a}{u-a}\right| + C$

20. $\displaystyle\int \frac{du}{u^2 - a^2} = \frac{1}{2a}\ln\left|\frac{u-a}{u+a}\right| + C$

21. $\displaystyle\int \sin^2 u\,du = \tfrac{1}{2}u - \tfrac{1}{4}\sin 2u + C$

22. $\displaystyle\int \cos^2 u\,du = \tfrac{1}{2}u + \tfrac{1}{4}\sin 2u + C$

23. $\displaystyle\int \tan^2 u\,du = \tan u - u + C$

24. $\displaystyle\int \cot^2 u\,du = -\cot u - u + C$

25. $\displaystyle\int \sin^3 u\,du = -\tfrac{1}{3}(2 + \sin^2 u)\cos u + C$

26. $\displaystyle\int \cos^3 u\,du = \tfrac{1}{3}(2 + \cos^2 u)\sin u + C$

27. $\displaystyle\int \tan^3 u\,du = \tfrac{1}{2}\tan^2 u + \ln|\cos u| + C$

28. $\displaystyle\int \cot^3 u\,du = -\tfrac{1}{2}\cot^2 u - \ln|\sin u| + C$

29. $\displaystyle\int \sec^3 u\,du = \tfrac{1}{2}\sec u \tan u + \tfrac{1}{2}\ln|\sec u + \tan u| + C$

30. $\displaystyle\int \csc^3 u\,du = -\tfrac{1}{2}\csc u \cot u + \tfrac{1}{2}\ln|\csc u - \cot u| + C$

31. $\displaystyle\int \sin^n u\,du = -\frac{1}{n}\sin^{n-1}u \cos u + \frac{n-1}{n}\int \sin^{n-2}u\,du$

32. $\displaystyle\int \cos^n u\,du = \frac{1}{n}\cos^{n-1}u \sin u + \frac{n-1}{n}\int \cos^{n-2}u\,du$

33. $\displaystyle\int \tan^n u\,du = \frac{1}{n-1}\tan^{n-1}u - \int \tan^{n-2}u\,du$

34. $\displaystyle\int \cot^n u\,du = \frac{-1}{n-1}\cot^{n-1}u - \int \cot^{n-2}u\,du$

35. $\displaystyle\int \sec^n u\,du = \frac{1}{n-1}\tan u \sec^{n-2}u + \frac{n-2}{n-1}\int \sec^{n-2}u\,du$

36. $\displaystyle\int \csc^n u\,du = \frac{-1}{n-1}\cot u \csc^{n-2}u + \frac{n-2}{n-1}\int \csc^{n-2}u\,du$

37. $\displaystyle\int \sin au \sin bu\,du = \frac{\sin(a-b)u}{2(a-b)} - \frac{\sin(a+b)u}{2(a+b)} + C$

38. $\displaystyle\int \cos au \cos bu\,du = \frac{\sin(a-b)u}{2(a-b)} + \frac{\sin(a+b)u}{2(a+b)} + C$

39. $\displaystyle\int \sin au \cos bu\,du = -\frac{\cos(a-b)u}{2(a-b)} - \frac{\cos(a+b)u}{2(a+b)} + C$

40. $\displaystyle\int u \sin u\,du = \sin u - u \cos u + C$

41. $\displaystyle\int u \cos u\,du = \cos u + u \sin u + C$

42. $\displaystyle\int u^n \sin u\,du = -u^n \cos u + n\int u^{n-1}\cos u\,du$

43. $\displaystyle\int u^n \cos u\,du = u^n \sin u - n\int u^{n-1}\sin u\,du$

44. $\displaystyle\int \sin^n u \cos^m u\,du = -\frac{\sin^{n-1}u \cos^{m+1}u}{n+m} + \frac{n-1}{n+m}\int \sin^{n-2}u \cos^m u\,du = \frac{\sin^{n+1}u \cos^{m-1}u}{n+m} + \frac{m-1}{n+m}\int \sin^n u \cos^{m-2}u\,du$